Cellular Biology: Structures and Functions

Cellular Biology: Structures and Functions

Edited by Gloria Doran

SYRAWOOD
PUBLISHING HOUSE

New York

Published by Syrawood Publishing House,
750 Third Avenue, 9th Floor,
New York, NY 10017, USA
www.syrawoodpublishinghouse.com

Cellular Biology: Structures and Functions
Edited by Gloria Doran

International Standard Book Number: 978-1-68286-647-4 (Hardback)

Cataloging-in-Publication Data

Cellular biology : structures and functions / edited by Gloria Doran.
    p. cm.
Includes bibliographical references and index.
ISBN 978-1-68286-647-4
1. Cytology. 2. Cells. 3. Biology. I. Doran, Gloria.
QH581.2 .C45 2019
571.6--dc23

TABLE OF CONTENTS

Preface VII

Chapter 1 *Drosophila* **MOF controls Checkpoint protein2 and regulates genomic
stability during early embryogenesis** 1
Sreerangam NCVL Pushpavalli, Arpita Sarkar, M Janaki Ramaiah,
Debabani Roy Chowdhury, Utpal Bhadra and Manika Pal-Bhadra

Chapter 2 **MKL1 inhibits cell cycle progression through p21 in podocytes** 16
Shuang Yang, Lingjia Liu, Pengjuan Xu and Zhuo Yang

Chapter 3 **MicroRNA *miR-378* promotes BMP2-induced osteogenic differentiation
of mesenchymal progenitor cells** 28
Marlinda Hupkes, Ana M Sotoca, José M Hendriks,
Everardus J van Zoelen and Koen J Dechering

Chapter 4 **Wingless gene cloning and its role in manipulating the wing dimorphism
in the white-backed planthopper, *Sogatella furcifera*** 43
Ju-Long Yu, Zhi-Fang An and Xiang-Dong Liu

Chapter 5 **Efficient isolation of specific genomic regions retaining molecular interactions
by the iChIP system using recombinant exogenous DNA-binding proteins** 52
Toshitsugu Fujita and Hodaka Fujii

Chapter 6 **Identification of a set of miRNAs differentially expressed in transiently
TIA-depleted HeLa cells by genome-wide profiling** 60
Carmen Sánchez-Jiménez, Isabel Carrascoso, Juan Barrero and José M Izquierdo

Chapter 7 **Efficient 5′-3′ DNA end resection by HerA and NurA is essential for cell
viability in the crenarchaeon *Sulfolobus islandicus*** 76
Qihong Huang, Linlin Liu, Junfeng Liu, Jinfeng Ni, Qunxin She and Yulong Shen

Chapter 8 **Downregulation of microRNA-100 protects apoptosis and promotes neuronal
growth in retinal ganglion cells** 90
Ning Kong, Xiaohe Lu and Bin Li

Chapter 9 **Pre-amplification in the context of high-throughput qPCR gene expression
experiment** 98
Vlasta Korenková, Justin Scott, Vendula Novosadová, Marie Jindřichová,
Lucie Langerová, David Švec, Monika Šídová and Robert Sjöback

Chapter 10 *Spdef* **deletion rescues the crypt cell proliferation defect in conditional *Gata6*
null mouse small intestine** 108
Boaz E Aronson, Kelly A Stapleton, Laurens ATM Vissers, Eva Stokhuijzen,
Hanneke Bruijnzeel and Stephen D Krasinski

Chapter 11   **The ICP22 protein selectively modifies the transcription of different kinetic classes of pseudorabies virus genes**..120
Irma F Takács, Dóra Tombácz, Beáta Berta, István Prazsák,
Nándor Póka and Zsolt Boldogkői

Chapter 12   **MicroRNA-19a regulates lipopolysaccharideinduced endothelial cell apoptosis through modulation of apoptosis signal-regulating kinase 1 expression**..132
Wei-Long Jiang, Yu-Feng Zhang, Qing-Qing Xia, Jian Zhu, Xin Yu,
Tao Fan and Feng Wang

Chapter 13   **Differential regulation of the *a-globin* locus by Krüppel-like factor 3 in erythroid and non-erythroid cells**..140
Alister PW Funnell, Douglas Vernimmen, Wooi F Lim, Ka Sin Mak,
Beeke Wienert, Gabriella E Martyn, Crisbel M Artuz, Jon Burdach,
Kate GR Quinlan, Douglas R Higgs, Emma Whitelaw,
Richard CM Pearson and Merlin Crossley

Chapter 14   **Screening and analysis of PoAkirin1 and two related genes in response to immunological stimulants in the Japanese flounder (*Paralichthys olivaceus*)**..152
Chang-Geng Yang, Xian-Li Wang, Bo Zhang, Bing Sun,
Shan-Shan Liu and Song-Lin Chen

Chapter 15   **Néstor-Guillermo Progeria Syndrome: a biochemical insight into Barrier-to-Autointegration Factor 1, alanine 12 threonine mutation**..165
Nicolas Paquet, Joseph K Box, Nicholas W Ashton, Amila Suraweera,
Laura V Croft, Aaron J Urquhart, Emma Bolderson, Shu-Dong Zhang,
Kenneth J O'Byrne and Derek J Richard

Chapter 16   **Identification of nucleotides and amino acids that mediate the interaction between ribosomal protein L30 and the SECIS element**..176
Abby L Bifano, Tarik Atassi, Tracey Ferrara and Donna M Driscoll

Chapter 17   **Mechanical stimulation of human tendon stem/ progenitor cells results in upregulation of matrix proteins, integrins and MMPs, and activation of p38 and ERK1/2 kinases**..188
Cvetan Popov, Martina Burggraf, Ludwika Kreja, Anita Ignatius,
Matthias Schieker and Denitsa Docheva

Chapter 18   **Nop17 is a key R2TP factor for the assembly and maturation of box C/D snoRNP complex**..199
Marcela B Prieto, Raphaela C Georg, Fernando A Gonzales-Zubiate,
Juliana S Luz and Carla C Oliveira

Permissions

List of Contributors

Index

PREFACE

Cell is the fundamental unit of life. The chemical, metabolic and physiological processes of cell organelles, as well as their interactions with their environment together are studied in cellular biology. Cell biology examines cell's life cycle which includes the study of the fundamental structures and functions of cell as well as an analysis of the different signaling pathways within the cell. Research in cellular biology is of significance in the field of genetics, biochemistry, molecular biology, immunology, etc. There has been rapid progress in this field and its applications are finding their way across multiple industries. This book covers in detail some existing theories and innovative concepts revolving around this field. It contains some path-breaking studies in the area of cell biology contributed by international experts. Cell biologists, geneticists, bioengineers, academicians and students will benefit alike from this book.

This book unites the global concepts and researches in an organized manner for a comprehensive understanding of the subject. It is a ripe text for all researchers, students, scientists or anyone else who is interested in acquiring a better knowledge of this dynamic field.

I extend my sincere thanks to the contributors for such eloquent research chapters. Finally, I thank my family for being a source of support and help.

Editor

Drosophila MOF controls Checkpoint protein2 and regulates genomic stability during early embryogenesis

Sreerangam NCVL Pushpavalli[1†], Arpita Sarkar[1†], M Janaki Ramaiah[1†], Debabani Roy Chowdhury[2], Utpal Bhadra[2] and Manika Pal-Bhadra[1*]

Abstract

Background: In *Drosophila* embryos, checkpoints maintain genome stability by delaying cell cycle progression that allows time for damage repair or to complete DNA synthesis. *Drosophila* MOF, a member of MYST histone acetyl transferase is an essential component of male X hyperactivation process. Until recently its involvement in G2/M cell cycle arrest and defects in ionizing radiation induced DNA damage pathways was not well established.

Results: *Drosophila* MOF is highly expressed during early embryogenesis. In the present study we show that haplo-insufficiency of maternal MOF leads to spontaneous mitotic defects like mitotic asynchrony, mitotic catastrophe and chromatid bridges in the syncytial embryos. Such abnormal nuclei are eliminated and digested in the yolk tissues by nuclear fall out mechanism. MOF negatively regulates *Drosophila* checkpoint kinase 2 tumor suppressor homologue. In response to DNA damage the checkpoint gene *Chk2* (*Drosophila mnk*) is activated in the *mof* mutants, there by causing centrosomal inactivation suggesting its role in response to genotoxic stress. A drastic decrease in the fall out nuclei in the syncytial embryos derived from *mof¹/+; mnk^{p6}/+* females further confirms the role of DNA damage response gene *Chk2* to ensure the removal of abnormal nuclei from the embryonic precursor pool and maintain genome stability. The fact that *mof* mutants undergo DNA damage has been further elucidated by the increased number of single and double stranded DNA breaks.

Conclusion: *mof* mutants exhibited genomic instability as evidenced by the occurance of frequent mitotic bridges in anaphase, asynchronous nuclear divisions, disruption of cytoskeleton, inactivation of centrosomes finally leading to DNA damage. Our findings are consistent to what has been reported earlier in mammals that; reduced levels of MOF resulted in increased genomic instability while total loss resulted in lethality. The study can be further extended using *Drosophila* as model system and carry out the interaction of MOF with the known components of the DNA damage pathway.

Keywords: *Mof*, Mitosis, Syncytial embryos, *Drosophila melanogaster*, *Chk2*, Anaphase bridges

Background

In eukaryotic organisms the individual identity of cells is determined by cell specific genes while a set of genes that are expressed in all cells functions as housekeeping genes. Eukaryotic DNA is highly packaged into chromatin structures, with core histone and non histone chromosomal proteins that regulate many cellular processes including DNA replication and repair of damaged DNA. Regulation of cell cycle involves processes that are crucial to the survival of a cell, wherein detection and repair of genetic damage occurs to control unwanted cell division and maintain genomic stability. Disruption of checkpoint function plays an important role in carcinogenesis and embryonic lethality [1,2]. Chromatin regulatory activities along with histone modifications facilitate the contact of repair proteins at the damaged sites and promote recruitment of components of signaling cascade. Acetylation of lysine16 on histone H4 (H4K16Ac) has the potential to create or

* Correspondence: manikapb@gmail.com
†Equal contributors
¹Centre for Chemical Biology, Indian Institute of Chemical Technology, Hyderabad 500607, India
Full list of author information is available at the end of the article

obscure binding platforms for chromatin modifying enzymes and transcriptional activators. Furthermore H4K16 acetylation can directly impact on higher order chromatin structure, thus creating an open highly accessible chromatin conformation. The major enzyme that acetylates H4K16 is MOF (Males Absent on First) which is highly conserved in mammals and *Drosophila*.

Drosophila histone acetyl transferase MOF is responsible for the interplay between the regulators of transcription and chromatin modifiers thereby governing the gene expression at transcriptional level. It belongs to the family of MYST histone acetyl transferases (HATs) which consists of a conserved catalytic MYST domain [3,4]. The members of this family display diverse roles in various nuclear processes and some of them have also been implicated in carcinogenesis [5]. MOF is an integral member in the *Drosophila melanogaster* dosage compensation process that ensures that males and females, despite unequal number of X chromosomes, express the same amount of X-linked gene products [6]. MOF has strict substrate specificity to H4K16 when compared to other HATs [7,8]. *Drosophila mof* was identified in a screen for ethyl methane sulfonate-induced male-specific lethal mutations and was shown to directly acetylate Histone H4 at K16 [9]. Deletion of *mof* in the case of both *Drosophila* and mammals caused substantial decrease in H4K16 acetylation indicating that *mof* is the major HAT for H4K16 [10,11].

Acetylation of H4K16 by MOF causes reduction in the chromatin compaction *in vitro* and decondensation of chromatin under *in vivo* conditions [12,13]. Hence MOF regulates chromatin based activities such as transcription and DNA damage repair by H4K16 acetylation. Moreover MOF is an important constituent of X-chromosome dosage compensation complex (DCC) resulting in two fold activation of X-linked genes in male flies. Males carrying loss of function *mof* mutation do not survive since they lack the H4K16Ac enrichment on the X-chromosome for transcription of the X-linked genes [6,14]. Interestingly mammalian MOF has high degree of sequence similarity to *Drosophila* MOF protein and H4K16 acetylation is also an epigenetic signature of cellular proliferation during embryogenesis and oncogenesis [15]. Further the role of MOF in ionizing radiation (IR) response is also conserved in *Drosophila* [11]. Recent studies in mammals suggest that the levels of H4K16 acetylation were reduced both in cancer cell lines and primary tumors [16]. Increased genomic instability, with high spontaneous chromosomal aberrations and reduced γ-H2AX foci formation after IR treatment are characteristic features of cultured $mMof^{+/-}$ cells. In the case of mammals total loss of function $(mof^{-/-})$ resulted in lethality [15].

Faithful transmission of genetic information in cellular organisms is carried out by two basic processes such as DNA replication and cell division. The first 13 syncytial nuclear divisions in *Drosophila* are maternally controlled and consist mainly of S and M phases with short or undetectable gap phases [17]. The syncytial cycles from 1–7 occur inside the embryos and nuclear migration to the cortex occurs during cycles 8 and 9 where further synchronous divisions take place before the onset of cellularisation at 14th nuclear cycle. During cycle 9 few nuclei migrate to the poles to form the pole cells that become the germ cells of the embryo [18]. After completion of 13 syncytial cycles, the embryo undergoes cellularization. Cell cycle checkpoints maintain genomic integrity and stability by regulating the progression of the cell cycle and inducing apoptosis in response to DNA damage to eliminate deleterious mutations from the genome. Defects in cell cycle checkpoints cause a wide variety of defects such as aging, genetic diseases, oncogenesis and neurodegeneration. Proper balance of cell cycle responses are critical for cell death or cell survival to occur. Though DNA damage and replication checkpoint induced apoptosis has been extensively studied, less is known about the cellular responses to stress during mitosis. Checkpoint failures lead to progression of mitosis without damage repair leading to mitotic catastrophe. Embryos exhibiting mitotic catastrophe have giant and fragmented nuclei lacking a regular pattern and 2N ploidy [19].

In the present study we report the identification and phenotypic characterization of *Drosophila* mutants which are haplo-insufficient for maternal MOF. During early embryogenesis mutation in *Drosophila mof* leads to spontaneous chromosomal aberrations and genomic instability leading to mitotic cell cycle progression without repair of damaged DNA. Most significantly, we found the activation of *Drosophila* homolog of the *checkpoint kinase 2* (*DmChk2* or *mnk*) in response to *mof* mutation causing centrosomal inactivation. We propose that *Drosophila* MOF, like its human counterpart, is required for maintaining genomic stability during embryogenesis.

Results

mof heterozygote embryos are haplo-insufficient for maternal MOF gene product

mof^1 is a EMS mutation having a single amino acid substitution in the acetyl co-enzyme motif [9]. Sequence analyses revealed that mof^3 results from a nonsense mutation at aminoacid 151 (Q151X) [11]. The nature of the *mof* alleles has been studied by quantifying the amount of maternal MOF gene product. For this purpose total protein was isolated from control (yw^{67c23}) mof^1 (Ethyl Methane Sulphonate mutation) and mof^3 (non-sense mutation) embryos (1–2 h) and western blot analysis was carried out using MOF antibody. A drastic decrease in MOF expression in *mof* heterozygote embryos compared to wild type controls indicated that *mof* mutation is haplo-insufficient for maternal gene product (Figure 1).

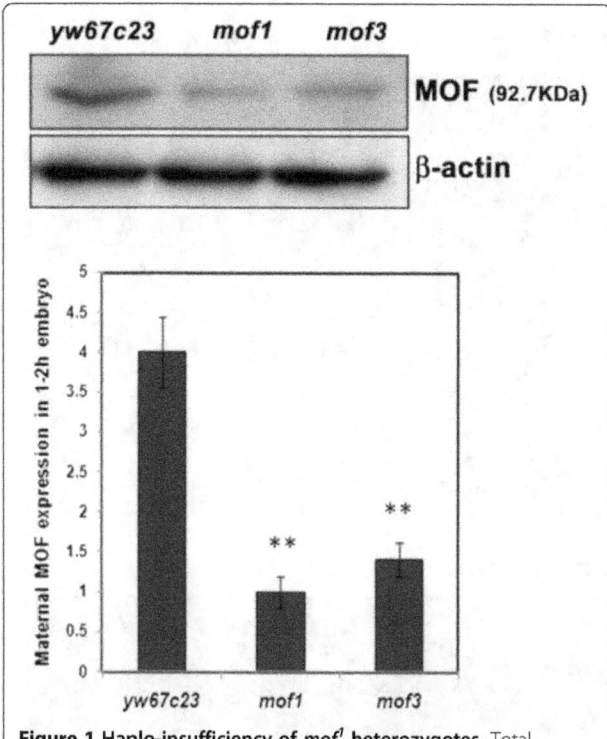

Figure 1 Haplo-insufficiency of *mof¹* heterozygotes. Total protein was isolated from early embryos of control (*yw^{67c23}*), *mof¹* and *mof³* (1–2 h) and western blot analysis carried out using mof antibody has shown 3-fold reduction in the maternal MOF protein. β-actin is used as an internal loading control. Statistical significance was assessed using student t-test. *** indicates P<0.001, ** indicates P<0.01, * indicates p<0.05.

Asynchronous cell cycle and Mitotic catastrophe in the *mof* embryos

MOF is a highly conserved MYST family HAT that acetylates histone at H4K16 and plays an important role in transcriptional activation. Studies of *mof* null mice showed delayed development with massive abnormal chromosomal aggregation, leading to death at an early stage [15]. *In vitro* and as well as *in vivo* studies in *Drosophila* has shown that MOF is required for efficient repair of DNA damage induced by ionizing radiation [11]. Since *mof¹* mutants are haplo-insufficient for the maternal gene product, we were interested to study the role of MOF during early mitotic divisions that are syncytial. Embryos derived from heterozygous mothers (haplo-insufficiency of maternal gene product) of *mof¹/FM7* (EMS mutagenesis), *mof³/FM7* (non-sense mutation) as well as *yw^{67c23}* (control) were collected. Early embryos (0–2 h) were fixed and mounted in propidium iodide (PI) to visualise the nuclei. *mof* heterozygote embryos exhibited mitotic catastrophe with fragmented nuclei that appear as large mass of chromatin compared to wild type control where the nuclei appeared normal (Figure 2A). During early embryogenesis the initial seven syncytial divisions occur at the interior of the embryo. During cycles 8 and 9 the nuclei migrate to

the cortex leaving only few yolk nuclei. We observed that abnormal nuclei in the *mof* heterozygote embryos are eliminated by nuclear fallout mechanism where in they are digested inside the yolk tissues. Nuclear fallout mechanism protects the organism by eliminating the abnormal nuclei from forming adult structures that might be deleterious. Nearly 70% of the *mof* heterozygotes exhibited a large number of fall out nuclei (high severity fall out nuclei=more than 5 fall out nuclei/embryo) (Figure 2B, 2C) compared to control nuclei (*yw^{67c23}*) where the number of fall out nuclei is negligible. Hence a decreased number of nuclei are present in the *mof* embryos compared to control (*yw^{67c23}*). The fall out nuclei in embryos were scored when they are 2-20μm below the cortex in the syncytial blastoderm stage as they are mis-interpreted in later stages where fall out co-exists with normal nuclear migration.

Abnormal mitosis in *mof* heterozygotes

Mof is a maternal effect gene and homozygotes for *mof* mutation do not survive (late larval lethal) till adult stage [11]. To study the role of MOF in early mitosis, we collected embryos derived from heterozygous mothers (haplo-insufficiency of maternal gene product) of *mof¹/FM7* and *yw^{67c23}* (control). During the early syncytial nuclear divisions, *mof¹* mutant embryos exhibited several mitotic defects such as chromatid bridges resulting in lagging chromosomes (Figure 3A), defects in sister chromatid separation (Figure 3B); telophase defects (Figure 3C) indicating that *mof* heterozygous embryos may be entering mitosis with damaged or incompletely replicated DNA. The lethality associated with *mof* homozygotes was fully rescued with the addition of *mof* transgene. Although the transgenic line expressing *mof* transgene was viable and fertile, it did not completely restore the chromosomal defects (only 60% of the defects were rescued) (Additional file 1). The embryos from *mof¹/+* display similar mitotic defects as that of *mof¹/FM7* while the embryos from *FM7/+* females do not show any mitotic defects indicating that *FM7* balancer has no role in causing the mitotic defects observed in the case of *mof¹/FM7* embryos (data not shown).

Mitotic asynchrony during early nuclear divisions in *mof* heterozygous embryos

In early embryos of wild type mitosis occurs synchronously and proceeds in the form of waves starting from the poles. Mitotic synchrony during pre-syncytial and syncytial divisons in *mof¹* and control embryos was studied by staining with antibody against Histone H3 Ser10 Phosphorylation (PH3) (the mitosis marker). Control embryos showed PH3 staining on all the chromsomes while in the case of *mof¹* heterozygotes both PH3 positive (dividing) and PH3 negative (non-dividing) chromosomes were observed. Thus the PH3 negative chromosomes in *mof¹*

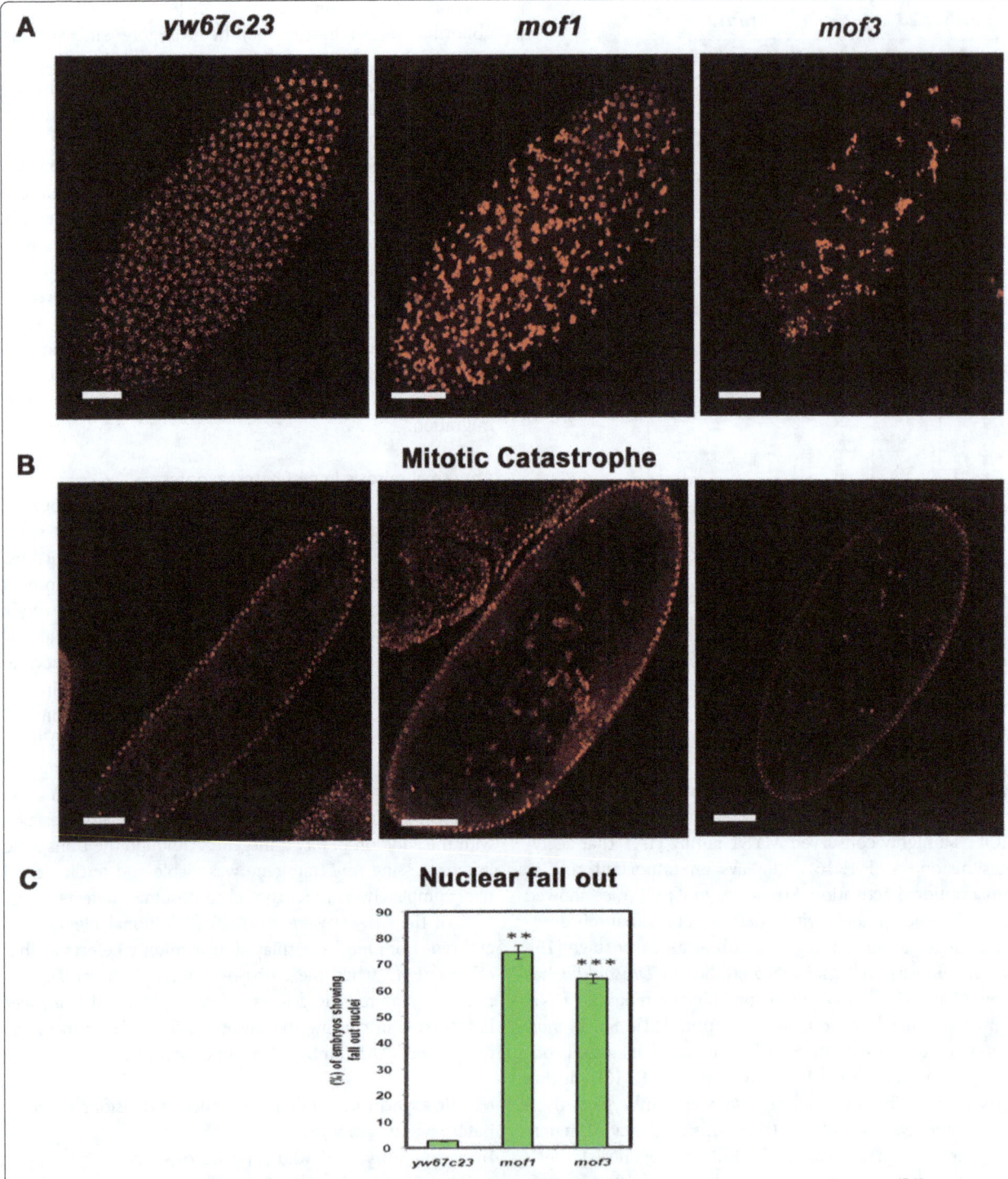

Figure 2 Loss of MOF causes asynchronous cell cycle, mitotic catastrophe and nuclear fallout. Early embryos (0–2 h) of yw^{67c23}, mof^1 and mof^3 were collected, fixed with DNA dye PI and visualized using confocal microscopy. (**A**) Large fragmented nuclei indicating the occurrence of mitotic catastrophe is seen in the early embryos of mof^1 and mof^3 mutants when compared to control embryos. (**B** & **C**) Increased number of fall out nuclei is observed in the mof^1 and mof^3 mutants when compared to control yw^{67c23} embryos. Bar indicates 10 μm scale. The data is represented in the form of bar diagram. Statistical significance was assessed using student t-test. *** indicates P<0.001, ** indicates P<0.01, * indicates p<0.05.

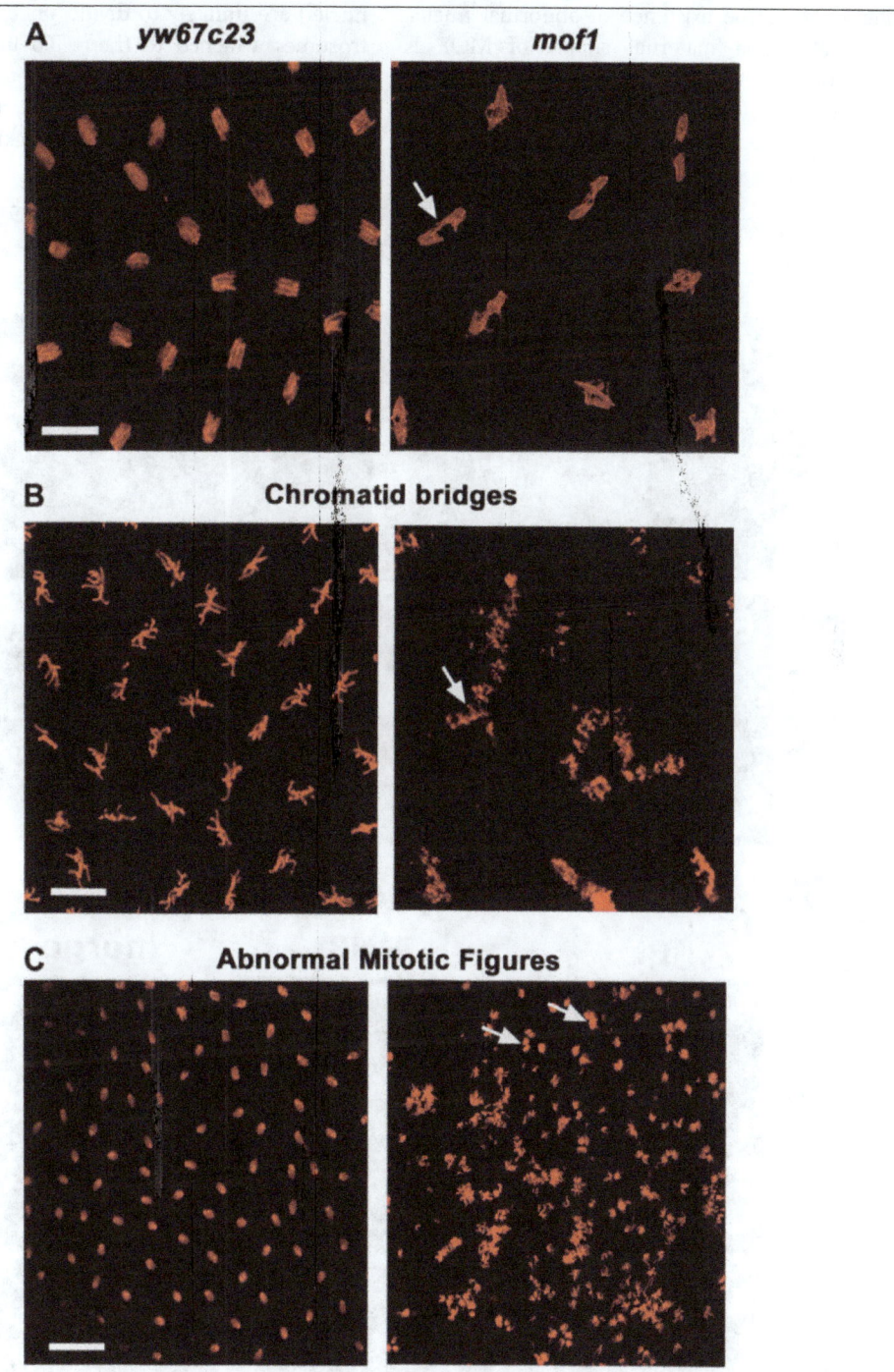

Figure 3 Loss of maternal MOF causes chromsomal defects in early embryos. Early embryos (0–2 h) from *mof*[1] and *yw*[67c23] were collected, fixed with DNA dye PI and visualised using confocal microscopy. The *mof*[1] mutants displayed a wide variety of chromosomal defects like (**A**) chromatid bridges which indicates the presence of lagging chromosomes (**B**) sister chromatid separation and (**C**) telophase defects. Bar indicates 10 µm scale.

heterozygotes indicate the existence of abnormal nuclei. Our data indicates that maternal supply of MOF is required for mitotic synchrony in pre-syncytial and syncytial blastoderm embryos (Figure 4A-A′, B-B′). These abnormal nuclei which loose association with cortex (fall out

nuclei) are unlikely to divide since they do not have centrosomes attached to them. To further confirm nuclear fallout early embryos of mof^1 heterozygotes and yw^{67c23} were immunostained with anti-centrosomin antibody. A number of free centrosomes lacking the chromosomes

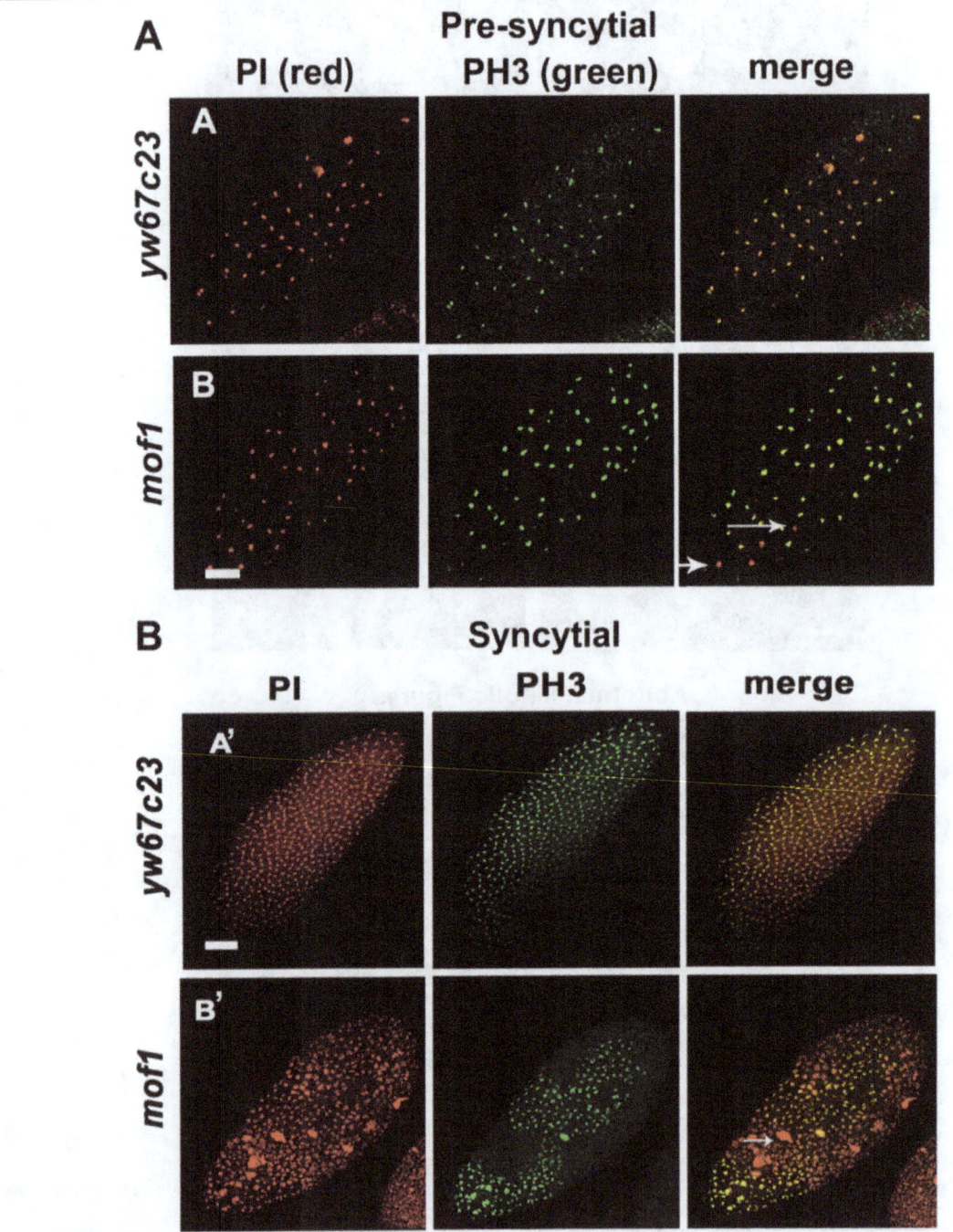

Figure 4 Mitotic asynchrony in mof^1 heterozygotes. Early embryos from 0–2 h were collected from control yw^{67c23} and mof^1 mutants and immunostaining was carried out using PH3 antibody (green) which is mitotic marker. DNA was stained with PI (red). **A** and A′ represents control yw^{67c23} while **B** and B′ represents mof^1 embryos during pre-syncytial and syncytial blastoderm stages respectively. In the case of mof^1 embryos the nuclei are unevenly spaced and not all chromsomes are stained with PH3 indicating the existance of mitotic asynchrony during pre-syncytial blastoderm stage. Bar indicates 10 μm scale.

were present in the *mof¹* heterozygous early embryos compared to control embryos (*yw⁶⁷ᶜ²³*). The free centrosomes in the embryo indicated the presence of abnormal nuclei that are eliminated by the nuclear fallout mechanism (Figure 5). In addition to free centrosomes we also observed chromosomes lacking centrosomes or with only one centrosome. These findings strongly suggest the involvement of centrosome inactivation in the *mof¹* early embryos.

Disruption of cytoskeleton in the *mof* heterozygous embryos

Cytoplasmic organization, nuclear division and nuclear migration in the syncytial embryos are modulated by the cytoskeletal proteins. Following the syncytial divisions individual cells are produced by a process called cellularization that occurs during interphase of nuclear cycle 14. Thus we were interested to study the changes in the organization of actin cytoskeleton and hence control *yw⁶⁷ᶜ²³* and *mof¹* heterozygous embryos were immunostained with ß-actin antibody. The typical honeycomb like structure of actin cytoskeleton observed in the control was lacking in the case of *mof¹* embryos. Moreover

in the *mof¹* embryos chromosomes were incompletely surrounded by the actin filaments along with few small cells that lack nuclei. These empty cells indicate the presence of abnormal nuclei which have been eliminated by nuclear fall out mechanism (Figure 6). In addition to the actin filaments the polymerization and depolymerization of microtubule network helps in mediating the coordinated nuclear movement (chromosomes) during syncytial stage of embryogenesis. Since polymerization and depolymerization of the microtubules is required for proper chromosome movement, we stained the *yw⁶⁷ᶜ²³* and *mof¹* embryos with alpha-tubulin antibody to visualize the organization of spindle fibres. Around 66% of *mof¹* embryos as opposed to only 7% of *yw⁶⁷ᶜ²³* embryos exhibited attachment of spindle fibres all over the chromsomes instead of the kinetochore, indicating disruption of the spindle fibre assembly and therefore leading to improper movement of chromosomes during anaphase resulting in lagging chromosomes. Number of embryos counted in the present study is 100 (Figure 7A, 7B).

The integrity of cell's cytoskeleton is crucial for the first occasion of vasa localization in the preplasmic cytoplasm as well as second occasion in the pole plasm.

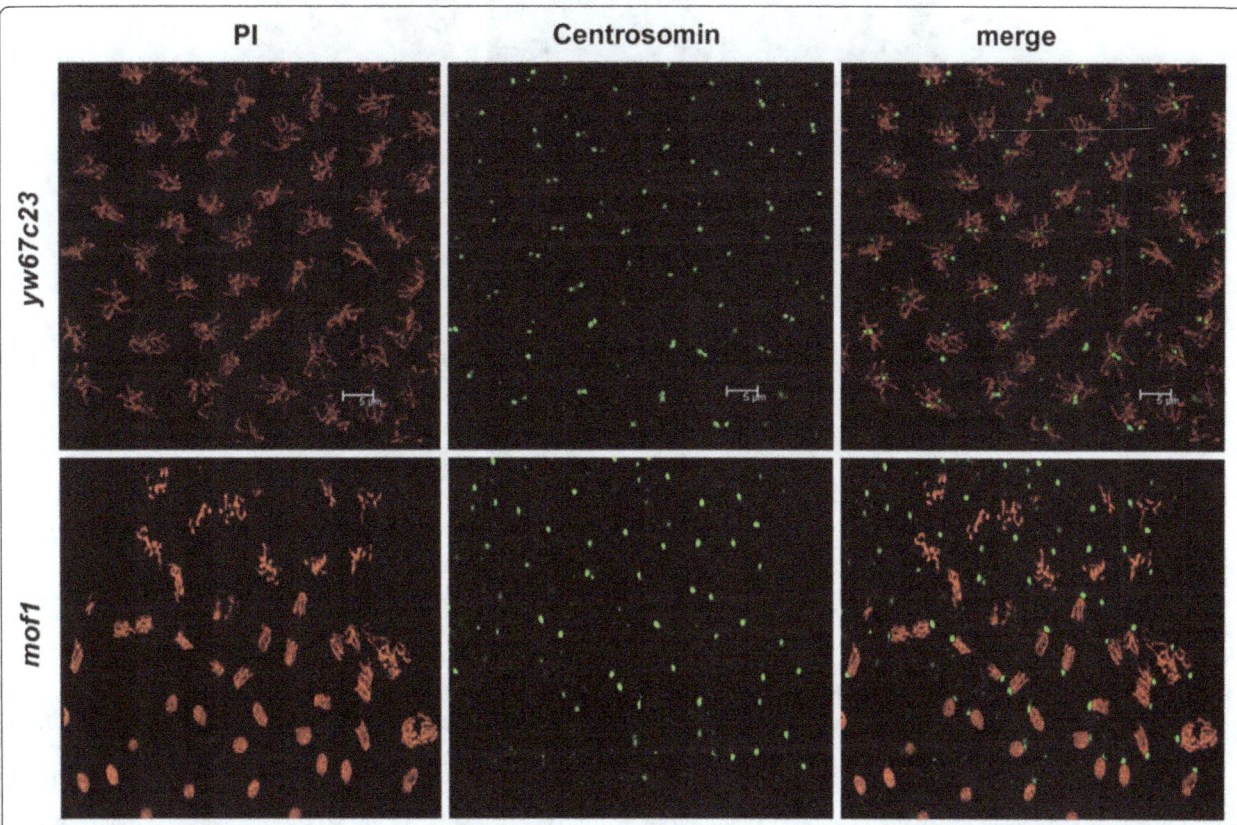

Figure 5 Loss of MOF results in free centrosomes. Embryos from 0–2 h from *yw⁶⁷ᶜ²³* and *mof¹* mutants were immunostained with anti-centrosomin antibody (green) and DNA was stained using PI (red). Free centrosomes without the chromosomes were present in the *mof¹* embryos indicating the presence of abnormal nuclei which have been removed by nuclear fallout mechanism. Bar indicates 5 µm scale.

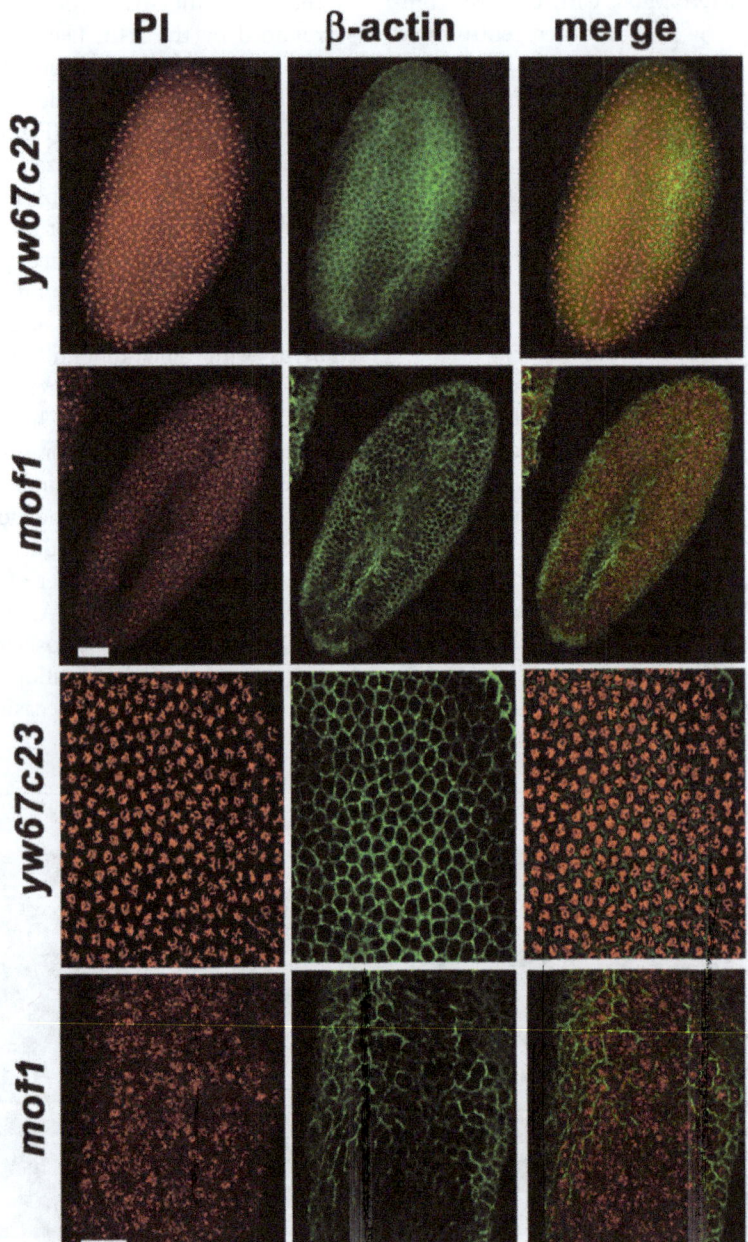

Figure 6 Defects in actin cytoskeleton in the *mof¹* heterozygotes. Early embryos (0–2 h) were stained with β-actin (green) antibody in both *yw⁶⁷ᶜ²³* and *mof¹* embryos. PI is used to stain the DNA. The honeycomb like structure which is characteristic of actin cytoskeleton is largely disrupted by the *mof* mutation. There are few cells which do not have chromsomes in them indicating abnormal nuclei which have been dropped into the cortex and digested by the yolk nuclei. Bar indicates 10 µm scale.

Proper function of the cytoskeleton is important for nuclear migration leading to formation of pole cells. Here *yw⁶⁷ᶜ²³* and *mof¹* embryos were immunostained with antibody against vasa which selectively stains the pole cells. As anticipated we observed drastic reduction in the number of pole cells in the *mof¹* embryos compared to control (*yw⁶⁷ᶜ²³*) indicating that MOF is required for proper nuclear migration and formation of pole cells (Figure 7C, 7D).

Elevated levels of DNA damage in *mof* heterozygous embryos

We next wanted to study more specific role of MOF in DNA damage. Thus *mof¹* heterozygote embryos as well as control embryos were used in an assay that determines the extent of DNA damage (single, double stranded DNA breaks) [20]. Genomic DNA was isolated from the embryos (0–2 h) and incubated with T4 DNA kinase and 32P ATP. The amount of incorporation of

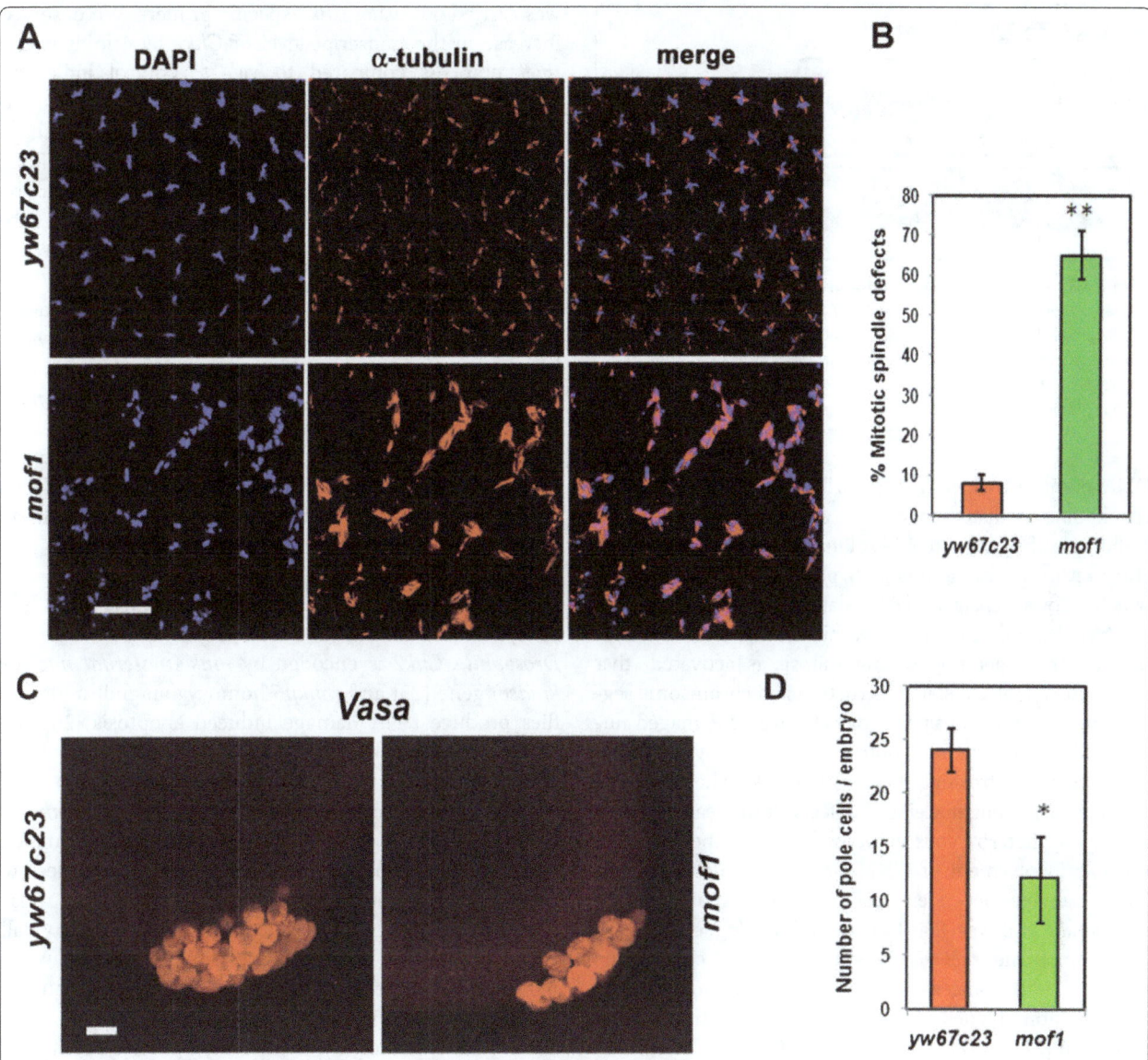

Figure 7 Loss of MOF causes defects in spindle fibre organisation and nuclear migration. (**A**) 0–2 h *mof¹* and *yw⁶⁷ᶜ²³* embryos were stained with alpha tubulin antibody (red) and DAPI is used as DNA dye. The chromosomes were not properly aligned in the metaphase plane with tubulin fibres attached all over the chromosmes. (**B**) The percentage of embryos exhibiting mitotic spindle defects is represented in the form of bar diagram. (**C**) The early embryos of *mof¹* and *yw⁶⁷ᶜ²³* were stained with vasa antibody to visualise the pole cells. There was drastic decrease in the number of pole cells in the *mof¹* embryos indicating that nuclear migration is affected. (**D**) The number of pole cells in control and *mof¹* mutant per embryo is represented in the form of bar diagram. Bar indicates 10 μm scale. Statistical significance was assessed using student t-test. *** indicates $P<0.001$, ** indicates $P<0.01$, * indicates $p<0.05$.

32PATP determined the number of exposed 5' phosphate groups in the DNA, indicating the number of single and double stranded lesions. Genomic DNA from *mof¹* embryos showed a highly elevated level of P32 incorporation of 6500 ± 370 cpm/ng compared to the controls which is 1011 ± 200 cpm/ng suggesting that the increase in lesions in *mof¹* embryos is due to progression through mitosis with damaged DNA or incompletely replicated DNA. To further confirm the double strand breaks observed in the *mof¹* heterozygote we carried out

western blot studies using phospho H2Av antibody. As expected in *mof¹* heterozygotes we observed an increase in the levels of H2Av-phosphorylation when compared to control (Figure 8).

Abnormal nuclei are eliminated by Chk2 activation

Late syncytial embryos of *Drosophila* exhibit two-stage response to DNA damage or replication defects [21]. Two different kinase pathways, ATM/Chk2 pathway and ATR/Chk1 pathway play a major role in response to

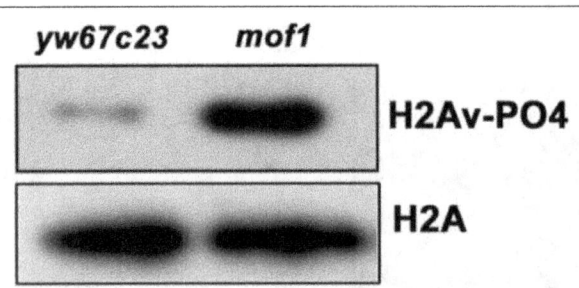

Figure 8 *mof¹* **mutation causes double strand breaks in DNA.** Total protein was isolated from *yw67c23* and *mof1* heterozygous embryos and western blot was carried out using H2Av-PO4 antibody. Here H2A antibody was used as loading control. An increase in the level of phosphorylation of H2Av was observed in the *mof1* heterozygotes.

DNA damage that is evolutionarily conserved. The DNA checkpoint mediated by *mei-41* and *grp*, the *Drosophila* orthologs of *ATR* and *Chk1* kinases, respectively, delay entry into mitosis via inhibitory phosphorylation of Cdk1, which allows repair of DNA damage or completion of DNA replication [22,23]. When this checkpoint fails, a second control operating during mitosis is activated, that results in changes in spindle structure and chromsome segregation to stop propagation of defective or damaged nuclei. This second step of control is mediated by activation of *Chk2* by centrosomal inactivation [24]. The increased number of fall out nuclei and defects during early mitosis in the *mof* heterozygous embryos led us to speculate the possible involvement of a DNA replication dependent or DNA damage dependent cell cycle checkpoint defect. *Drosophila* embryos that lack *Chk1* homologue (*grp*) and *ATR* homologue (*mei-41*) show inactivation of centrosome during the late stages of syncytial division proving that both the homologues are not required for centrosome inactivation [21].

To show that DNA replication checkpoint is intact in the *mof¹* embryos, we studied the levels of *grp* (*Chk1*) and *mei-41* (*ATR*) and we found that their levels remained the same in syncytial cycles 10–13 (Figure 9A). Also the levels of cyclins remained the same in the wild type and *mof¹* embryos in syncytial cycles 10–13 indicating that our data do not support a role of *Drosophila mof* in the control of cell cycle in the syncytial embryos through regulation of cyclins and *grp*. Thus our data do not support a role for *Drosophila* MOF in control of cell cycle timing in syncytial embryos via regulation of cyclins or *grp* levels.

The defective mitotic spindles that are short, anastral and associated with poorly aligned chromosomes in the *mof* embryos exhibited key features reminiscent of *Chk2* mediated centrosomal inactivation. This led us to investigate the possible role of checkpoint gene *Chk2* in this event. Total RNA was isolated from syncytial cycles 10–13 of control *yw⁶⁷ᶜ²³*, *mof¹* heterozygotes and RT-PCR

was carried out using *Chk2* specific primers. We observed increase in the transcript level of *Chk2* by 4-folds in the *mof¹* embryos compared to *yw⁶⁷ᶜ²³* control indicating *Chk2* mediated centrosome inactivation (Figure 9B). To further confirm MOF mediated *Chk2* regulation we have conducted the *mof* knock down experiment using dsRNA in S2 cells. The expression of *Chk2* was found to be enhanced upon *mof* depletion. *GFP* dsRNA did not cause any significant change in levels of mof and thus used as control in the RNAi study (Figure 9B').

Since *Chk2* is a major target of ataxia telangiectasia-mutated (ATM), the expression pattern of ATM in *mof¹* heterozygous embryos was also studied. We observed that there was pronounced increase in levels of ATM in *mof¹* heterozygotes. As reported earlier [11] we also observed increased expression of *p53* in *mof¹* heterozygous embryos indicating that *mof* mutation causes spontaneous DNA damage leading to the activation of ATM-Chk2 pathway (Figure 9A).

Centrosome inactivation in asynchronous nuclei of syncytial *mof* heterozygous embryos

Drosophila Chk2 is encoded by *mnk* (*maternal nuclear kinase*) gene [25] and *mnkp6* homozygous null mutation flies produce DNA damage induced apoptosis [26]. To further confirm *ATM/Chk2* mediated centrosomal inactivation, we crossed *mof¹/FM7* virgin females with *mnkp6* males to produce heterozygous *mof¹/+; mnkᵖ⁶/+* flies. The *mof¹/+; mnkᵖ⁶/+* females were further mated with wild type males and 0–2 h embryos were collected. These embryos were stained with PI to check for nuclear fallout. The *mof¹* embryos exhibited high severity fall out nuclei (more than 5 fall out nuclei/embryo) while *mof¹/+; mnkᵖ⁶/+* embryos had only low severity fall out nuclei (less than 5) (Figure 9C). Thus Chk2 activation contributes significantly to the *mof¹* phenotype in syncytial embryos.

Discussion

MOF is a member of MYST family of histone acetyl transferases and is the essential component of the X-chromosome dosage compensation system in *Drosophila*. All MOF deficient mouse embryos fail to develop the expanded blastocyst stage and die at implantation *in vivo*. In the present study we observed loss of maternal MOF in the *Drosophila* embryos caused mitotic defects during early syncytial cycles and chromosomal aberrations as visualized by the presence of chromatid bridges and lagging chromosomes. These defects occur spontaneously in *mof* heterozygotes without stress and resemble the defects induced by X-ray irradiation or chemical treatment of wild type embryos [27]. Our data has shown that endogenous DNA damage occurs during the process of development as shown by the presence of single and double

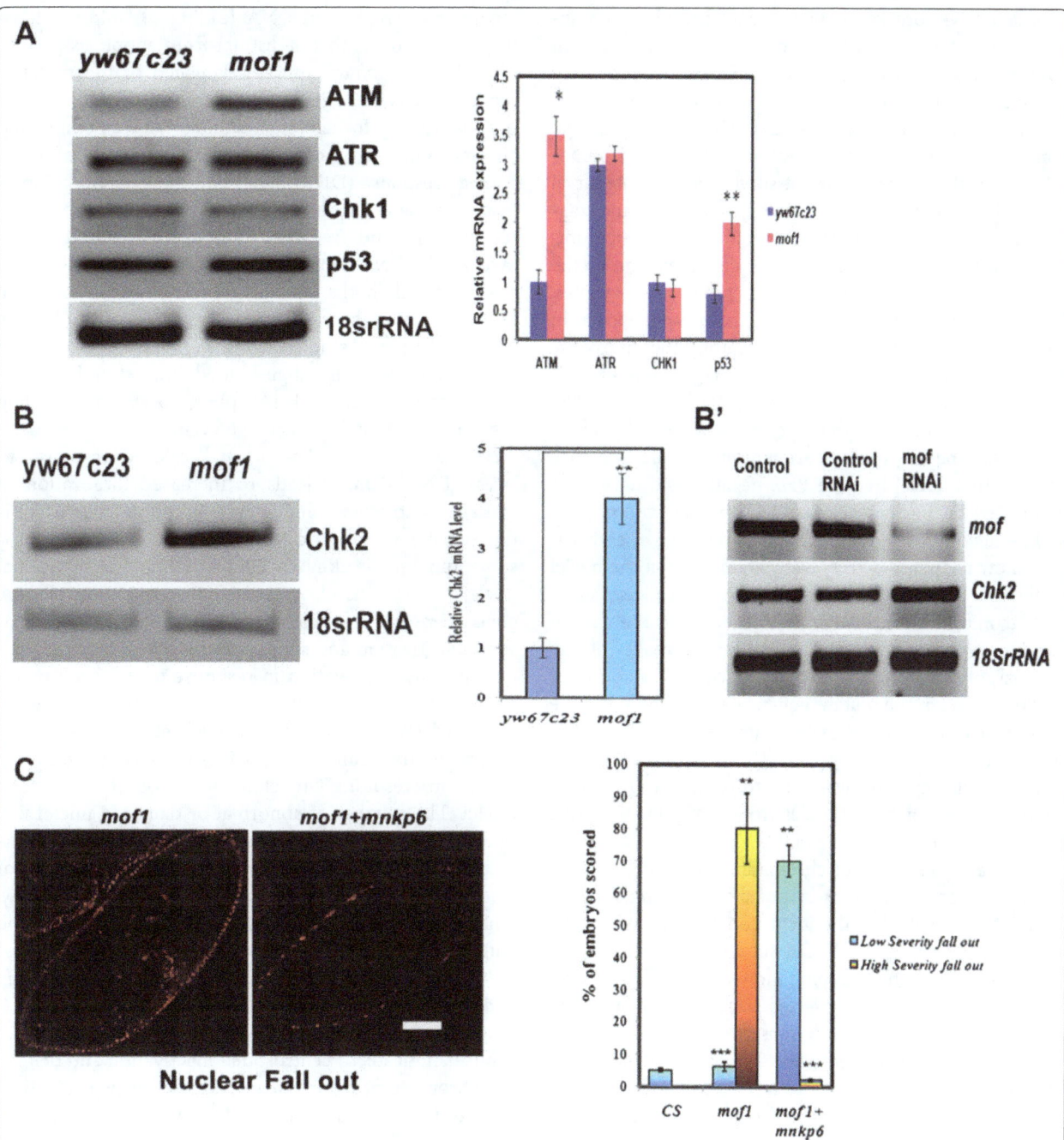

Figure 9 Activation of *Chk2* in the *mof* heterozygous embryos. (**A**) Total RNA was isolated from embryos of syncytial cycles of 10–13 from control *yw⁶⁷ᶜ²³* and *mof¹* embyros and semi-quantitative RT-PCR was carried out using primers against *ATM, ATR, Chk1* and *p53* to study their expression pattern. (**B**) Semi-quantitative RT-PCR was carried out using *chk2* primers in *yw⁶⁷ᶜ²³* and *mof¹* embryos (syncytial cycles 10–13). We observed 4-fold increase in the level of *chk2* in *mof¹* embryos compared to control indicating activation of *chk2*. (**B'**) Depletion of MOF by RNAi was performed in S2 cells (*Drosophila* Schneider cells) by incubation with dsRNA against *mof*. The transfection was conducted for 72 h time period. Total RNA isolated was subjected to RT-PCR analysis against *Chk2* and *mof*. Here dsRNA against GFP was used as control (control RNAi). The depletion of mof leads to an increase in the levels of *Chk2* mRNA. (**C**) Early embryos from *mof¹* and *mof¹/+; mnk^{p6}/+* females were collected and stained with DNA dye PI to study the severity of nuclear fallout. The high severity of nuclear fallout in *mof¹* embryos was drastically reduced in the presence of *mnk^{p6}* (*chk2*) mutation and is graphically represented. Statistical significance was assessed using student t-test. *** indicates P<0.001, ** indicates P<0.01, * indicates p<0.05. Bar indicates 10 µm scale.

stranded DNA breaks. Similar phenotypes like spontaneous mitotic defects and chromosomal aberrations in *Drosophila* were also observed in *recq5* DNA helicase mutants that are involved in DNA replication and maintainance of genomic integrity [28]. MOF is an essential component of dosage compensation complex and homozygotes for the mutation do not survive beyond late larval stages. The nuclei in *mof* early embryos are large and fragmented resembling mitotic catastrophe. The typical mitotic wave in wild type embryos is disrupted due to the presence of abnormal nuclei in the *mof* heterozygotes. Such abnormal nuclei are PH3 negative and appear in clusters [29]. Cellularisation occurs during the interphase of 14th nuclear division. During this stage the actin filaments in *mof^1* embryos loose the typical honeycomb like structures often leading to empty cages without the chromosomes. The empty cages are indicative of the presence of abnormal nuclei that are PH3 negative and have been eliminated by nuclear fallout while the centrosomes are still retained in the cortex. A small percentage of embryos also show telophase defects like rounding off of the nuclei while the chromosomes are still in the anaphase stage. Similar mitotic defects were also observed in the case of another *mof* allele (*mof^3*). The late larval lethality of *mof* homozygotes were rescued to 100% with the addition of *mof* transgene in the mutant genetic background while the mitotic defects in the *mof* heterozygote embryos were partially rescued.

In general every organism tries to protect itself by preventing these abnormal nuclei from being incorporated into forming adult structures [30]. Mutation of PcG genes which are components of chromatin remodeling that aid in the maintenance of transcriptional state during embryogenesis also resulted in the formation of abnormal nuclei [19]. Also the severity of the fall out nuclei in the *mof^1* heterozygotes was high containing more than 5 fall out nuclei per embryo which was reverted back to normal in the presence of wild type *mof* transgene. The response to DNA damage and mitotic defects maintain genomic stability by blocking chromosome segregation and removing the abnormal nuclei by nuclear fallout mechanism. Staining with PH3 antibody and DNA dye in the syncytial embryos is a good system to detect irregular or damaged DNA in *Drosophila* and also for studying maternal genes required for mitosis and genomic stability. The abnormal nuclei that stain negatively for PH3 are asynchronous and are seen during nuclear cycles 11–13. Following mitotic failure the defective nuclei drop into the interior of embryos and free centrosomes are seen in the cortex [29]. In the wild type embryos all the nuclei are in actively dividing phase compared to *mof* heterozygote where in nondividing nuclei are also present.

In a variety of systems, cell cycle checkpoint defects lead to progression into mitosis with damaged DNA or incompletely replicated DNA leading to "mitotic catastrophe", a process that is distinct from apoptosis [31]. In syncytial *Drosophila* embryos damaged or incompletely replicated DNA triggers centrosome inactivation during mitosis leading to defects in spindle fibre assembly and chromosome segregation [24]. The hallmark of DNA damage response (DDR) involves the phosphorylation of histone variant H2AX that play an essential role in the recruitment and retention of downstream proteins involved in DNA repair. In addition to this γ–H2AX is also involved in the transduction and amplification of DDR from megabase domains surrounding the damage site [32]. In our studies using *mof^1* heterozygotes we observed increase in single and double stranded DNA breaks and as well as H2Av phosphorylation revealing the occurrence of DNA damage event.

Drosophila Chk2 plays a vital role in response to stress. DNA damage leads to increased localization of Chk2 to centrosomes and spindle fibres and also Chk2 is the signal for mitotic catastrophe that disrupts centrosome function leading to elimination of the abnormal nuclei [24]. It was also reported that mutation of *Mnk* gene (*Drosophila* homolog of *Chk2*) prevents centrosome inactivation and suppresses defects associated with chromosome segregation in response to damaged or incompletely replicated DNA. In our study we observed increased level of Chk2 in the *mof* heterozygotes in response to the damaged nuclei causing centrosome inactivation resulting in elimination of the damaged nuclei. The number of abnormal or damaged nuclei was reduced in the embryos of *mof^1/+; mnk^{p6}/+* mothers indicating that mutation of *mnk^{p6}* prevents inactivation of centrosomes and hence loss of nuclei from the cortex. Unlike the cell cycle delays that occur to repair the damaged DNA or incompletely replicated DNA, *Drosophila* embryonic system utilizes the delay to identify and discard those abnormal nuclei. When the DNA lesions enter into mitosis, Chk2 is activated as a response and leads to centrosomal inactivation and delinks the chromosomes from their centrosomes which ultimately results in loss of the nuclei [29]. It has been proposed that Chk2 functions at two points during early embryogenesis in response to genotoxic stress. At the onset of mitosis DNA lesions leads to activation of Chk2 that target proteins involved in centrosomal spindle activity and in maintaining γTURC localisation. This causes failure in anaphase chromosome segregation. Once failure of mitotic division occurs, Chk2 causes centrosomal inactivation and disrupts the link between centrosomes and nuclei. Since centrosomes anchor nuclei to the cortex, Chk2 response to DNA damage results in loss of nuclei from the cortex [33]. The sensing of DNA lesions by DDR machinery occurs in a complex and heterogeneous chromatin environment [34,35].

Earlier reports also emphasized on the alteration in the chromatin structure that helps in the sensing and as well as spreading of the DNA damage response apart from double strand breaks [36-38].

DNA damage induces the activation of chromatin bound Chk2 by a chromatin derived signal resulting in the dissociation of the activated Chk2 from the chromatin. Chk2 is phosphorylated at T68 by ataxia telangiectasia-mutated (ATM) and transmits the DNA damage signals from the upstream phosphatidylinositol 3'-kinase like kinases to the effector substrates including p53, Braca1, Cdc25A and Cdc25C [39]. In addition Chk2 has been reported to phosphorylate p53, thereby enhancing the transcriptional activity of p53 responsive genes [40]. Further the functional link between p53 and Chk2 during DNA damage occurs through the phosphorylation and acceleration of degradation of Hdmx, a negative regulator of p53 [41].

This study revealed for the first time the role of MOF during early embryogenesis in *Drosophila* apart from dosage compensation and response to ionizing radiation.

Conclusions

Recent investigations have clearly demonstrated the role of MOF in response to ionizing radiation is conserved in *Drosophila melanogaster*. In human cells knockdown of hMOF results in loss of H4K16Ac and destabilization of nucleosomes that correlates with regions of chromatin decondensation. While acetylated H4 K16 appears to 'open up' the *Drosophila* male X chromosome to make it more accessible to transcription, which is an important part of the dosage compensation mechanism in the fly. Reduced levels of MOF in mammals correlate with decreased H4K16Ac, cell proliferation, cell survival and increased genomic instability. *Drosophila* haplo-insufficency of maternal MOF causes several mitotic defects in the syncytial embryos and a large number of abnormal nuclei have been removed through the process of nuclear fallout. The increased number of abnormal or fall out nuclei correlated with reduced nuclear density in syncytial blastoderm embryos of *mof* heterozygotes. Our study demonstrates that in response to spontaneous DNA damage (increased number of single and double stranded DNA breaks) in *mof* heterozygotes, Chk2 is activated leading to centrosomal inactivation and loss of damaged nuclei from the cortex of the syncytial embryos. Furthermore removal of one copy of *Chk2* in the *mof* mutant background considerably reduced the number of fall out nuclei in the syncytial embryos indicating the restoration of genomic stability. Hence MOF seems to play a crucial role in ensuring genomic stability during early embryogenesis both in mammals and *Drosophila*.

Methods

Fly stocks

Flies were cultured at standard corn meal agar food at 25°C. yw^{67c23} was obtained from flybase. mof^1, mof^3, mof^3+ *18H1*and mnk^{p6} fly stocks were kind gift from John C. Lucchesi, Joel C. Eissenberg and William E. Theurkauf respectively. Both mof^1 and mof^3 are null alleles of *mof* [11].

Immunostaining of *Drosophila* embryos

Drosophila embryos were dechorionised with 50% commercial bleach for 2–3 minutes and fixed with paraformaldehyde/heptane mix for 15–20 min. Primary antibodies for immunostaining were used in the following dilution: goat anti-vasa (1:20), rat anti-α-tubulin (1:20), rabbit anti-PH3 (1:50), mouse anti-β-actin (1:30). Goat or donkey raised FITC, Cy3 and Cy5 conjugated secondary antibodies were used at a dilution of 1:50. Embryos were then mounted in Vectashield mounting media containing PI or DAPI and viewed using confocal microscope (FV1000, Olympus, Japan).

Total RNA isolation and RT- PCR analysis

Total cellular RNA from embryos was isolated by Trizol (Invitrogen). RNase-Free DNase treatment was subsequently carried out to remove DNA contaminants and further cleaned using RNeasy Mini Kit (Qiagen, Germany). Three micrograms of RNA was used for first strand cDNA synthesis using SuperScript™ (Invitrogen, USA). PCR analysis was carried out using the following primers

18SrRNA
FP- 5' CCTTATGGGACGTGTGCTTT 3',
RP- 5' CCTGCTGCCTTCCTTAGATG 3'
Chk2
FP-5' CAAGCTGCTGATCAACCAAA 3'
RP-5' GCCTCGACCCTCACGTATTA 3'
ATM
FP-5' ATCATAGCTTGGGCATACGG3'
RP-5' TTTGTTCTCCTTCGCGATCT3'
p53
FP-5'AAT GCC CAT CCA ACC ACT TA3',
RP-5'AAG GCT CAA CGC TAA GGT GA3'
Mof
FP-5'CTGGGTAGGCTGAGCTATCG3',
RP-5' CCAGACGAGGTAATCGGTGT3'
Chk1/grp
FP-5'CCGGACTCAATTACCTGCAT3',
RP-5'GTTTGCTCCAAGGAGTCTGC 3'
ATR/Mei-41
FP-5' TCAGGAGACGCTAGCCATTT3'
RP-5' TGCAGAACTGCCATGAACTC 3'

Total protein isolation and western blot analysis

Nearly 0.1 gm of embryos were collected and thoroughly homogenized in lysis buffer (6% SDS, 1 mM EDTA, 2 mM

PMSF, 10 μg/ml Aprotinin, 10 μg/ml Leupeptin, 10 μg/ml Pepstatin). The lysed samples were boiled at 95°C for 5 minutes and centrifuged at 12,000 rpm for 10 minutes at 4°C. The supernatant was collected and western blot analysis was carried out by standard protocols using MOF (1:300), ß-actin (1:500) (Abcam) and H2Av-PO4 (1:500) antibodies (LS biosciences).

DNA damage assay

In order to assess DNA damage, genomic DNA was isolated from 0–2 h methanol-fixed wild type control and mof^1 embryos as described earlier with slight modifications [21]. NaOH lysis is followed by incubation at 65°C for 40 min. The embryos were gently homogenised and treated with proteinase K for 8 h at 55°C, followed by RNAse A treatment and phenol-chloroform extraction. Each sample of DNA (100 ng) was incubated at 37°C for 20 min in 25 μl of a reaction mixture containing 50 mM imidazole pH 6.4, 12 mM MgCl2, 1 mM β-mercaptoethanol, 100 μM ADP, 20 units T4 DNA Kinase (Gibco BRL) and 2.5 μl of 3000 Ci / mmol [32P]ATP (Amersham). Column purification was carried out using G-50 spin columns to remove unincorporated labeled nucleotides. The level of p32 incorporated into the genomic DNA was measured by liquid scintillation counter. Over 300 embryos were used to generate the DNA for each experiment and the experiments were repeated thrice.

Knock down studies

MOF knockdown by RNAi was conducted in S2 cells using Schnieder's insect medium supplemented with 10% fetal bovine serum at 25°C. Here for the effective knock down of mof, we transfected 50 μg dsRNA in S2 cells and were incubated for 72 h as described earlier [11]. Control RNAi experiments using GFP was also employed.

Statistical analysis

Statistical analysis was performed using the graph pad software to evaluate the significant difference between the control and treated samples. The results obtained were expressed as mean ± SD. All the experiments were conducted in triplicates. Statistical significance was assessed using student t-test. *** indicates P<0.001, ** indicates P<0.01, * indicates p<0.05.

Abbreviations

Mof: Males absent on the First; HAT: Histone acetyl transferase; DCC: Dosage compensation complex; PI: Propidium Iodide; DDR: DNA damage response; mnk: Maternal nuclear kinase; PH3: Anti-phospho-Histone H3 (Ser10).

Competing interests

The authors declare that they have no competing interests.

Authors' contributions

SNCVLP, AS, MJR and DRC carried out the experiments. SNCLVP and MJR designed the experiments. UB and MPB conceived the idea, designed the experiment and wrote the manuscript. All authors read and approved the final manuscript.

Acknowledgements

The authors thank John C. Lucchesi, Joel C. Eissenberg, William E. Theurkauf and Bloomington Stock Centre for the fly stocks. The authors also thank M. Kuroda for the MOF antibody and Jordon Raff for the centrosomin antibody. This work was supported by Wellcome Trust, UK to authors MBP [GAP 0158] and UB [GAP 0065].

Author details

[1]Centre for Chemical Biology, Indian Institute of Chemical Technology, Hyderabad 500607, India. [2]Functional Genomics and Gene Silencing Group, Centre for Cellular and Molecular Biology, Hyderabad 500007, India.

References

1. Zhou BB, Elledge SJ: The DNA damage response: putting checkpoints in perspective. Nature 2000, 408:433–443.
2. Jaklevic BR, Su TT: Relative contribution of DNA repair, cell cycle check points and cell death to survival after DNA damage in Drosophila larvae. Curr Biol 2004, 14:23–32.
3. Voss AK, Thomas T: MYST family histone acetyltransferases take center stage in stem cells and development. Bioessays 2009, 31:1050–1061.
4. Morales V, Straub T, Neumann MF, Mengus G, Akhtar A, Becker PB: Functional integration of the histone acetyltransferase MOF into the dosage compensation complex. EMBO J 2004, 23:2258–2268.
5. Yang XJ: The diverse superfamily of lysine acetyltransferases and their roles in leukemia and other diseases. Nucleic Acids Res 2004, 32:959–976.
6. Smith ER, Pannuti A, Gu W, Steurnagel A, Cook RG, Allis CD, Lucchesi JC: The Drosophila MSL complex acetylates histone H4 at lysine 16, a chromatin modification linked to dosage compensation. Mol Cell Biol 2000, 20:312–318.
7. Smith ER, Cayrou C, Huang R, Lane WS, Cote J, Lucchesi JC: A human protein complex homologous to the Drosophila MSL complex is responsible for the majority of histone H4 acetylation at lysine 16. Mol Cell Biol 2005, 25:9175–9188.
8. Li X, Wu L, Corsa CA, Kunkel S, Dou Y: Two mammalian MOF complexes regulate transcription activation by distinct mechanisms. Mol Cell 2009, 36:290–301.
9. Hilfiker A, Hilfiker-Kleiner D, Pannuti A, Lucchesi JC: Mof, a putative acetyl transferase gene related to the Tip60 and MOZ human genes and to the SAS genes of yeast, is required for dosage compensation in Drosophila. EMBO J 1997, 16:2054–2060.
10. Taipale M, Rea S, Richter K, Vilar A, Lichter P, Imhof A, Akhtar A: hMOF histone acetyltransferase is required for histone H4 lysine 16 acetylation in mammalian cells. Mol Cell Biol 2005, 25:6798–6810.
11. Pal- Bhadra M, Horikoshi N, Pushpavalli Sreerangam NCVL, Sarkar A, Bag I, Krishnan A, Lucchesi JC, Kumar R, Yang Q, Pandita RJ, Singh M, Bhadra U, Eissenberg JC, Pandita TK: The role of MOF in the ionizing radiation response is conserved in Drosophila melanogaster. Chromosoma 2011, http://dx.doi.org/10.1007/s00412-011-0344-7.
12. Shogren-Knaak M, Ishii H, Sun JM, Pazin MJ, Davie JR, Peterson CL: Histone H4-K16 acetylation controls chromatin structure and protein interactions. Science 2006, 311:844–847.
13. Corona DF, Clapier CR, Becker PB, Tamkun JW: Modulation of ISWI function by site-specific histoneacetylation. EMBO Rep 2002, 3:242–247.
14. Akhtar A, Becker PB: Activation of transcription through histone H4 acetylation by MOF, an acetyltransferase essential for dosage compensation in Drosophila. Mol Cell 2000, 5:367–375.
15. Gupta A, Guerin-Peyrou TG, Sharma GG, Park C, Agarwal M, Ganju RK, Pandita S, Choi K, Sukumar S, Pandita RK, Ludwig T, Pandita TK: The mammalian ortholog of Drosophila MOF that acetylates histone H4

lysine 16 is essential for embryogenesis and oncogenesis. *Mol Cell Biol* 2008, **28**:397–409.

16. Fraga MF, Ballestar E, Villar-Garea A, Boix-Chornet M, Espada J, Schotta G, Bonaldi T, Haydon C, Ropero S, Petrie K, Iyer NG, Perez-Rosado A, Calvo E, Lopez JA, Cano A, Calasanz MJ, Colomer D, Piris MA, Ahn N, Imhof A, Caldas C, Jenuwein T, Esteller M: **Loss of acetylation at Lys16 and trimethylation at Lys20 of histone H4 is a common hallmark of human cancer.** *Nat Genet* 2005, **37**:391–400.

17. Tram U, Riggs B, Sullivan W: **Cleavage and gastrulation in *Drosophila* embryos.** In *Encyclopedia of Life Sciences*. London: Macmillan Reference Ltd; 2002.

18. Foe VE, Odell GM Edgar BA, Bate M, Martinez Arias A: **Mitosis and morphogenesis in the *Drosophila* embryo, Point and counterpoint.** In *The Development of Drosophila melanogaster vol 1*. Cold Spring Harbor, New York: Cold Spring Harbor Laboratory Press; 1993:149–300.

19. Dor E, Beck SA, Brock HW: **Polycomb group mutants exhibit mitotic defects in syncytial cell cycles of *Drosophila* embryos.** *Dev Biol* 2006, **29**:312–322.

20. Fogarty P, Campbell SD, Abu-Shumays R, Phalle BS, Yu KR, Uy GL, Goldberg ML, Sullivan W: **The Drosophila grapes gene is related to checkpoint gene chk1/rad27 and is required for late syncytial division fidelity.** *Curr Biol* 1997, **7**:418–426.

21. Sibon OC, Kelkar A, Lemstra W, Theurkauf WE: **DNA replication/DNA-damage-dependent centrosome inactivation in Drosophila embryos.** *Nat Cell Biol* 2000, **2**:90–95.

22. Sibon OC, Stevenson VA, Theurkauf WE: **DNA-replication checkpoint control at the Drosophila midblastula transition.** *Nature* 1997, **388**:93–97.

23. Sibon OC, Laurencon A, Hawley R, Theurkauf WE: **The Drosophila ATM homologue Mei-41 has an essential checkpoint function at the midblastula transition.** *Curr Biol* 1999, **9**:302–312.

24. Takada S, Kelkar A, Theurkauf WE: **Drosophila checkpoint kinase 2 couples centrosome function and spindle assembly to genomic integrity.** *Cell* 2003, **113**:87–99.

25. Oishi I, Sugiyama S, Otani H, Yamamura H, Nishida Y, Minami Y: **A novel Drosophila nuclear protein serine/threonine kinase expressed in the germline during its establishment.** *Mech Dev* 1998, **71**:49–63.

26. Xu J, Xin S, Du W: **Drosophila Chk2 is required for DNA damage-mediated cell cycle arrest and apoptosis.** *FEBS Lett* 2001, **508**:394–398.

27. Callaini G, Dallai R, Riparbelli MG: **Cytochalasin induces spindle fusion in the syncytial blastoderm of the early *Drosophila* embryo.** *Biol Cell* 1992, **74**:249–254.

28. Nakayama M, Yamaguchi S, Sagisu Y, Sakurai H, Ito F, Kawasaki K: **Loss of RecQ5 leads to spontaneous mitotic defects and chromosomal aberrations in *Drosophila melanogaster*.** *DNA Repair* 2009, **8**:232–241.

29. Sakurai H, Okado M, Fumiaki I, Kawasaki K: **Anaphase DNA bridges induced by lack of RecQ5 in Drosophila syncytial embryos.** *FEBS Lett* 2011, **585**:1923–1928.

30. Sullivan W, Daily DR, Fogarty P, Yook KJ, Pimpinelli S: **Delays in anaphase initiation occur in individual nuclei of the syncytial *Drosophila* embryo.** *Mol Biol Cell* 1993, **4**:885–896.

31. Roninson IB, Broude EV, Chang BD: **If not apoptosis, then what? Treatment-induced senescence and mitotic catastrophe in tumor cells.** *Drug Resist Update* 2001, **4**:303–313.

32. Polo SE, Jackson SP: **Dynamics of DNA damage response proteins at DNA breaks: a focus on protein modifications.** *Genes Dev* 2011, **25**:409–433.

33. Raff JW, Glover DM: **Centrosomes, and not nuclei, initiate pole cell formation in Drosophila embryos.** *Cell* 1989, **57**:611–619.

34. Misteli T, Soutoglou E: **The emerging role of nuclear architecture in DNA repair and genome maintenance.** *Nat Rev Mol Cell Biol* 2009, **10**:243–254.

35. Shi L, Oberdoerffer P: **Chromatin dynamics in DNA double-strand break repair.** *Biochim Biophys Acta* 2012, **1819**:811–819.

36. Bakkenist CJ, Kastan MB: **DNA damage activates ATM through intermolecular autophosphorylation and dimer dissociation.** *Nature* 2003, **421**:499–506.

37. Bencokova Z, Kaufmann MR, Pires IM, Lecane PS, Giaccia AJ, Hammond EM: **ATM activation and signaling under hypoxic conditions.** *Mol Cell Biol* 2009, **29**:526–537.

38. Hunt CR, Pandita RK, Laszlo A, Higashikubo R, Agarwal M, Kitamura T, Gupta A, Rie N, Horikoshi N, Baskaran R, Lee JH, Lobrich M, Paull TT, Roti JL: **Hyperthermia activates a subset of ataxia-telangiectasia mutated effectors** independent of DNA strand breaks and heat shock protein 70 status. *Cancer Res* 2007, **67**:3010–3017.

39. Li J, Stern DF: **DNA damage regulates Chk2 association with chromatin.** *J Biol Chem* 2005, **280**:37948–37956.

40. Takai H, Naka K, Okada Y, Watanabe M, Harada N, Saito S, Anderson CW, Appella E, Nakanishi M, Suzuki H, Nagashima K, Sawa H, Ikeda K, Motoyama N: **Chk2-deficient mice exhibit radioresistance and defective p53-mediated transcription.** *EMBO J* 2002, **21**:5195–5205.

41. Chen LD, Gilkes M, Pan Y, Lane WS, Chen J: **ATM and Chk2-dependent phosphorylation of MDMX contribute to p53 activation after DNA damage.** *EMBO J* 2005, **24**:3411–3422.

MKL1 inhibits cell cycle progression through p21 in podocytes

Shuang Yang[1*], Lingjia Liu[1], Pengjuan Xu[2] and Zhuo Yang[1*]

Abstract

Background: The glomerular podocyte is a highly specialized cell type with the ability to ultrafilter blood and support glomerular capillary pressure. However, little is known about the genetic programs leading to this functionality or the final phenotype.

Results: In the current study, we found that the expression of a myocardin/MKL family member, MKL1, was significantly upregulated during cell cycle arrest induced by a temperature switch in murine podocyte clone 5 (MPC5) cells. Further investigation demonstrated that overexpression of MKL1 led to inhibition of cell proliferation by decreasing the number of cells in S phase of the cell cycle. In contrast, MKL1 knockdown by RNA interference had the opposite effect, highlighting a potential role of MKL1 in blocking G1/S transition of the cell cycle in MPC5 cells. Additionally, using an RT^2 Profiler PCR Array, p21 was identified as a direct target of MKL1. We further revealed that MKL1 activated p21 transcription by recruitment to the CArG element in its promoter, thus resulting in cell cycle arrest. In addition, the expression of MKL1 is positively correlated with that of p21 in podocytes in postnatal mouse kidney and significantly upregulated during the morphological switch of podocytes from proliferation to differentiation.

Conclusions: Our observations demonstrate that MKL1 has physiological roles in the maturation and development of podocytes, and thus its misregulation might lead to glomerular and renal dysfunction.

Keywords: Kidney development, Podocyte, Cell growth arrest, MKL1

Background

Podocytes, also called visceral glomerular epithelial cells, are terminally differentiated cells overlaying the outer region of the glomerular basement membrane of renal glomeruli. These cells have several key functions including the prevention of proteinuria, synthesis of basement membrane components, regulation of glomerular filtration, and counteraction of intraglomerular hydrostatic pressure [1]. Podocyte injury is typically associated with proteinuria and progressive glomerulosclerosis [2].

Podocytes are derived from epithelial cells originating in the metanephric mesenchyme, which develop into postmitotic terminally differentiated cells, and therefore have similarities to neurons [3,4]. During glomerulogenesis, podocytes proliferate until the S-shape body stage

and exit the cell cycle at the capillary loop stage [5,6]. Podocytes then acquire their fully differentiated phenotype, a process that is not complete until 1 week after birth in the mouse. Mature podocytes tightly regulate and maintain their quiescent and differentiated phenotype, and therefore lost podocytes cannot be replaced by proliferation of neighboring undamaged cells. Indeed, studies have shown that the inability to proliferate contributes to glomerular scarring [7]. Thus, the mechanism responsible for the cell cycle arrest that occurs during nephrogenesis may also participate in maintenance of cell cycle quiescence in mature podocytes.

Megakaryoblastic leukemia 1 (MKL1) was originally found in a study of a chromosomal translocation, t (1;22), which is closely related to the incidence of acute megakaryoblastic leukemia in infants and children [8,9]. MKL1 has been recently shown to be a member of a three-protein family that includes MKL2 and myocardin. These myocardin/MKL proteins serve as serum response factor (SRF) coactivators by binding to SRF and

* Correspondence: yangshuang@nankai.edu.cn; zhuoyang@nankai.edu.cn
[1]Medical School, Tianjin Key Laboratory of Tumor Microenvironment and Neurovascular Regulation, Nankai University, 94 Weijin Road, Tianjin 300071, China
Full list of author information is available at the end of the article

strongly activating SRF target genes [10-13]. In contrast to myocardin, which has cardiac and smooth muscle-specific expression [10,14-16], MKL1 and MKL2 are expressed in a wide range of embryonic and adult tissues [8,11,17-20]. MKL1 regulates many processes including muscle cell differentiation [17], cardiovascular development [18], remodeling of neuronal networks in the developing and adult brain [19], megakaryocytic differentiation and migration [21,22], modulation of cellular motile functions, and epithelial-mesenchymal transition [23-25]. Notably, there is increasing evidence of the involvement of the myocardin/MKL family in suppression of cell proliferation and cell cycle progression. Both myocardin and MKL1 exert anti-proliferative effects in various cell lines [26-28]. Therefore, unraveling the functional pathways in which these proteins have a role and furthering our comprehension of the cellular mechanisms intrinsic to their regulation of cell proliferation will become increasingly important.

In the present study, we found that MKL1 expression was upregulated during temperature-switched growth arrest in murine podocyte clone 5 (MPC5) cells. Overexpression of MKL1 resulted in inhibition of G1/S cell cycle progression in cell viability and EdU cell proliferation assays, whereas MKL1 knockdown had the opposite effect. We further demonstrated that MKL1 induced expression of the cyclin-dependent kinase inhibitor (CKI) p21 during the regulation of cell cycle arrest. Importantly, MKL1 expression was observed in podocytes of the mouse kidney during postnatal development, which was upregulated during the morphological switch of podocytes from proliferation to differentiation.

Results

Expression of MKL1 is upregulated during temperature-switched cell growth arrest in MPC5 cells

To assess the possible role of myocardin/MKL proteins in podocyte growth arrest, MPC5 cells were respectively maintained at the permissive temperature of 33°C and at the nonpermissive temperature of 37°C for 10 days. As shown in Figure 1A, a cell viability assay indicated that MPC5 cells cultured at 37°C showed a significant decrease in cell number compared with those cultured at 33°C. The results of immunofluorescence staining in the EdU cell proliferation assay further revealed a marked decrease in the number of cells in S phase by the temperature switch to 37°C (Figure 1B and C). The percentage of cells in S phase decreased from 46.37% to 20.03% as early as 2 days after the temperature switch. At 4–6 days, the percentage of cells in S phase further decreased from 14.97% to 6.70%, which is consistent with a previous report indicating that the temperature switch induces growth arrest of podocytes *in vitro* [29].

The expression of myocardin/MKL proteins was then measured during temperature-switched growth arrest in MPC5 cells. qPCR analysis indicated that the temperature switch to 37°C induced an approximate 1.8-fold increase in MKL1 mRNA expression compared with the basal level at 2 days (Figure 1D). At 4–10 days, MKL1 expression showed a 2-4-fold increase at the mRNA level. Western blotting was used to confirm the upregulation of MKL1 expression at the protein level (Figure 1E). However, the alteration in myocardin and MKL2 expression was not as evident (Figure 1D). Considering the dominant presence of MKL1 over its other family members, we focused on the effects of MKL1 in subsequent experiments.

MKL1 functions as an effective inducer of cell growth arrest in MPC5 cells

Next, a mouse MKL1 expression plasmid [11] was transiently transfected into MPC5 cells. Overexpression of MKL1 was assessed by western blotting (Figure 2A). Compared with control cells, the cell viability assay indicated that ectopic expression of MKL1 inhibited MPC5 cell proliferation (Figure 2B). Results of DNA analysis by flow cytometry further confirmed that MKL1-overexpressing MPC5 cells had a lower population of S phase cells and a higher population of G0/G1 phase cells (Additional file 1: Figure S1). The EdU cell proliferation assay revealed a marked decrease in the number of S phase cells after MKL1 overexpression (Figure 2C). The percentage of cells in S phase decreased from 55.56% to 28.39% at 72 h after transfection of the MKL1 expression plasmid. Furthermore, MPC5 cells were stably transfected with either the MKL1 expression plasmid (ΔMKL1) or the empty vector (ΔControl). Overexpression of MKL1 was then examined by western blotting (Figure 2D). Cell viability and EdU cell proliferation assays confirmed that MKL1 overexpression induced a delay in G1/S phase transition of MPC5 cells (Figure 2E and F).

Therefore, we hypothesized that knockdown of MKL1 by RNA interference would result in an increase in the number of cells in S phase. To test our hypothesis, a MKL1-targeting shRNA plasmid (shMKL1) or a scrambled control shRNA plasmid (shControl) were transiently transfected into MPC5 cells. Knockdown of MKL1 expression was confirmed by western blotting (Figure 3A). Compared with shControl cells, the cell viability assay indicated that depletion of MKL1 promoted MPC5 cell proliferation (Figure 3B). The results of DNA analysis by flow cytometry further showed that MKL1 knockdown MPC5 cells had a higher population of cells in S phase and a lower population of cells in G0/G1 phase compared with control cells (Additional file 2: Figure S2). The EdU cell proliferation assay revealed that

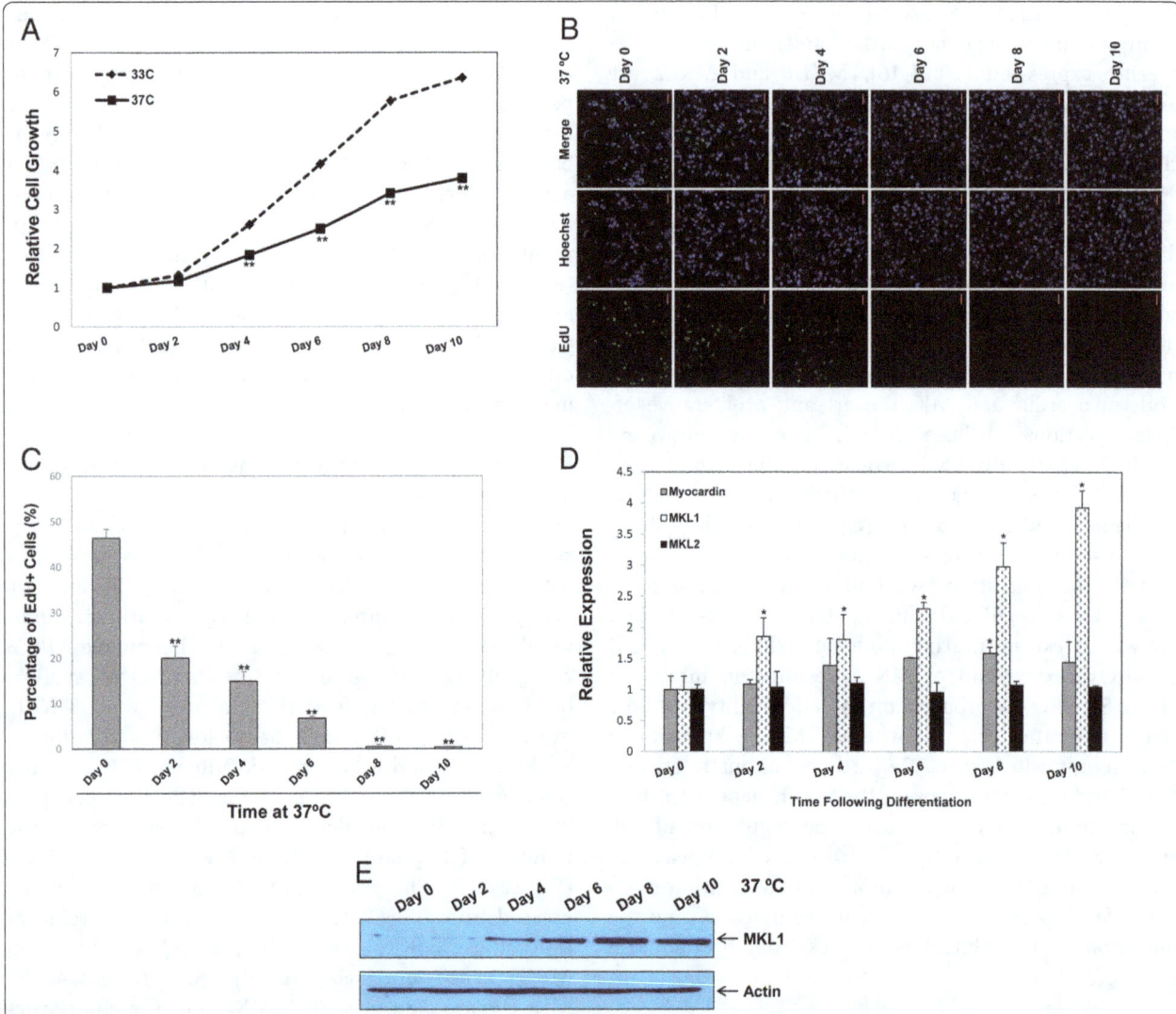

Figure 1 MKL1 is upregulated during temperature-switched cell cycle arrest in MPC5 cells. A) MPC5 cells were cultured at the permissive temperature of 33°C or the nonpermissive temperature of 37°C. At the indicated time points, cell growth was measured using a CCK-8 assay. *******p* < 0.01 compared with the control (one-way ANOVA followed by Tukey's HSD test). **B)** MPC5 cells were shifted from 33°C to 37°C and cultured for the indicated times. Cell proliferation was measured by immunofluorescence analysis of EdU incorporation. Scale bars, 25 μm. **C)** The percentage of proliferating cells was calculated as EdU-positive cells/Hoechst-stained cells × 100%. **p < 0.01 compared with the control (one-way ANOVA followed by Tukey's HSD test). **D)** MPC5 cells were shifted from 33°C to 37°C and cultured for the indicated times. The mRNA expression levels of myocardin, MKL1 and MKL2 were verified by qPCR. GAPDH was used to normalize expression levels. *p < 0.05 compared with the control (one-way ANOVA followed by Tukey's HSD test). **E)** MPC5 cells were shifted from 33°C to 37°C and cultured for the indicated times. Protein expression levels of MKL1 were examined by western blotting. Actin was used to normalize MKL1 levels.

repression of MKL1 resulted in a significant increase in the number of cells in S phase from 47.40% to 77.07% at 72 h after MKL1 knockdown (Figure 3C). In addition, MPC5 cells were stably transfected with either shMKL1 (ΔshMKL1) or shControl (ΔshControl). Knockdown of MKL1 expression was then confirmed by western blotting (Figure 3D). The cell viability and EdU cell proliferation assays confirmed that repression of MKL1 remarkably promoted cell cycle progression through S phase in MPC5 cells (Figure 3E and F).

MKL1 regulates podocyte proliferation by targeting p21
To identify the potential cellular pathways regulated by MKL1, differences in the mRNA levels of selected signaling molecules were examined using an RT^2 Profiler PCR Array by comparing MKL1-expressing MPC5 cells with the control cells. We observed alterations in the expression of several cell cycle regulators, including p21, Gadd45a, Ddit3, E2F2, and cyclin A1 (Table 1). qPCR and western blotting were performed to verify these findings (Figure 4A and Additional file 3: Figure S3).

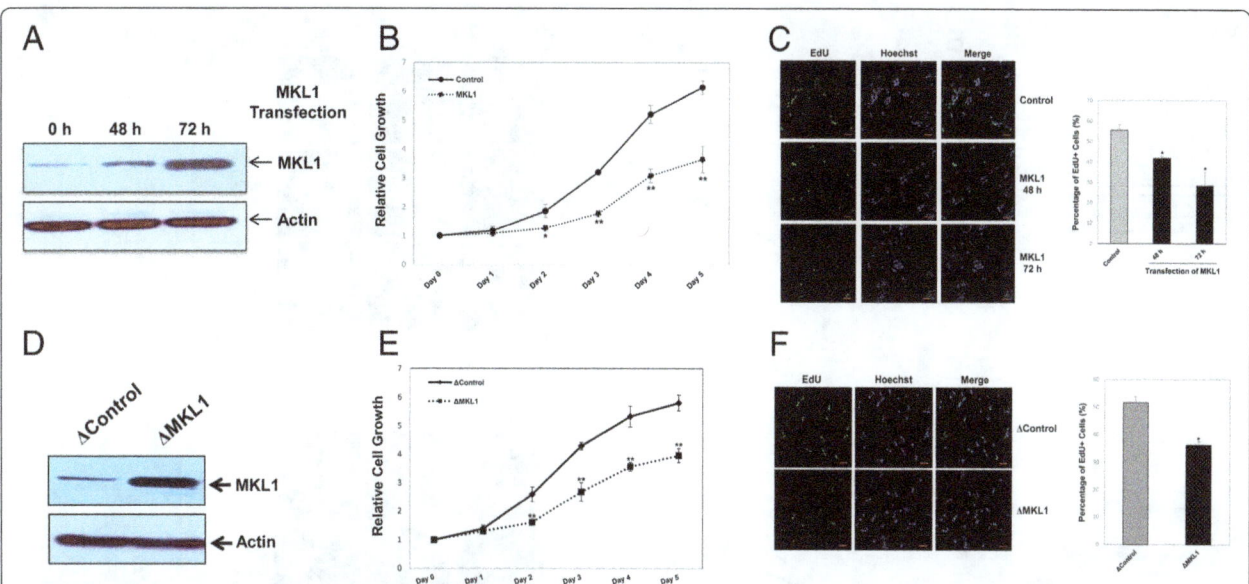

Figure 2 Overexpression of MKL1 induces MPC5 cell growth arrest. A) MPC5 cells were transiently transfected with a mouse MKL1 expression plasmid and cultured at 33°C. Expression of MKL1 protein was verified by western blotting. Actin was used to normalize MKL1 levels. **B)** At the indicated time points, cell growth was measured using the CCK-8 assay. *$p < 0.05$ and **$p < 0.01$ compared with the control (one-way ANOVA followed by Tukey's HSD test). **C)** Cell proliferation was measured by immunofluorescence analysis of EdU incorporation. Scale bars, 25 μm. The percentage of proliferating cells was calculated as EdU-positive cells/Hoechst-stained cells × 100%. *$p < 0.05$ compared with the control (one-way ANOVA followed by Tukey's HSD test). **D)** MPC5 cells were stably transfected with a mouse MKL1 expression plasmid and cultured at 33°C. Expression of MKL1 protein was verified by western blotting. Actin was used to normalize MKL1 levels. **E)** At the indicated time points, cell growth was measured using the CCK-8 assay. **$p < 0.01$ compared with the control (one-way ANOVA followed by Tukey's HSD test). **F)** Cell proliferation was measured by immunofluorescence analysis of EdU incorporation. Scale bars, 25 μm. The percentage of proliferating cells was calculated as EdU-positive cells/Hoechst-stained cells × 100%. *$p < 0.05$ compared with the control (one-way ANOVA followed by Tukey's HSD test).

Considering that MKL1 functions with its co-factor SRF by binding to the CArG box in the promoter region of target genes [12,13], we performed a search of the transcription factor database TRANSFAC and identified a CArG box (CCTTTTCTGG) at position −316/-307 in the mouse p21 promoter (Figure 4B). Thus, we assessed whether MKL1 was a bona fide activator of p21 transcription using reporter gene assays. As shown in Figure 4C, MKL1 significantly increased mouse p21 promoter activity of the wild-type −1562/+200 reporter by approximately 49% relative to the control without MKL1 transfection. Furthermore, we found that MKL1 activated the promoter activity of p21 in a dose-dependent manner (Additional file 4: Figure S4). A series of truncated p21 promoter-reporter constructs were thus generated for analysis, as shown in Figure 4B. The results showed that deletion of the CArG box significantly abolished MKL1-induced transactivation of the p21 promoter compared with that in the control without MKL1 transfection (Figure 4C). Next, we prepared mutants of the CArG box (CCTTTTCTGG to CCTTTTCTTT) by site-directed mutagenesis. We found that mutation of the CArG box was sufficient to interfere with MKL1-activated transcription of the p21 promoter (Figure 4D). Chromatin immunoprecipitation

(ChIP) assays were then performed using an anti-MKL1 antibody, anti-SRF antibody, or control IgG in MPC5 cells. The results indicated that both MKL1 and SRF were able to bind to the p21 promoter during basal conditions in a CArG-dependent manner (Figure 4E). Overexpression of MKL1 resulted in a 1.9-fold increase in its binding to the endogenous p21 promoter in qChIP analysis (Figure 4F). These results suggested that the overexpressed MKL1 in conjunction with SRF promotes p21 transcription by binding to the CArG box in its promoter.

Importantly, to further show that MKL1 regulates podocyte proliferation by targeting p21, MKL1-overexpressing MPC5 cells (ΔMLK1) were transfected with a p21-targeting shRNA plasmid (shp21). Expression of MKL1 and p21 was assessed by western blotting (Figure 4G). The EdU cell proliferation assay revealed a marked decrease in the number of cells in S phase after MKL1 overexpression, whereas p21 interference remarkably attenuated the MKL1-inhibited cell cycle progression through S phase. The percentage of cells in S phase increased from 41.00% to 67.19% at 48 h after transfection of shp21 in MKL1-overexpressing MPC5 cells (Figure 4H). These observations confirmed that MKL1 inhibits MPC5 cell proliferation, which is effectively mediated by targeting p21.

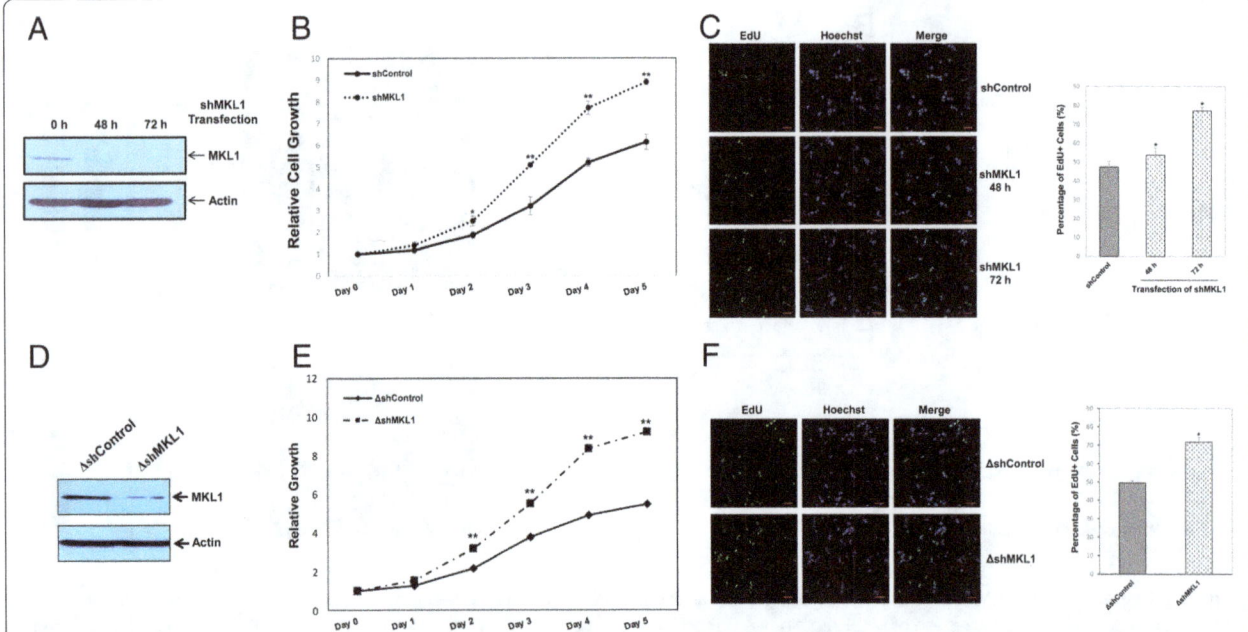

Figure 3 Knockdown of MKL1 promotes cell cycle progression through S phase. A) MPC5 cells were transiently transfected with a MKL1-specific shRNA plasmid (shMKL1) or a scrambled shRNA control plasmid (shControl) and cultured at 33°C. The efficiency of MKL1 protein knockdown was examined by western blotting. Actin was used to normalize MKL1 levels. **B)** At the indicated time points, cell growth was measured using the CCK-8 assay. * $p < 0.05$ and ** $p < 0.01$ compared with the control (one-way ANOVA followed by Tukey's HSD test). **C)** Cell proliferation was measured by immunofluorescence analysis of EdU incorporation. Scale bars, 25 μm. The percentage of proliferating cells was calculated as EdU-positive cells/Hoechst-stained cells × 100%. * $p < 0.05$ compared with the control (one-way ANOVA followed by Tukey's HSD test). **D)** MPC5 cells were stably transfected with a MKL1-specific shRNA plasmid (ΔshMKL1) or a scrambled shRNA control plasmid (ΔshControl) and cultured at 33°C. Expression of MKL1 protein was verified by western blotting. Actin was used to normalize MKL1 levels. **E)** At the indicated time points, cell growth was measured using the CCK-8 assay. ** $p < 0.01$ compared with the control (one-way ANOVA followed by Tukey's HSD test). **F)** Cell proliferation was measured by immunofluorescence analysis of EdU incorporation. Scale bars, 25 μm. The percentage of proliferating cells was calculated as EdU-positive cells/Hoechst-stained cells × 100%. * $p < 0.05$ compared with the control (one-way ANOVA followed by Tukey's HSD test).

The expression of MKL1 and p21 is positively correlated in podocytes *in vivo*

Next, we detected the appearance of MKL1 in developing podocytes by examining its expression in the newborn mouse kidney that displays glomeruli at various stages of development from the S-shaped body through the capillary loop stage to mature glomeruli [5,6]. Immunofluorescence was used to detect MKL1 expression in the mouse kidney at postnatal day (P) 1, 3, 5, 7, 14, 21, 28 and 49. As seen in Figure 5A b-e, S-shaped and comma-shaped bodies were observed in the renal cortex of immature mice at P1–5, ultimately vascularizing into

a capillary loop of mature nephron at P7. Moreover, MKL1 expression was found at all stages of renal glomerulus and tubule formation during postnatal development (Figure 5A b-i). Importantly, there was an increase in the expression of MKL1 in glomeruli between P5 and P7 (Figure 5A d-e).

Consequently, to address the correlation between MKL1 and p21 expression in newborn mouse kidney, immunofluorescent staining was used to detect p21 expression at P3-28. The results showed a relatively weak expression of p21 in glomeruli at P3-7 during the postnatal development (Figure 5B b-d). A remarkable

Table 1 Genes regulated by MKL1

Unigene	GeneBank™ accession no.	Symbol	Description	Fold change (MKL1 vs. Control)	P value
Mm.195663	NM_007669	Cdkn1a	Cyclin-dependent kinase inhibitor 1A (p21)	4.8402	0.0317
Mm.72235	NM_007836	Gadd45a	Growth arrest and DNA-damage-inducible 45 alpha	2.9501	0.0268
Mm.110220	NM_007837	Ddit3	DNA-damage inducible transcript 3	2.5385	0.0282
Mm.307932	NM_177733	E2F2	E2F transcription factor 2	-8.3195	0.0146
Mm.4815	NM_007628	Ccna1	Cyclin A1	-6.1312	0.0195

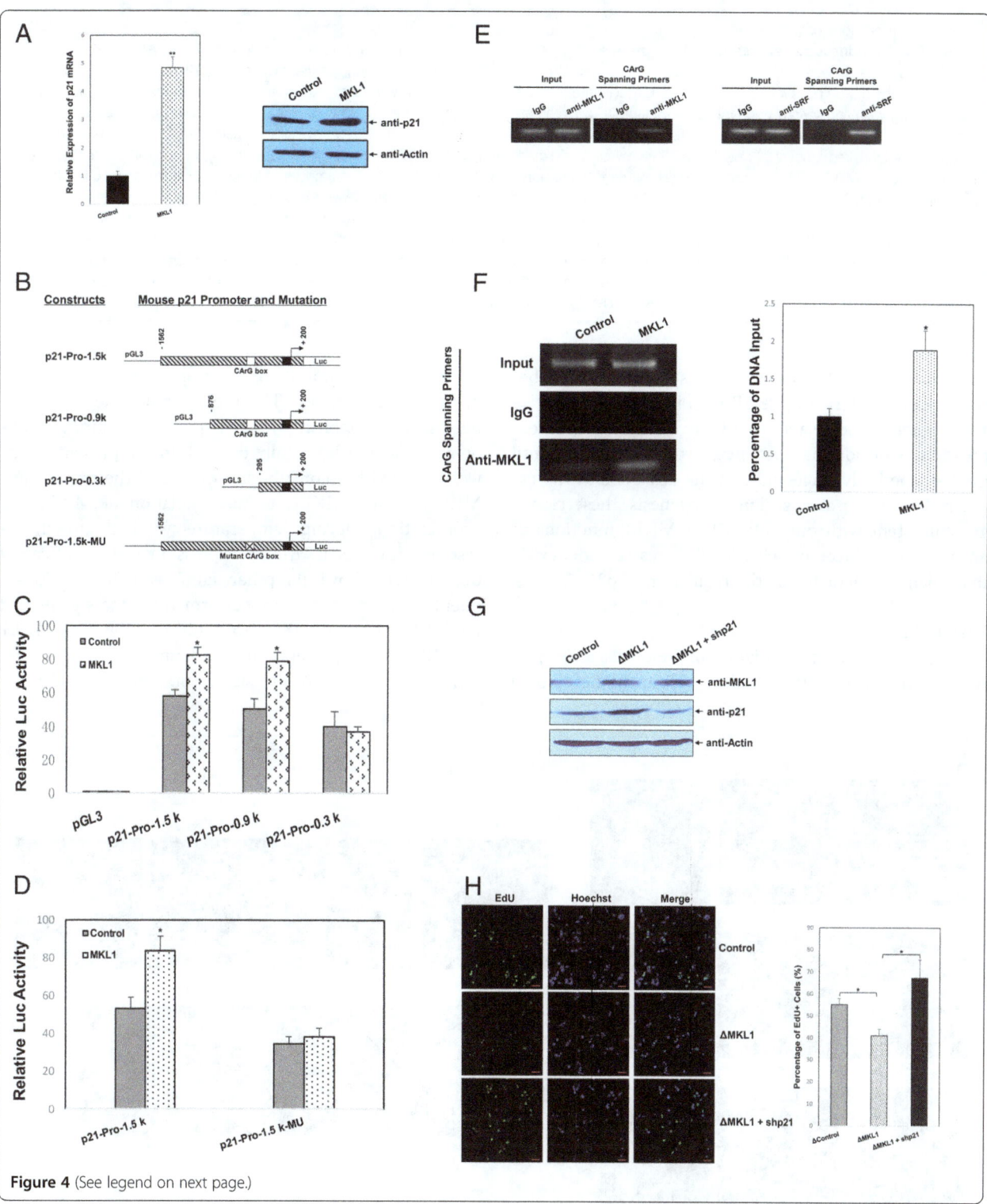

Figure 4 (See legend on next page.)

(See figure on previous page.)

Figure 4 MKL1 induces p21 expression. A) The expression of p21 was examined by qPCR and western blotting in MKL1-overexpressing MPC5 cells. GAPDH and actin were used to normalize p21 levels. **$p < 0.01$ compared with the control (unpaired Student's t-test). **B)** Sequential deletions and mutations of the mouse p21 promoter were fused to a luciferase reporter. MPC5 cells in 24-well plates were co-transfected with the MKL1 expression plasmid (1 μg/well) and various wild-type **C)** or mutant **D)** p21 promoter luciferase reporters (1 μg/well). The luciferase activity of the extracts was determined at 24 h after transfection using a Betascope analyzer. Luciferase values were normalized to Renilla activities. *$p < 0.05$ compared with the empty vector (unpaired Student's t-test). **E)** ChIP assays were performed using an anti-MKL1 antibody, anti-SRF antibody, or control IgG in MPC5 cells. The association of MKL1 or SRF with the proximal mouse p21 promoter was analyzed by PCR. The amount of input was confirmed by equal loading of chromatin. **F)** ChIP assays were performed using the anti-MKL1 antibody or control IgG in MKL1-overexpressing MPC5 cells. The association of MKL1 with the proximal mouse p21 promoter was analyzed by PCR or qPCR. The amount of input DNA was confirmed by equal loading of chromatin. *$p < 0.05$ compared with the empty vector (unpaired Student's t-test). **G)** ΔMKL1 cells were transiently transfected with a p21-specific shRNA plasmid (shp21) and cultured at 33°C. The efficiency of p21 knockdown was examined by western blotting. Actin was used to normalize MKL1 and p21 levels. **H)** Cell proliferation was measured by immunofluorescence analysis of EdU incorporation. Scale bars, 25 μm. The percentage of proliferating cells was calculated as EdU-positive cells/Hoechst-stained cells × 100%. *$p < 0.05$ compared with the control (one-way ANOVA followed by Tukey's HSD test).

upregulation of p21 expression was then observed between P14 and P21 (Figure 5B e-f), indicating a positive and sequential correlation between MKL1 and p21 expression. Considering that podocytes proliferate until the S-shape body stage and exit the cell cycle at the capillary loop stage during glomerulogenesis, these results are consistent with our notion that MKL1 functions as an effective inducer of cell growth arrest in podocytes, that might be mediated by the regulation of p21.

Discussion

Growth arrest and differentiation of podocytes are essential for normal formation of glomeruli in the developing kidney and paramount for normal glomerular function in the mature kidney. The precise cell cycle regulation necessary to establish podocyte quiescence during development has not been fully defined. In the present study, we identified the contribution of one of the myocardin/MKL proteins, MKL1, to the regulation of MPC5 cell proliferation. During temperature-switched growth arrest of MPC5 cells, MKL1 expression was significantly upregulated above its other family members. Consequently, experiments were performed to assess gain- and loss-of-function of MKL1 to study the effect of MKL1 on MPC5 cell proliferation. We found that overexpression of MKL1 resulted in significant repression of G1/S

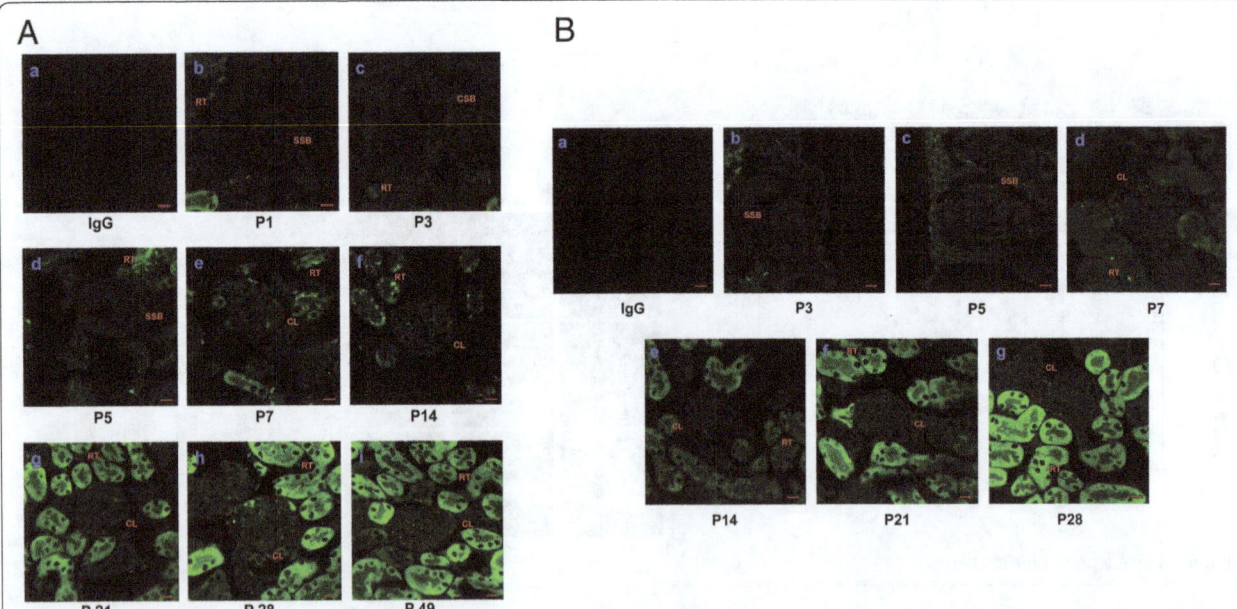

Figure 5 The expression of MKL1 and p21 is correlated in podocytes *in vivo*. A) MKL1 expression was examined by immunofluorescence staining of the mouse renal cortex at P1 (b), P3 (c), P5 (d), P7 (e), P14 (f), P21 (g), P28 (h), and P49 (i). The negative control image shows the renal cortex stained with a species-appropriate IgG (a). Scale bar, 50 μm. **B)** The expression of p21 was examined by immunofluorescent staining in the mouse renal cortex at P3 (b), P5 (c), P7 (d), P14 (e), P21 (f) and P28 (g). The negative control image was from the renal cortex, in which the primary antibody was a species-appropriate IgG (a). Scale bar, 50 μm. The S-shaped body (SSB), comma-shaped body (CSB), capillary loop (CL), and renal tubule (RT) are indicated.

phase progression of the cell cycle, whereas MKL1 knockdown had the opposite effect. Importantly, we demonstrated that MKL1 activated transcription of the cell cycle inhibitor p21 by recruitment to its promoter in a CArG element-dependent manner. In addition, MKL1 expression in the mouse kidney during postnatal development showed that MKL1 is expressed in podocytes *in vivo*, which was significantly upregulated during the morphological switch of podocytes from proliferation to differentiation.

Recent evidence supports an important and complex role of MKL1 in maintenance of proper differentiation in various cell lines including smooth muscle cells [30], myofibroblasts [31], megakaryocytes [22], and adipocytes [32]. Whether MKL1 functions in cell proliferation, however, remains poorly understood. Shaposhnikov *et al.* reported that MKL1 regulates the expression levels of two proapoptotic Bcl-1 family members, Bok and Noxa, and thus is involved in apoptotic signaling in NIH3T3 fibroblasts [33]. Here, we extended the study to show that elevated expression of MKL1 significantly blocked cell cycle progression by decreasing the number of MPC5 cells in S phase. On the other hand, RNA interference of MKL1 expression resulted in the opposite effect to promote G1/S transition of the cell cycle, confirming a potential role of MKL1 in regulation of cell proliferation. These observations collectively position MKL1 as a potential regulatory factor at the crossroad between cell proliferation and differentiation.

Indeed, myocardin family members have been previously implicated in the regulation of cell proliferation. Myocardin impairs the proliferation of vascular smooth muscle cells, Chinese hamster ovarian (CHO) cells, and leiomyosarcoma cells [27,28]. In HT1080 fibrosarcoma cells, myocardin inhibits proliferation at a low cell density and abrogates colony formation in soft agar [34]. However, the suggested molecular mechanisms underlying myocardin/MKL-induced cell growth arrest are conflicting. In human uterine leiomyosarcoma cells, myocardin in conjunction with SRF directly binds to the p21 promoter and induces its expression, thus resulting in G1/S arrest [27]. In contrast, myocardin has been shown to downregulate expression of c-myc, CDK1, CDK2, and S6K, but not p21 and p27, which leads to G2/M arrest and accumulation of polyploidy cells [35]. Our data presented here support the notion that MKL1 functions as an effective regulator to inhibit cell proliferation by altering p21 expression in MPC5 cells. Significant upregulation of p21 expression at both mRNA and protein levels was observed after transfecting the MKL1 expression plasmid, whereas MKL1 interference resulted in the opposite effect to downregulate p21 expression (Additional file 5: Figure S5), suggesting that a direct mechanism is involved in this regulation. We further

demonstrated that MKL1 induced promoter activity of the *p21* gene in a dose-dependent manner. Importantly, we found that deletion or mutation of the CArG element in the mouse p21 promoter remarkably abolished the stimulatory effect on p21 transcription induced by MKL1. Transfection of the MKL1 expression plasmid led to a marked increase in the binding affinity of MKL1 for the endogenous p21 promoter, indicating a significant role of the CArG element in mediating MKL1-induced expression of p21. In addition to p21, we identified obvious candidates involved in MKL1-regulated MPC5 cell proliferation, such as Gadd45a, Ddit3, E2F2, and cyclin A1. However, these genes are not potential targets of myocardin/MKLs/SRF (unpublished data). These results indicate that an SRF-independent mechanism might contribute to MKL-mediated G1/S arrest of the cell cycle.

In the present study, we found that MKL1 was expressed in podocytes of the mouse kidney during postnatal development. Moreover, a significant increase in MKL1 expression was observed between P5 and P7 during postnatal development of the kidney, highlighting a potential role of MKL1 in the physiological and morphological switch of podocytes from proliferation to differentiation. Therefore, using the conditionally immortalized mouse podocyte cell line MPC5, we further revealed that MKL1 functioned as an effective inducer to inhibit cell proliferation and trigger cell cycle arrest at G1/S transition. Several studies have also demonstrated the presence of an intrinsic barrier to replication associated with activation of the cell cycle in podocytes. Re-expression of cell cycle proteins has been reported during glomerular disorders. *De novo* cyclin A staining is observed in podocytes of children with collapsing glomerulopathy [36] and focal segmental glomerulosclerosis (FSGS) [37]. Positive signals for cyclin D have also been reported in the cellular lesions of FSGS [38]. Recently, strong upregulation of CKIs p21 and p27 was reported in podocytes during Heymann nephritis and in diabetic ZDF-fa/fa rats [39,40]. Moreover, the glomerular tufts in crescentic glomerulonephritis strongly express CKIs [41], suggesting that podocytes upregulate CKIs to maintain cell cycle quiescence and preserve normal physiological functions. Here, we extended the study showing that MKL1 acted as an upstream regulator of a variety of cell cycle factors, such as p21 and cyclin A1, to control cell cycle progression in podocytes. In addition, we found significant upregulation of MKL1 expression in the renal tubular cells of newborn mouse kidneys. Recent reports have revealed a potential role of MKL1 in the regulation of renal tubular diseases. For example, Xu *et al.* reported that MKL1 is induced by glucose and synergizes with glucose to induce collagen expression in cultured renal tubular epithelial cells and the kidneys of mice with diabetic nephropathy, eventually leading to tubulointerstitial fibrosis [42]. Moreover, suppressor of cancer cell invasion (SCAI) has been demonstrated to negatively regulate epithelial-mesenchymal transition and renal fibrosis in LLC-PK1 (CL4) proximal tubular epithelial cells, which is at least partially mediated by repression of MKL1 and MKL2 [43]. Taken together, these observations indicate that MKL1 performs physiological roles in maturation and development of the kidney, and thus its dysfunction might lead to glomerular and renal diseases.

Conclusion
In the present study, we found a potential mechanism of MKL1/p21-mediated cell cycle quiescence in podocytes. Therefore, these findings reveal a novel function of MKL1 during podocyte proliferation and differentiation, furthering our understanding of kidney development and the mechanisms of kidney diseases.

Methods
Plasmid construction
The mouse p21 promoter sequence (−1562/+200) was obtained by PCR from mouse blood genomic DNA and cloned into the pGL3-basic vector (Promega) using the forward primer, 5′-AGCAAGAATTCACAGACCGATG-3′, and reverse primer, 5′-GTACCTGACACATACA CACC-3′. The mutagenesis of the CArG box in the mouse p21 promoter was performed using the Quick-Change Site-Directed Mutagenesis Kit (Stratagene) with the forward primer: 5′-gtactccctgtCCTTTTCT*TT* gaagtggtgatt-3′ and reverse primer: 5′-aatcaccacttc *AA*AGAAAAGGacaggggagtac-3′.

Cell culture and transfection
MPC-5 cells were propagated in collagen I-coated dishes at 33°C (permissive temperature) in RPMI supplemented with 10% FBS and 20 U/ml of recombinant mouse IFN-γ (R&D). To induce differentiation, the medium was changed to RPMI with 5% FBS without IFN-γ, and the cells were shifted to 37°C (nonpermissive temperature) for 10 days. Under these conditions, cells underwent growth arrest, increased in size, and developed elongated cell processes. The cells were transfected using TurboFect™ Transfection Reagent (Fermentas) according to the manufacturer's protocols.

The mouse MKL1 expression plasmid was introduced into MPC5 by transient transfection. G418-resistant clones were isolated over a period of 3–4 weeks. The overexpression of MKL1 was confirmed by western blotting assay.

Cell viability assay
MPC5 cells transfected with the mouse MKL1 expression plasmid or empty vector were seeded onto 96-well plates at a density of 2×10^3 cells/well and incubated in RPMI containing 5% FBS at 33°C for 6 days. The cell viability was assessed using the CCK-8 assay according to the manufacturer's protocols (Dojindo). Six parallel replicates were measured for each sample.

5-ethynyl-2′-deoxyuridine (EdU) cell proliferation assay
MPC5 cells transfected with the mouse MKL1 expression plasmid or empty vector were seeded onto 24-well plates at 50-60% confluence. Cells were incubated with 50 µM EdU for 2 h at 48–72 h after transfection. After the 2-h pulse, the cells were washed twice with PBS and fixed with 4% paraformaldehyde at room temperature for 30 min. The cells were subsequently washed with glycine (2 mg/ml) for 5 min, added 0.2% Trion X-100 for 10 min, washed with PBS twice, and added click reaction buffer (Tris–HCl, pH 8.5, 100 mM; CuSO4, 1 mM; Apollo 550 fluorescent azide, 100 mM; ascorbic acid, 100 mM) for 30 min while protecting from light. The cells were then washed again with 0.5% Triton X-100 for three times, stained with Hoechst (5 mg/ml) for 30 min at room temperature, washed with 0.5% Triton X-100 for three times. Images were taken and analyzed using Confocal FV1000 (Olympus). EdU positive cells were calculated with (EdU add-in cells/Hoechst stained cells) × 100%. At least 200 cells were counted per well.

RNA extraction and quantitative RT-PCR
Using TRIzol Reagent (Invitrogen), total RNA was extracted from MPC5 cells that were cultured at the nonpermissive temperature of 37°C for 10 days. Total RNA (0.5 µg) from each sample was used for first-strand cDNA synthesis using M-MLV Reverse Transcriptase (Promega). The specific products of mouse myocardin, MKL1 and MKL2 were amplified by quantitative PCR using the following primers: myocardin, 5′-GATGGG CTCTCTCCAGATCAG-3′ (forward) and 5′-GGCTGC ATCATTCTTGTCACTT-3′ (reverse); MKL1, 5′-CCCA AAGGTAGCAGACAGTTC-3′ (forward) and 5′-GAGT GGGTGATATGGAGGTGG-3′ (reverse); and MKL2, 5′-GAGCGAGCCAGAACTGAGAAT-3′ (forward) and 5′-ACTCGAATCCACAGGAAGGATG-3′ (reverse). The verification of gene expression levels was performed by quantitative RT-PCR using EvaGreen (Biotium). GAPDH was used as an internal control.

Western blotting and antibodies
Whole cell extract preparation and western blotting with the appropriate antibodies were performed as previously described [44]. The following antibodies (Abs) were used: rabbit polyclonal Ab against MKL1 at 1:800 dilution (ab49311; Abcam), rabbit polyclonal Ab against p21 at 1:500 dilution (10355-1-AP; Proteintech) and mouse monoclonal Ab against actin at 1:1000 dilution (A-4700; Sigma).

RT^2 Profiler™ PCR array

Total RNA was extracted from MPC5 cells stably transfected with the mouse MKL1 expression plasmid (ΔMKL1) or empty vector (ΔControl). For PCR array experiments, an RT^2 Profiler custom PCR array was used to simultaneously examine the mRNA levels of 84 genes closely associated with cell cycle, including 5 "housekeeping genes", in 96-well plates following the manufacturer's protocol (PAMM-020Z, QIAGEN). Briefly, first-strand cDNAs were synthesized from 1 μg of total RNA using the TaqMan RT reagent kit (QIAGEN) according to the manufacturer's protocol. The reaction mixtures (25 μl) were incubated at 25°C for 10 min, followed by incubation at 48°C for 30 min and 95°C for 5 min, then cooled on ice. Arrays were performed independently at least three times for each cell line; values were obtained for the threshold cycle (Ct) for each gene and normalized using the average of five housekeeping genes on the same array (Actb, B2m, Gapdh, Gusb, and Hsp90ab1). The Ct values for housekeeping genes and a dilution series of ACTB were monitored for consistency between the arrays. The change (ΔCt) between ΔMKL1 and ΔControl was determined using the formula ΔCt = Ct(ΔMKL1) − Ct (ΔControl), and the fold change was determined using the formula fold change = $2^{(-\Delta Ct)}$. The resulting values were reported as fold change; only genes showing twofold or greater change were considered. The negative controls ensured the absence of DNA contamination and set the threshold for determining absence versus presence of expression.

Luciferase assay

MPC5 cells were co-transfected with the wild-type or mutant mouse p21 promoters and MKL1 expression plasmid in 24-well plates. Lysates were prepared at 24 h after transfection, and luciferase activities were measured using the Dual-Luciferase Reporter Assay System (Promega) according to the manufacturer's protocols. The luciferase activities were normalized to the values for Renilla luciferase.

Chromatin immunoprecipitation (CHIP)

ChIP assays were performed using reagents commercially obtained from Upstate, essentially according to the manufacturer's instructions. The antibodies used in these experiments were rabbit polyclonal Ab against MKL1 (ab49311; Abcam) and anti-rabbit normal IgG (sc-2345, Santa Cruz). The amounts of each specific DNA fragment in immunoprecipitates were determined by PCR or quantitative PCR reactions. The fragment of mouse p21 promoter, containing the CArG box, was amplified using the forward primer: 5′-CCCTCGTGCTTA GACCA-3′, and reverse primer: 5′-GCTGTTGCTGC TACCCA-3′.

Immunofluorescent analysis

Tissue samples were placed in 4% paraformaldehyde in PBS over 2 hours at 4°C, immersed in 30% sucrose overnight at 4°C, and then were embedded in OCT compound (Tissue-Tek, Miles) and sectioned at 5 mm (Leica CM 1850, Leica Instruments) for the morphological and immunohistochemistry study. Sections were washed three times in PBS for five minutes each and incubated in blocking buffer with 10% serum of the secondary antibody host species for 1 hour at room temperature. After that, sections were incubated with rabbit polyclonal Ab against MKL1 (ab49311; Abcam) and p21 (10355-1-AP; Proteintech) at 1:100 dilution overnight at 4°C. After being washed three times for 10 minutes with PBS at room temperature, sections were incubated with Alexa 488-conjugated anti-rabbit IgG for 3 hours at room temperature. Negative control samples were treated with species-appropriate IgG instead of primary antibody. Images were taken and analyzed using Confocal FV1000 (Olympus).

Statistical analysis

SPSS 17.0 software (SPSS) was used for statistical analysis. The data from all the experiments are presented as means ± SD and represent three independent experiments. One-way analysis of variance (ANOVA) was used to compare means between treatment groups and Tukey's HSD (honestly significant difference) test was used to evaluate the statistically significant differences between groups. Where appropriate, Student's t-test for unpaired observations was applied. A p-value < 0.05 was considered significant.

Additional files

Additional file 1: Figure S1. Overexpression of MKL1 induces cell cycle delay at the G1-S phase transition.

Additional file 2: Figure S2. Knockdown of MKL1 promotes cell cycle progression at the G1-S phase transition.

Additional file 3: Figure S3. MKL1 inhibits cell proliferation by altering the levels of cell cycle regulators.

Additional file 4: Figure S4. MKL1 activates the transcriptional activity of p21 in a dose-dependent manner.

Additional file 5: Figure S5. Knockdown of MKL1 results in upregulation of p21 expression.

Competing interests
The authors declare that no competing interests exist.

Authors' contributions
Contributed reagents/materials/analysis tools: SY ZY. Wrote the paper: SY. Conceived and designed the experiments (Figures 1, 2, 3 and 4 and supporting materials): SY LJL ZY. Conceived and designed the experiments (Figure 5): PJX ZY. Acquisition of data (Figures 1, 2, 3 and 4 and supporting materials): SY LJL. Acquisition of data (Figure 5): PJX. All authors read and approved the final manuscript.

Acknowledgements

This work was partly supported by grant from the National Basic Research Program of China (2011CB944003), the China National Nature Science Foundation (No. 81072153, No. 81272184), the New Century Excellent Talents Supporting Program (No. NCET-11-0260), and the Tianjin Natural Science Foundation (No. 13JCZDJC30200). The mouse MKL1 expression plasmid is a kind gift from Dr. Da-Zhi Wang, Children's Hospital Boston.

Author details

[1]Medical School, Tianjin Key Laboratory of Tumor Microenvironment and Neurovascular Regulation, Nankai University, 94 Weijin Road, Tianjin 300071, China. [2]Tianjin University of Traditional Chinese Medicine, Tianjin 300193, China.

References

1. Pavenstädt H, Kriz W, Kretzler M. Cell biology of the glomerular podocyte. Physiol Rev. 2003;83(1):253–307.
2. Greka A, Mundel P. Cell biology and pathology of podocytes. Annu Rev Physiol. 2012;74:299–323.
3. Kreidberg JA. Podocyte differentiation and glomerulogenesis. J Am Soc Nephrol. 2003;14(3):806–14.
4. Kobayashi N, Gao SY, Chen J, Saito K, Miyawaki K, Li CY, et al. Process formation of the renal glomerular podocyte: is there common molecular machinery for processes of podocytes and neurons? Anat Sci Int. 2004;79(1):1–10.
5. Marshall CB, Shankland SJ. Cell cycle regulatory proteins in podocyte health and disease. Nephron Exp Nephrol. 2007;106(2):e51–9.
6. Price PM. A role for novel cell-cycle proteins in podocyte biology. Kidney Int. 2010;77(8):660–1.
7. Kriz W, Shirato I, Nagata M, LeHir M, Lemley KV. The podocyte's response to stress: the enigma of foot process effacement. Am J Physiol Renal Physiol. 2013;304(4):F333–47.
8. Ma Z, Morris SW, Valentine V, Li M, Herbrick JA, Cui X, et al. Fusion of two novel genes, RBM15 and MKL1, in the t(1;22)(p13;q13) of acute megakaryoblastic leukemia. Nat Genet. 2001;28(3):220–1.
9. Mercher T, Coniat MB, Monni R, Mauchauffe M, Nguyen Khac F, Gressin L, et al. Involvement of a human gene related to the Drosophila spen gene in the recurrent t(1;22) translocation of acutemegakaryocytic leukemia. Proc Natl Acad Sci U S A. 2001;98(10):5776–9.
10. Wang D, Chang PS, Wang Z, Sutherland L, Richardson JA, Small E, et al. Activation of cardiac gene expression by myocardin, a transcriptional cofactor for serum response factor. Cell. 2001;105(7):851–62.
11. Wang DZ, Li S, Hockemeyer D, Sutherland L, Wang Z, Schratt G, et al. Potentiation of serum response factor activity by a family of myocardin-related transcription factors. Proc Natl Acad Sci U S A. 2002;99(23):14855–60.
12. Cen B, Selvaraj A, Burgess RC, Hitzler JK, Ma Z, Morris SW, et al. Megakaryoblastic leukemia 1, a potent transcriptional coactivator for serum response factor (SRF), is required forserum induction of SRF target genes. Mol Cell Biol. 2003;23(18):6597–608.
13. Selvaraj A, Prywes R. Expression profiling of serum inducible genes identifies a subset of SRF target genes that are MKL dependent. BMC Mol Biol. 2004;5:13.
14. Wang Z, Wang DZ, Pipes GC, Olson EN. Myocardin is a master regulator of smooth muscle gene expression. Proc Natl Acad Sci U S A. 2003;100(12):7129–34.
15. Yoshida T, Sinha S, Dandré F, Wamhoff BR, Hoofnagle MH, Kremer BE, et al. Myocardin is a key regulator of CArG-dependent transcription of multiple smooth muscle marker genes. Circ Res. 2003;92(8):856–64.
16. Chen J, Kitchen CM, Streb JW, Miano JM. Myocardin: a component of a molecular switch for smooth muscle differentiation. J Mol Cell Cardiol. 2002;34(10):1345–56.
17. Selvaraj A, Prywes R. Megakaryoblastic leukemia-1/2, a transcriptional co-activator of serum response factor, is required for skeletal myogenic differentiation. J Biol Chem. 2003;278(43):41977–87.
18. Parmacek MS. Myocardin-related transcription factors: critical coactivators regulating cardiovascular development and adaptation. Circ Res. 2007;100(5):633–44.
19. Mokalled MH, Johnson A, Kim Y, Oh J, Olson EN. Myocardin-related transcription factors regulate the Cdk5/Pctaire1 kinase cascade to control neurite outgrowth, neuronal migration and brain development. Development. 2010;137(14):2365–74.
20. Ly DL, Waheed F, Lodyga M, Speight P, Masszi A, Nakano H, et al. Hyperosmotic stress regulates the distribution and stability of myocardin-related transcription factor, a key modulator of the cytoskeleton. Am J Physiol Cell Physiol. 2013;304(2):C115–27.
21. Cheng EC, Luo Q, Bruscia EM, Renda MJ, Troy JA, Massaro SA, et al. Role for MKL1 in megakaryocytic maturation. Blood. 2009;113(12):2826–34.
22. Smith EC, Teixeira AM, Chen RC, Wang L, Gao Y, Hahn KL, et al. Induction of megakaryocyte differentiation drives nuclear accumulation and transcriptional function of MKL1 via actin polymerization and RhoA activation. Blood. 2013;121(7):1094–101.
23. Fan L, Sebe A, Péterfi Z, Masszi A, Thirone AC, Rotstein OD, et al. Cell contact-dependent regulation of epithelial-myofibroblast transition via the rho-rho kinase-phospho-myosin pathway. Mol Biol Cell. 2007;18(3):1083–97.
24. Elberg G, Chen L, Elberg D, Chan MD, Logan CJ, Turman MA. MKL1 mediates TGF-beta1-induced alpha-smooth muscle actin expression in human renal epithelial cells. Am J Physiol Renal Physiol. 2008;294(5):F1116–28.
25. Mihira H, Suzuki HI, Akatsu Y, Yoshimatsu Y, Igarashi T, Miyazono K, et al. TGF-β-induced mesenchymal transition of MS-1 endothelial cells requires Smad-dependent cooperative activation of Rho signals and MRTF-A. J Biochem. 2012;151(2):145–56.
26. Descot A, Hoffmann R, Shaposhnikov D, Reschke M, Ullrich A, Posern G. Negative regulation of the EGFR-MAPK cascade by actin-MAL-mediated Mig6/Errfi-1 induction. Mol Cell. 2009;35(3):291–304.
27. Kimura Y, Morita T, Hayashi K, Miki T, Sobue K. Myocardin functions as an effective inducer of growth arrest and differentiation in human uterine leiomyosarcoma cells. Cancer Res. 2010;70(2):501–11.
28. Tang RH, Zheng XL, Callis TE, Stansfield WE, He J, Baldwin AS, et al. Myocardin inhibits cellular proliferation by inhibiting NF-kappaB(p65)-dependent cell cycle progression. Proc Natl Acad Sci U S A. 2008;105(9):3362–7.
29. Saleem MA, O'Hare MJ, Reiser J, Coward RJ, Inward CD, Farren T, et al. A conditionally immortalized human podocyte cell line demonstrating nephrin and podocin expression. J Am Soc Nephrol. 2002;13(3):630–8.
30. Jeon ES, Park WS, Lee MJ, Kim YM, Han J, Kim JH. A Rho kinase/myocardin-related transcription factor-A-dependent mechanism underlies the sphingosylphosphorylcholine-induced differentiation of mesenchymal stem cells into contractile smooth musclecells. Circ Res. 2008;103(6):635–42.
31. Velasquez LS, Sutherland LB, Liu Z, Grinnell F, Kamm KE, Schneider JW, et al. Activation of MRTF-A-dependent gene expression with a small molecule promotes myofibroblast differentiationand wound healing. Proc Natl Acad Sci U S A. 2013;110(42):16850–5.
32. Nobusue H, Onishi N, Shimizu T, Sugihara E, Oki Y, Sumikawa Y, et al. Regulation of MKL1 via actin cytoskeleton dynamics drives adipocyte differentiation. Nat Commun. 2014;5:3368.
33. Shaposhnikov D, Descot A, Schilling J, Posern G. Myocardin-related transcription factor A regulates expression of Bok and Noxa and is involved in apoptotic signaling. Cell Cycle. 2012;11(1):141–50.
34. Milyavsky M, Shats I, Cholostoy A, Brosh R, Buganim Y, Weisz L, et al. Inactivation of myocardin and p16 during malignant transformation contributes to a differentiation defect. Cancer Cell. 2007;11(2):133–46.
35. Shaposhnikov D, Kuffer C, Storchova Z, Posern G. Myocardin related transcription factors are required for coordinated cell cycle progression. Cell Cycle. 2013;12(11):1762–72.
36. Barisoni L, Mokrzycki M, Sablay L, Nagata M, Yamase H, Mundel P. Podocyte cell cycle regulation and proliferation in collapsing glomerulopathies. Kidney Int. 2000;58(1):137–43.
37. Wang S, Kim JH, Moon KC, Hong HK, Lee HS. Cell-cycle mechanisms involved in podocyte proliferation in cellular lesion of focal segmental glomerulosclerosis. Am J Kidney Dis. 2004;43(1):19–27.
38. Srivastava T, Garola RE, Whiting JM, Alon US. Cell-cycle regulatory proteins in podocyte cell in idiopathic nephrotic syndrome of childhood. Kidney Int. 2003;63(4):1374–81.
39. Shankland SJ, Floege J, Thomas SE, Nangaku M, Hugo C, Pippin J, et al. Cyclin kinase inhibitors are increased during experimental membranous nephropathy: potential role in limiting glomerular epithelial cell proliferation in vivo. Kidney Int. 1997;52(2):404–13.

40. Hoshi S, Shu Y, Yoshida F, Inagaki T, Sonoda J, Watanabe T, et al. Podocyte injury promotes progressive nephropathy in zucker diabetic fatty rats. Lab Invest. 2002;82(1):25–35.

41. Nitta K, Horita S, Honda K, Uchida K, Watanabe T, Nihei H, et al. Glomerular expression of cell-cycle-regulatory proteins in human crescentic glomerulonephritis. Virchows Arch. 1999;435(4):422–7.

42. Xu H, Wu X, Qin H, Tian W, Chen J, Sun L, et al. Myocardin-Related Transcription Factor A Epigenetically Regulates Renal Fibrosis in Diabetic Nephropathy. J Am Soc Nephrol. 2014; [Epub ahead of print].

43. Fintha A, Gasparics Á, Fang L, Erdei Z, Hamar P, Mózes MM, et al. Characterization and role of SCAI during renal fibrosis and epithelial-to-mesenchymal transition. Am J Pathol. 2013;182(2):388–400.

44. Yang S, Du J, Wang Z, Yuan W, Qiao Y, Zhang M, et al. BMP-6 promotes E-cadherin expression through repressing deltaEF1 in breast cancer cells. BMC Cancer. 2007;7:211.

MicroRNA *miR-378* promotes BMP2-induced osteogenic differentiation of mesenchymal progenitor cells

Marlinda Hupkes[1*], Ana M Sotoca[1], José M Hendriks[1], Everardus J van Zoelen[1] and Koen J Dechering[1,2,3]

Abstract

Background: MicroRNAs (miRNAs) are a family of small, non-coding single-stranded RNA molecules involved in post-transcriptional regulation of gene expression. As such, they are believed to play a role in regulating the step-wise changes in gene expression patterns that occur during cell fate specification of multipotent stem cells. Here, we have studied whether terminal differentiation of C2C12 myoblasts is indeed controlled by lineage-specific changes in miRNA expression.

Results: Using a previously generated RNA polymerase II (Pol-II) ChIP-on-chip dataset, we show differential Pol-II occupancy at the promoter regions of six miRNAs during C2C12 myogenic versus BMP2-induced osteogenic differentiation. Overexpression of one of these miRNAs, miR-378, enhances Alp activity, calcium deposition and mRNA expression of osteogenic marker genes in the presence of BMP2.

Conclusions: Our results demonstrate a previously unknown role for miR-378 in promoting BMP2-induced osteogenic differentiation.

Background

The generation of distinct populations of terminally differentiated, mature specialized cell types from multipotent stem cells, via progenitor cells, is characterized by a progressive restriction of differentiation potential that involves a tightly controlled, coordinated activation and repression of specific subsets of genes. This process depends on the orchestrated action of key regulatory transcription factors in combination with changes in epigenetic modifications that regulate which regions in the genome are accessible for transcription [1]. The more recently discovered family of microRNAs (miRNAs) is thought to provide an additional layer of gene control that integrates with these transcriptional and epigenetic regulatory processes to further modulate the final gene expression profile of a specific cell type [2].

MicroRNAs (miRNAs) are a class of small, evolutionarily conserved non-coding RNA molecules (~19-25 nucleotides) involved in post-transcriptional gene silencing and as such play important roles in diverse biological processes such as developmental timing [3], insulin secretion [4], apoptosis [5], oncogenesis [6] and organ development [7,8]. MiRNAs are transcribed from the genome as long primary transcripts (pri-miRNA) encoding one or more miRNAs, which are processed in the nucleus by the so-called 'microprocessor' complex consisting of DGCR8 (DiGeorge Syndrome Critical Region 8) and the ribonuclease III (RNase III) enzyme DROSHA [9]. This liberates the precursor-miRNA (pre-miRNA), a hairpin-type structure, which has a characteristic 3' overhang of two nucleotides and is subsequently exported from the nucleus by Exportin-5, a RAN GTPase protein [10]. Inside the cytoplasm, the pre-miRNA hairpin loop is removed by a second RNase III enzyme, DICER, yielding a ~22 nucleotide long imperfect RNA duplex. This duplex contains two potentially functional mature miRNAs termed the 5p and 3p strands, referring to which end of the pre-miRNA they are derived from [8]. One of these strands

* Correspondence: m.hupkes@science.ru.nl
[1]Department of Cell & Applied Biology, Faculty of Science, Nijmegen Centre for Molecular Life Sciences (NCMLS), Radboud University Nijmegen, Heyendaalseweg 135, 6525 AJ, Nijmegen, The Netherlands
Full list of author information is available at the end of the article

is then incorporated into the RNA-induced silencing complex (RISC), which guides the mature miRNA to its target mRNA. In general, one strand is inserted into the RISC complex at much higher efficiency than the other, whereby the strand choice may be based on a variety of factors including thermodynamic instability, strength of the base-pairing and position of the stem-loop [11]. The strand that is incorporated into RISC with lowest efficiency is referred to with an asterisk (miRNA*) and, since non-incorporated strands are thought to be degraded, is less-abundant than its counterpart [12].

The RISC-incorporated miRNA regulates gene expression through sequence-specific interactions with its target site, which is typically located within the 3' untranslated region (3'UTR) of an mRNA transcript. Animal miRNAs usually exhibit only partial complementarity to their mRNA targets, whereby nucleotides 2–8 at the 5' end of the miRNA, referred to as the 'seed region', are thought to be the primary determinant of target specificity [11,13]. Interaction of the miRNA with its target mRNA can interfere with protein translation and/or induce mRNA degradation through a variety of different mechanisms [14], thereby decreasing the protein output. The mechanism and level of effect are thought to be influenced by the degree of complementarity between the miRNA and its mRNA target, the surrounding sequences in the target 3'UTR [12] and their relative abundance [15].

Estimated numbers of miRNA genes amount to nearly 1% of the number of predicted protein-coding genes in the genome of higher eukaryotes, a percentage similar to that of other large gene families with regulatory roles, such as the homeodomain transcription factor family [11]. In addition, miRNAs are estimated to target the expression of approximately one-third of all mammalian genes [8]. Due to the imperfect complementarity between a miRNA and its target, most miRNAs are predicted to be able to bind to and regulate a large number of different mRNA targets [2]. In addition, multiple different miRNAs can synergistically target and control a single mRNA target [2], providing the potential for complex regulatory networks. Many miRNAs studied so far are differentially expressed during development and differentiation, suggesting that each cell type might have its own unique miRNA profile that could affect the utilization of thousands of mRNAs and thus 'micromanage' the output of the transcriptome [2,8]. Several studies have indeed provided examples of miRNAs that play a role in the regulation of cellular differentiation, including hematopoietic cell differentiation [16], adipogenesis [17], osteogenesis [18] and myogenesis [12]. In addition, it has been shown that expression of only three miRNAs (miR-200c, miR-302 and miR-369) is sufficient to induce pluripotency in mouse cells, demonstrating that miRNAs can act as major determinants of cell fate [19]. Since miRNAs have been discovered relatively recently, however, much still remains to be learned about their role in cellular programming, including the identification and detailed analysis of their targets.

In the present study, we took advantage of the robust and homogeneous differentiation characteristics of the mouse C2C12 myoblast cell line to investigate whether lineage-specific changes in miRNA expression might underlie their terminal differentiation. C2C12 cells were originally derived from regenerating muscle tissue [20] and are considered to represent the transit amplifying progenitor population that is derived from muscle satellite stem cells [21]. When cultured under low-serum conditions, C2C12 cells terminally differentiate and fuse into multi-nucleated myotubes upon reaching confluence, which is preceded by upregulation of the key myogenic transcription factors Myod1 and Myog. However, treatment of C2C12 cells with bone morphogenetic protein (BMP) 2 induces these cells to differentiate into osteoblasts, which involves the upregulation of key osteogenic transcription factors Dlx5, Sp7 and Runx2 [22-24], subsequently leading to the expression of late osteoblast marker genes, such as Alpl and Bglap [25,26]. These characteristics make C2C12 progenitor cells an excellent model system to study the molecular mechanisms that underlie cell-fate specification and terminal differentiation. Using a previously generated RNA polymerase II (Pol-II) ChIP-on-chip dataset [27], we show that several miRNAs have differential Pol-II occupancy during C2C12 myogenic versus osteogenic differentiation and that overexpression of one of these miRNAs, miR-378, promotes BMP2-induced osteogenic differentiation of C2C12 cells.

Results

C2C12 lineage-specific miRNA expression

To identify miRNAs that are differentially expressed during C2C12 myogenic versus BMP2-induced osteogenic differentiation, and thereby might play a role in lineage restriction, we made use of our previously generated Pol-II ChIP-on-chip dataset [27]. This dataset contains Pol-II occupancy data for undifferentiated C2C12 cells (d0) and cells treated with (osteogenesis) or without (myogenesis) BMP2 for 1, 3 and 6 days, whereby changes in Pol-II occupancy are considered to reflect changes in transcriptional activity. Since miRNA genes are usually also regulated by Pol-II promoters [28], this dataset formed a good starting point to search for lineage-specific miRNA expression profiles. Our selection criteria (see Methods) thus led to the identification of 6 miRNA genes, namely miR-21, miR-34b/c, miR-99b, miR-365-2, miR-378 and miR-675, located in the vicinity of enriched regions with differential Pol-II occupancy profiles during myogenic versus osteogenic differentiation within our dataset (Figure 1A).

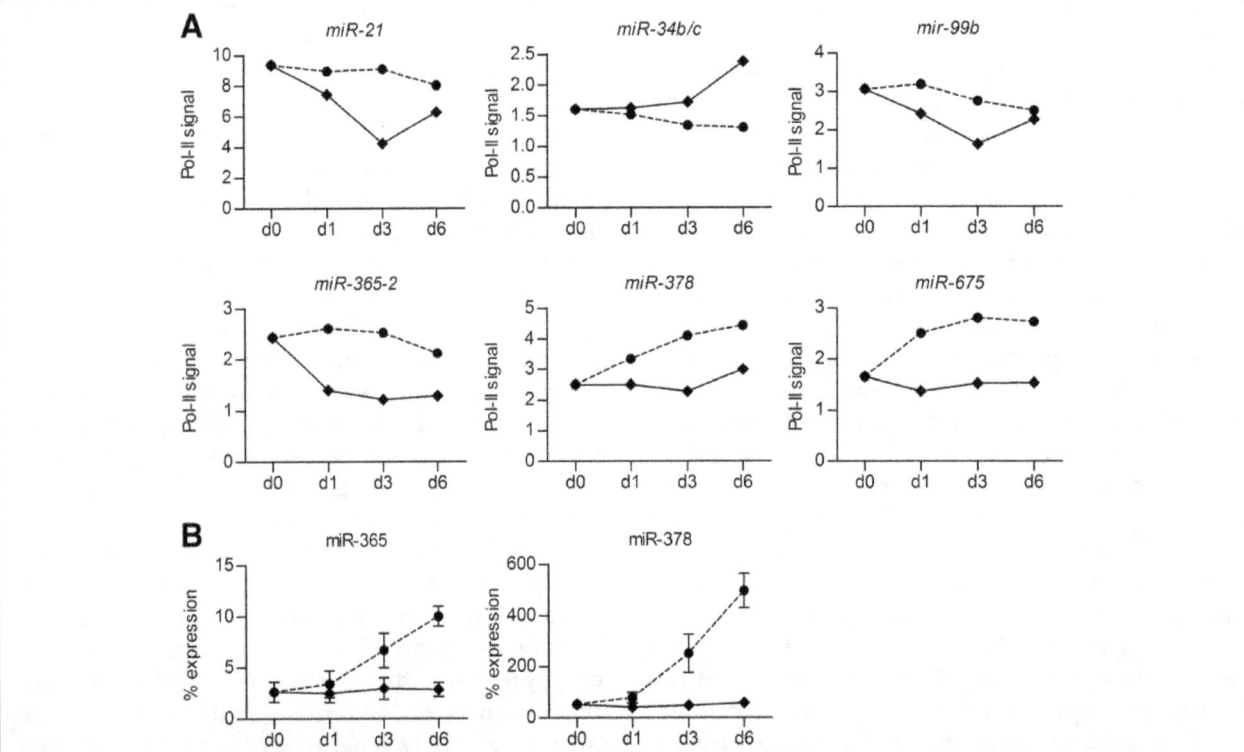

Figure 1 C2C12 lineage-specific miRNA expression. C2C12 cells were treated with (diamonds) or without (circles) 300 ng/ml BMP2 for 6 days, during which chromatin and RNA were harvested in parallel at d0, d1, d3 and d6. **A)** Pol-II enrichment [27] at regions associated with miRNA genes *miR-21* (RefSeq Accession NR_029738; enriched region from −1492 to +1661 bp), *miR-34b/c* (RefSeq Accessions NR_029655 and NR_029654; enriched regions from −287 to +236 bp (*miR-34b*) and −822 to −299 bp (*miR-34c*)), *miR-99b* (RefSeq Accession NR_029536; enriched region from −1113 to −399 bp), *miR-365-2* (RefSeq Accession NR_029959; enriched region from +187 to +1282 bp), *miR-378* (RefSeq Accession NR_029879.1; enriched region from −2068 to −130 bp) and *miR-675* (RefSeq Accession NR_030416.1; enriched region from −1055 to −235 bp) as determined by Pol-II ChIP-on-chip analysis of single biological samples [27]. **B)** Mature miRNA levels of *miR-365 (3p)* and *miR-378 (3p)* as determined by real-time PCR and expressed as a percentage of the control small, non-coding RNA *snoRNA202*. The mean values +/− SD from duplicate measurements are shown for all data points.

Since most of these enriched Pol-II regions could alternatively be associated to other surrounding (predicted) genes, we subsequently validated whether the identified Pol-II occupancy profiles correspond to the actual expression profile of two of these miRNAs, *miR-365* and *miR-378*, by quantitative PCR analysis of the mature miRNAs (Figure 1B). For miR-365, the higher levels of Pol-II occupancy on the associated enriched region during myogenesis versus osteogenesis is reflected by higher levels of mature miRNA expression. While Pol-II occupancy appears to be specifically downregulated during osteogenesis and does not change during myogenesis, however, mature miR-365 levels do not change during osteogenesis and are upregulated during myogenesis. For miR-378, the associated Pol-II occupancy profile and the mature miRNA expression pattern are very similar. These results confirm a lineage-specific difference in the expression of both miR-365 and miR-378. Given the high expression levels of mature miR-378 relative to miR-365, we subsequently focused on this miRNA to further investigate its potential role in C2C12 lineage-specific differentiation.

Effect of miR-378 overexpression on genome-wide mRNA expression levels

To gain more understanding of the role and putative target of miR-378 in C2C12 differentiation, we first created a stably transduced C2C12 cell line overexpressing miR-378 (C2C12-pMirn378) and a control cell line transduced with the parent vector (C2C12-pMirn0). We subsequently examined the effect of miR-378 overexpression on gene expression levels during C2C12 lineage-specific differentiation by means of genome-wide mRNA profiling of undifferentiated (d0) C2C12-pMirn378 and control C2C12-pMirn0 cells and of both cell lines treated with or without BMP2 for 3 and 6 days.

We first explored changes in gene expression levels during differentiation of the control C2C12-pMirn0 cells. Comparison of expression levels in differentiating cells (d3 and d6 time points) versus undifferentiated (d0) cells in this control group revealed a significant upregulation of 4521 probes during C2C12-pMirn0 treatment without BMP2. Functional gene annotation of this set of probes according to Gene Ontology (GO: biological processes category) revealed significant enrichment of many GO terms related to muscle development (including for example 'muscle organ development', 'striated muscle cell development', 'muscle contraction' and 'muscle cell development'; data not shown), consistent with an upregulation of the muscle transcription program under these culture conditions. This is illustrated by the expression profiles of several myogenic marker genes in our control C2C12-pMirn0 cells (Additional file 1A).

Similarly, we observed a significant upregulation of 4664 probes during C2C12-pMirn0 treatment with BMP2 (d3 and d6 time points) as compared to undifferentiated (d0) cells in the control group. Functional gene annotation of these probes according to GO (biological processes) revealed significant enrichment of GO terms related to bone development (including 'skeletal system development', 'bone development', 'extracellular matrix organization' and 'ossification', 'skeletal system morphogenesis', 'osteoblast differentiation', 'bone mineralization'; data not shown), consistent with the expected osteogenesis-inducing effect of BMP2 on our control C2C12-pMirn0 cells. The expression profiles of several osteogenic marker genes are presented in Additional file 1B.

Finally, control C2C12-pMirn0 cultures treated both with and without BMP2 showed a clear cell cycle withdrawal signature as common functional gene annotation of the sets of probes significantly downregulated during myogenic (5396 probes) and osteogenic (4550 probes) differentiation. To illustrate, the expression profiles of several cell-cycle regulators are shown in Additional file 1C.

We thus conclude that treatment of our control C2C12-pMirn0 cells with and without BMP2 had induced the expected changes in transcription patterns corresponding to osteogenic and myogenic differentiation, respectively.

We next examined the effect of miR-378 overexpression on these gene expression profiles. MiR-378 is expressed approximately 11-fold higher in C2C12-pMirn378 cells than in C2C12-pMirn0 cells at the d0 time point (Figure 2A). Similar to C2C12-pMirn0 cells, miR-378 expression increases during myogenic differentiation of C2C12-pMirn378 cells (Figure 2A). While miR-378 levels remain higher in C2C12-pMirn378 versus C2C12-pMirn0 cells during myogenesis, the fold overexpression decreases to approximately 3-fold at d3 and 2-fold at d6 (Figure 2A). The fold overexpression of miR-378 in C2C12-pMirn378 versus C2C12-pMirn0 cells also decreases to approximately 8-fold at d3 and 3-fold at d6 during BMP2-induced osteogenesis (Figure 2A).

Gene expression levels in C2C12-pMirn378 cells were compared to those in control C2C12-pMirn0 cells for each time point during each treatment separately. The Venn diagrams in Figure 2B-C, Figure 3A and Figure 4A demonstrate the number of probes found to be significantly higher- or lower expressed in the C2C12-pMirn378 cells versus C2C12-pMirn0 cells at each indicated time point during myogenesis (Figure 2B-C) and osteogenesis (Figures 3A and 4A). We subsequently focused on the sets of probes that are consistently expressed at either higher or lower levels at at least two consecutive time points during differentiation.

The Venn diagram in Figure 2C shows that during myogenic differentiation hardly any probes are consistently higher expressed in C2C12-pMirn378 cells than in the C2C12-pMirn0 cells. However, we did observe a significantly lower expression of 53 probes at two or more consecutive time points (Figure 2B). GO-analysis of this set of probes (Figure 2D) revealed a significant enrichment of GO terms associated with various alternative differentiation pathways, including osteogenesis, blood vessel development, neuron differentiation and cartilage development. Most of these genes are, however, upregulated during (both C2C12-pMirn378 and C2C12-pMirn0) myogenic differentiation, so they do not appear to be specific for a particular lineage. We did not observe any significant differences between C2C12-pMirn378 and C2C12-pMirn0 cells in the expression of muscle marker genes, such as for example the myogenic transcription factors *Myog* and *Mef2c*, *Ckm*, *Chrng* and the sarcomeric proteins *Actn3* and *Tnnc2* during myogenesis (Figure 2E), suggesting that miR-378 overexpression does not have an effect on C2C12 muscle differentiation.

Compared to myogenesis, many more probes are differentially expressed in C2C12-pMirn378 cells versus C2C12-pMirn0 cells during osteogenic differentiation (Figure 3A and Figure 4A). We observed a consistent (at at least two consecutive time points) lower expression of 253 probes (Figure 3A) and higher expression of 286 probes (Figure 4A) in the C2C12-pMirn378 cells. GO-analysis showed that the set of lower expressed probes was significantly enriched for numerous GO terms associated with muscle differentiation (including for example 'striated muscle development', 'muscle tissue development' and 'muscle organ development': Figure 3B), and includes genes such as the myogenic transcription factor *Myog* and the sarcomeric proteins *Tnnt2* and *Actc1*. These genes, which are upregulated during (both C2C12-pMirn378 and C2C12-pMirn0) myogenesis, are downregulated during BMP2-induced osteogenesis of C2C12-pMirn0 cells, which is further enhanced in C2C12-pMirn378 cells (Figure 3C). Besides terms associated with muscle differentiation, GO-analysis also revealed significant enrichment of GO terms associated with Wnt signaling ('Wnt receptor signaling pathway' and 'Wnt receptor signaling pathway through beta-catenin'), which include genes for the Wnt proteins *Wnt5a* and *Wnt10a*. In control C2C12-pMirn0 cells, *Wnt10a* is upregulated specifically during myogenesis, while *Wnt5a* is upregulated specifically during BMP2-induced osteogenesis (Figure 3C).

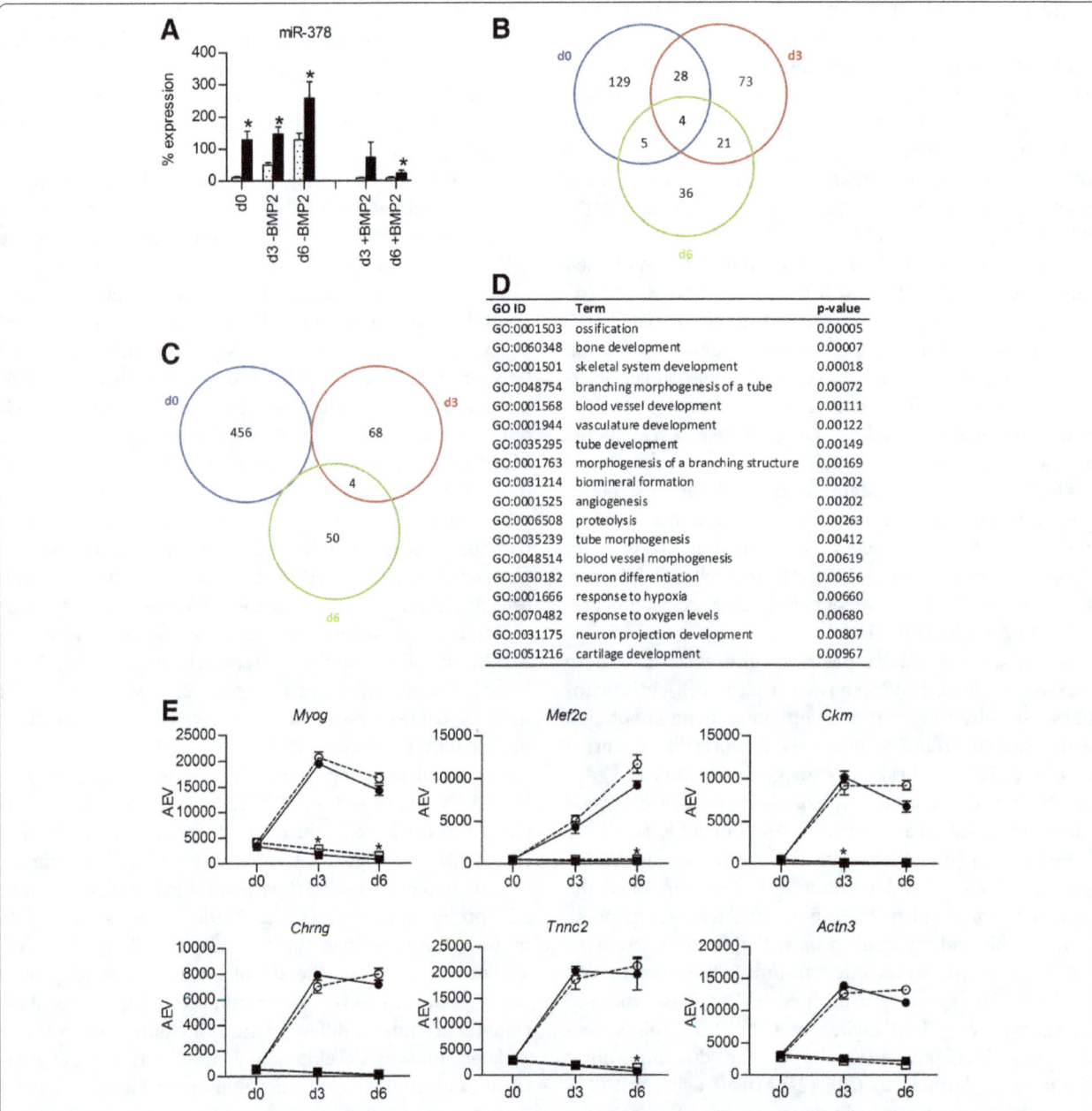

Figure 2 Effect of miR-378 overexpression on C2C12 muscle transcription program. A) Mature miR-378 levels in C2C12-pMirn0 (light bars) and C2C12-pMirn378 (dark bars) cells at indicated time points during treatment with or without 300 ng/ml BMP2. MiR-378 levels are expressed as a percentage of the control small, non-coding RNA *snoRNA202*. Mean values +/− SD of three biological replicates, whereby each measurement was made in duplicate, are shown. *p < 0.05 when compared to the C2C12-pMirn0 sample at the same time point and treatment. **B-C)** Venn diagrams representing the number of probes on the microarray that are significantly lower **(B)** or higher **(C)** expressed in C2C12-pMirn378 cells than in control C2C12-pMirn0 cells at each time point during differentiation in the absence of BMP2. **D)** Enriched (p < 0.01) GO terms within the set of probes that are significantly lower expressed in C2C12-pMirn378 versus C2C12-pMirn0 cells at at least two consecutive time points during differentiation in the absence of BMP2. **E)** mRNA expression profiles of the muscle transcription factors myogenin (*Myog*; 1419391_at) and myocyte enhancer factor 2C (*Mef2c*; 1421027_a_at) and other muscle marker genes muscle creatine kinase (*Ckm*; 1417614_at), the acetylcholine receptor subunit gamma (*Chrng*; 1449532_at) and the sarcomeric genes fast troponin C2 (*Tnnc2*; 1417464_at) and actinin alpha 3 (*Actn3*; 1418677_at) at indicated time points during differentiation of C2C12-pMirn0 (light bullets) and C2C12-pMirn378 (dark bullets) cells treated with (squares) or without (circles) 300 ng/ml BMP2 as revealed from microarray analysis. Mean expression values +/− SD from triplicate microarray experiments are shown for all data points. When the error bar is not visible, the SD falls within the printed data point. All SD values are, however, listed in Additional file 2. AEV = average expression value; *significantly different expression in C2C12-pMirn378 compared to control C2C12-pMirn0 at the same time point and treatment.

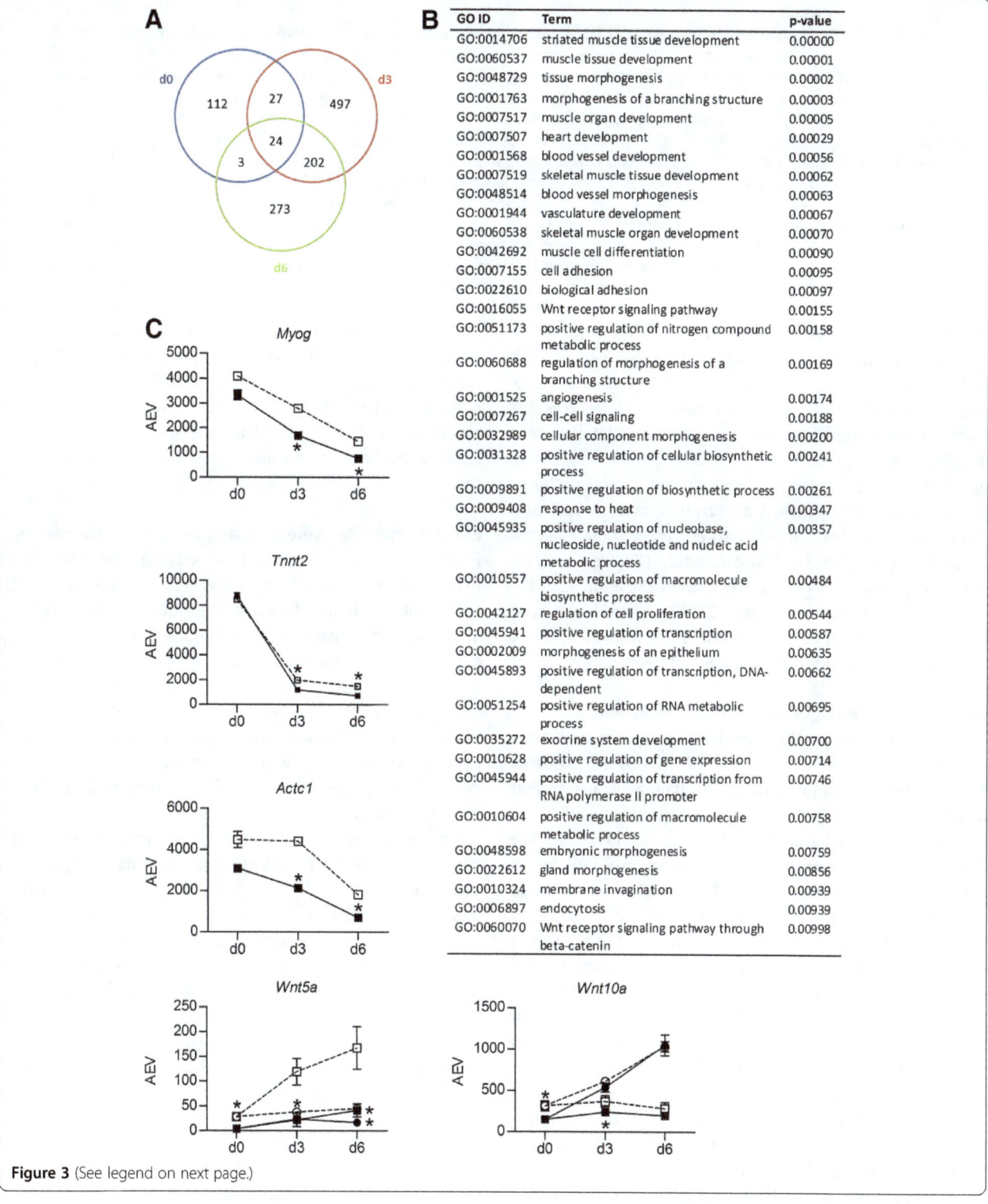

Figure 3 (See legend on next page.)

Figure 3 Effect of miR-378 overexpression on C2C12 bone transcription program: downregulation. A) Venn diagram representing the number of probes on the microarray that are significantly (as defined in Methods) lower expressed in C2C12-pMirn378 cells than in control C2C12-pMirn0 cells at each time point during differentiation in the presence of 300 ng/ml BMP2. **B)** Enriched (p < 0.01) GO terms (in biological processes category) within the set of probes that are significantly lower expressed in C2C12-pMirn378 versus C2C12-pMirn0 cells at at least two consecutive time points during differentiation in the presence of 300 ng/ml BMP2. **C)** Microarray profiles of representative genes that are expressed at significantly lower levels in C2C12-pMirn378 cells than in C2C12-pMirn0 cells at at least two consecutive time points during differentiation in the presence of BMP2 (light bullets: C2C12-pMirn0, dark bullets: C2C12-pMirn378, squares: treatment with BMP2, circles: treatment without BMP2); *Myog*, troponin T2, cardiac (*Tnnt2*; 1418726_a_at), actin alpha cardiac muscle 1 (*Actc1*; 1415927_at), wingless-related MMTV integration site 5a (*Wnt5a*; 1436791_at) and 10a (*Wnt10a*; 1460657_at). For *Myog*, *Tnnt2* and *Actc1*, expression levels are only shown for cells treated with BMP2. Mean expression values +/– SD from triplicate microarray experiments are shown for all data points. When the error bar is not visible, the SD falls within the printed data point. All SD values are, however, listed in Additional file 2. AEV = average expression value; *significantly different expression in C2C12-pMirn378 compared to control C2C12-pMirn0 at the same time point and treatment (see Methods for definition).

Interestingly, GO-analysis of the set of 286 probes that are consistently expressed higher in C2C12-pMirn378 cells than in C2C12-pMirn0 cells during BMP2 treatment revealed significant enrichment of GO terms related to bone differentiation (including 'skeletal system development', 'extracellular matrix organization', 'bone development' and 'ossification': Figure 4B), and includes genes for the osteogenic transcription factors *Sp7* and *Dlx5* and other osteogenic marker genes such as *Alpl*, *Vdr*, *Col1a1*, *Pdgfra*, *Fgfr3* and *Kazald1* (Figure 4C). The higher expression of osteogenic marker genes in C2C12-pMirn378 cells versus control C2C12-pMirn0 cells suggests that overexpression of miR-378 has a positive effect on C2C12 BMP2-induced osteogenic differentiation.

Putative miR-378 target selection and validation

While our mRNA profiling analysis revealed that a large number of genes are affected by miR-378 overexpression, we expected the majority of these changes in expression to be the result of indirect, downstream events following the initial effect of miR-378 on its direct target(s). We therefore set out next to identify direct miR-378 target genes. Given the general effect of miR-378 overexpression on osteogenesis, we hypothesized that miR-378 might target signaling pathways involved in the initial activation of the osteogenic transcription program. We therefore focused on genes that were downregulated by miR-378 overexpression early during BMP2 treatment (i.e. at at least the d0 and d3 time point) and had at least one predicted miR-378 target site in their 3'UTR (see Methods). From this group, we selected three candidate target genes that are known to play a role in the regulation of osteoblast differentiation: the Wnt signaling proteins *Wnt5a* and *Wnt10a* and the BMP-inhibitor *Grem1* (gremlin 1).

To determine whether these candidates are indeed directly targeted by miR-378, we used an *in vitro* luciferase reporter assay. Reporter constructs containing the 3'UTRs of *Wnt5a*, *Wnt10a* and *Grem1*, as well as a positive control containing the miR-378 target sequence, fused to a luciferase reporter gene were co-transfected into HEK293 cells together with the miR-378 overexpression pMirn378

or control plasmid pMirn0 to examine changes in luciferase activity (see Methods). Overexpression of miR-378 significantly suppressed luciferase activity of the positive control, but had no significant effect on the 3'UTR-luciferase reporter constructs (data not shown). Our selected candidates therefore do not appear to be direct targets of miR-378.

Effect of miR-378 overexpression on C2C12 differentiation

Finally, we examined the overall effect of miR-378 overexpression on C2C12 myogenesis and osteogenesis by means of biochemical assays for differentiation markers. The effect on myogenic differentiation was assessed by comparing creatine kinase (Ck) activity in C2C12-pMirn0 and C2C12-pMirn378 cells after treatment with DM in the absence of BMP2 (Figure 5A). Consistent with the lack of effect on myogenic marker gene expression, no significant differences in Ck activity were observed between the two cell lines, again indicating that overexpression of miR-378 does not affect C2C12 myogenesis.

The effect of miR-378 overexpression on osteogenesis was assessed by comparing alkaline phosphatase (Alp) activity in and calcium release by C2C12-pMirn0 and C2C12-pMirn378 cells after treatment with BMP2 (Figure 5B-C). The results demonstrate both an increase in Alp activity and a significant enhancement of calcium deposition by the C2C12-pMirn378 cells. In agreement with the higher expression levels of osteogenic marker genes observed in this cell-line, these results further indicate that overexpression of miR-378 enhances C2C12 BMP2-induced osteogenesis.

Discussion

In this study we used a previously generated Pol-II ChIP-on-chip dataset to identify miRNAs that are differentially expressed during C2C12 myogenic versus osteogenic differentiation and thus possibly play a role in lineage specification. Overexpression of one of these miRNAs, miR-378, had no apparent effect on myogenesis while enhancing BMP2-induced osteogenesis, suggesting a positive role for this miRNA in the osteogenic differentiation program.

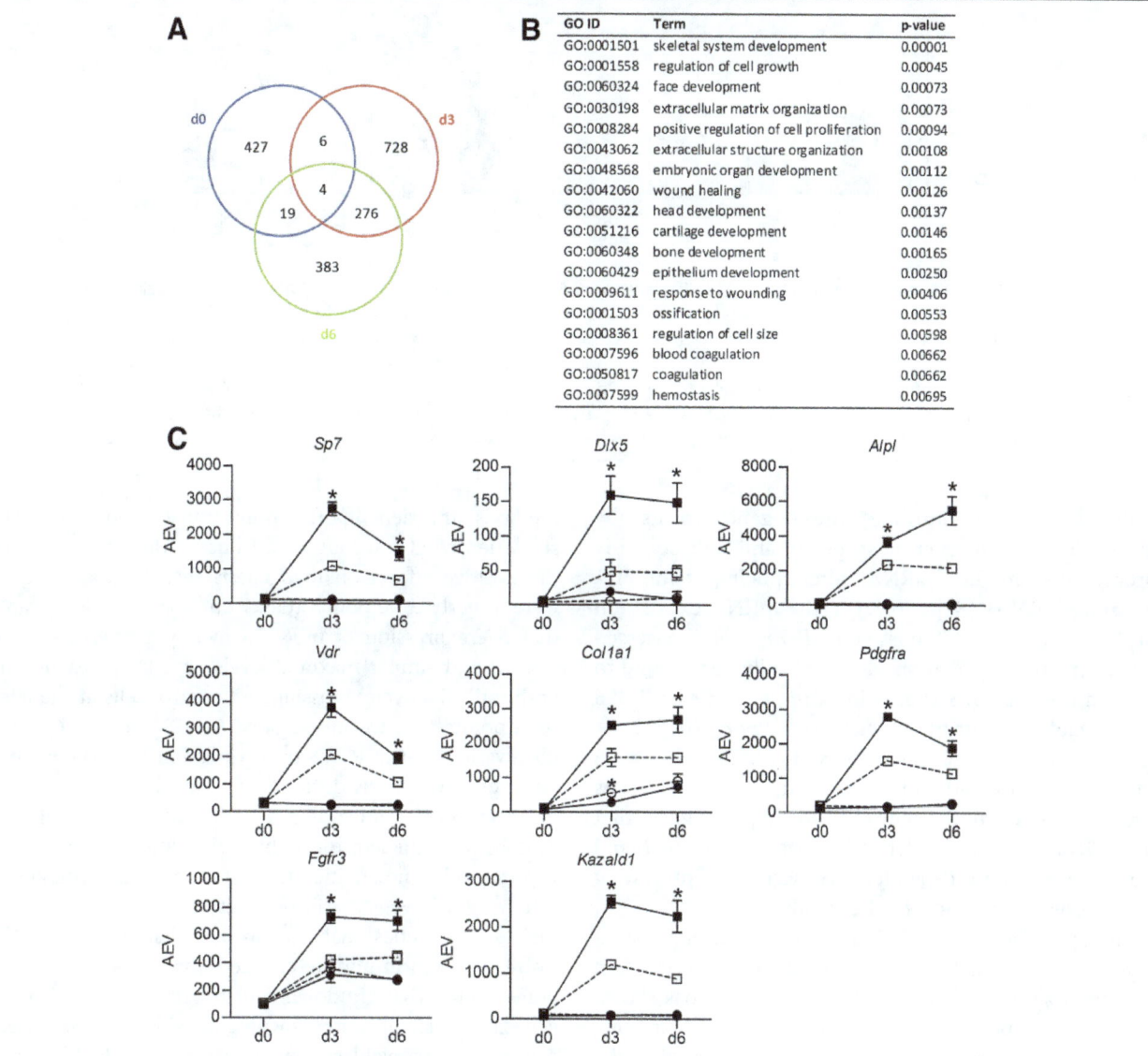

Figure 4 Effect of miR-378 overexpression on C2C12 bone transcription program: upregulation. A) Venn diagram representing the number of probes on the microarray that are significantly (as defined in Methods) higher expressed in C2C12-pMirn378 cells than in control C2C12-pMirn0 cells at each time point during differentiation in the presence of 300 ng/ml BMP2. **B)** Enriched (p < 0.01) GO terms (in biological processes category) within the set of probes that are significantly higher expressed in C2C12-pMirn378 versus C2C12-pMirn0 cells at least two consecutive time points during differentiation in the presence of 300 ng/ml BMP2. **C)** Microarray profiles of bone markers Sp7 transcription factor 7 (*Sp7*; 1418425_at), distal-less homeobox 5 (*Dlx5*; 1449863_a_at), alkaline phosphatase (*Alpl*; 1423611_at), vitamin D receptor (*Vdr*; 1418175_at), collagen type I alpha 1 (*Col1a1*; 1423669_at), platelet derived growth factor receptor alpha polypeptide (*Pdgfra*; 1421917_at), fibroblast growth factor receptor 3 (*Fgfr3*; 1421841_at) and Kazal-type serine peptidase inhibitor domain 1 (*Kazald1*; 1436528_at) at indicated time points during differentiation of C2C12-pMirn0 (light bullets) and C2C12-pMirn378 (dark bullets) cells treated with (squares) or without (circles) 300 ng/ml BMP2 as revealed from microarray analysis. Mean expression values +/– SD from triplicate microarray experiments are shown for all data points. When the error bar is not visible, the SD falls within the printed data point. All SD values are, however, listed in Additional file 2. AEV = average expression value; *significantly different expression in C2C12-pMirn378 compared to control C2C12-pMirn0 at the same time point and treatment (see Methods for definition).

Our finding that miR-378 is strongly upregulated during C2C12 myogenic differentiation corresponds well to other reports demonstrating miR-378 upregulation during myogenesis and high levels of this miRNA in skeletal muscle [29,30]. This upregulation of mature miR-378 matches an increase in Pol-II occupancy at a region located within the first intron of the *Ppargc1b* gene, just upstream of the miR-378 gene. This Pol-II enriched area lies adjacent to an E-box containing Myod-binding region previously shown to be important for miR-378 upregulation during myogenesis [29]. Approximately a third of all miRNA genes, including miR-378, lie within introns of protein-

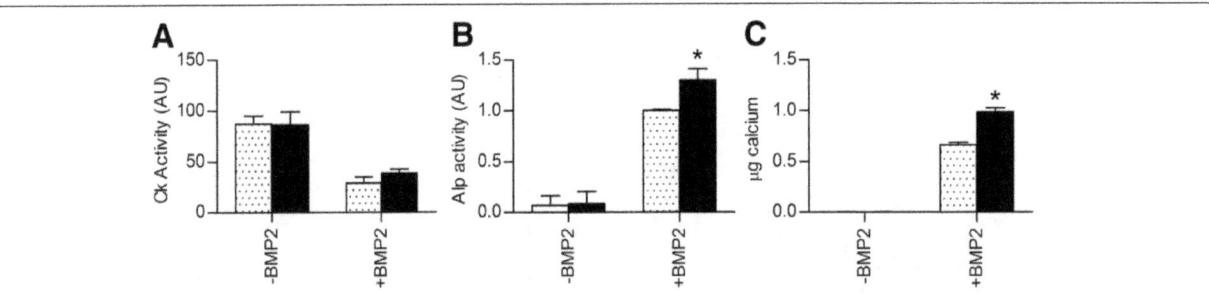

Figure 5 Effect of miR-378 overexpression on C2C12 differentiation. A-C) Creatine kinase **(A)** and Alp **(B)** activity in and calcium deposition **(C)** by C2C12-pMirn0 (light bars) and C2C12-pMirn378 (dark bars) cells after 6 **(A-B)** or 10 **(C)** days of culture in the presence or absence of 300 ng/ml BMP2. **A)** The mean values +/− SD of two biological replicates, whereby each measurement was made in duplicate, are shown. **B)** Mean values +/− SD of three independent experiments with biological duplicates each. For each independent experiment, Alp activity values were normalized to the average of the C2C12-pMirn0 samples treated with BMP2. **C)** Data shows calcium deposition in 24-well plates and is representative of 2 independent experiments. Mean values +/− SD of two biological replicates in one experiment, whereby each measurement was made in duplicate, are shown. *p < 0.05 when compared to the C2C12-pMirn0 sample with the same treatment. AU: arbitrary units.

coding genes. Such intronic miRNA genes are usually co-regulated with their host genes and subsequently processed to mature miRNAs after splicing of the pre-messenger RNAs [31]. However, the mRNA expression profile of the miR-378 host gene, *Ppargc1b*, as assessed by our microarray analysis, does not fully correspond to the mature miR-378 expression profile; while miR-378 is upregulated during myogenesis, *Ppargc1b* mRNA levels do not change (data not shown). Together with the increase in Pol-II and Myod occupancy seen at sites within the first *Ppargc1b* intron, this might suggest that miR-378 is regulated independently from *Ppargc1b* and transcribed as an independent transcript, an interesting hypothesis that requires further study.

The upregulation of miR-378 specifically during C2C12 myogenic differentiation suggests a role for this miRNA in this pathway. Indeed, a study by Gagan *et al.* has shown that miR-378 promotes C2C12 myogenesis by targeting *Msc* (musculin, also known as myogenic repressor; *MyoR*), a repressor of myogenic differentiation that inhibits Myod activity by binding to its co-activators or binding directly to Myod target sequences [29]. In addition, miR-378 has been shown to target mitogen-activated protein kinase 1 (*Mapk1*) and *Bmp2*, which are relevant to myoblast proliferation and differentiation, respectively, in pigs [30]. Similarly, miR-378 has also been shown to play a role in the repression of cardiac hypertrophy by targeting *Mapk1*, *Igf1r* (insulin-like growth factor 1 receptor), *Grb2* (growth factor receptor-bound protein 2) and *Ksr1* (kinase suppressor of ras 1), components of the MAP kinase pathway, in rat cardiomyocytes [32]. In contrast, we did not observe any significant effect of overexpression of miR-378 on C2C12 myogenesis, as assessed by the expression of several myogenic marker genes and Ck activity. The discrepancy with the work of Gagan *et al.* might be attributed to a difference in levels of miR-378 overexpression resulting from the use of different overexpression

methods (transient lipofectamine transfection versus our stable lentiviral-transduced cell lines). Alternatively, since the positive effects on myogenesis seen by Gagan *et al.* were at early time points (day 1 and day 3), it is possible that overexpression of miR-378 merely accelerates myogenesis and similar maximal levels have been reached by both miR-378 overexpressing and control cells at the later time points that we investigated (day 3 and day 6). Our observation that miR-378 overexpression has no apparent effect on myogenesis does not rule out that it plays a role in this process; most likely, endogenous levels of this miRNA are sufficient for its biological function, and overexpression has no additional effect on myogenic markers.

It would, however, still be interesting to take a closer look at the genes that are downregulated by miR-378 overexpression in undifferentiated myoblasts (day 0 time point); genes that are downregulated during C2C12 myogenesis, and significantly downregulated by miR-378 overexpression in myoblasts, such as for example (data not shown) *Fgf7* (fibroblast growth factor 7), *Crlf1* (cytokine receptor-like factor 1), *Ereg* (epiregulin) and *Cck* (cholecystokinin), are potential targets of this miRNA and interesting candidates for further study on the role of miR-378 in myogenesis. Unfortunately, we did not observe a significant effect of miR-378 overexpression on mRNA levels of its published targets *Msc*, *Mapk1*, *Igf1r*, *Grb2* and *Ksr1* [29,30,32]. This does not contradict the findings in these publications, since it is possible that miR-378 exerts its effect on these targets at the level of protein translation and not by inducing mRNA degradation (see below).

Besides its putative role in myogenesis, we clearly demonstrate an effect of miR-378 on C2C12 bone differentiation. Our observation that miR-378 overexpression promotes C2C12 osteogenesis in the presence of BMP2, as assessed by Alp activity, calcium deposition and expression of osteogenic marker genes, was surprising considering the lack of changes in its expression level during

BMP2-induced osteogenic differentiation. Since this effect of miR-378 overexpression is limited only to BMP2-treated cells, we believe that miR-378 on its own is not a major determinant of the osteogenic cell fate, but more likely plays a role in fine-tuning osteogenic gene expression within the BMP2-induced cellular environment.

A role for miR-378 in modulating osteogenic differentiation has previously been described by Kahai et al. in the context of a nephronectin (*Npnt*)-3'UTR over-expressing MC3T3-E1 osteo-progenitor cell line [33]. Npnt is an extracellular matrix protein that, when overexpressed, enhances MC3T3-E1 osteoblast differentiation. Npnt secretion depends on its glycosylation by glycosylation-associated enzymes including Galnt7 (UDP-N-acetyl-alpha-D-galactosamine:polypeptide N-acetylgalactosaminyltransferase 7). The 3'UTR of both *Npnt* and *Galnt7* contain a miR-378 binding site. Kahai et al. demonstrated that, during late stages of MC3T3-E1 development (in the presence of ascorbic acid, β-glycerophosphate and dexamethasone), stable cell lines overexpressing *Npnt* containing its 3'UTR (*Npnt-3'UTR*) have a higher rate of osteoblast differentiation and bone nodule formation than cell lines overexpressing *Npnt* without its 3'UTR; this is further enhanced by co-transfection with *miR-378*. Interestingly, co-transfection of *Npnt-3'UTR* with *miR-378* enhanced production of Npnt and promoted Npnt glycosylation. It was suggested that interaction of the *Npnt* 3'UTR with miR-378 sequestered this miRNA away from *Galnt7*, leading to enhanced Galnt7 activity, a subsequent increase in Npnt glycosylation and secretion and, as a result, a higher rate of osteogenesis. In addition, it was proposed that binding of miR-378 to the *Npnt* 3'UTR resulted in preventing access of other miRNAs, thereby protecting the *Npnt* mRNA from post-transcriptional regulation and resulting in the observed increase in Npnt synthesis [33]. In line with these findings, we observed significantly higher levels of *Npnt* mRNA in our C2C12-pMirn378 versus control cells after 6 days of osteogenic differentiation (data not shown). It would therefore be interesting to determine whether a similar Npnt/Galnt7 –mediated mechanism might also play a role in the effect miR-378 overexpression has on BMP2-induced C2C12 osteogenesis. However, the positive effect of miR-378 overexpression on MC3T3-E1 osteoblast differentiation described by Kahai et al. was only observed when co-transfected with *Npnt-3'UTR* and only during later stages of development. In fact, stable transfection of MC3T3-E1 cells with *miR-378* alone actually inhibited osteogenesis [33]. This is in direct contrast with our observation in BMP2-induced C2C12 cells and indicates that the effect of miR-378 may depend on the osteogenic model system used and/or the signaling pathways involved in inducing differentiation: for example, it is conceivable that miR-378 acts specifically on the BMP2

signaling pathway to positively reinforce the BMP2 effect on our C2C12 model system, while this mechanism might not play a role in the differentiation of MC3T3-E1 cells by Kahai et al., which occurred in the absence of BMP2. Further exploration of the mechanism underlying the positive effect of miR-378 on our BMP2-induced C2C12 system may help shed light on this issue.

We were as yet unable to identify the genes that are directly targeted by miR-378 during BMP2-induced C2C12 osteogenesis. Most effects seen in our mRNA microarray analysis are likely to be secondary to the initial effect of miR-378, making it difficult to identify its direct target(s). Given the overall positive effect of miR-378 on the expression of osteogenic markers, and negative effect on myogenic markers, we expected the initial targeting event to take place early during the differentiation process. To identify direct miR-378 targets, we therefore selected genes *a)* that were downregulated by miR-378 overexpression early (day 0) and consistently during osteogenesis, *b)* that contained a predicted miR-378 target site in their 3'UTR and *c)* that were known to play a role in the regulation of osteoblast differentiation. This led to the selection of *Grem1*, *Wnt5a* and *Wnt10a* as putative targets. Grem1 is a secreted glycoprotein that binds BMP2 and prevents BMP2 signaling and activity in cells of the osteoblast lineage [34]. Targeting of *Grem1* by miR-378 could thus increase the levels of BMP2 available for inducing osteogenesis. Wnts are a family of 19 secreted glycoproteins that activate their cell surface receptors to induce specific intracellular signaling cascades controlling gene expression and play a critical role in embryonic development, postnatal development and adult tissue homeostasis [35]. Wnt signaling regulates cellular processes including proliferation, differentiation, and apoptosis through β-catenin-dependent canonical and β-catenin-independent non-canonical pathways and has been shown to play an important role in bone formation [36]. Wnt5a has been found to be the most dominant Wnt expressed during osteogenesis of human mesenchymal stem cells (hMSCs) both *in vitro* and *in vivo* [37] and Wnt5a signaling has been shown to be important for BMP2-mediated osteogenesis in MC3T3-E1 cells, though the exact signaling pathways involved remain unclear [38]. Wnt10a has also been shown to stimulate osteogenesis [39]. Given their important role in osteoblast formation, it was interesting to determine whether these Wnts were indeed targeted by miR-378 and subsequently how this could relate to the observed increase in osteogenic differentiation. However, our luciferase-reporter assay demonstrated that miR-378 did not directly target the 3'UTR of any of these selected candidates and further work is thus required to identify the mechanism by which miR-378 exerts its effect.

The imperfect complementarity that may exist between a miRNA and its target, the possibility for combinatorial regulation that depends on the presence of other miRNAs to observe an effect, and the various mechanisms by which miRNAs may act, pose a great challenge common to all studies of miRNA function. In our approach we assumed that miR-378 exerts its effect by mRNA destabilization and/or degradation, resulting in a decrease in mRNA levels of its target(s). It is possible, however, for a miRNA to have only very subtle effects on (multiple) targets that cannot be observed as a change in mRNA levels of its direct targets, or to affect protein translation without affecting mRNA levels [14,40]. In addition, miRNAs have been shown to be able to affect mRNA levels of their target genes via alternative mechanisms than binding to their 3'UTR, which would not be detected using a luciferase-3'UTR reporter assay. For instance, it has been shown that miRNAs can affect gene transcription by inducing histone modifications at target promoter sites [41]. Interestingly, a study by Gerin et al. has shown that miR-378 can specifically increase the transcriptional activity of Cebpa and Cebpb (CCAAT/enhancer binding protein, alpha and beta) on adipocyte gene promoters, though it could not be excluded that this was an indirect effect through e.g. inhibition of a co-repressor [42]. Given the role of Cebpb in synergizing with Runx2 to regulate bone-specific gene expression [43], it would be very interesting to investigate whether a similar mechanism underlies the effect of miR-378 on BMP2-induced osteogenesis.

So far, we have attributed the effects seen in C2C12 cells transduced with the miR-378 precursor expression construct to mature miR-378, the 3p strand of the precursor miRNA. However, it should be noted that these cells also overexpress miR-378*, the less-abundant 5p strand. Although present at 10–30 times lower levels than miR-378 (data not shown), it cannot be excluded that the effects seen are (in part) the result of miR-378* overexpression, and it would thus be interesting to also search for putative miR-378* targets within the group of affected genes.

In this study, we used our previous Pol-II ChIP-on-chip dataset to identify lineage-specific miRNA expression. Since the probes on the arrays used for this dataset were restricted to promoter sequences of protein coding genes, the results of this approach do not represent the full picture of Pol-II occupancy at all miRNA gene promoters in the genome. This could explain why several miRNAs known to be specifically upregulated during myogenesis, the so-called myomiRs (*miR-1/206* and *miR-133* families) [12], were not identified. However, our approach did provide a first means to identify several miRNAs with differential Pol-II occupancy during myogenic versus osteogenic differentiation. Most of these miRNAs, including miR-21, miR-34b/c, miR-99b, miR-365 and miR-675, have an as yet unknown role in these differentiation pathways and are thus attractive candidates for further investigation.

Conclusions

In the present study we have identified a list of miRNAs that potentially play a role in C2C12 lineage specification and demonstrated a previously unknown role for miR-378 in enhancing BMP2-induced osteogenic differentiation. Future studies will focus on further exploring the precise function of these miRNAs during cellular differentiation, including the challenging task of identifying their targets and mechanisms of action.

Methods
Cell culture and treatment
Murine C2C12 myoblasts (as well as C2C12-derived stable cell lines: see below) and Human Embryonic Kidney (HEK) 293 cells (American Type Culture Collection, Manassas, VA) were maintained at subconfluent densities in DMEM (Invitrogen, Carlsbad, CA), supplemented with 10% newborn calf serum (NCS; Thermo Fisher Scientific, Waltham, MA), antibiotics (100 U/ml penicillin, 100 µg/ml streptomycin: Sigma-Aldrich, St. Louis, MO), and 2 mM L-glutamine (Invitrogen), further designated as growth medium (GM), at 37°C in a humidified atmosphere containing 7.5% CO_2. To study C2C12 differentiation, cells were plated at 2.5×10^4 cells per cm^2 in GM and grown for 24 hours to sub-confluence. Subsequently, medium was replaced by DMEM containing 5% NCS (referred to as differentiation medium (DM)) in the presence or absence of 300 ng/ml recombinant human bone morphogenetic protein 2 (BMP2; R&D Systems, Minneapolis, MN). For calcium deposition studies, 0.2 mM ascorbate and 10 mM β-glycerophosphate were added to the DM. Medium was replaced every 3–4 days.

Pol-II ChIP-on-chip and selection of differentially enriched microRNA genes
Generation of the RNA polymerase II (Pol-II) ChIP-on-chip dataset used in this study has been described in Hupkes et al. [27]. Enriched regions within 500 base pairs (bp) upstream from a miRNA transcription start site (TSS), within the miRNA gene, or up to 500 bp downstream from the gene end were assigned to that associated miRNA. MiRNA-associated active regions with an absolute average log2 fold > 0.4 of untreated over BMP2-treated Pol-II enrichment values at each time point were selected as differentially expressed during myogenic versus BMP2-induced osteogenic C2C12 differentiation.

RNA isolation and miRNA real-time polymerase chain reaction (PCR)
RNA (including miRNA) was extracted using TRIzol® according to the manufacturer's instructions (Invitrogen). RNA was precipitated with isopropanol and, after air-drying, dissolved in DEPC-treated H_2O. Total RNA

concentrations were quantified by measuring absorbance at 260 nm.

The TaqMan® miRNA Reverse Transcription Kit (Applied Biosystems, Carlsbad, CA), including TaqMan® stem-loop primers miR-378 and miR-365 (cat. # 4427975_002243 and 4427975_001020, respectively; TaqMan® miRNA Assays (Applied Biosystems)) and snoRNA202 (cat. # 4427975_001232; TaqMan® Small RNA Control Assays (Applied Biosystems)) were used for reverse transcription (RT) of miR-378 (3p), miR-365 (3p) and the small, non-coding control RNA snoRNA202 from 100 ng of total RNA each, according to the manufacturer's protocol.

TaqMan® PCR primers and probes for miR-378, miR-365 and snoRNA202, included in the above-mentioned TaqMan® miRNA and small RNA Control assays, together with the TaqMan® Universal PCR Master Mix II, no uracil N-glycosylase (UNG) (Applied Biosystems) were subsequently used for quantitative PCR analysis, also according to the manufacturer's instructions. MiR-378 and miR-365 expression levels were expressed as a percentage of the control small, non-coding RNA snoRNA202.

Expression constructs

The lentivector-based miR-378 precursor expression construct PMIRH378PA-1 (referred to as 'pMirn378') and its parent lentivector pCDH-CMV-MCS-EF1-copGFP (referred to as 'pMirn0') were purchased from System Biosciences (Mountain View, CA). Both vectors contain an expression module for the copGFP fluorescent marker gene to enable monitoring of cells positive for transfection and transduction. MiTarget™ 3'UTR miRNA target clones were purchased from GeneCopoeia (Rockville, MD) and consisted of the *Grem1*, *Wnt5a* or *Wnt10a* (accession numbers NM_011824.3, NM_009524.2 and NM_009518.1 respectively) 3'UTR sequence, the miR-378 target sequence (5'-CCTTCTGACTCCAAGTCCAGT-3'; positive control) or no additional sequence (negative control) inserted in the pEZX-MT01 vector downstream of the firefly luciferase reporter gene (constructs are referred to as Grem1-luc, Wnt5a-luc, Wnt10a-luc, Pos-luc and Neg-luc, respectively). The firefly luciferase gene, driven by an SV40 promoter, resulted in the transcription of a chimeric transcript consisting of luciferase and the inserted target sequence. The pEZX-MT01 vector also contained the *Renilla* luciferase gene under the control of a CMV promoter to allow dual analysis of firefly and *Renilla* luciferase activities in individual samples. Firefly luciferase activity was thus normalized to account for potential differences in transfection efficiencies between different samples.

Stable C2C12-pMirn cell lines

Lentiviruses were produced from pMirn378 and pMirn0 as described previously [44]. For infection, C2C12 cells were initially seeded in a 24-wells plate in GM at a density of 3.0×10^3 cells per well. The next day, cells were infected for 48 hours with 800 ng of virus in GM containing 8 µg/ml of polybrene, whereby the infection medium was refreshed after 24 hours. The cells were then washed twice with GM and maintained in GM for another 24 hours. Subsequently, cells were transferred to T75 flasks and maintained in GM until a confluency of approximately 60% was reached. Finally, copGFP-positive cells were sorted by FACS, resulting in the stably transduced cell lines C2C12-pMirn0 and C2C12-pMirn378.

Microarray processing and identification of significantly regulated genes

For mRNA expression profiling analysis, total RNA samples were purified using the RNeasy Mini Kit (Qiagen, Venlo, the Netherlands), according to the manufacturer's RNA cleanup protocol. Quality of RNA samples was evaluated by capillary electrophoresis on an Agilent 2100 Bioanalyzer (Agilent Technologies, Santa Clara, CA). In total, 30 RNA samples were obtained from triplicate experiments of C2C12-pMirn0 or C2C12-pMirn378 cells cultured for 0, 3 or 6 days in DM with or without 300 ng/ml BMP2. Following purification, 200 ng of total RNA were amplified, labeled, and fragmented using the GeneChip 3' IVT Express Kit (Affymetrix, Santa Clara, CA) according to the manufacturer's instructions. Fragmented amplified RNA (10 µg) was subsequently applied to the GeneChip Mouse Genome 430 2.0 array (Affymetrix) and hybridized for 16 hours at 45°C at 60 rpm in a GeneChip Hybridization Oven 640 (Affymetrix). Following hybridization, the arrays were washed and stained with a GeneChip Fluidics Station 450 (Affymetrix) using the Affymetrix Hybridization Wash Stain (HWS) kit. The arrays were laser scanned with a GeneChip Scanner 3000 7G (Affymetrix). Data was saved as raw image file and quantified using Affymetrix GeneChip Command Console v 1.0 (Affymetrix). These data were imported to R 2.4.1 using the Bioconductor (www.bioconductor.org) Affymetrix package. The model-based Robust Multiarray Average (RMA) algorithm was used to generate the probe set summary based on the full annotation on gene level and normalization was done according to the quantile method. To identify genes that are differentially expressed in C2C12-pMirn378 versus C2C12-pMirn0 samples, expression ratios were calculated for each time point and treatment using the Limma algorithm in R, applying moderated t-tests. A similar approach was taken to identify genes that are up- or downregulated during differentiation of C2C12-pMirn0 cells, whereby expression ratios were calculated for each time point during each treatment versus the d0 base line value. To correct for multiple hypothesis testing, the q value [45] was calculated for each p value using Benjamini-Hochberg correction, indicating the significance of the corresponding ratio.

Genes with a q value < 0.005 and an absolute log2 expression ratio between C2C12-pMirn378 and C2C12-pMirn0 > 0.6 were considered to be significantly differentially expressed at the corresponding time point and treatment. Genes with a q value < 0.005 for the d6 vs d0 time point and an average log2 expression ratio of the d3 vs d0 and d6 vs d0 time points < –0.6 or > 0.6 for the same treatment were considered to be significantly down- or upregulated, respectively, during that particular treatment. Results are listed in Additional file 2. In addition, raw and processed microarray data were submitted to the U.S. National Center for Biotechnology Information Gene Expression Omnibus (GEO) database (GSE51883). The Web-based platform DAVID Bioinformatics Resources [46] was used to identify enriched Gene Ontology (GO) terms of the biological process category [47] in the sets of significantly differentially expressed genes relative to all probes represented on the array, whereby a p value < 0.01 was considered a significant enrichment.

Target prediction

TargetScan version 4.0, PITA, DIANA, PicTar, FINDTAR3 and Miranda databases were used to identify potential miR-378 target sites in genes that were downregulated in C2C12-pMirn378 cells as compared to C2C12-pMirn0 cells.

Transfections and luciferase reporter assays

HEK293 cells were seeded in 24-well plates in GM and medium was refreshed after 24 hours. One hour prior to transfection, medium was replaced by GM lacking antibiotics. 3'UTR miRNA target clones (0.4 µg) were subsequently co-transfected with pMirn0 or pMirn378 (0.4 µg) using Lipofectamine 2000 (Invitrogen) according to the manufacturer's instructions. After 5 hours of incubation with transfection reagents, medium was replaced by GM. Twenty-four hours later, firefly and *Renilla* luciferase activities were measured from the same samples using the LucPair™ miR Duo-Luciferase Assay Kit according to the manufacturer's instructions (Genecopoeia). Firefly luciferase activity was then normalized for transfection efficiency using the *Renilla* luciferase activity in the same sample. Normalized luciferase values are presented as percentage of the control samples co-transfected with the Neg-luc vector.

Creatine kinase assay

Creatine kinase (Ck) enzymatic activity was measured in cell lysates using the EnzyChrom™ Creatine Kinase Assay Kit (ECPK-100, BioAssay Systems, Hayward, CA) according to the manufacturer's protocol. Cell lysates were obtained from cells seeded in 48-well plates: cells

were washed twice with PBS, lysed by incubation in 50 µl lysis buffer (Promega, Madison, WI) on ice for 10 minutes, scraped loose and spun down to remove cellular debris. Supernatant was then collected and diluted 2.5 times in H_2O, of which 10 µl was used for each Ck measurement. Results of the Ck assay were normalized for protein content, as measured using the Bio-Rad Protein assay (Bio-Rad, Hercules, CA) according to the manufacturer's protocol ("Microassay Procedure for Microtiter Plates") and thus expressed as 'arbitrary units (AU)'. Samples were diluted such that absorbance at 595 nm for each sample fell within the linear range of a bovine serum albumin (BSA) standard curve.

Alkaline phosphatase and mineralization assays

Alkaline phosphatase (Alp) enzymatic activity was measured as described previously [48] and normalized for neutral red staining to correct for potential differences in cell number [49].

Calcium deposition in the extracellular matrix (calcium release) was measured as described by Piek *et al.* [44].

Statistical analysis

For miRNA real-time PCR analysis, Ck, Alp, calcium and luciferase assays, Student's 2-tailed *t* test was used to compare miR-378-overexpressing samples with their controls whereby a difference with p < 0.05 was considered significant.

Additional files

Additional file 1: Figure S1. Microarray expression profiles of control C2C12-pMirn0 cells. mRNA expression profiles of A) the muscle transcription factors myogenin (*Myog;* 1419391_at) and myocyte enhancer factor 2C (*Mef2c;* 1421027_a_at) and other muscle marker genes muscle creatine kinase (*Ckm;* 1417614_at), the acetylcholine receptor subunit gamma (*Chrng;* 1449532_at) and the sarcomeric genes fast troponin C2 (*Tnnc2;* 1417464_at) and actinin alpha 3 (*Actn3;* 1418677_at), B) the osteogenic transcription factors Sp7 transcription factor 7 (*Sp7;* 1418425_at), distal-less homeobox 5 (*Dlx5;* 1449863_a_at) and runt-related transcription factor 2 (*Runx2;* 1424704_at), and other osteogenic marker genes alkaline phosphatase (*Alpl;* 1423611_at), bone gamma-carboxyglutamate (gla) protein (*Bglap;* 1449880_s_at) and vitamin D receptor (*Vdr;* 1418175_at) and C) the cell-cycle regulators cyclins A2 (*Ccna2;* 1417910_at) and B1 (*Ccnb1;* 1419943_s_at), cell division cycle 7 (*Cdc7;* 1426002_a_at) and 20 (*Cdc20;* 1416664_at) and the cyclin-dependent kinases 1 (*Cdk1;* 1448314_at) and 4 (*Cdk4;* 1422441_x_at) at indicated time points during differentiation of C2C12-pMirn0 cells treated with (diamonds) or without (circles) 300 ng/ml BMP2 as revealed from microarray analysis. Mean expression values +/– SD from triplicate microarray experiments are shown for all data points. When the error bar is not visible, the SD falls within the printed data point. All SD values are, however, listed in Additional file 2. AEV = average expression value.

Additional file 2: Table S1. Results of mRNA expression profiling. Gene expression profiling results, listing normalized values in C2C12-pMirn0 and C2C12-pMirn378 cells after 0 (d0), 3 (d3) and 6 (d6) days of treatment with or without 300 ng/ml BMP2 as average and standard deviation of three biological replicates, including q values for indicated combinations. Genes that are significantly up- or downregulated during myogenic (column AF) and osteogenic (column AG) differentiation of C2C12-pMirn0 control cells are indicated with 'SU' and 'SD' , respectively, in the appropriate columns. Genes that are significantly up (SU)- or downregulated (SD) in C2C12-pMirn378 cells as compared to C2C12-pMirn0 cells during myogenesis (column AH (SD) and AI (SU)) or osteogenesis (column AJ (SD) and AK (SU)) are grouped into 7 groups; 1-significant difference only on d0; 2-significant difference only on d3; 3-significant difference only on d6; 4-significant difference on d0 and d3; 5-significant difference on d3 and d6; 6-significant difference on d0 and d6; 7-significant difference on d0, d3 and d6.

Competing interests
The authors declare that they have no competing interests.

Authors' contributions
MH conceived of the study, participated in its design and coordination, carried out part of the molecular and cellular studies and drafted the manuscript. AS participated in the design, carried out the luciferase assays, helped to draft the manuscript and performed the statistical analysis. JH carried out the miRNA qPCR studies. JvZ and KD participated in the design and coordination of the study and helped to draft the manuscript. All authors read and approved the final manuscript.

Acknowledgements
The authors would like to acknowledge Eddy van der Struik, Roselinde I. van Ravestein-van Os and Susanne Bauerschmidt from the Merck Research Laboratories for their assistance with the mRNA expression microarrays. This work was supported by a Casimir grant from NWO (project number 018.002.035) and by Merck Sharp & Dohme (Oss, the Netherlands).

Author details
[1]Department of Cell & Applied Biology, Faculty of Science, Nijmegen Centre for Molecular Life Sciences (NCMLS), Radboud University Nijmegen, Heyendaalseweg 135, 6525 AJ, Nijmegen, The Netherlands. [2]Merck Research Laboratories, PO Box 20, 5340 BH, Oss, The Netherlands. [3]Current affiliation: TropIQ Health Sciences, PO Box 9101, 6500 HB, Nijmegen, The Netherlands.

References
1. Bernstein BE, Meissner A, Lander ES: **The mammalian epigenome.** *Cell* 2007, **128**(4):669–681.
2. Bartel DP, Chen CZ: **Micromanagers of gene expression: the potentially widespread influence of metazoan microRNAs.** *Nat Rev Genet* 2004, **5**(5):396–400.
3. Reinhart BJ, Slack FJ, Basson M, Pasquinelli AE, Bettinger JC, Rougvie AE, Horvitz HR, Ruvkun G: **The 21-nucleotide let-7 RNA regulates developmental timing in Caenorhabditis elegans.** *Nature* 2000, **403**(6772):901–906.
4. Poy MN, Eliasson L, Krutzfeldt J, Kuwajima S, Ma X, Macdonald PE, Pfeffer S, Tuschl T, Rajewsky N, Rorsman P, Stoffel M: **A pancreatic islet-specific microRNA regulates insulin secretion.** *Nature* 2004, **432**(7014):226–230.
5. Brennecke J, Hipfner DR, Stark A, Russell RB, Cohen SM: **Bantam encodes a developmentally regulated microRNA that controls cell proliferation and regulates the proapoptotic gene hid in Drosophila.** *Cell* 2003, **113**(1):25–36.
6. Lu J, Getz G, Miska EA, Alvarez-Saavedra E, Lamb J, Peck D, Sweet-Cordero A, Ebert BL, Mak RH, Ferrando AA, Downing JR, Jacks T, Horvitz HR, Golub TR: **MicroRNA expression profiles classify human cancers.** *Nature* 2005, **435**(7043):834–838.
7. Yekta S, Shih IH, Bartel DP: **MicroRNA-directed cleavage of HOXB8 mRNA.** *Science* 2004, **304**(5670):594–596.
8. Kim VN: **MicroRNA biogenesis: coordinated cropping and dicing.** *Nat Rev Mol Cell Biol* 2005, **6**(5):376–385.
9. Lee Y, Ahn C, Han J, Choi H, Kim J, Yim J, Lee J, Provost P, Radmark O, Kim S, Kim VN: **The nuclear RNase III Drosha initiates microRNA processing.** *Nature* 2003, **425**(6956):415–419.
10. Yi R, Qin Y, Macara IG, Cullen BR: **Exportin-5 mediates the nuclear export of pre-microRNAs and short hairpin RNAs.** *Genes Dev* 2003, **17**(24):3011–3016.
11. Bartel DP: **MicroRNAs: genomics, biogenesis, mechanism, and function.** *Cell* 2004, **116**(2):281–297.
12. Goljanek-Whysall K, Sweetman D, Munsterberg AE: **microRNAs in skeletal muscle differentiation and disease.** *Clin Sci (Lond)* 2012, **123**(11):611–625.
13. Lewis BP, Shih IH, Jones-Rhoades MW, Bartel DP, Burge CB: **Prediction of mammalian microRNA targets.** *Cell* 2003, **115**(7):787–798.
14. Morozova N, Zinovyev A, Nonne N, Pritchard LL, Gorban AN, Harel-Bellan A: **Kinetic signatures of microRNA modes of action.** *RNA* 2012, **18**(9):1635–1655.
15. Mukherji S, Ebert MS, Zheng GX, Tsang JS, Sharp PA, van Oudenaarden A: **MicroRNAs can generate thresholds in target gene expression.** *Nat Genet* 2011, **43**(9):854–859.
16. Chen CZ, Li L, Lodish HF, Bartel DP: **MicroRNAs modulate hematopoietic lineage differentiation.** *Science* 2004, **303**(5654):83–86.
17. McGregor RA, Choi MS: **microRNAs in the regulation of adipogenesis and obesity.** *Curr Mol Med* 2011, **11**(4):304–316.
18. Dong S, Yang B, Guo H, Kang F: **MicroRNAs regulate osteogenesis and chondrogenesis.** *Biochem Biophys Res Commun* 2012, **418**(4):587–591.
19. Miyoshi N, Ishii H, Nagano H, Haraguchi N, Dewi DL, Kano Y, Nishikawa S, Tanemura M, Mimori K, Tanaka F, Saito T, Nishimura J, Takemasa I, Mizushima T, Ikeda M, Yamamoto H, Sekimoto M, Doki Y, Mori M: **Reprogramming of mouse and human cells to pluripotency using mature microRNAs.** *Cell Stem Cell* 2011, **8**(6):633–638.
20. Yaffe D, Saxel O: **Serial passaging and differentiation of myogenic cells isolated from dystrophic mouse muscle.** *Nature* 1977, **270**(5639):725–727.
21. Lee S, Shin HS, Shireman PK, Vasilaki A, Van Remmen H, Csete ME: **Glutathione-peroxidase-1 null muscle progenitor cells are globally defective.** *Free Radic Biol Med* 2006, **41**(7):1174–1184.
22. Lee MH, Kim YJ, Kim HJ, Park HD, Kang AR, Kyung HM, Sung JH, Wozney JM, Ryoo HM: **BMP-2-induced Runx2 expression is mediated by Dlx5, and TGF-beta 1 opposes the BMP-2-induced osteoblast differentiation by suppression of Dlx5 expression.** *J Biol Chem* 2003, **278**(36):34387–34394.
23. Lee MH, Kwon TG, Park HS, Wozney JM, Ryoo HM: **BMP-2-induced Osterix expression is mediated by Dlx5 but is independent of Runx2.** *Biochem Biophys Res Commun* 2003, **309**(3):689–694.
24. Katagiri T, Yamaguchi A, Komaki M, Abe E, Takahashi N, Ikeda T, Rosen V, Wozney JM, Fujisawa-Sehara A, Suda T: **Bone morphogenetic protein-2 converts the differentiation pathway of C2C12 myoblasts into the osteoblast lineage.** *J Cell Biol* 1994, **127**(6 Pt 1):1755–1766.
25. Vaes BL, Dechering KJ, Feijen A, Hendriks JM, Lefevre C, Mummery CL, Olijve W, van Zoelen EJ, Steegenga WT: **Comprehensive microarray analysis of bone morphogenetic protein 2-induced osteoblast differentiation resulting in the identification of novel markers for bone development.** *J Bone Miner Res* 2002, **17**(12):2106–2118.
26. Vaes BL, Dechering KJ, van Someren EP, Hendriks JM, van de Ven CJ, Feijen A, Mummery CL, Reinders MJ, Olijve W, van Zoelen EJ, Steegenga WT: **Microarray analysis reveals expression regulation of Wnt antagonists in differentiating osteoblasts.** *Bone* 2005, **36**(5):803–811.
27. Hupkes M, van Someren EP, Middelkamp SH, Piek E, van Zoelen EJ, Dechering KJ: **DNA methylation restricts spontaneous multi-lineage differentiation of mesenchymal progenitor cells, but is stable during growth factor-induced terminal differentiation.** *Biochim Biophys Acta* 2011, **1813**(5):839–849.
28. Lee Y, Kim M, Han J, Yeom KH, Lee S, Baek SH, Kim VN: **MicroRNA genes are transcribed by RNA polymerase II.** *EMBO J* 2004, **23**(20):4051–4060.
29. Gagan J, Dey BK, Layer R, Yan Z, Dutta A: **MicroRNA-378 targets the myogenic repressor MyoR during myoblast differentiation.** *J Biol Chem* 2011, **286**(22):19431–19438.
30. Hou X, Tang Z, Liu H, Wang N, Ju H, Li K: **Discovery of MicroRNAs associated with myogenesis by deep sequencing of serial developmental skeletal muscles in pigs.** *PLoS One* 2012, **7**(12):e52123.
31. Baskerville S, Bartel DP: **Microarray profiling of microRNAs reveals frequent coexpression with neighboring miRNAs and host genes.** *RNA* 2005, **11**(3):241–247.
32. Ganesan J, Ramanujam D, Sassi Y, Ahles A, Jentzsch C, Werfel S, Leierseder S, Loyer X, Giacca M, Zentilin L, Thum T, Laggerbauer B, Engelhardt S:

MiR-378 controls cardiac hypertrophy by combined repression of mitogen-activated protein kinase pathway factors. *Circulation* 2013, **127**(21):2097–2106.

33. Kahai S, Lee SC, Lee DY, Yang J, Li M, Wang CH, Jiang Z, Zhang Y, Peng C, Yang BB: MicroRNA miR-378 regulates nephronectin expression modulating osteoblast differentiation by targeting GalNT-7. *PLoS One* 2009, **4**(10):e7535.

34. Gazzerro E, Smerdel-Ramoya A, Zanotti S, Stadmeyer L, Durant D, Economides AN, Canalis E: Conditional deletion of gremlin causes a transient increase in bone formation and bone mass. *J Biol Chem* 2007, **282**(43):31549–31557.

35. Kim JH, Liu X, Wang J, Chen X, Zhang H, Kim SH, Cui J, Li R, Zhang W, Kong Y, Zhang J, Shui W, Lamplot J, Rogers MR, Zhao C, Wang N, Rajan P, Tomal J, Statz J, Wu N, Luu HH, Haydon RC, He TC: Wnt signaling in bone formation and its therapeutic potential for bone diseases. *Ther Adv Musculoskelet Dis* 2013, **5**(1):13–31.

36. Milat F, Ng KW: Is Wnt signalling the final common pathway leading to bone formation? *Mol Cell Endocrinol* 2009, **310**(1–2):52–62.

37. Guo J, Jin J, Cooper LF: Dissection of sets of genes that control the character of wnt5a-deficient mouse calvarial cells. *Bone* 2008, **43**(5):961–971.

38. Nemoto E, Ebe Y, Kanaya S, Tsuchiya M, Nakamura T, Tamura M, Shimauchi H: Wnt5a signaling is a substantial constituent in bone morphogenetic protein-2-mediated osteoblastogenesis. *Biochem Biophys Res Commun* 2012, **422**(4):627–632.

39. Cawthorn WP, Bree AJ, Yao Y, Du B, Hemati N, Martinez-Santibanez G, MacDougald OA: Wnt6, Wnt10a and Wnt10b inhibit adipogenesis and stimulate osteoblastogenesis through a beta-catenin-dependent mechanism. *Bone* 2012, **50**(2):477–489.

40. Williams AH, Liu N, van Rooij E, Olson EN: MicroRNA control of muscle development and disease. *Curr Opin Cell Biol* 2009, **21**(3):461–469.

41. Kim DH, Saetrom P, Snove O Jr, Rossi JJ: MicroRNA-directed transcriptional gene silencing in mammalian cells. *Proc Natl Acad Sci USA* 2008, **105**(42):16230–16235.

42. Gerin I, Bommer GT, McCoin CS, Sousa KM, Krishnan V, MacDougald OA: Roles for miRNA-378/378* in adipocyte gene expression and lipogenesis. *Am J Physiol Endocrinol Metab* 2010, **299**(2):E198–E206.

43. Gutierrez S, Javed A, Tennant DK, van Rees M, Montecino M, Stein GS, Stein JL, Lian JB: CCAAT/enhancer-binding proteins (C/EBP) beta and delta activate osteocalcin gene transcription and synergize with Runx2 at the C/EBP element to regulate bone-specific expression. *J Biol Chem* 2002, **277**(2):1316–1323.

44. Piek E, Sleumer LS, van Someren EP, Heuver L, de Haan JR, de Grijs I, Gilissen C, Hendriks JM, van Ravestein-van Os RI, Bauerschmidt S, Dechering KJ, van Zoelen EJ: Osteo-transcriptomics of human mesenchymal stem cells: accelerated gene expression and osteoblast differentiation induced by vitamin D reveals c-MYC as an enhancer of BMP2-induced osteogenesis. *Bone* 2010, **46**(3):613–627.

45. Storey JD: A direct approach to false discovery rates. *J Roy Stat Soc B* 2002, **64**:479–498.

46. Huang DW, Sherman BT, Lempicki RA: Systematic and integrative analysis of large gene lists using DAVID bioinformatics resources. *Nat Protoc* 2009, **4**(1):44–57.

47. Ashburner M, Ball CA, Blake JA, Botstein D, Butler H, Cherry JM, Davis AP, Dolinski K, Dwight SS, Eppig JT, Harris MA, Hill DP, Issel-Tarver L, Kasarskis A, Lewis S, Matese JC, Richardson JE, Ringwald M, Rubin GM, Sherlock G, Consortium GO: Gene Ontology: tool for the unification of biology. *Nat Genet* 2000, **25**(1):25–29.

48. van der Plas A, Aarden EM, Feijen JH, de Boer AH, Wiltink A, Alblas MJ, de Leij L, Nijweide PJ: Characteristics and properties of osteocytes in culture. *J Bone Miner Res* 1994, **9**(11):1697–1704.

49. Lowik CW, Alblas MJ, van de Ruit M, Papapoulos SE, van der Pluijm G: Quantification of adherent and nonadherent cells cultured in 96-well plates using the supravital stain neutral red. *Anal Biochem* 1993, **213**(2):426–433.

Wingless gene cloning and its role in manipulating the wing dimorphism in the white-backed planthopper, *Sogatella furcifera*

Ju-Long Yu, Zhi-Fang An and Xiang-Dong Liu[*]

Abstract

Background: Wingless gene (*Wg*) plays a fundamental role in regulating the segment polarity and wing imaginal discs of insects. The rice planthoppers have an obvious wing dimorphism, and the long- and short-winged forms exist normally in natural populations. However, the molecular characteristics and functions of *Wg* in rice planthoppers are poorly understood, and the relationship between expression level of *Wg* and wing dimorphism has not been clarified.

Results: In this study, wingless gene (*Wg*) was cloned from three species of rice planthopper, *Sogatella furcifera*, *Laodelphgax striatellus* and *Nilaparvata lugens*, and its characteristics and role in determining the wing dimorphism of *S. furcifera* were explored. The results showed that only three different amino acid residuals encoded by *Wg* were found between *S. furcifera* and *L. striatellus*, but more than 10 residuals in *N. lugens* were different with *L. striatellus* and *S. furcifera*. The sequences of amino acids encoded by *Wg* showed a high degree of identity between these three species of rice planthopper that belong to the same family, Delphacidae. The macropterous and brachypterous lineages of *S. furcifera* were established by selection experiment. The *Wg* mRNA expression levels in nymphs were significantly higher in the macropterous lineage than in the brachypterous lineage of *S. furcifera*. In macropterous adults, the *Wg* was expressed mainly in wings and legs, and less in body segments. Ingestion of 100 ng/μL double-stranded RNA of *Wg* from second instar nymphs led to a significant decrease of expression level of *Wg* during nymphal stage and of body weight of subsequent adults. Moreover, RNAi of *Wg* resulted in significantly shorter and deformative wings, including shrunken and unfolded wings.

Conclusion: *Wg* has high degree of identity among three species of rice planthopper. *Wg* is involved in the development and growth of wings in *S. furcifera*. Expression level of *Wg* during the nymphal stage manipulates the size and pattern of wings in *S. furcifera*.

Keywords: Rice planthopper, *Sogatella furcifera*, *Wingless* gene, Wing deformation, Wing length

Background

Wingless/Wnt signaling pathway is a complicated protein-protein interaction network that regulates important developmental processes, such as cell proliferation, polarity and fate specification [1]. The wingless gene (*Wg*) in *Drosophila* is homologous with the mammalian *Wnt*-1 gene [2], which manipulates embryonic development [3], as well as the limb formation of adults [4]. *Wg* encodes a kind of cysteine-rich secreted protein [5], and the secreted location and concentration gradients mediate the composition of the midgut morphology, development of central nervous system and the formation of imaginal discs of wings, eyes and legs [6-8].

Wg protein belongs to a kind of morphogen which acts as a signaling molecule to directly control specific cellular responses [9]. Different concentrations of *Wg* protein lead to different cell reactions. It can stimulate some specific target genes when its concentration reaches a threshold gradient [10,11]. It is well known that *Wg* influences development of *Drosophila* wing imaginal disc [12-14]. *Wg* gene is expressed in a narrow stripe along the dorsal/ventral (D/V) boundary during the development of wing disc of *Drosophila* [15], and it has been considered as one of important signal transduction pathways in the

* Correspondence: liuxd@njau.edu.cn
Department of Entomology, Nanjing Agricultural University, Nanjing 210095, China

formation of wing pattern [16]. During the formation of wings, *Wg* regulates the expression of downstream target genes through changing its own concentration, and then results in the morphological differentiation of wing imaginal disc cells [9]. When *Wg* gene in *Tribolium castaneum* was knocked down in the late instar of larvae, the wing width of adults decreased [17]. *Wg* gene of *Bombyx mori* exhibited the highest expression level in wing primordium of the fifth instar larvae, and its expression level decreased gradually after pupation. When the expression level of *Wg* was reduced in the fifth instar larvae by RNA interference method, the resulting adults showed a partial or even complete loss of wings [18].

The white-backed planthopper, *Sogatella furcifera* (Hemiptera: Delphacidae), is one of the most devastating pests in rice fields in Asia. It sucks phloem sap of rice and causes a decrease of grain weight [19]. *Sogatella furcifera* has a wing dimorphism which females have either macropterous or brachypterous wings, but males usually are macropterous in China. In their natural environments, the macropterous rice planthoppers can make long-distance migrations to expand their occurrence regions [20]. Although the genetic basis of wing polymorphism in insects is generally not well understood, it has been verified that the wing forms of rice planthopper have a genetic component and are not purely environmentally determined [21]. The titer of juvenile hormone and DNA methylation were also thought to be involved in the determination of wing forms [22-24]. There are some differentially expressed genes between the long- and short-winged brown planthoppers, such as *flightin*, *troponin C4*, *titin* and *myosin heavy chain* [25,26]. However, the molecular characteristics, expression and biological function of *Wg*, an important gene relating to growth and wing imaginal disc development of insects, in rice planthoppers are still not well understood. Moreover, it is unknown whether the *Wg* plays a key role in determining the development of wings and in manipulating the wing dimorphism in rice planthoppers. In the present study, therefore, the full-length cDNAs encoding *Wg* were cloned and characterized from the three common species of rice planthopper, *S. furcifera*, *Laodelphgax striatellus* and *Nilaparvata lugens*. And then the expression differences of *Wg* between the macropterous and brachypterous lineages of *S. furcifera* which wing forms were selected for more than 20 generations under a constant condition were examined using the quantitative real-time PCR method. Finally, the survival of nymphs, body weight, wing length and wing pattern of adults in the macropterous lineage were measured when the *Wg* expression was knocked down by ingestion of dsRNA of *Wg* in nymphs. This study will illustrate the role of *Wg* in determining the wing dimorphism of rice planthoppers.

Results

Selection response of wing forms of *S. furcifera*

All the adults from the parents (macropterous female × macropterous male, M♀ × M♂ lineage) were macropterous after seven generations of selection. All the males from the parents (brachypterous female × macropterous male, B♀ × M♂ lineage) were macropterous during all these 40 generations of selection, and more than 95% of the females were brachypterous after 25 generations of selection (Figure 1). The *S. furcifera* had significant selection response in wing forms. The macropterous pure line had been obtained by seven continuous generations of selection from M♀ × M♂ lineage (Figure 1).

Characteristics of *Wg* from three species of rice planthoppers

The full-length cDNA clone encoding *Wg* was isolated from *S. furcifera* and *L. striatellus* by the 3′ and 5′ RACE, as well as the open reading frame (ORF) of *Wg* from *N. lugens*. The cDNA length of *Wg* from *S. furcifera* was 1571 bp which contains 31 bp 5′-untranslated region (UTR), 367 bp 3′-UTR with a consensus polyadenylation sequence and 1173 bp ORF. The deduced protein consisted of 390 amino acid residues. A full-length cDNA of *Wg* from *L. striatellus* was 1443 bp containing 34 bp 5′-UTR,1173 bp ORF, and 236 bp 3′-UTR, which also encoded 390 amino acid residues. The cDNA sequence of *Wg* from *N. lugens* had a complete 1185 bp ORF encoding 394 amino acid residues. The deduced amino acid sequences encoded by *Wg* had high level of identity among the three species of rice planthopper. There were only three different amino acid residues between *S. furcifera* and *L. striatellus*. *Nilaparvata lugens* *Wg* had ten and eleven different amino acid residues compared to *S. furcifera* and *L. striatellus*, respectively.

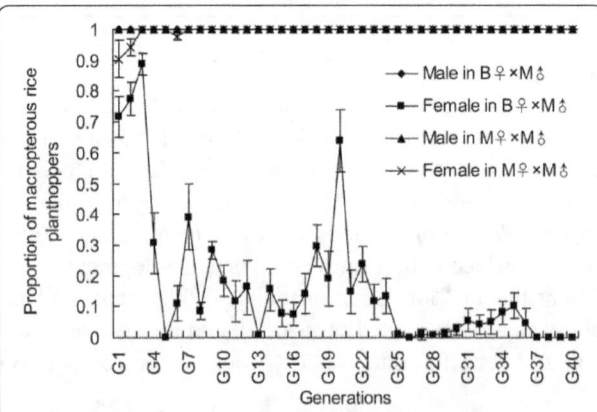

Figure 1 The proportion of macropterous rice planthoppers in the offspring from the M♀ × M♂ and B♀ × M♂ lineages of *Sogatella furcifera* at different generations of selection.

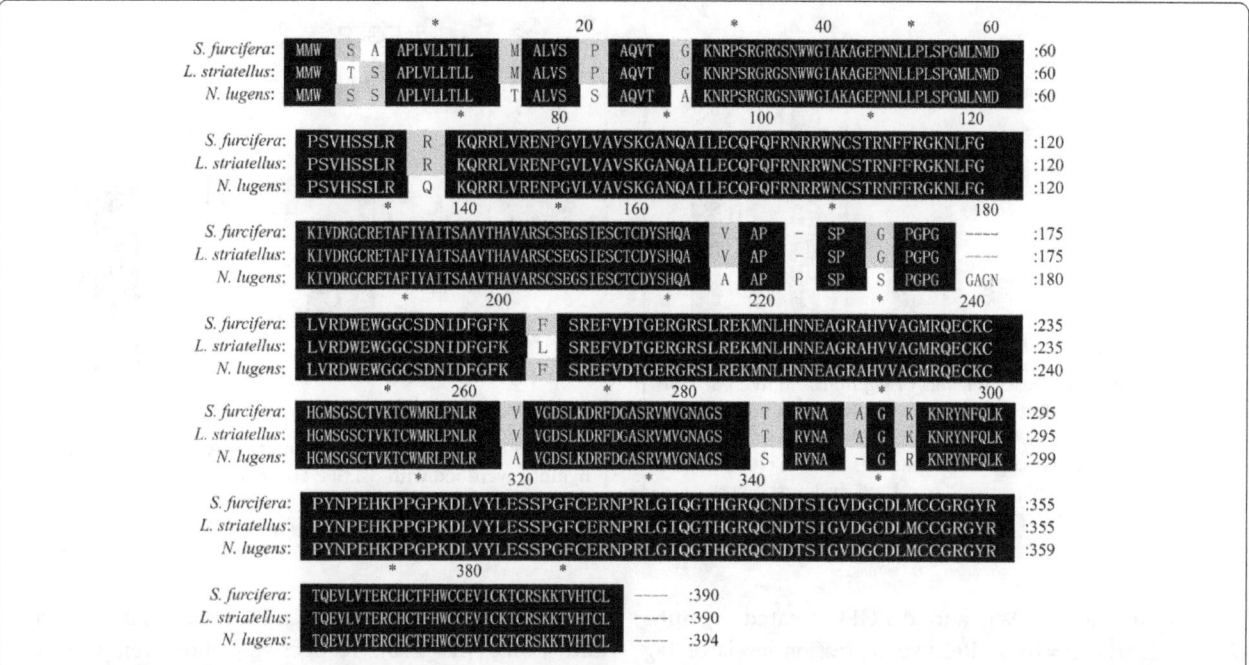

Figure 2 Alignment of the deduced amino acid sequences encoded by *Wg* from three species of rice planthopper, *Nilaparvata lugens*, *Sogatella furcifera* and *Laodelphax striatellus*. Amino acids covered by black, grey, and white bars are 100%, 80%, and below 80% identity among the three species of rice planthoppers, respectively.

The result implied that *Wg* was highly conservative in rice planthoppers (Figure 2).

Expression of *Wg* in the macropterous and brachypterous lineages of *S. furcifera*

Relative expression levels of *Wg* mRNA were significantly higher in nymphs than in adults regardless of the macropterous and brachypterous lineages (for the macropterous lineage: $F_{4,\ 25} = 36.876$, $P < 0.0001$; for the brachypterous lineage: $F_{4,\ 25} = 51.908$, $P < 0.0001$). Moreover, the expression levels were significantly higher in the macropterous lineage than in the brachypterous lineage during the 3rd instar ($t = 4.238$, $df = 5$, $P = 0.0082$), 4th instar ($t = 3.554$, $df = 5$, $P = 0.0163$), and 5th instar nymphs ($t = 5.946$, $df = 5$, $P = 0.0019$), whereas there were no significant differences between the two lineages during adults (male: $t = 0.870$, $df = 5$, $P = 0.4243$; female: $t = 2.018$, $df = 5$, $P = 0.0997$, Figure 3).

Relative expression levels of *Wg* mRNA in different parts of the macropterous adults of *S. furcifera*, such as head, thorax, abdomen, wings and legs, showed that *Wg* was expressed mainly in wings and legs, and at much lower levels in body segments: head, thorax, and abdomen (males: $F_{4,\ 10} = 73.556$, $P < 0.0001$; females: $F_{4,\ 10} = 81.891$, $P < 0.0001$). Relative expression levels between females and males were not significantly different (Figure 4).

Survival and body weight of *S. furcifera* after *Wg* RNAi

The results showed that mortality rates of *S. furcifera* nymphs ingested artificial diet with 100 ng/μL dsWg and dsEGFP were $38.81 \pm 3.77\%$ and $30.35 \pm 6.54\%$, respectively, and there were no significant differences ($t = 1.209$, $df = 8$, $P = 0.261$). Mortality of nymphs in the control was $6.41 \pm 1.73\%$ which was significantly

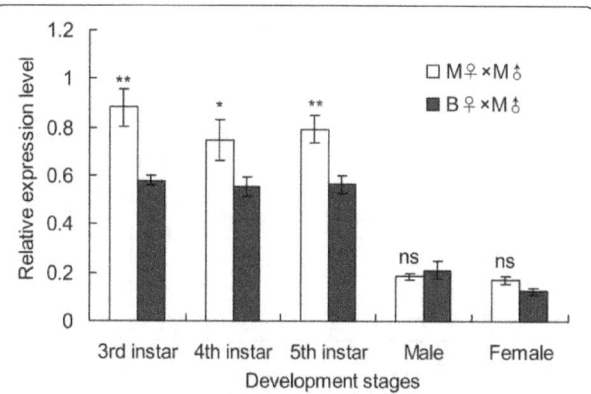

Figure 3 Relative expression levels of *Wg* mRNA in the macropterous and brachypterous lineages of *Sogatella furcifera*. Error bars indicate standard error (SE, n = 6). * and ** above the bars indicate significant differences at $P < 0.05$ and 0.01 level between the macropterous and brachypterous lineages, respectively.

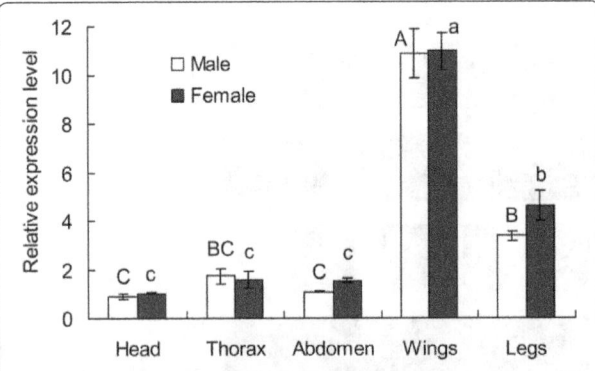

Figure 4 Relative expression levels of *Wg* mRNA in five parts of the macropterous adults of *Sogatella furcifera*. Error bars indicate standard error (SE, n = 3). Different capital and lowercase letters above the bars indicate significant differences at *P* < 0.05 level among different parts of male and female adults, respectively.

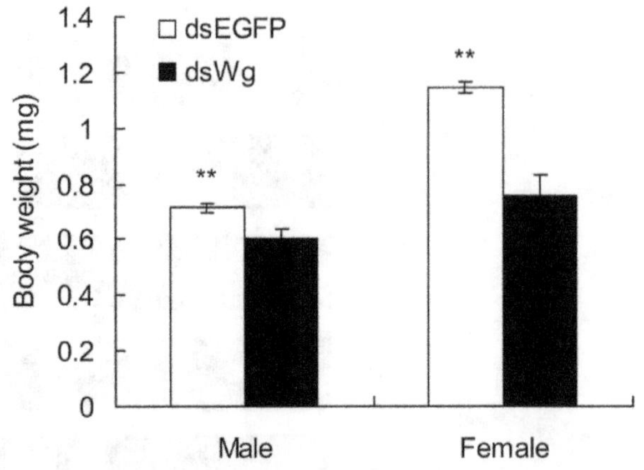

Figure 6 Body weights of male and female adults when their nymphs were fed with dsEGFP and dsWg. ** indicates the significant differences between the treatments of dsEGFP and dsWg by a student's *t* test at *P* < 0.01 level.

lower than these dsWg and dsEGFP treated nymphs ($F_{2,\ 10} = 11.041$, $P = 0.03$). Relative expression levels of *Wg* mRNA in nymphs fed on 100 ng/μL dsWg for seven days significantly decreased ($F_{2,\ 6} = 8.836$, $P = 0.018$, Figure 5). Body weights of these resulting adults were reduced when they were fed on artificial diet with dsWg during nymphs in comparison with those fed on diet with dsEGFP (males: $t = 2.934$, $P = 0.008$; females: $t = 5.360$, $P < 0.0001$, Figure 6).

Wings of *S. furcifera* treated by *Wg* RNAi

Wg RNAi resulted in seriously deformative wings in *S. furcifera*. The abnormal rates of wings in the blank (CK)

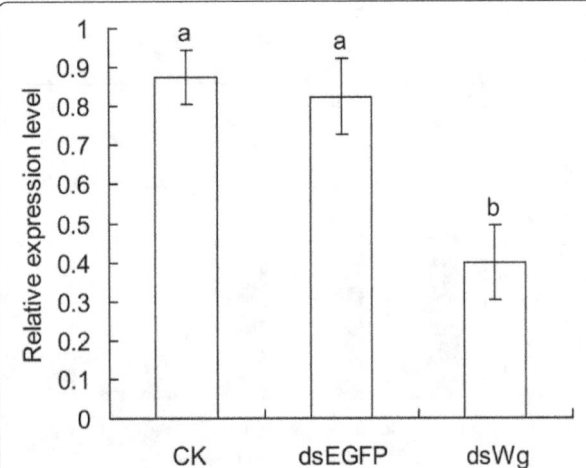

Figure 5 Relative expression levels of *Wg* mRNA in nymphs of *Sogatella furcifera* feeding on artificial diet with 100 ng/μL dsEGFP and dsWg for seven days. Error bars indicate standard error (SE, n = 3). Different lowercase letters above the bars indicate significant differences among the *Wg* RNAi and controls at *P* < 0.05 level.

and negative controls (dsEGFP) were 1.90% (n = 105) and 1.82% (n = 110), respectively, however, the value increased to 18.55% (n = 124) when the nymphs were fed on diet with dsWg. There were significant differences among dsWg, dsEGFP and CK in the abnormal rates of wings ($x^2 = 29.88$, df = 2, P < 0.0001). The normal wings of *S. furcifera* were smooth and folded on the back (Figure 7 A and B), but the adults had shrunken (Figure 7C and D) or unfolded wings on the back (Figure 7E and F) when their nymphs were ingested artificial diet with 100 ng/μL dsWg. The wings from nymphs ingested dsWg had obvious patterning defects in the vines and pigmentation, and the background and veins of wings were depigmented (Figure 7). The lengths of fore- and hind-wing from base to tip became significantly shorter when nymphs were fed with 100 ng/μL dsWg than that fed with 100 ng/μL dsEGFP (female fore-wing: $t = 3.68$, $P < 0.0001$; female hind-wing: $t = 3.68$, $P < 0.0001$; male fore-wing: $t = 4.74$, $P < 0.0001$; male hind-wing: $t = 5.88.$, $P < 0.0001$, Figure 8). The *Wg* was involved in the determination of wing length and pattern in *S. furcifera*.

Discussion

The white-backed planthopper is one of insects with wing-dimorphism which adult was either macropterous or brachypterous. It has been reported that the development of wing was controlled by multiple genes [27-30]. The wing pattern genes are mainly involved in determining three kinds of development of insect wings: proximal-distal, dorsal-ventral and anterior-posterior. The *Wg* occupies a key position in the genetic determination system of the development of dorsal-ventral axis, and it directly impacts the expression of downstream genes controlling the wing pattern [9]. In this study, we found the wing forms of *S. furcifera* exhibited a significant selection response,

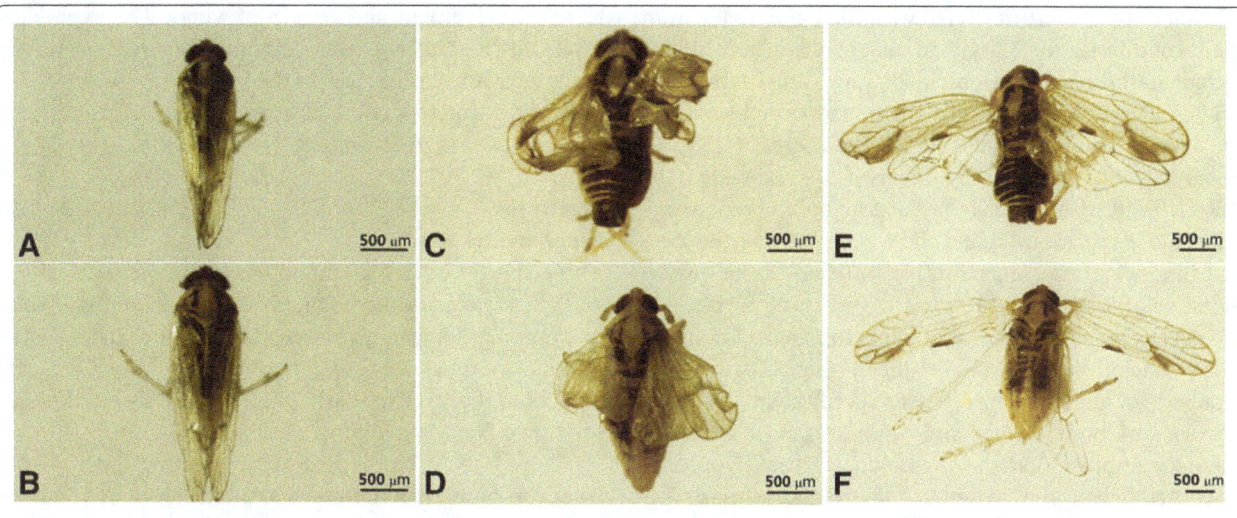

Figure 7 Wings of adult *Sogatella furcifera*. A and **B**: nymphs ingested 100 ng/µL dsEGFP. **C-F**: nymphs ingested 100 ng/µL dsWg.

and the macropterous pure line and brachypterous near-pure line were obtained after seven and twenty-five generations of selection, respectively. Additionally, we cloned the full-length cDNA of *Wg* from *S. furcifera* and *L. striatellus*, and the ORF from *N. lugens*. The amino acid residues encoded by *Wg* had more than 99% identity between *S. furcifera* and *L. striatellus*, whereas the value was 94% between *S. furcifera* and *N. lugens*. The *Wg* is highly conservative within the Delphacidae. Among the three species of rice planthopper, *S. furcifera*, *L. striatellus* and *N. lugens*, the *Wg* from *S. furcifera* had higher similarity with *L. striatellus*, but relatively lower with *N. lugens*. Peng et al. [17] presumed that the genetic basis of wing forms in *S. furcifera* was as similar as *L. striatellus* and both were controlled by two pairs of alleles, one locating on a euchromosome and the other on a sex chromosome, but it was significantly different from *N. lugens* which was controlled by an allele locating on a euchromosome

[28]. The variations of *Wg* among these three species of rice planthopper supported the opinion of Peng et al. [17].

In this study, we found that the *Wg* mRNA expressed highly in third, fourth and fifth instar nymphs, but surprisingly, in adult, the expression level of *Wg* mRNA was significantly lower. The nymphs from third to fifth instar are in the process of development and growth of wings, therefore the *Wg* mRNA expression shows a high level. In the adult, the wings are fully formed, so the expression level becomes lower. In pea aphid, an insect that also has the wing dimorphism, there is a trend towards higher *Wg* mRNA expression in the winged morphs than the wingless ones [27]. In the present study, we found that *Wg* mRNA expression levels in the 3rd-5th instar nymphs of the macropterous pure line of *S. furcifera* were significantly higher than that in the brachypterous line, but no differences were found in adults between these two lines. Classic *Wingless/Wnt* signaling pathway in *Drosophila* is related to wing formation process, including wing growth and morphogenesis [31]. In our study, we found that the higher expression of *Wg* mRNA in the nymphal stage was beneficial to form the long-winged rice planthoppers. The relative expression of *Wg* mRNA was more abundant in wings than that in the other parts of adult. These results revealed that the *Wg* gene indeed played an important role in the wing development and growth of *S. furcifera*.

Studies had already illustrated that *Wg* was required for wing cell survival, particularly during the rapid growth phase of wing development. The determination of the final wing size of wings in *Drosophila* was only a part of functions of *Wg* [32]. The lower expression level of *Wg* mRNA in the nymphs of the brachypterous *S. furcifera* lineage might be resulted from the requirement of the development of short wings or from other functions of *Wg*.

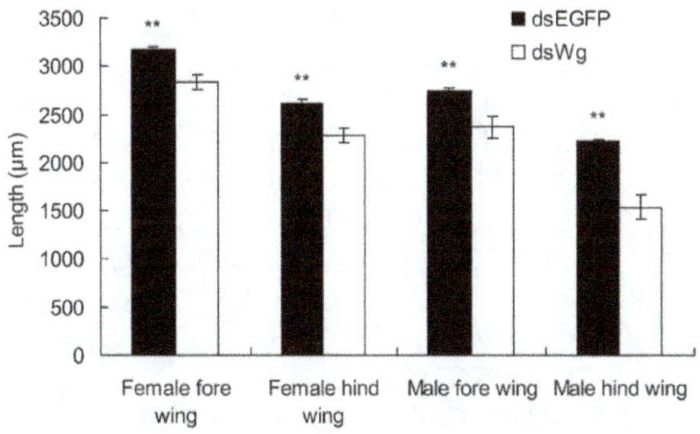

Figure 8 Lengths of the fore- and hind-wing of *Sogatella furcifera* adults which nymphs were fed on diet with 100 ng/uL dsEGFP and dsWg. ** indicates the significant differences between dsEGFP and dsWg treatments by a student's *t* test at *P* < 0.01 level.

Nymphs of *S. furcifera* treated with 100 ng/μL ds*Wg* from the second instar for 7 days had abnormal wings as adults, including shrunken and unfolded wings. In the previous studies, it has been found that the mutation of *Wg* caused the loss of wings in adults of *D. melanogaster* [33]. In *Bombyx mori*, the knocking-down of *Wg* expression by dsRNA resulted in partially missing or even absent wings [18]. In *S. furcifera*, we found the wing lengths of the fore- and hind-wing were significantly shortened caused by the ingestion of dsWg. Therefore, the *Wg* was really involved in the development and growth of wings in rice planthoppers. The β-catenin pathway, one of the *Wg* signal pathway, is required for growth and cell fate specification [34]. The decrease of body weight and wing length of adults *S. furcifera* implied that the β-catenin pathway would be disrupted when the nymphs fed on diet with dsWg. The *Wg* performs the function to determine the growth of wings.

The genetic determination of wing morphs in the rice planthoppers is complex. It has been illustrated that the wing form of rice planthoppers is a quantitative trait controlled by many genes [35-37]. We found here that the wing lengths of female and male adults became significantly shorter when the silencing of the *Wg* was performed at nymphal stage, however, the wings were still longer than that of the naturally branchypterous ones. The forewing lengths were 1.2-1.8 mm and 3.2-3.8 mm in the brachypterous and macropterous females of *S. furcifera*, respectively [38]. In this study the wing length from the dsWg nymphs was 2.84 ± 0.07 mm which was an intermediate wing form. We also found that the relative expression level of *Wg* mRNA in third instar nymphs of the brachypterous lineage was as about 66 percent as the macropterous lineage, and wing length of brachypterous female in natural field was as 47 percent as the macropterous one [38]. However, when the *Wg* mRNA expression level in the macropterous lineage was interfered 50 percent by dsWg in nymphs, the wing length of female was only shortened 10 percent. The expression levels of *Wg* mRNA in nymphs were not linearly related with the wing lengths of rice planthoppers. We presumed that the wing forms of rice planthoppers might be a threshold response to *Wg*. When the expression level of *Wg* was less than a lower threshold gradient, the adults would be short-winged. Otherwise, when it was more than a higher threshold, the adults would be long-winged. Fortunately, we have attained an adult which wings only covered its abdomen when the nymph was fed with a higher concentration 400 ng/uL of dsWg. Of course, this comprehensive process of *Wg* manipulating the wing size and pattern of rice planthoppers still needs the further study.

Conclusions

The wingless gene has a high level of identity among the three species of rice planthopper, and was strongly related to the development and growth of wings. The expression levels of *Wg* mRNA were significantly higher in the macropterous lineage than in the brachypterous lineage of *S. furcifera*. Interference of *Wg* resulted in the shorter and abnormal wings in *S. furcifera*.

Methods
Experimental insects

Three species of rice planthoppers, *S. furcifera*, *N. lugens* and *L. striatellus*, were originally collected from rice fields in Nanjing, Jiangsu province, China, and then reared in laboratory at $25 \pm 1°C$, relative humidity $75 \pm 10\%$, and 14 h light/10 h dark photoperiod using the rice seedlings (var. Wuyunjing 7).

Selection of the macropterous and brachypterous lineages of *S. furcifera*

Due to shortage of the brachypterous males in *S. furcifera* in Jiangsu province, only two mating combinations: macropterous male × macropterous female (M♂ × M♀, called the macropterous lineage) and macropterous male × brachypterous female (M♂ × B♀, called the brachypterous lineage) were established, and reared under a constant condition ($25 \pm 1°C$, RH $75 \pm 10\%$, and 14 h light/10 h dark). In their offspring generations, only the females and males with the same wing form as their parents were chosen to mate and propagate. The detail method for selection experiment of wing forms was described by Peng et al. [28]. Continuous 40 generations of selection were performed, and the macropterous and brachypterous lineages were used in the following experiments.

Cloning of the wingless gene cDNA from three species of rice planthopper

Total RNA was isolated from the homogenization of 10 nymphs of rice planthopper (two individuals from each instar) and four adults (two macropterous males, and one macropterus and one brachypterous female) with Trizol (Invitrogen Co., USA). Total RNA 1 μg was used to synthesize the templates for cloning the *Wg* cDNA using the BD SMART™ RACE cDNA Amplification Kit (Clontech). The PCR primers were SfGSP1-5′: 5′-AGAAGCCAGGCGACGACTCCAGGTAG-3′, SfNGSP1-5′: 5′-CGTCGAACCTGTCCTTGAGGCTGTC-3′, SfGSP2-3′: 5′-ACTTTGGATTCAAGTTCTCGCGGGAG-3′, and SfNGSP2-3′: 5′-AAGCCCTACAACCCGGAGCACAAGC-3′ for *S. furcifera*, and LsGSP3-5′:5′-CGTCGAACCTGTCCTTGAGGCTGTC-3′, LsNGSP3-5′:5′-AAGCCCTACAACCCGGAGCACAAGC-3′, LsGSP4-3′:5′-CACAACACATCAGGTCGCAGCCGT-3′, and LsNGSP4-3′:5′-TTGTGCTCCGGGTTGTAGGGCTTCAG-3′ for *L. striatellus*. PCR reaction system 25 μL was used which contained 1 μL template, 2.5 μL 10 × *Taq* buffer (Mg^{2+} free), 2 μL $MgCl_2$

(25 mM), 2 μL dNTP mixture (2.5 mM/each), 1 μL forward and 1 μL reverse primers, 0.25 μL *Taq* polymerase (Takara Bio.) and 15.25 μL double distilled H_2O. PCR was preceded by denaturation at 94°C for 2 min, followed by 5 cycles at 94°C for 30 s and 72°C for 5 min, and by another 5 cycles at 94°C for 30 s, 70°C for 30s, and 72°C for 4 min, and then by 25 cycles at 94°C for 30 s, 68°C for 30 s, 72°C for 4 min and finishing with chain extension at 72°C for 10 min. The amplified product was separated by 1.5% agarose gel, and the target product was recycled and purified with the Wizard DNA Gel Extraction Kit (Unigenes Promega, Madison, Wis., USA), cloned into the EASY-T3 vector (TransGen Biotech, Beijing, China), and transformed into the DH5α competent cells. Positive clones were chosen to sequence. According to the sequences of *Wg* in *S. furcifera* and *L. striatellus*, the end to end primers *sfwg*-F2 5′-GATGGCAGCGCAATGATG-3′ and *Sfwg*-R2 5′-CGCTGTGGGTTTGGGTAA-3′ for *S. furcifera*, *Lswg*-F2 5′-AGTACAAGCCTGGTGATGG-3′ and *Lswg*-R2 5′-GGCAATAAAGTGAATTGAATAA-3′ for *L. striatellus* were designed to check the full-length of *Wg*. Unfortunately, We did not clone the full-length of *Wg* from *N. lugens*, and only cloned its ORF used the end to end primers for *S. furcifera*. The ORF sequences of *Wg* gene in the three species of rice planthopper were found using the ORF Finder online service in NCBI (http://www.ncbi.nlm.nih.gov/gorf/orfig.cgi).

Expression level of *Wg* mRNA in the macropterous and brachypterous lineages of *S. furcifera*

The 3rd, 4th and 5th instar nymphs and adults from the 29th generation of the macropterous lineage and the 30th generation of the brachypterous lineage of *S. furcifera* were collected and frozen in 300 μL Trizol for 24 hours at −70°C. The total RNA from six individuals was isolated using Trizol. The samples were grinded well by mixing some quartz rocks. The total RNA was adjusted to 1 μg/μL with DEPC-treated H_2O, and 2 μg RNA was used for RT-PCR in the 20 μL reaction mixture using the One Step SYBR PrimerScript RT-PCR Kit (Takara) to synthesize the first-strand cDNA. The real-time qPCR was performed on an ABI 7500 Real-Time PCR System (Applied Biosystems, Foster City, CA, USA) to quantify the expression levels of *Wg*. The first-strand cDNA (2 μL) diluted 50-fold was used as templates in each 20 μL reaction mixture. The qRT-PCR was performed at 95°C for 30 s, and then for 40 cycles at 95°C for 5 s and 60°C for 34 s. The ribosomal 18 s rRNA was used as an internal control. Relative expression level of *Wg* mRNA was computed based on the internal control gene using the $2^{-\triangle\triangle Ct}$ method. The special primer sets for quantifying *Wg* in *S. furcifera* were the sense primer (*SFwg*-F) 5′-TCGAATGCCAATTCCAGTTTAG -3′ and the antisense

primer (*SFwg*-R) 5′-CCCCTATCCACAATCTTTCCA -3′. And in the internal control, the primers for *S. furcifera* 18S rRNA gene were 18S rRNA-F 5′-ACAAGTATCAATTG GAGGGCAAGTCTGG-3′ and 18S rRNA-R 5′-ATGCA CACAGTATACAGGCGTGACAAG -3′. The expression levels of *Wg* mRNA in head, thorax, abdomen, wings and legs were also detected in the macropterous lineage of *S. furcifera* which was selected for 37 generations. For each sample, we carried out six biological replicates.

Synthesis of the double stranded RNA (dsRNA)

A 246 bp of nucleotide sequence of *S. furcifera Wg* was cloned using the sense primer ds*Wg* F: 5′- GCTGCC CAACCTGCGCGT-3′ and antisense primer ds*Wg* R: 5′- CCGTGCGTGCCCTGGATG-3′ designed by the coding region sequence of *Wg*, and inserted into the EASY-T3 vector (Trans Gen Biotech, Beijing, China) to confirm the sequences. The plasmid containing the specific nucleotide sequences was extracted by the MinBEST plasmid Purification Kit (Takara). The diluted plasmid was used as templates for amplification of the target sequence (ds*Wg*) by PCR. The sense primer was T7 *Wg* F: 5′-TAA TACGACTCACTATAGGG GCTGCCCAACCTGCGCGT-3′ and the antisense primer was T7 *Wg* R: 5′- TAAT ACGACTCACTATAGGG CCGTGCGTGCCCTGGATG-3′. The PCR procedures for synthesizing dsWg were performed at 94°C for 2 min followed by 35 cycles at 94°C for 30 s, 55°C for 20 s and 72°C for 20 s, and the last cycle was followed by final extension at 72°C for 5 min. The T7 RiboMAX™ Express RNAi System (Promega USA) was used to produce the specific dsRNA. The dsRNA synthesized here was washed twice using 70% ethanol, dried and suspended into an appropriate amount of the nuclease-free water, and then quantified by the Nano Drop spectrophotometer (Thermo Scientific, Wilmington, DE, USA) at 260 nm. The quality and size of the dsRNA products were verified by 1.5% agarose gel electrophoresis. Enhanced green fluorescent protein gene dsRNA (dsEGFP) was used as the negative control. The sense primer was T7 EGFP-R 5′-TAATACGACTCACTATAGGGAAGTTCAGCGTGTC CG-3′ and the antisense primer was T7 EGFP-F 5′-TAAT ACGACTCACTATAGGGCACCTTGATGCCGTTC-3′ for synthesizing the dsEGFP. The synthesized dsRNA was stored at −70°C.

Nymph RNAi

To determine the function of *Wg* in *S. furcifera*, the second instar nymphs were fed on artificial diet with the dsWg [39-41]. Groups of 30 second-instar nymphs were collected from the 37th generation selection of the macropterous lineage of *S. furcifera*, and placed into a feeding chamber constructed from a 3 cm diameter and 12 cm height cylindrical glass tube with one end covered by insect-proof nylon mesh net (48-micron) and the other end covered by two-layer of Parafilm sandwiched together [42]. Seventy-microliter liquid diets containing 100 ng/ul dsWg were put into the two layers. The nymphs feeding on artificial diet containing 100 ng/ul dsEGFP and water were the negative and blank control, respectively. The artificial diet was replaced daily. After seven days on the diet, the nymphs were counted and collected to examine the expression levels of *Wg* mRNA by qPCR method. When the nymphs grew into adults, their survival, wing forms, wing lengths, and body weights were measured. Curled wings were unfolded by dipping in absolute ethanol for 48 hours before measurement. The lengths of wings were measured from the base to tip under a stereomicroscope using the CellSens V 1.5 system (Olympus ®). The dsWg and dsEGFP treatments were repeated at least three times.

Data analysis

Wg protein sequences in three species of rice planthopper were aligned in a multiple sequence alignment using CLUSTAL X [43]. The differences between two means, such as the relative expression levels of *Wg* mRNA between the macropterous and brachypterous lineages, were compared by a student's *t* test. The expression levels of *Wg* mRNA, body weight and wing length among different tissues or RNAi treatments were analyzed by the ANOVA followed by the Tukey test. The proportions of adults with abnormal wings among RNAi treatments by dsWg and dsEGFP and the control were compared by the Chi square analysis using the Crosstabs method. All these statistic analyses were performed using SPSS 13.0 software (SPSS, Chicago, IL, USA).

Abbreviations

Wg: Wingless gene; *EGFP*: Enhanced green fluorescent potein; PCR: Polymerase chain reaction; RT-PCR: Reverse transcriptase PCR; qRT-PCR: Quantitative real-time PCR; cDNA: Complementary DNA; RACE: Rapid-amplification of cDNA ends; dsRNA: Double-stranded RNA; RNAi: RNA interference; ORF: Open reading frame; UTR: Untranslated region; SE: Standard error; ANOVA: Analysis of variance.

Competing interests

The authors declare that they have no competing interests.

Authors' contributions

JLY and ZFA carried out the experiments. XDL and JLY participated in the data analysis and drafted the manuscript. XDL conceived and supervised the research. All authors read and approved the final manuscript.

Acknowledgements

This study was supported by the grant from Major Project of Chinese National Programs for Fundamental Research and Development (973 Program, No. 2010CB126201).

References

1. Seto ES, Bellen HJ: **The ins and outs of wingless signaling.** *Trends Cell Biol* 2004, **14**(1):45–53.

2. Rijsewijk F, Schuermann M, Wagenaar E, Parren P, Weigel D, Nusse R: **The** *Drosophila* **homolog of the mouse mammary oncogene int-1 is identical to the segment polarity gene wingless.** *Cell* 1987, **50**(4):649–657.

3. Gonsalves FC, DasGupta R: **Function of the wingless signaling pathway in** Drosophila. *Methods Mol Biol* 2008, **469**:115–125.

4. Schubiger M, Sustar A, Schubiger G: **Regeneration and transdetermination: The role of wingless and its regulation.** *Dev Biol* 2010, **347**(2):315–324.

5. Cadigan KM, Nusse R: **Wnt signaling: a common theme in animal development.** *Genes Dev* 1997, **11**(24):3286–3305.

6. Chu-LaGraff Q, Doe CQ: **Neuroblast specification and formation regulated by wingless in the Drosophila CNS.** *Science* 1993, **261**(5128):1594–1597.

7. Siegfried E, Perrimon N: **Drosophila wingless: A paradigm for the function and mechanism of Wnt signaling.** *Bioessays* 1994, **16**(6):395–404.

8. Takashima S, Paul M, Aghajanian P, Younossi-Hartenstein A, Hartenstein V: **Migration of** *Drosophila* **intestinal stem cells across organ boundaries.** *Development* 2013, **140**(9):1903–1911.

9. Swarup S, Verheyen EM: **Wnt/Wingless signaling in** *Drosophila*. *Cold Spring Harb Perspect Biol* 2012, **4**:a007930.

10. Cadigan KM: **Regulating morphogen gradients in the** *Drosophila* **wing.** *Semin Cell Dev Biol* 2002, **13**(2):83–90.

11. Tabata T, Takei Y: **Morphogens, their identification and regulation.** *Development* 2004, **131**(4):703–712.

12. Phillips RG, Whittle JRS: **Wingless expression mediates determination of peripheral nervous system elements in late stages of** *Drosophila* **wing disc development.** *Development* 1993, **118**(2):427–438.

13. Simcox AA, Roberts IJH, Hersperger E, Gribbin MC, Shearn A, Whittle JRS: **Imaginal discs can be recovered from cultured embryos mutant for the segment-polarity genes engrailed naked and patched but not from wingless.** *Development* 1989, **107**(4):715–722.

14. Williams JA, Paddock SW, Carroll SB: **Pattern formation in a secondary field: a hierarchy of regulatory genes subdivides the developing** *Drosophila* **wing disc into discrete subregions.** *Development* 1993, **117**(2):571–584.

15. Campbell G, Weaver T, Tomlinson A: **Axis specification in the developing** *Drosophila* **appendage: the role of wingless, decapentaplegic, and the homeobox gene aristaless.** *Cell* 1993, **74**(6):1113–1123.

16. Yang L, Meng F, Ma D, Xie W, Fang M: **Bridging decapentaplegic and wingless signaling in** *Drosophila* **wings through repression of naked cuticle by brinker.** *Development* 2013, **140**(2):413–422.

17. Peng YN, Li CJ, Li B, Li ZY: **Knocking-down of Wingless/Wnt1 influences the development of** *Tribolium castaneum*. *Chin J Biotechnol Mol Biol* 2012, **28**(8):733–738.

18. Tong XL: *Study on the wing pattern genes in the silkworm, Bombyx mori*. PhD thesis. China: Southwest University; 2008.

19. Zhu ZR, Cheng J: **Sucking rates of the white-backed planthopper** *Sogatella furcifera* **(Horv.) (Homoptera, Delphacidae) and yield loss of rice.** *J Pest Science* 2002, **75**(5):113–117.

20. Otuka A, Matsumura M, Watanabe T: **The search for domestic migration of the white-backed planthopper,** *Sogatella furcifera* **(Horvath) (Homoptera: Delphacidae), in Japan.** *Appl Entomol Zool* 2009, **44**(3):379–386.

21. Denno RF, Roderick GK: **Population biology of planthoppers.** *Ann Rev Entomol* 1990, **35**:489–520.

22. Roff DA: **The evolution of wing dimorphism in insects.** *Evolution* 1986, **40**(5):1009–1020.

23. Zhang QX, Sun ZX, Li GH, Wang FH: **Effects of three kinds of exogenous hormones on wing dimorphism of** *Sogatella furcifera* **(Horvath).** *Acta Ecol Sin* 2008, **28**(12):5994–5998.

24. Zhou X, Chen J, Zhang M, Liang S, Wang F: **Differential DNA methylation between two wing phenotypes adults of** *Sogatella furcifera*. *Genesis* 2013, **51**(12):819–826.

25. Xue JA, Bao YY, Li BL, Cheng YB, Peng ZY, Liu H, Xu HJ, Zhu ZR, Lou YG, Cheng JA, Zhang CX: **Transcriptome analysis of the brown planthopper** *Nilaparvata lugens*. *PLoS One* 2010, **5**:e14233.

26. Xue J, Zhang XQ, Xu HJ, Fan HW, Huang HJ, Ma XF, Wang CY, Chen JG, Cheng JA, Zhang CX: **Molecular characterization of the** *flightin* **gene in the wing-dimorphic planthopper,** *Nilaparvata lugens*, **and its evolution in Pancrustacea.** *Insect Biochem Mol Biol* 2013, **43**(5):433–443.

27. Brisson JA, Ishiawa A, Miura T: **Wing development genes of the pea aphid and differential gene expression between winged and unwinged morphs.** *Insect Mol Biol* 2010, **19**(Suppl 2):63–73.

28. Peng J, Zhang C, An ZF, Yu JL, Liu XD: **Genetic analysis of wing-form determination in three species of rice planthoppers (Hemiptera: Delphacidae).** *Acta Entomol Sin* 2012, **55**(8):971–980.

29. Weatherbee SD, Halder G, Kim J, Hudson A, Carroll S: **Ultrabithorax regulates genes at several levels of the wing-patterning hierarchy to shape the development of the *Drosophila* haltere.** *Genes Dev* 1998, **12**(10):1474–1482.

30. Braendle C, Davis GK, Brisson JA, Stern DL: **Wing dimorphism in aphids.** *Heredity* 2006, **97**(3):192–199.

31. Logan CY, Nusse R: **The *Wnt* signaling pathway in development and disease.** *Annu Rev Cell Dev Biol* 2004, **20**:781–810.

32. Johnston LA, Sanders AL: **Wingless promotes cell survival but constrains growth during *Drosophila* wing development.** *Nat Cell Biol* 2003, **5**(9):827–833.

33. Sharma RP, Chopra VL: **Effect of *wingless* (*wg*[1]) mutation on wing and haltere development in *Drosophila melanogaster*.** *Dev Biol* 1976, **48**(2):461–465.

34. Kohn AD, Moon RT: **Wnt and calcium signaling: β-Catenin-independent pathways.** *Cell Calcium* 2005, **38**(3–4):439–446.

35. Mahmud FS: **Alary polymorphism in the small brown planthopper *Laodelphax striarellus* (Homoptera: Delphacidae).** *Ent Exp Appl* 1980, **28**(1):47–53.

36. Mori K, Nakasuji F: **Genetic analysis of the wing-form determination of the small brown planthopper, *Laodelphax striatellus* (Hemiptera: Delphacidae).** *Res Popul Ecol* 1990, **32**(2):279–287.

37. Liu JN, Gui FR, Li ZY: **Factors of influencing the development of wing dimorphism in the rice white-backed planthopper, *Sogatella furcifera* (Horváth).** *Acta phytophylacica Sinica* 2010, **37**(6):511–516.

38. Cook AG, Perfect TJ: **Determining the wing-morph of adult *Nilaparvata lugens* and *Sogatella furcifera* from morphometric measurements on the fifth-instar nymphs.** *Ent Exp Appl* 1982, **31**:159–164.

39. Chen J, Zhang D, Yao Q, Zhang J, Dong X, Tian H, Chen J, Zhang W: **Feeding-based RNA interference of a trehalose phosphate synthase gene in the brown planthopper, Nilaparvata lugens.** *Insect Mol Biol* 2010, **19**(6):777–786.

40. Fu Q, Zhang ZT, Hu C, Lai FX, Sun ZX: **A chemically defined diet enables continuous rearing of the brown planthopper, *Nilaparvata lugens* (Stål) (Homoptera: Delphacidae).** *Appl Entomol Zool* 2001, **36**(1):111–116.

41. Jia S, Wan PJ, Zhou LT, Mu LL, Li GQ: **Molecular cloning and RNA interference-mediated functional characterization of a Halloween gene spook in the white-backed planthopper *Sogatella furcifera*.** *BMC Mol Biol* 2013, **14**:19.

42. Yao J, Rotenberg D, Afsharifar A, Barandoc-Alviar K, Whitfield AE: **Development of RNAi methods for *Peregrinus maidis*, the corn planthopper.** *PLoS One* 2013, **8**:e70243.

43. Jeanmougin F, Thompson JD, Gouy M, Higgins DG, Gibson TJ: **Multiple sequence alignment with Clustal X.** *Trends Biochem Sci* 1998, **23**(10):403–405.

Efficient isolation of specific genomic regions retaining molecular interactions by the iChIP system using recombinant exogenous DNA-binding proteins

Toshitsugu Fujita and Hodaka Fujii[*]

Abstract

Background: Comprehensive understanding of mechanisms of genome functions requires identification of molecules interacting with genomic regions of interest *in vivo*. We previously developed the insertional chromatin immunoprecipitation (iChIP) technology to isolate specific genomic regions retaining molecular interactions and identify their associated molecules. iChIP consists of locus-tagging and affinity purification. The recognition sequences of an exogenous DNA-binding protein such as LexA are inserted into a genomic region of interest in the cell to be analyzed. The exogenous DNA-binding protein fused with a tag(s) is expressed in the cell and the target genomic region is purified with antibody against the tag(s). In this study, we developed the iChIP system using recombinant DNA-binding proteins to make iChIP more straightforward than the conventional iChIP system using expression of the exogenous DNA-binding proteins in the cells to be analyzed.

Results: In this system, recombinant 3xFNLDD-D (r3xFNLDD-D) consisting of the 3xFLAG-tag, a nuclear localization signal (NLS), the DNA-binding domain plus the dimerization domain of the LexA protein, and the Dock-tag is used for isolation of specific genomic regions. r3xFNLDD-D was expressed using a silkworm-baculovirus expression system and purified by affinity purification. iChIP using r3xFNLDD-D could efficiently isolate the single-copy chicken *Pax5* (*cPax5*) locus, in which LexA binding elements were inserted, with negligible contamination of other genomic regions. In addition, we could detect RNA associated with the *cPax5* locus using this form of the iChIP system combined with RT-PCR.

Conclusions: The iChIP system using r3xFNLDD-D can isolate specific genomic regions retaining molecular interactions without expression of the exogenous DNA-binding protein in the cell to be analyzed. iChIP using r3xFNLDD-D would be more straightforward and useful for analysis of specific genomic regions to elucidate their functions as compared to the previously published iChIP protocol.

Keywords: iChIP, Locus-specific ChIP, r3xFNLDD-D, ChIP, Chromatin immunoprecipitation

Background

Genome functions are mediated by various molecular complexes in the context of chromatin [1]. Comprehensive understanding of mechanisms of genome functions requires identification of molecules interacting with genomic regions of interest *in vivo*. To this end, we recently developed the locus-specific chromatin immunoprecipitation (ChIP) technologies consisting of insertional ChIP (iChIP) [2-5] and engineered DNA-binding molecule-mediated ChIP (enChIP) [6-8] to isolate genomic regions of interest retaining molecular interactions. The functions of the genomic regions can be comprehensively understood by analysis of DNA, RNA, proteins, or other molecules interacting with the genomic regions.

In principle, iChIP is based on locus-tagging by inserting recognition sequences of an exogenous DNA-binding

---

* Correspondence: hodaka@biken.osaka-u.ac.jp
Chromatin Biochemistry Research Group, Combined Program on Microbiology and Immunology, Research Institute for Microbial Diseases, Osaka University, Suita, Osaka, Japan

protein to isolate specific genomic regions using the exogenous DNA-binding molecule. In contrast, enChIP is based on recognition of endogenous DNA sequences by engineered DNA-binding molecules such as transactivator-like (TAL) proteins and the clustered regularly interspaced short palindromic repeats (CRISPR) system. The scheme of iChIP is as follows: (i) The recognition sequences of an exogenous DNA-binding protein such as a bacterial protein, LexA, are inserted into the genomic region of interest in the cell to be analyzed. (ii) The DNA-binding domain (DB) of the exogenous DNA-binding protein is fused with a tag(s) and a nuclear localization signal(s) (NLS(s)) and expressed in the cell to be analyzed. (iii) The resultant cell is stimulated and crosslinked with formaldehyde or other crosslinkers, if necessary. (iv) The cell is lysed, and the chromatin DNA is fragmented by sonication or enzymatic digestion. (v) The complexes including the exogenous DB are immunoprecipitated with antibody (Ab) against the tag(s). (vi) The isolated complexes which retain molecular interactions are reverse crosslinked, if necessary, and subsequent purification of DNA, RNA, proteins, or other molecules allows their identification and characterization. We successfully identified proteins and RNA components of an insulator, which functions as boundaries of chromatin domains [9], by using iChIP combined with mass spectrometry (iChIP-MS) or RT-PCR (iChIP-RT-PCR) [3]. iChIP has also been used for identification of proteins or DNA interacting with specific genomic regions by other researchers [10-13]. Thus, iChIP is a useful technology for elucidation of molecular mechanisms of genome functions.

We recently developed 3xFNLDD, the second-generation tagged LexA DB consisting of 3xFLAG-tag, an NLS, and DB plus the dimerization domain of LexA, to utilize in iChIP [4]. 3xFNLDD is expressed in the cell to be analyzed for binding to the inserted LexA BE and subsequent purification of target genomic regions in the iChIP technology. If target genomic regions inserted with LexA BE can be pulled down using recombinant 3xFNLDD conjugated to Ab against the tag (Figure 1), expression of 3xFNLDD in the cell to be analyzed would not be necessary. In addition, it is not necessary to consider unexpected side effects of expression of 3xFNLDD on cell behavior, if any, making the procedure more straightforward than the conventional iChIP system using expression of the exogenous DNA-binding proteins in the cells to be analyzed.

In this study, we developed the iChIP system using the recombinant C-terminally Dock-tagged 3xFNLDD (r3xFNLDD-D). r3xFNLDD-D was expressed using a silkworm-baculovirus expression system and purified by affinity purification. iChIP using r3xFNLDD-D could effectively isolate the single-copy chicken *Pax5* (c*Pax5*) locus from a chicken B cell line, DT40. In addition, we could detect RNA associated with the c*Pax5* locus using

this form of the iChIP system combined with RT-PCR. Thus, iChIP using r3xFNLDD-D would be more straightforward and useful than the conventional iChIP system using expression of the exogenous DNA-binding proteins in the cells to be analyzed to isolate specific genomic regions for their biochemical analysis.

Results and discussion

Expression and purification of r3xFNLDD-D

For preparation of the purified r3xFNLDD-D, we utilized a silkworm-baculovirus expression system [14]. In this system, r3xFNLDD-D was expressed in a silkworm pupa by infection of baculoviruses expressing r3xFNLDD-D. The expressed protein was purified from the pupal homogenates using Dock Catch Resin, which specifically binds to Dock-tag in a calcium-dependent manner [14]. As shown in Figure 2A, SDS-PAGE followed by Coomassie Brilliant Blue (CBB) staining detected a single protein band at 35 kDa in the elution fraction. This protein was confirmed as r3xFNLDD-D by immunoblot analysis with anti-Dock Ab (Figure 2B). Thus, r3xFNLDD-D could be expressed in a silkworm pupa and purified without visible degradation.

Efficient isolation of a target genomic region by iChIP using r3xFNLDD-D

Next, we examined whether the purified r3xFNLDD-D could be utilized for isolation of genomic regions of interest from vertebrate cells. To this end, we used the chicken DT40-derived cell line, DT40#205-2, in which 8 × repeats of LexA BE were inserted 0.3 kbp upstream of the exon 1A of the single-copy endogenous c*Pax5* gene [15] (Figure 3A). The crosslinked chromatin prepared from the cell line was subjected to iChIP using r3xFNLDD-D as shown in Figure 1. After purification of the immunoprecipitated genomic DNA, the yield of the c*Pax5* 1A promoter region was evaluated by detection of the LexA BE site (LexA BE) and the region 0.2 kbp upstream of LexA BE (i.e., 0.7 kbp upstream of the transcription start site (TSS) of c*Pax5* exon 1A) (−0.7 k) by real-time PCR (Figure 3A). As shown in Figure 3B, the yields of LexA BE and −0.7 k were more than 20% and 5% of input, respectively, when 10 μg of each r3xFNLDD-D and anti-FLAG Ab were used. In contrast, the yield of the genomic region 10 kbp upstream of the TSS of the exon 1A (−10 k) was less than 0.01%. These results suggested that r3xFNLDD-D can bind to LexA BE even in the crosslinked chromatin and iChIP using r3xFNLDD-D is able to specifically purify target genomic regions. The specific isolation of the c*Pax5* 1A promoter region was completely blocked when we inhibited binding of r3xFNLDD-D to anti-FLAG Ab with excessive amounts of 3xFLAG peptide (Figure 3C). The c*Pax5* 1A promoter region was not isolated when parental DT40 was used instead of DT40#205-2 (Figure 3C). These results clearly

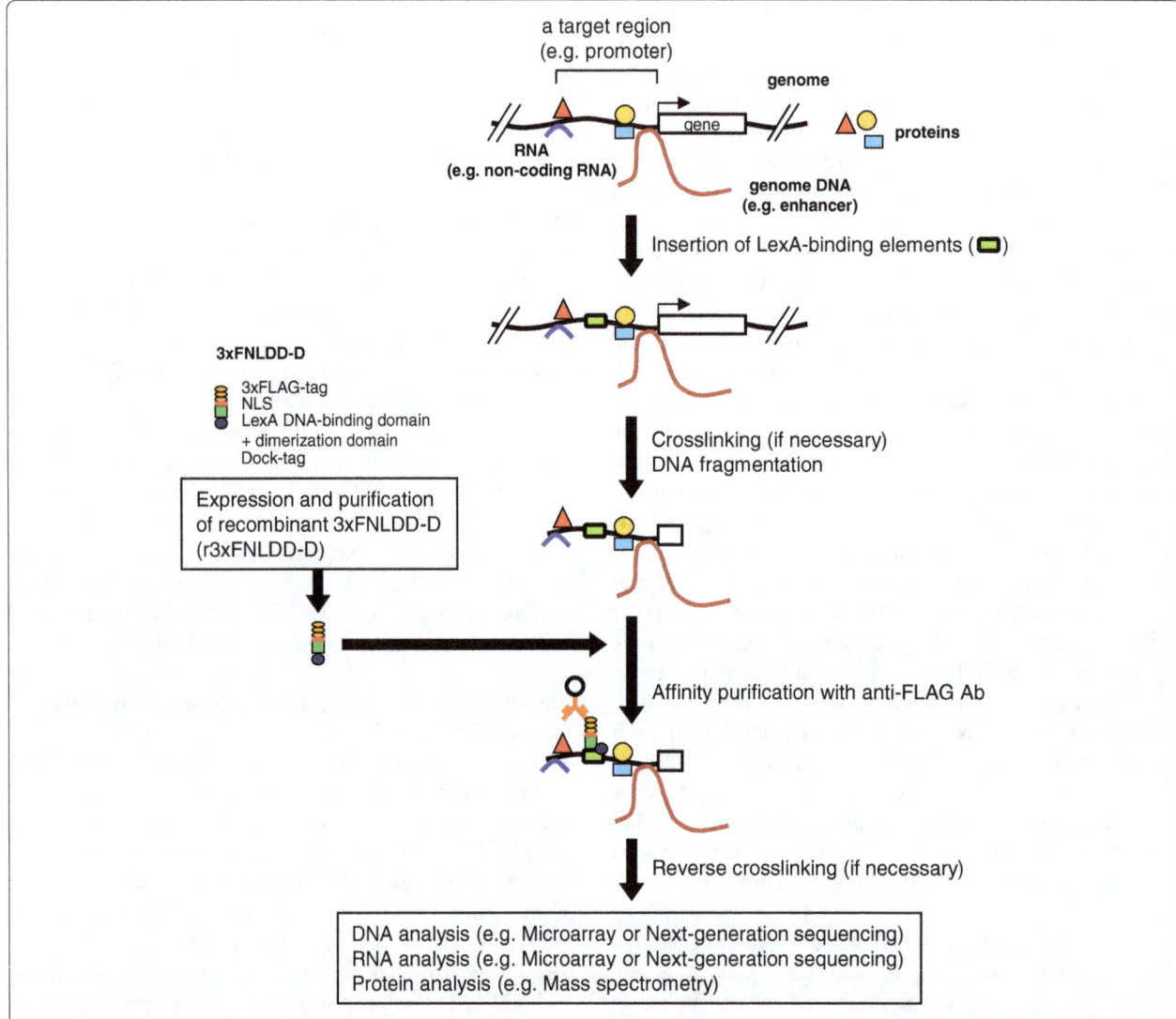

Figure 1 Scheme of iChIP using r3xFNLDD-D. 3xFNLDD-D consisting of 3xFLAG-tag, a nuclear localization signal (NLS), the DNA-binding domain (DB) plus the dimerization domain of the LexA protein, and Dock-tag, is expressed and purified. The recognition sequences of the LexA protein (LexA BE) are inserted into a genomic region of interest, usually by homologous recombination, in the cell to be analyzed. The resultant cell is stimulated and crosslinked with formaldehyde or other crosslinkers, if necessary. The cell is lysed, and the genomic DNA is fragmented. The target genomic region is affinity purified with r3xFNLDD-D conjugated with anti-FLAG antibody (Ab). After reverse crosslinking, if necessary, purification of the chromatin components (DNA, RNA, proteins, other molecules) allows their identification and characterization.

demonstrated that isolation of target genomic regions is mediated by binding of r3xFNLDD-D to LexA BE. The yield of the target genomic region by the modified iChIP system was comparable with that of the previously reported iChIP protocol [4].

Optimization of iChIP using r3xFNLDD-D

Next, we titrated amounts of r3xFNLDD-D and anti-FLAG Ab to optimize the system (Figure 4A). The yield of the LexA BE site in the cPax5 1A promoter was comparable when 0.5 - 10 µg of each r3xFNLDD-D and anti-FLAG Ab was used with chromatin prepared from 1×10^7 of DT40#205-2 cells. In contrast, use of 0.01 -

0.1 µg of each protein showed lower yield, suggesting that 0.5 µg of each r3xFNLDD-D and anti-FLAG Ab are sufficient for 1×10^7 cells. The yield of iChIP using 0.5 µg of r3xFNLDD-D was 20% of input for LexA BE and less than 0.01% for −10 k, which is comparable with that using 10 µg of r3xFNLDD-D (Figures 3B and 4B). In this regard, we observed the yield of 15% of input for the same locus when 3xFLNDD was expressed in DT40#205-2 and the conventional iChIP protocol was used (T.F., H.F., unpublished observation). These results showed that the iChIP using r3xFNLDD-D could purify the target region with efficiency comparable to the conventional iChIP. We also examined whether it would

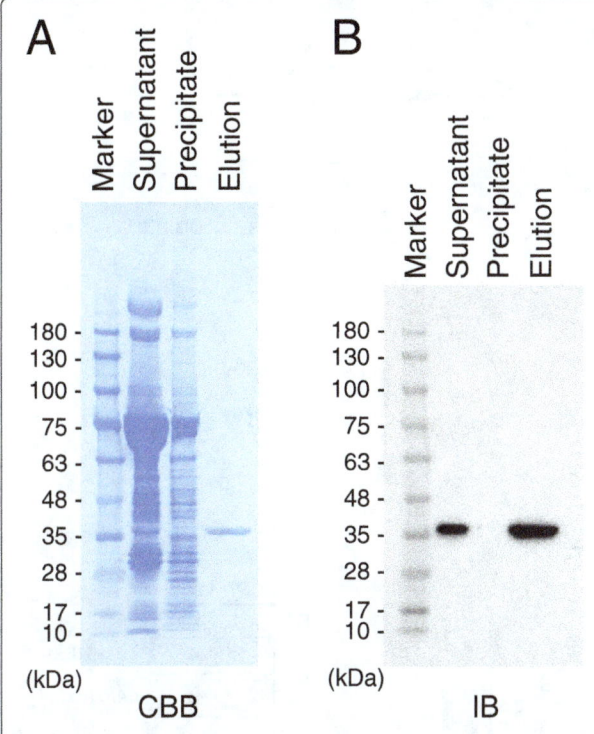

Figure 2 Expression and purification of r3xFNLDD-D. (A)
Coomassie Brilliant Blue (CBB) staining of the recombinant protein.
The purified proteins were subjected to SDS-PAGE and CBB staining.
(B) Immunoblot analysis (IB) of r3xFNLDD-D. The purified proteins
were subjected to SDS-PAGE and IB with anti-Dock Ab. Supernatant:
the supernatant prepared from the silkworm pupal homogenates.
Precipitant: the insoluble precipitate prepared from the silkworm
pupal homogenates. Elution: the eluate after affinity purification with
Dock Catch Resin.

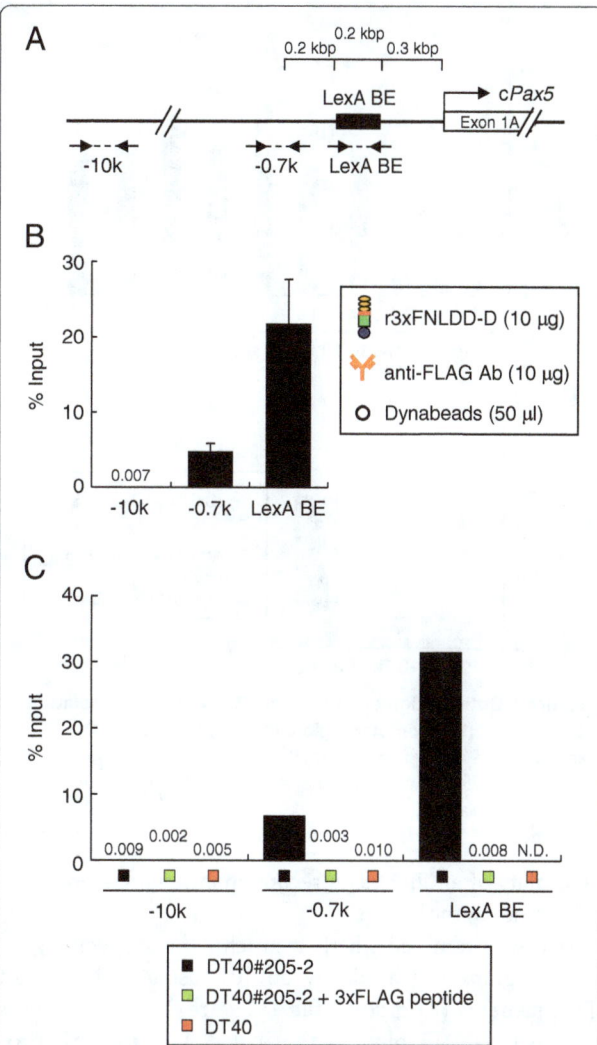

**Figure 3 Isolation of the *cPax5* 1A promoter region by iChIP
using r3xFNLDD-D. (A)** Scheme of the LexA BE-inserted *cPax5* 1A
promoter region with primer positions. The positions of PCR primers
with distances from the transcription start site (TSS) are indicated.
(B) The results of iChIP using 10 μg of r3xFNLDD-D. % of input is
shown (mean +/− SEM, n = 3). **(C)** Specific isolation of the target
genomic region by iChIP using r3xFNLDD-D. N.D.: not detected.

be possible to purify the c*Pax5* 1A promoter region using
r3xFNLDD-D with Dock Catch Resin, which binds to
the C-terminal Dock-tag of r3xFNLDD-D. As shown in
Additional file 1: Figure S1, 2% of input of the LexA
BE site could be isolated with negligible contamination
of −10 k, indicating that the C-terminal Dock-tag of
r3xFNLDD-D can also be utilized for iChIP, although the
yield was much lower than that using anti-FLAG Ab.

Isolation of RNA associated with the *cPax5* locus by iChIP using r3xFNLDD-D

Next, we examined whether iChIP using r3xFNLDD-D
could be utilized to isolate genomic regions of interest and
identify molecules interacting with those genomic regions
in cells. To this end, we attempted to isolate the c*Pax5*
locus including the exon 1A region by iChIP using
r3xFNLDD-D and detect the nascent RNA transcribed
from the TSS of the c*Pax5* exon 1A by RT-PCR (Figure 5A).
Transcription from the c*Pax5* exon 1A was not disrupted
by the presence of LexA BE inserted in the 1A promoter
region (Additional file 1: Figure S2). As shown in Figure 5B,
iChIP using r3xFNLDD-D isolated the c*Pax5* exon 1A

region but not the exon 3 region of the irrelevant c*AID*
gene, which encodes an enzyme essential for B cell-specific
immunoglobulin somatic hypermutation and class switch
recombination [16]. After purification of the associated
RNA, RT-PCR analysis detected RNA transcribed
from the exon 1A of the c*Pax5* gene but not that
from the c*AID* gene in the iChIP sample (Figure 5C)
(the full-length images with size markers are shown in
Additional file 1: Figure S3). These results suggested
that iChIP using r3xFNLDD-D is able to isolate specific
genomic regions retaining molecules interacting with
the genomic regions.

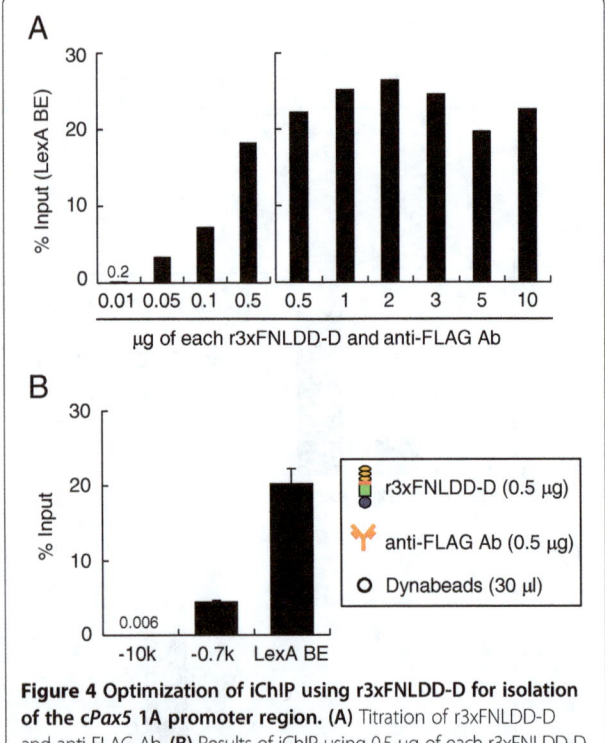

Figure 4 Optimization of iChIP using r3xFNLDD-D for isolation of the *cPax5* 1A promoter region. (A) Titration of r3xFNLDD-D and anti-FLAG Ab. **(B)** Results of iChIP using 0.5 μg of each r3xFNLDD-D and anti-FLAG Ab. % of input is shown. The error bar represents the range of duplicate experiments.

Feasibility of enChIP using recombinant engineered DNA-binding molecules

Lastly, we examined whether enrichment of specific genomic regions is feasible by enChIP using recombinant TAL proteins (Additional file 1: Figure S4). enChIP uses engineered DNA-binding molecules such as TAL proteins or the CRISPR system consisting of a catalytically inactive form of Cas9 (dCas9) and guide RNA (gRNA) for locus-tagging and affinity purification of the target loci [6-8]. We generated a construct expressing a fusion protein, r3xFN-5′HS5-TAL-G, consisting of 3xFLAG-tag, an NLS, a recombinant TAL protein recognizing human 5′ HS5 region, which functions as an insulator to regulate transcription of the *β-globin* genes (Additional file 1: Figure S5A) [17], and the glutathione S-transferase (GST)-tag, and prepared the recombinant protein by using the silkworm-baculovirus expression system (Additional file 1: Figure S5B and C). We used the recombinant protein for enChIP analysis of the 5′HS5 locus. As shown in Additional file 1: Figure S5D, the 5′HS5 site was enriched several-fold compared to the irrelevant *interferon regulatory factor 1* (*IRF-1*) promoter region when non-crosslinked native chromatin prepared from the 293T cell line was used. These results suggest that enChIP using recombinant TAL proteins is feasible. However, we found that the r3xFN-5′HS5-TAL-G (ca. 160 kDa) showed massive degradation (Additional file 1:

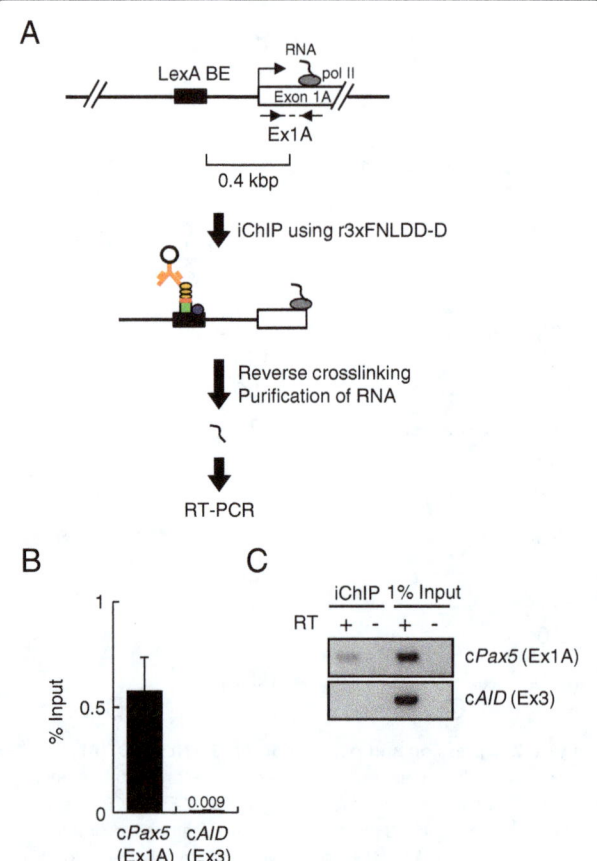

Figure 5 Detection of RNA associated with the *cPax5* locus. (A) Scheme of iChIP. After isolation of the *cPax5* locus by iChIP using r3xFNLDD-D, the nascent RNA transcribed on the exon 1A of the *cPax5* gene was detected by RT-PCR. **(B)** Results of iChIP using 0.5 μg of each r3xFNLDD-D and anti-FLAG Ab. % of input is shown. The error bar represents the range of duplicate experiments. **(C)** Detection of RNA corresponding to the exon 1A of the *cPax5* gene but not that corresponding to the exon 3 of the *cAID* gene by RT-PCR.

Figure S5B and C). In addition, we failed to detect enrichment of the target 5′HS5 site when we used crosslinked chromatin (data not shown). These results also suggest that improvement in production of recombinant TAL proteins and their access to target loci might be required for efficient isolation of target regions and identification of associated molecules.

Applications and advantages of iChIP using r3xFNLDD-D

In this study, we applied RT-PCR to detection of RNA interacting with a genomic region of interest in cells. Next-generation sequencing or microarray analysis can be combined with iChIP using r3xFNLDD-D for non-biased identification of interacting RNA as well as DNA. Moreover, mass spectrometry can be combined for non-biased identification of interacting proteins.

Because iChIP using r3xFNLDD-D does not require expression of 3xFNLDD in cells, it is of great use in the iChIP analysis of primary cells isolated from organisms, especially higher eukaryotes such as mice. In the case of application of the standard iChIP technology to mice, it is time-consuming to establish mouse lines expressing 3xFNLDD in the cells to be analyzed as well as those possessing LexA BE in specific genomic regions. In this regard, iChIP using r3xFNLDD-D is able to skip the mating steps between mice expressing 3xFNLDD and those possessing LexA BE, substantially accelerating iChIP analysis using organisms.

Compared to enChIP or proteomics of isolated chromatin (PICh), which uses specific biotinylated nucleic acid probes such as locked nucleic acids (LNAs) that hybridize target genomic regions for their isolation [18], iChIP requires insertion of LexA BE, which takes time and effort. However, recent advancement of genome editing technologies using TALEN and CRISPR makes insertion of exogenous sequences in target loci much more easily. In addition, insertion of such exogenous sequences may abrogate function of genomic regions through changes in nucleosome positioning or other mechanisms. Therefore, it is necessary to confirm that the insertion of LexA BE does not abrogate function of genomic regions before isolating the genomic regions by iChIP. On the other hand, the locus-tagging system used in iChIP can be used for isolation of a specific target allele such as a maternal or paternal allele. Feasibility of such allele-specific analysis is one of advantages of iChIP over enChIP and PICh when allele-specific analysis is required, for example, in the analysis of genome imprinting.

Conclusions

In this study, we established the iChIP system using r3xFNLDD-D to make the iChIP technology much more straightforward than the conventional iChIP system using expression of the exogenous DNA-binding proteins in the cells to be analyzed. Using this system, we were able to isolate target genomic regions; % of input reached more than 20% for the cPax5 1A promoter region. In addition, we could detect RNA associated with the cPax5 locus, suggesting that iChIP using r3xFNLDD-D can isolate target genomic regions retaining molecular interactions. Thus, the modified iChIP protocol described here using r3xFNLDD-D has advantages over the previously published protocol in that it is more straightforward and useful for analysis of specific genomic regions to elucidate their functions.

Methods

Expression and purification of r3xFNLDD-D

Expression of 3xFNLDD-D was performed using the silkworm-baculovirus expression system (ProCube) (Sysmex Corporation, http://procube.sysmex.co.jp/eng/) as described previously [14]. Briefly, the coding sequence of 3xFNLDD [4] was inserted into the transfer vector pM31a (Sysmex Corporation) to fuse the Dock-tag at its C-terminus and co-transfected with linearized genomic DNA of ABv baculovirus (*Bombyx mori* nucleopolyhedrovirus; CPd strain, Sysmex Corporation) into the *B. mori*-derived cell line, BmN, to generate the recombinant baculovirus. The generated baculovirus was infected into a silkworm pupa to express 3xFNLDD-D. The expressed 3xFNLDD-D was

Table 1 Primers used in this study

Number	Name	Sequence (5′ → 3′)	Experiments
26572	LexA-N2	ttctctatcgataggtacctcg	Real-time PCR in Figures 3, 4 and Additional file 1: Figure S1 (LexA BE)
27428	LexA-C-for-Pax5	cgctgcgtggtcgagcgtactg	Real-time PCR in Figures 3, 4 and Additional file 1: Figure S1 (LexA BE)
27134	cPax5-ChIP-UP(−0.2 k)-F	gggctcttatttcgttttttcttgtt	Real-time PCR in Figures 3 and 4 (−0.7 k)
27135	cPax5-ChIP-UP(−0.2 k)-R	gtgcttatttgtcagcgtggttg	Real-time PCR in Figures 3 and 4 (−0.7 k)
27013	cPax5-ChIP-UP(−10 k)-F	tccacatcgttacattgtcacttct	Real-time PCR in Figures 3, 4 and Additional file 1: Figure S1 (−10 k)
27014	cPax5-ChIP-UP(−10 k)-R	taaaagccctcagttcgatttattg	Real-time PCR in Figures 3, 4 and Additional file 1: Figure S1 (−10 k)
26552	cPax5-inExon1A-F	cctaaaacgtttagtttcagctcagt	RT-PCR in Figure 5 and Additional file 1: Figure S2 (cPax5 Ex1A)
26553	cPax5-inExon1A-R	ttcgtggctctctcaggtca	RT-PCR in Figure 5 and Additional file 1: Figure S2 (cPax5 Ex1A)
27571	cAID-Ex3-F	catgtggaggttctcttcctacg	RT-PCR in Figure 5 and Additional file 1: Figure S2 (cAID Ex3)
27572	cAID-Ex3-R	caagtttgggtaggcacgaag	RT-PCR in Figure 5 and Additional file 1: Figure S2 (cAID Ex3)
26773	18SrRNA-F2	cttagagggacaagtggcg	RT-PCR in Additional file 1: Figure S2 (18S)
26774	18SrRNA-R2	acgctgagccagtcagtgta	RT-PCR in Additional file 1: Figure S2 (18S)
27420	hHS5-TAL-Target-F	ccagtttcttccagtttcccttt	Real-time PCR in Additional file 1: Figure S5 (5′HS5)
27421	hHS5-TAL-Target-R	ttttcaaaatgcaaggtgatgtc	Real-time PCR in Additional file 1: Figure S5 (5′HS5)
27310	hIRF1-prom-F	cgcctgcgttcgggagatatac	Real-time PCR in Additional file 1: Figure S5 (IRF-1)
27312	hIRF1-prom-R1 + 2	ctgtcctctcactccgccttgt	Real-time PCR in Additional file 1: Figure S5 (IRF-1)

purified with Dock Catch Resin (Sysmex Corporation) as described previously [14]. The immunoblot analysis was performed with anti-Dock Ab (Sysmex Corporation).

Cell lines
The chicken B cell line DT40 was provided by the RIKEN BioResource Center through the National Bio-Resource Project of the Ministry of Education, Science, Sports and Culture of Japan. DT40 and DT40#205-2, in which LexA BE was inserted in the 1A promoter region of the cPax5 gene (Fujita and Fujii, manuscript submitted), were maintained in RPMI-1640 (Wako) with 4 mM glutamine, 10% (v/v) fetal bovine serum, 1% chicken serum, and 50 μM 2-mercaptoethanol at 39.5°C.

Chromatin preparation
Cells (2×10^7) were fixed with 1% formaldehyde at 37°C for 5 min. The chromatin fraction was extracted and fragmented (2 kbp-long on average) by sonication as described previously [19] except for using 800 μl of *in vitro* Modified Lysis Buffer 3 (10 mM Tris pH 8.0, 150 mM NaCl, 1 mM EDTA, 0.5 mM EGTA) and Ultrasonic disruptor UD-201 (TOMY SEIKO). After sonication, Triton X-100 was added to final concentration at 0.1%.

iChIP using r3xFNLDD-D
The sonicated chromatin (400 μl) was pre-cleared with 0.01 - 10 μg of normal mouse IgG (Santa Cruz Biotechnology) conjugated to 30 - 50 μl of Dynabeads-Protein G (Invitrogen) and subsequently incubated with 0.01 - 10 μg of r3xFNLDD-D and anti-FLAG M2 Ab (Sigma-Aldrich) conjugated to 30 - 50 μl of Dynabeads-Protein G at 4°C for 20 h. 100 μg of 3xFLAG peptide was added to inhibit binding of r3xFNLDD-D to anti-FLAG Ab. The Dynabeads were washed four times with 1 ml of *in vitro* Wash Buffer (20 mM Tris pH 8.0, 150 mM NaCl, 2 mM EDTA, 0.1% Triton X-100) and once with 1 ml of TBS-IGEPAL-CA630 (50 mM Tris pH 7.5, 150 mM NaCl, 0.1% IGEPAL-CA630). The isolated chromatin complexes were eluted with 120 μl of Elution Buffer (500 μg/ml 3xFLAG peptide (Sigma-Aldrich), 50 mM Tris pH 7.5, 150 mM NaCl, 0.1% IGEPAL-CA630) at 37°C for 30 min. After reverse cross-linking at 65°C, DNA was purified with ChIP DNA Clean & Concentrator (Zymo Research) and used as template for real-time PCR with SYBR Select PCR system (Applied Biosystems) using the Applied Biosystems 7900HT Fast Real-Time PCR System. PCR cycles were as follows: heating at 50°C for 2 min followed by 95°C for 10 min; 40 cycles of 95°C for 15 sec and 60°C for 1 min. The primers used in this experiment are shown in Table 1.

Isolation of interacting RNA and RT-PCR
Chromatin preparation and iChIP using r3xFNLDD-D were performed as described above except for addition of RNasin Plus RNase Inhibitor (Promega) in all buffers. After reverse crosslinking at 65°C, RNA was isolated with Isogen II (Nippon gene) combined with Direct-zol RNA Mini Prep (Zymo Research). The purified RNA was used as template for reverse transcription with ReverTra Ace qPCR RT Master Mix with gDNA Remover (Toyobo). The cDNA was used as template for PCR with AmpliTaq Gold 360 Master Mix (Applied Biosystems). PCR cycles were as follows: heating at 95°C for 10 min; 40 cycles of 95°C for 30 sec, 60°C for 30 sec, 72°C for 1 min; and the final extending 72°C for 2 min. The primers used in this experiment are shown in Table 1.

Competing interests
Patents on iChIP are already registered ("Method for isolating specific genomic regions", August 20, 2010, PCT/JP2010/064052; April 9, 2013, US patent 8415098; November 22, 2013, Japanese patent 5413924). The authors filed a patent application on enChIP ("Method for isolating specific genomic regions using DNA-binding molecules recognizing endogenous DNA sequences", September 6, 2013, PCT/JP2013/74107). This does not alter the authors' adherence to all the BMC Molecular Biology policies on sharing data and materials.

Authors' contributions
TF and HF conceived this form of the iChIP and enChIP technologies, designed and performed experiments, and wrote the manuscript. HF directed and supervised the study. Both authors read and approved the final manuscript.

Acknowledgements
We thank F. Kitaura for technical assistance. This work was supported by Takeda Science Foundation (TF), the Uehara Memorial Foundation (HF), the Kurata Memorial Hitachi Science and Technology Foundation (TF and HF), Adaptable & Seamless Technology Transfer Program through Target-driven R&D (A-STEP) by the Japan Science and Technology Agency (JST) (#AS251Z01861Q) (HF), Grant-in-Aid for Young Scientists (B) (#25830131) (TF), "Transcription Cycle" (#25118512) (HF) from the Ministry of Education, Culture, Sports, Science and Technology of Japan.

References
1. van Driel R, Fransz PF, Verschure PJ: **The eukaryotic genome: a system regulated at different hierarchical levels.** *J Cell Sci* 2003, 116:4067–4075.
2. Hoshino A, Fujii H: **Insertional chromatin immunoprecipitation: a method for isolating specific genomic regions.** *J Biosci Bioeng* 2009, 108(5):446–449.
3. Fujita T, Fujii H: **Direct idenification of insulator components by insertional chromatin immunoprecipitation.** *PLoS One* 2011, 6(10):e26109.
4. Fujita T, Fujii H: **Efficient isolation of specific genomic regions by insertional chromatin immunoprecipitation (iChIP) with a second-generation tagged LexA DNA-binding domain.** *Adv Biosci Biotechnol* 2012, 3(5):626–629.
5. Fujita T, Fujii H: **Locus-specific biochemical epigenetics/chromatin biochemistry by insertional chromatin immunoprecipitation.** *ISRN Biochem* 2013, 2013:913273.
6. Fujita T, Fujii H: **Efficient isolation of specific genomic regions and identification of associated proteins by engineered DNA-binding molecule-mediated chromatin immunoprecipitation (enChIP) using CRISPR.** *Biochem Biophys Res Commun* 2013, 439:132–136.

7. Fujita T, Asano Y, Ohtsuka J, Takada Y, Saito K, Ohki R, Fujii H: **Identification of telomere-associated molecules by engineered DNA-binding molecule-mediated chromatin immunoprecipitation (enChIP).** *Sci Rep* 2013, **3**:3171.

8. Fujita T, Fujii H: **Identification of proteins associated with an IFNγ-responsive promoter by a retroviral expression system for enChIP using CRISPR.** *PLoS One* 2014, **9**(7):e103084.

9. Gaszner M, Felsenfeld G: **Insulators: exploiting transcriptional and epigenetic mechanisms.** *Nat Rev Genet* 2006, **7**:703–713.

10. McCullagh E, Seshan A, El-Samad H, Madhani HD: **Coordinate control of gene expression noise and interchromosomal interactions in a MAP kinase pathway.** *Nat Cell Biol* 2010, **12**(10):954–962.

11. Agelopoulos M, McKay DJ, Mann RS: **Developmental regulation of chromain conformation by Hox proteins in Drosophila.** *Cell Rep* 2012, **1**(4):350–359.

12. Byrum SD, Raman A, Taverna SD, Tackett A: **ChAP-MS: a method for identification of proteins and histone posttranslational modifications at a single genomic locus.** *Cell Rep* 2012, **2**(1):198–205.

13. Pourfarzad F, Aghajanirefah A, de Boer E, Ten Have S, Bryn van Dijk T, Kheradmandkia S, Stadhouders R, Thongjuea S, Soler E, Gillemans N, von Lindern M, Demmers J, Philipsen S, Grosveld F: **Locus-specific proteomics by TChP: targeted chromatin purification.** *Cell Rep* 2013, **4**:589–600.

14. Kamezaki Y, Enomoto C, Ishikawa Y, Koyama T, Nara S, Suzuki T, Sakka K: **The Dock tag, an affinity tool for the purification of recombinant proteins, based on the interaction between dockerin and cohesin comains from Clostridium josui cellulosome.** *Protein Expr Purif* 2010, **70**:23–31.

15. Fujita T, Fujii H: **Species-specific 5'-genomic structure and multiple transcription start sites in the chicken Pax5 gene.** *Gene* 2011, **477**:24–31.

16. Arakawa H, Buerstedde JM: **Immunoglobulin gene conversion: insights from bursal B cells and the DT40 cell line.** *Dev Dyn* 2004, **229**:458–464.

17. Kukreti S, Kaur H, Kaushik M, Bansal A, Saxena S, Kaushik S, Kukreti R: **Structural polymorphism at LCR and its role in beta-globin gene regulation.** *Biochimie* 2010, **92**:1199–1206.

18. Déjardin J, Kingston RE: **Purification of proteins associated with specific genomic loci.** *Cell* 2009, **136**(1):175–186.

19. Fujita T, Ryser S, Tortola S, Piuz I, Schlegel W: **Gene-specific recruitment of positive and negative elongation factors during stimulated transcription of the MKP-1 gene in neuroendocrine cells.** *Nucleic Acids Res* 2007, **35**:1007–1017.

Identification of a set of miRNAs differentially expressed in transiently TIA-depleted HeLa cells by genome-wide profiling

Carmen Sánchez-Jiménez[1], Isabel Carrascoso[1], Juan Barrero[2] and José M Izquierdo[1][*]

Abstract

Background: T-cell intracellular antigen (TIA) proteins function as regulators of cell homeostasis. These proteins control gene expression globally at multiple levels in response to dynamic regulatory changes and environmental stresses. Herein we identified a micro(mi)RNA signature associated to transiently TIA-depleted HeLa cells and analyzed the potential role of miRNAs combining genome-wide analysis data on mRNA and miRNA profiles.

Results: Using high-throughput miRNA expression profiling, transient depletion of TIA-proteins in HeLa cells was observed to promote significant and reproducible changes affecting to a pool of up-regulated miRNAs involving miR-30b-3p, miR125a-3p, miR-193a-5p, miR-197-3p, miR-203a, miR-210, miR-371-5p, miR-373-5p, miR-483-5p, miR-492, miR-498, miR-503-5p, miR-572, miR-586, miR-612, miR-615-3p, miR-623, miR-625-5p, miR-629-5p, miR-638, miR-658, miR-663a, miR-671-5p, miR-769-3p and miR-744-5p. Some up-regulated and unchanged miRNAs were validated and previous results confirmed by reverse transcription and real time PCR. By target prediction of the miRNAs and combined analysis of the genome-wide expression profiles identified in TIA-depleted HeLa cells, we detected connections between up-regulated miRNAs and potential target genes. Gene Ontology (GO) and Kyoto Encyclopedia of Genes and Genomes (KEGG) database analysis suggest that target genes are related with biological processes associated to the regulation of DNA-dependent transcription, signal transduction and multicellular organismal development as well as with the enrichment of pathways involved in cancer, focal adhesion, regulation of actin cytoskeleton, endocytosis and MAPK and Wnt signaling pathways, respectively. When the collection of experimentally defined differentially expressed genes in TIA-depleted HeLa cells was intersected with potential target genes only 7 out of 68 (10%) up- and 71 out of 328 (22%) down-regulated genes were shared. GO and KEGG database analyses showed that the enrichment categories of biological processes and cellular pathways were related with innate immune response, signal transduction, response to interleukin-1, glomerular basement membrane development as well as neuroactive ligand-receptor interaction, endocytosis, lysosomes and apoptosis, respectively.

Conclusion: All this considered, these observations suggest that individual miRNAs could act as potential mediators of the epigenetic switch linking transcriptomic dynamics and cell phenotypes mediated by TIA proteins.

Keywords: TIA1, TIAR, miRNAs, Gene regulatory networks

* Correspondence: jmizquierdo@cbm.uam.es
[1]Centro de Biología Molecular Severo Ochoa, Consejo Superior de Investigaciones Científicas, Universidad Autónoma de Madrid (CSIC/UAM), C/ Nicolás Cabrera 1, Cantoblanco, Madrid 28049, Spain
Full list of author information is available at the end of the article

Background

Nowadays, the central dogma of Molecular Biology — developed from classic research works aimed at determining the biology of prokaryotic organisms— is known to reflect only a part of the agenda containing the genetic information that gives rise to the complexity of eukaryotic organisms. The characterization of post-transcriptional events leading to the generation of multiple RNAs, proteins and functions from only one RNA precursor shows up the existence of multiple overlapping regulatory networks and mechanisms for the control of biological functions beyond transcriptional regulation. There is increasing evidence to support the idea that transcriptome and proteome regulation and heterogeneity are key stages to understand differences in the protein diversity observed in organisms of similar genetic complexity. It's therefore necessary to fully characterize the modulators linking and synchronizing multiple layers of gene expression regulation.

T-cell intracellular antigen 1 (TIA1) and TIA1 related/like (TIAR/TIAL1) proteins are two DNA/RNA binding proteins that regulate many aspects of RNA metabolism at different levels. These multifunctional regulators can modulate: i) DNA-dependent transcription through its interaction with DNA and RNA polymerase II [1-4]; ii) alternative splicing of pre-mRNA through the selection of atypical 5′ spliced sites [5-8], and iii) stability and/or translation of eukaryotic mRNAs through the interaction with 5′ and/or 3′ untranslatable regions [9-16]. Some of these regulatory layers operate in the control of main biological programmes so as to maintain cellular homeostasis; this programmes include apoptosis, inflammation, cell responses to stress or viral infections ([10] and references included). Furthermore, mice lacking either TIA1 [16] or TIAR [17], as well as ectopically over-expressing TIAR [18], show higher rates of embryonic lethality.

MicroRNAs (miRNAs) are a class of 19-25 nt long non-coding RNAs that regulate post-transcriptionally gene expression by binding with partially complementary sequences on target mRNAs and inhibiting translation or affecting stability of these mRNAs [19]. Multiple lines of evidence indicate that they are key regulators of numerous critical functions in developmental, cell differentiation and disease processes, including tumorigenesis and cancer progression [19]. However, defining the place and function of miRNAs in complex regulatory networks is not straightforward. Systems' approaches such as the inference of a module network from expression data can help to achieve this goal [19].

We have previously described specific changes of transcriptomic dynamics associated to inflammation, angiogenesis, metabolism, and cell proliferation-related genes upon TIA1/TIAR-RNA interference-based silencing in HeLa cells [10]. In the present study, we test the hypothesis whether there are specific changes associated to the pattern of miRNA expression which may interfere/modulate with target genes and, therefore, contribute to the phenotypes described in TIA-depleted HeLa cells. Herein, we identified a miRNA signature that is concomitant and coherent with biological processes and pathways associated to the phenotypes observed in HeLa cells lacking TIA proteins.

Methods

Cell culture and RNA interference (RNAi)analysis

HeLa cells were grown and transfected with 20 nM of either a control siRNA (non-silencing siRNA duplex fluorescein labeled 27-6411-02FL from Gene Link) or two siRNAs against TIA1 (5′-AAGCTCTAATTCTGCAA CTCTTT-3′; 5′-AACAACTAA TGCGTCAGACTTTT-3′) and TIAR (5′-AAGTCCTTATACTTCAGTTGTTC-3′; 5′-AACCATGGAATCAACAAGGATTT-3′) directed to the positions 59-81/647-669 and 65-87/971-993 to the coding regions of TIA1 and TIAR mRNAs, respectively, as described previously [7,10].

Cell extract preparation, western blot analysis and RNA purification

Whole-cell extracts were prepared by resuspensing the cells in lysis buffer (50 mM Tris–HCl, pH 8.0, 140 mM NaCl, 1.5 mM MgCl$_2$, 0.5% Nonidet P-40 plus a mixture of protease inhibitors), freeze-thawing three times, and centrifugation at 10,000 rpm for 5 min. in a microfuge at 4°C [10]. Resulting supernatants were recovered and stored at -70°C [10]. Immunoblots were carried out loading equal amounts of protein (15 μg) on 10% SDS-PAGE and using the following antibodies: anti-TIA1 and anti-TIAR from Santa Cruz Biotechnology (CA, USA) and anti-α-tubulin from Sigma (UK). RNA was extracted by using a miRVANA kit (Ambion, TX, USA) according to the manufacturer's protocol.

MicroRNA expression profiling analysis

The quality of the total RNA was verified by an Agilent 2100 Bioanalyzer profile. 2000 ng total RNA from each sample was labeled with Hy3™ or Hy5™ fluorescent label, using the miRCURY LNA™ microRNA Labeling Kit Hy3™/Hy5™ (Exiqon, Denmark), following the procedure described by the manufacturer. A Hy3™- and a Hy5™-labeled RNA sample were mixed pair-wise and hybridized to the miRCURY LNA™ microRNA Array (Exiqon, Denmark), which contains capture probes targeting all microRNAs for human registered in the miRBASE version 9.1. The hybridization was performed according to the miRCURY LNA™ microRNA Array Instruction manual using a Tecan HS4800™ hybridization station (Tecan, Austria). After hybridization the microarray slides were scanned and stored in an ozone free environment (ozone level below 2.0 ppb) in order to prevent potential bleaching of the fluorescent dyes. The miRCURY

LNA™ microRNA Array slides were scanned using the Agilent G2565BA Microarray Scanner System (Agilent Technologies, Inc., USA) and the image analysis was carried out using the ImaGene® 7.0 (miRCURY LNA™ microRNA Array Analysis Software, Exiqon, Denmark). The quantified signals were background corrected (local background subtraction) and normalized using the global Lowess (LOcally WEighted Scatterplot Smoothing) regression3 algorithm.

Local background was corrected by normexp method with an offset of 50. Background corrected intensities were transformed to log scale (base 2) and normalized by Lowess for each array [20]. Finally, to have similar intensity distribution across all arrays, Lowess-normalized-intensity values were scaled by adjusting their quantiles [21]. After data processing each probe was tested for changes in expression over replicates using an empirical Bayes moderated t statistic [22]. To control the false discovery rate (FDR), P values were corrected using the method of Benjamini and Hochberg (1995) [23]. FIESTA viewer (http://bioinfogp.cnb.csic.es/tools/ FIESTA) was used to visualize all microarray results and to evaluate the numerical thresholds (-2> fold change >2; FDR < 0.0001) applied for selecting differentially expressed genes [24].

The miRPlus sequences are licensed human sequences (Exiqon, Denmark). Some of them are already annotated in the miRBase database version 18. Microarray data discussed in this publication have been deposited in the NCBI Gene Expression Omnibus database (http://www.ncbi.nlm.nih.gov/geo/info/linking.html) and are accessible through the GEO Series accession number GSE41213.

QPCR
The method was optimized for microRNA, and reagents, primers, and probes were obtained from Applied Biosystems. Reverse transcriptase (RT) reactions and real-time PCR (PCR) were performed according to manufacturer protocols at the Genomic PCR Core Facility at Universidad Autónoma de Madrid in Madrid Scientific Park. Analyses were performed in two independent samples by triplicate, including no-template and RT-minus controls. U6 RNA and miR-200 expression were used as endogenous reference controls. Relative miRNA expression was calculated using the comparative cycle threshold method.

Generation of miRNA targets dataset
In silico targets predicted for each of the differentially expressed miRNAs by three different algorithms: TargetScan 5.2 ([25] and references included), PicTar-Vert ([26] and references included) and miRDB [27,28] were downloaded using web-app miRBase [29] (http://www.mirbase.org). Given that each algorithm focus on different aspects of miRNA-mRNA pairing, and the lack of experimental validation of most miRNAs targets do not allow a false-positive elimination, we kept the datasets by considering the following score values: TargetScan 5.2 (aggregate PCT > 0.1), PicTar-Vert (PicTar score > 2) and miRBD (target score > 70).

Gene ontology, pathway and network analyses
The Gene Ontology (GO) and Kyoto Encyclopedia of Genes and Genomes (KEGG) database analysis were conducted using software programmes provided by GenCodis3 (http://genecodis.cnb.csic.es) [30,31]. Networks and regulatory topologies of functional relationships between gene clusters were created using the CytoScape [32] (http://www.cytoscape.org).

Results
To analyze the putative role of TIA proteins in the control of gene expression on microRNAs, we transfected HeLa cells with double-stranded small interfering RNAs (siRNAs) targeting TIA1 and TIAR mRNAs or with control siRNA (C), as previously reported [10]. The effect of siRNAs on TIA1/TIAR expression was analyzed by Western blotting (Figure 1A). Upon TIA1/TIAR-RNA interference-based silencing, 80-90% and 70-80% depletions of TIA1 and TIAR proteins, respectively (Figure 1A) were achieved 72 h after transfection, thus in agreement with previous observations [10]. By contrast, α-tubulin protein (Figure 1A) was used as control of siRNA specificity, and its expression level was not significantly affected by gene interference approach.

To test the microRNA (miRNA) expression profiles resulting from the reduction of TIA1 and TIAR protein levels, the differences in global miRNA expression patterns in control and TIA1/TIAR-depleted HeLa cells were determined by means of a miRCURY™ LNA Array with specific probes for simultaneous analysis of at least 600 different miRNAs. After having passed sample QC on the Bioanalyzer2100 and RNA measurement on the Nanodrop instrument, the samples were labelled using the miRCURY™ Hy3™/Hy5™ labelling kit and hybridized on the miRCURY™ LNA Array (v.8.1) (Figure 1B). Analysis of the scanned slides showed that the labelling was successful as all capture probes for the control spike-in oligo nucleotides produced signals in the expected range. The quantified signals were normalized using the global Lowess (Locally WEighted Scatterplot Smoothing) regression algorithm, which we have found produces the best within-slide normalization to minimize the intensity-dependent differences between the dyes (Figure 1B). The positive effect of this normalization is illustrated in 3 different plots for each experimental conditions analyzed (Additional file 1). It is interesting to note the high reproducibility of individual miRNA expression levels and their correlation across the different miRNA pools (Additional file 1). On the other hand, an appropriate statistical test analysis was made using a linear model (as implemented in

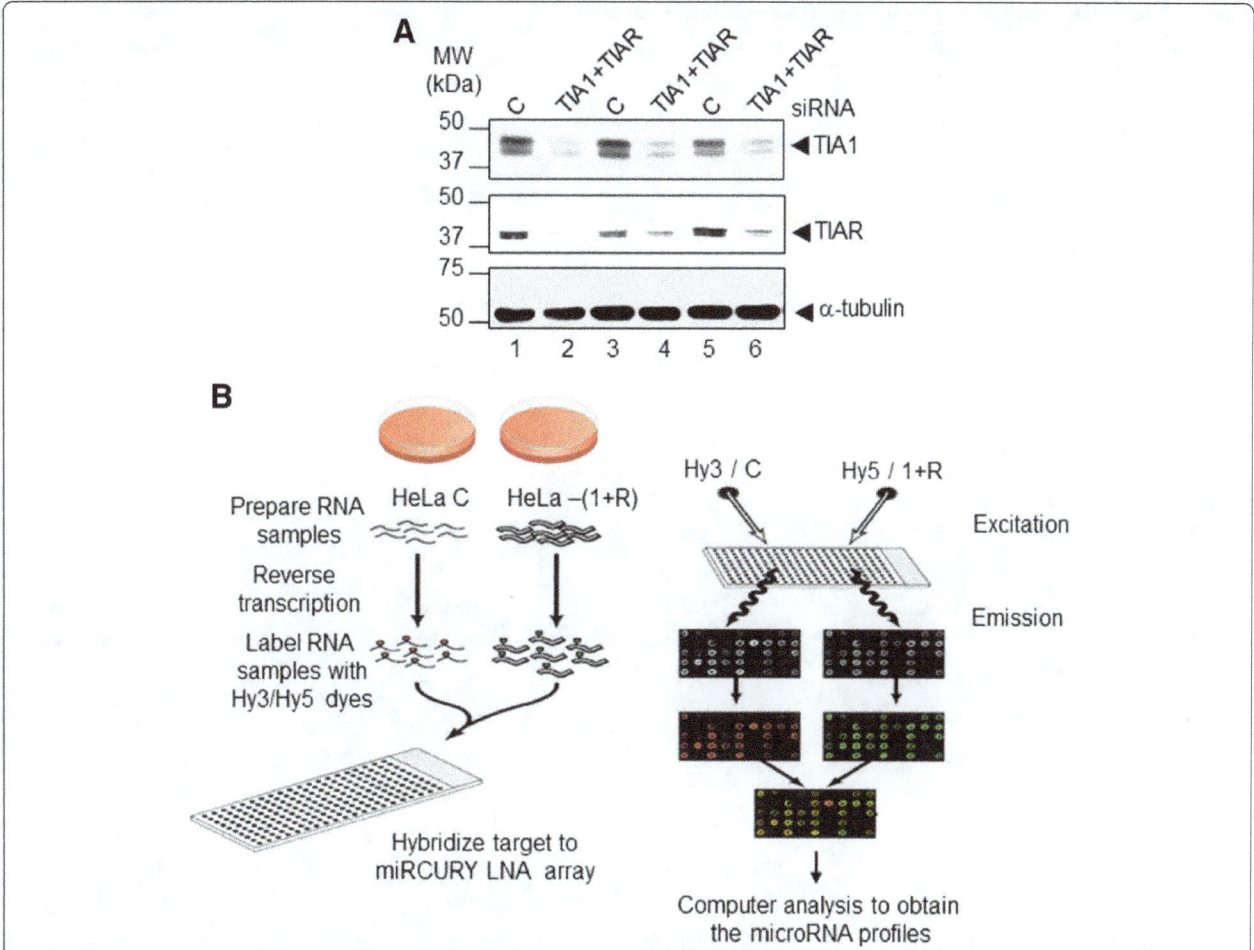

Figure 1 Small interfering RNA (siRNA)-mediated depletion of TIA proteins. (**A**) Western blot analysis of HeLa cell lysates prepared 72 h after transfection with siRNAs against control (C; lanes 1, 3 and 5) and TIA1 plus TIAR (lanes 2, 4 and 6). The blot was probed with antibodies against TIA1, TIAR, and α-tubulin proteins, as indicated. Molecular weight markers and the identities of protein bands are shown. (**B**) A schematic representation depicting the miRNA array strategy used in this study. The preparation of the RNA samples from control and TIA-depleted HeLa cells, labeling of above RNA samples with Hy3 and Hy5 dyes, hybridization and analyses of the resulting miRNA profiles are illustrated.

the limma R/Biocounductor package) to compare miRNA expression patterns for three two-color arrays performed from three independent biological replicates (Figure 1 and Additional file 1).

As shown in Table 1, depletion of TIA-proteins resulted in significantly altered miRNA expression profiling. The expression level of 29 out of 600 miRNAs was found to more than double when comparing channels Hy5 and Hy3. The genome-wide profiling analysis identified 17 miRNAs and 12 putative in-silico miRNAs (identified as miRPlus). These were the well-established miRNAs: miR-197-3p, miR-210, miR-373-5p, miR-492, miR-498, miR-503-5p, miR-572, miR-586, miR-612, miR-615-3p, miR-623, miR-625-5p, miR-638, miR-658, miR-663a, miR-671-5p and miR-769-3p, which were differentially up-regulated at least 2-fold (FDR<0.0001). The identified miRPlus sequences were in licensed human sequences; now a great number of these sequences have been

annotated in the corresponding miRNA database. This information is available in Table 1. Regarding the differentially expressed miRPlus in our experimental conditions, an update of these miRPlus sequences in the miRBase 18 indicates that miRPlus-17836 is miR-30b-3p, miRPlus-17864 is miR-744-5p, miRPlus-17867 is miR-203a, miRPlus-17877 and miRPlus-17960 are miR-483-5p, miRPlus-17878 is miR-193a-5p, miRPlus-17942 is miR-125a-3p, miRPlus-17950 is miR-371-5p and miRPlus-17961 is miR-629-5p (Table 1). The remaining miRPlus sequences have not been yet assigned to the putative specific miRNAs (Additional file 2).

To validate previous results on identified miRNAs and their relative expression levels, four up-regulated and two unchanged miRNAs were confirmed by TaqMan reverse transcription and polymerase chain reaction (QPCR) analyses. As shown in Figure 2, the results obtained by quantitative amplification were fully comparable and the relative

Table 1 MicroRNA expression profiling in TIA-depleted HeLa cells

Fold change	pval (LiMMA)	FDR (LiMMA)	miRNA ID	miRNA ID update
2.21	4.13E-06	8.26E-05	hsa-miR-197_MM2	hsa-miR-197-3p
3.84	5.00E-07	1.70E-05	hsa-miR-210	-
2.3	4.84E-06	9.38E-05	hsa-miR-373*	hsa-miR-373-5p
2.75	3.78E-06	7.77E-05	hsa-miR-492	-
2.79	1.38E-06	3.55E-05	hsa-miR-498	-
2.5	5.22E-06	9.89E-05	hsa-miR-503	hsa-miR-503-5p
4.29	1.00E-07	4.05E-06	hsa-miR-572	-
3.78	3.60E-07	1.27E-05	hsa-miR-586	-
2.24	5.01E-06	9.59E-05	hsa-miR-612	-
2.82	4.50E-06	8.81E-05	hsa-miR-615	hsa-miR-615-3p
3.18	1.91E-06	4.50E-05	hsa-miR-623	-
2.84	1.23E-06	3.31E-05	hsa-miR-625	hsa-miR-625-5p
2.11	3.63E-06	7.54E-05	hsa-miR-638	-
2.44	1.64E-06	3.96E-05	hsa-miR-658	-
3.61	2.57E-06	5.63E-05	hsa-miR-663	hsa-miR-663a
2.23	2.75E-06	5.96E-05	hsa-miR-671	hsa-miR-671-5p
3.65	3.18E-06	6.70E-05	hsa-miR-769-3p	-
3.44	2.60E-07	9.88E-06	miRPlus_17832	n.d.
2.73	9.60E-07	2.77E-05	miRPlus_17836	hsa-miR-30b-3p
4.38	5.70E-07	1.91E-05	miRPlus_17856	n.d.
3.34	2.26E-06	5.13E-05	miRPlus_17864	hsa-miR-744-5p
2.5	2.91E-06	6.23E-05	miRPlus_17867	hsa-miR-203a
2.7	6.30E-07	2.04E-05	miRPlus_17877/17960	hsa-miR-483-5p
2.33	4.69E-06	9.13E-05	miRPlus_17878	hsa-miR-193a-5p
2.84	1.87E-06	4.43E-05	miRPlus_17881	n.d.
3.63	5.50E-07	1.85E-05	miRPlus_17890	n.d.
3.3	2.16E-06	4.94E-05	miRPlus_17942	hsa-miR-125a-3p
2.15	2.85E-06	6.14E-05	miRPlus_17950	hsa-miR-371-5p
2.6	1.34E-06	3.45E-05	miRPlus_17961	hsa-miR-629-5p

miRNA cluster defining a signature of up-regulated miRNAs in TIA-depleted HeLa cells. The microarray data were analyzed by the limma R method. Fold is an average measure of the fold change in differential expression and the false discovery rate (FDR) indicates the expected percentage of false positives (FDR < 0.0001). miRNA ID update are miRNAs and miRPlus renamed in agreement with miRBase 18 database. n.d. means non-determined.

fold changes in miRNA expression were consistent with the data detected by hybridization in the corresponding microarrays (compare Table 1 and Figure 2).

As a first attempt to understand the relevance of 29 differentially up-regulated miRNAs, in silico methods for predicting human miRNA target genes were used (Figure 3). The potential target genes were identified by searching the TargetScan, PicTar and miRBD databases [25-28]. By comparing and selecting alone non-repeated target genes, 2683 miRNA target genes were identified (Additional file 3). All these potential target genes were tested using computer tool GeneCodis3 [20,21] to elucidate the biological processes and cellular pathways assigned using Gene Ontology (GO) and Kyoto Encyclopedia of Genes and Genomes (KEGG) database analyses (Figure 3 and Additional file 3).

A total of 253 biological processes from GO database and 50 cellular pathways from KEGG categories were estimated with a significance hypergeometric test (corrected hypergeometric p value < 0.01). Target genes corresponding to the biological processes from GO categories were mainly involved in regulation of DNA-dependent transcription, signal transduction, multicellular organismal development, positive/negative regulation of transcription from RNA polymerase II promoter, cell adhesion, transmembrane transport, apoptotic process and nervous system development (Table 2 and Additional file 3). In addition, target genes associated to the main biological pathways were also identified using KEGG database involving pathways in cancer, MAPK signalling pathway, focal adhesion, endocytosis, regulation of

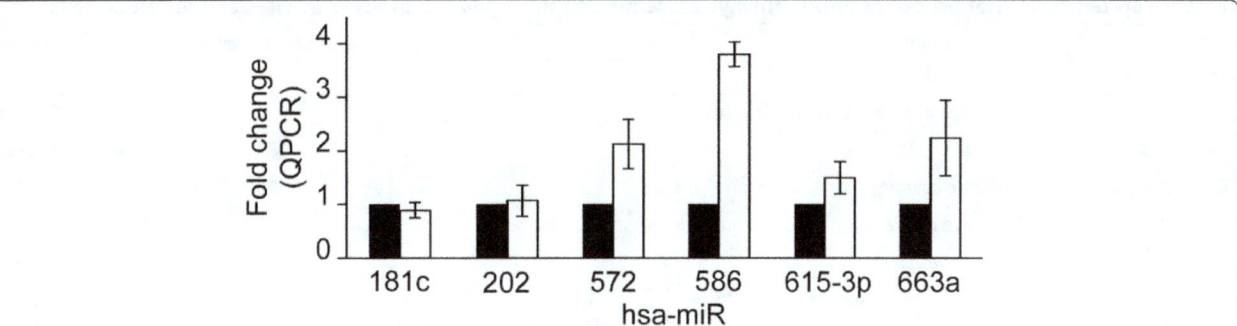

Figure 2 Validation of miRNA array-predicted changes by quantitative RT-PCR (QPCR). Quantitative miRNA expression analysis was carried out using TaqMan probes by QPCR. The represented values were normalized and expressed relative to control values (whose value was fixed arbitrarily to 1) as ratios and are means ± SD (n = 2).

actin cytoskeleton, Wnt signalling pathway, neuroactive ligand-receptor interaction, glutamatergic synapse, ubiquitin mediated proteolysis and tuberculosis (Table 2 and Additional file 3). Collectively, these results suggest that TIA-protein depletion promotes the induction of a miRNA

signature which may directly and indirectly contribute to the establishment of the observed cellular phenotypes in TIA-deficient HeLa cells [10].

On the other hand, the collection of differentially expressed genes previously identified by expression microarray Human

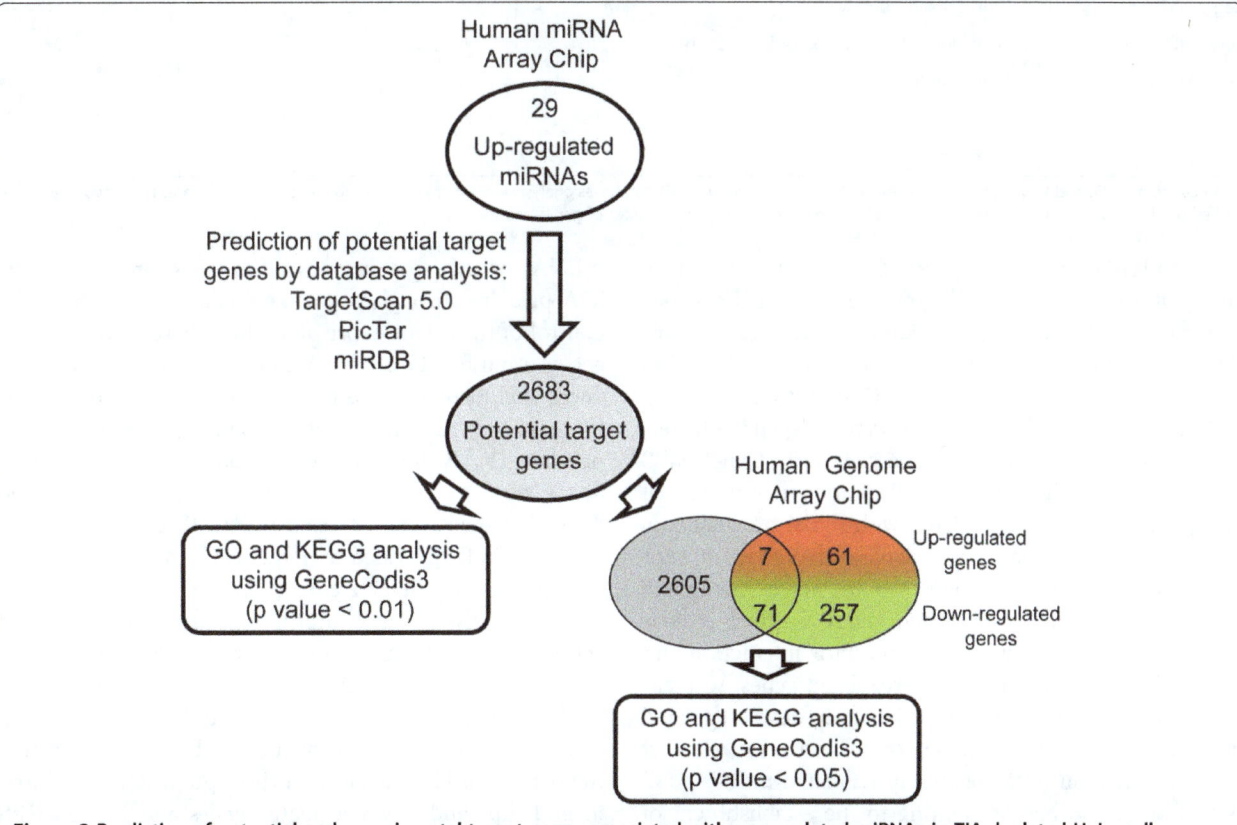

Figure 3 Prediction of potential and experimental target genes associated with up-regulated miRNAs in TIA-depleted HeLa cells. Integrative analyses of potential target genes and miRNAs regulated in TIA-depleted HeLa cells using TargetScan 5.2, PicTar-Vert and miRDB software tools (see corresponding section in Methods). Diagrams depicting the number of potential target genes and up-regulated miRNAs (Additional file 3) by TIA silencing are shown. Venn diagram depicting the numbers of genes that were intersected among putative target genes (highlighted in gray) associated to miRNAs and experimentally defined differentially expressed genes (see additional data files in [10]) in HeLa cells lacking TIA proteins. The Gene Ontology (GO) and Kyoto Encyclopedia of Genes and Genomes (KEGG) database analyses were conducted using software programmes provided by GenCodis3.

Table 2 Top ten biological processes and pathways associated to up-regulated miRNAs in TIA-depleted HeLa cells

Term	Description	Number of genes	p-value
GO			
GO:0006355	Regulation of transcription, DNA-dependent	269	2.49E-33
GO:0007165	Signal transduction	188	1.00E-18
GO:0007275	Multicellular organismal development	151	5.17E14
GO:0045944	Positive regulation of transcription from RNA pol II promoter	124	2.69E-21
GO:0007155	Cell adhesion	103	4.06E-12
GO:0055085	Transmembrane transport	102	5.89E-08
GO:0045893	Positive regulation of transcription, DNA-dependent	96	5.89E-14
GO:0006915	Apoptotic process	96	4.13E-07
GO:0000122	Negative regulation of transcription from RNA pol II promoter	95	4.66E-17
GO:0007399	Nervous system development	93	1.98E-16
KEGG			
Kegg:05200	Pathways in cancer	68	5.72E-11
Kegg:04010	MAPK signaling pathway	58	3.68E-08
Kegg:04510	Focal adhesion	49	2.75E-08
Kegg:04144	Endocytosis	46	7.60E-07
Kegg:04810	Regulation of actin cytoskeleton	44	1.48E-04
Kegg:04310	Wnt signaling pathway	37	3.88E-05
Kegg:04080	Neuroactive ligand-receptor interaction	35	9.43E-04
Kegg:04724	Glutamatergic synapse	32	2.34E04
Kegg:04120	Ubiquitin mediated proteolysis	31	5.96E-03
Kegg:05152	Tuberculosis	28	1.17E-04

Gene Ontology (GO) and Kyoto Encyclopedia of Genes and Genomes (KEGG) databases analysis were carried out using GeneCodis3 software. The categories were ranked on their numbers of associated genes and the ten with the highest number of genes are shown.

Genome U133 Plus 2.0 (Affymetrix) in TIA-depleted HeLa cells [10] were intersected with potential target genes identified using in silico target prediction tools. The results show that only 7 out of 68 (10%) up- and 71 out of 328 (22%) down-regulated genes were shared (Figure 3 and Table 3). GO and KEGG database analyses were independently performed for up- and down-regulated genes with GeneCodis3 software. The results suggest that the enrichment functional categories (p < 0.05) are related with signal transduction, innate immune response and response to interleukin-1 for up-regulated genes and glomerular basement membrane development for down-regulated genes. Further, cellular pathways of neuroactive lingand-receptor interaction and apoptosis were associated to up-regulated genes, whereas endocytosis and lysosomes were linked to down-regulated genes (Table 4). Taken together, these results suggest that at least a fraction (40%) of the up-regulated miRNAs (12 out of 30) could be contributing to the establishment of differential expression profiles associated to the HeLa cells lacking TIA proteins.

Based on above observations an interesting question emerges: why should many specific genes be up- or down-regulated by induced miRNAs in TIA-depleted HeLa cells? A simple answer to this question might be that these mRNAs or their precursors (pre-mRNAs) are targeted by TIA proteins through one or multiple layers to exert the control of their gene expression; thus, these regulators can act as multifunctional RNA binding proteins (see references included in Background). To approach this issue, the experimental profiles of the binding patterns of TIA proteins (i.e., the RNA map corresponding to TIA proteins) at the up- or down-regulated pre-mRNAs (Figure 4 and Table 3) were examined using the iCLIP database of TIA proteins kindly provided by Jernej Ule's laboratory [8]. For example, as shown in Figure 4 and in Table 3, the pre-mRNAs analyzed show multiple sequence sites located across the full-length pre-mRNAs, both exons and introns, and which we have classified as genes with either low (+), medium (++) or high (+++) density of TIA binding sites. The greatest relevance of this observation is the fact that the density of binding sites on these pre-mRNAs is found in both up- and down-regulated genes and located with frequency on the last exons of these pre-mRNAs and particularly on the sequences located at the 3' untranslated regions of the mature mRNAs. In this regard, it is reasonable to think that the existence of a feedback loop that represses the expression of many genes, which could be activated in the absence of TIA proteins, for example at

Table 3 Intersection between potential and experimentally defined target genes related to miRNAs in TIA-depleted HeLa cells

Gene symbol	Description	Associated miRNA	TIA-iCLIP
Up-regulated genes			
CNR1	Cannabinoid receptor 1 (brain)	miR-30b-3p	+
EIF4A2	Eukaryotic translation initiation factor 4A, isoform 2	miR-586	+++
EREG	Epiregulin	miR-586	+
F2RL2	Coagulation factor II (thrombin) receptor-like 2	miR-30b-3p	+
IL1R1	Interleukin 1 receptor, type I	miR-498	++
IRAK2	Interleukin-1 receptor-associated kinase 2	miR-503-5p	++
SELI	Selenoprotein I	miR-197-3p	++
Down-regulated genes			
ACADSB	Acyl-Coenzyme A dehydrogenase, short/branched chain	miR-203	++
ACOX1	Acyl-Coenzyme A oxidase 1, palmitoyl	miR-373-5p	++
ALS2CR4	Amyotrophic lateral sclerosis 2 (juvenile) chromosome region, candidate 4	miR-203	+++
ANKH	Ankylosis, progressive homolog (mouse)	miR-203	+++
AP1S2	Adaptor-related protein complex 1, sigma 2 subunit	miR-203	++
AP2B1	Adaptor-related protein complex 2, beta 1 subunit	miR-203	+++
APPBP2	Amyloid beta precursor protein (cytoplasmic tail) binding protein 2	miR-612	++
BBS1	Bardet-Biedl syndrome 1	miR-612	+
BRIP1	BRCA1 interacting protein C-terminal helicase 1	miR-373-5p	+
C16orf72	Chromosome 16 open reading frame 72	miR-671-5p	+++
C18orf54	Chromosome 18 open reading frame 54	miR-625-5p	++
C1orf96	Chromosome 1 open reading frame 96	miR-373-5p	++
C20orf108	Chromosome 20 open reading frame 108	miR-30b-3p	+
CCDC50	Coiled-coil domain containing 50	miR-203	+
CENPH	Centromere protein H	miR-612	+
CLCC1	Chloride channel CLIC-like 1	miR-373-5p and 30b-3p	++
COL4A4	Collagen, type IV, alpha 4	miR-203	+++
CTDSPL2	CTD (carboxy-terminal domain, RNA polymerase II, polypeptide A) small phosphatase like 2	miR-203	+++
CTSC	Cathepsin C	miR-203	++
DAB2	Disabled homolog 2, mitogen-responsive phosphoprotein (Drosophila)	miR-203	+++
DDIT4	DNA-damage-inducible transcript 4	miR-30b-3p	++
EEF1A1	Eukaryotic translation elongation factor 1 alpha 1	miR-373-5p	+
EIF4EBP2	Eukaryotic translation initiation factor 4E binding protein 2	miR-373-5p	++
ELMOD2	ELMO/CED-12 domain containing 2	miR-30b-3p	++
EPHA7	EPH receptor A7	miR-503-5p	+++
FAM129A	Family with sequence similarity 129, member A	miR-373-5p and 586	++
FBXO9	F-box protein 9	miR-203	++
IQCE	IQ motif containing E	miT-483-5p	+
KCTD12	Potassium channel tetramerisation domain containing 12	miR-373-5p and 586	+
KLF9	Kruppel-like factor 9	miR-373-5p	++
KRIT1	KRIT1, ankyrin repeat containing	miR-373-5p	+
LAMP2	Lysosomal-associated membrane protein 2	miR-373-5p	++
MCM4	Minichromosome maintenance complex component 4	miR-373-5p	++

Table 3 Intersection between potential and experimentally defined target genes related to miRNAs in TIA-depleted HeLa cells (Continued)

MECP2	Methyl CpG binding protein 2 (Rett syndrome)	miR-203	+
MIB1	Mindbomb homolog 1 (Drosophila)	miR-373-5p, 503 and 203	++
MOBKL2B	MOB1, Mps One Binder kinase activator-like 2B (yeast)	miR-503-5p and 203	+++
MTAP	Methylthioadenosine phosphorylase	miR-125a-3p	++
MTHFD2L	Methylenetetrahydrofolate dehydrogenase (NADP+ dependent) 2-like	miR-373-5p	++
NAV1	Neuron navigator 1	miR-503-5p	++
NDRG3	NDRG family member 3	miR-203	++
NDST1	N-deacetylase/N-sulfotransferase (heparan glucosaminyl) 1	miR-30b-3p	+
NHLRC2	NHL repeat containing 2	miR-373-5p	++
NID1	Nidogen 1	miR-30b-3p	++
PAWR	PRKC, apoptosis, WT1, regulator	miR-30b-3p	+++
PCGF6	Polycomb group ring finger 6	miR-203	+
PDE1A	Phosphodiesterase 1A, calmodulin-dependent	miR-373-5p	+++
PGM2	Phosphoglucomutase 2	miR-498	+++
PLD1	Phospholipase D1, phosphatidylcholine-specific	miR-203	+++
RAB22A	RAB22A, member RAS oncogene family	miR-498	++
RBM8A	RNA binding motif protein 8A	miR-373-5p	+
RECK	Reversion-inducing-cysteine-rich protein with kazal motifs	miR-503-5p	++
RGC32	Regulator of cell cycle	miR-30b-5p	++
SFPQ	Splicing factor proline/glutamine-rich (polypyrimidine tract binding protein associated)	miR-586	++
SGPL1	Sphingosine-1-phosphate lyase 1	miR-373-5p	++
SLC12A2	Solute carrier family 12 (sodium/potassium/chloride transporters), member 2	miR-503-5p, 586 and 203	+
SLC35B3	Solute carrier family 35, member B3	miR-203	++
SNAPC3	Small nuclear RNA activating complex, polypeptide 3, 50kDa	miR-373-5p and 671	+++
STXBP4	Syntaxin binding protein 4	miR-625-5p	+
SUDS3	Suppressor of defective silencing 3 homolog (S. cerevisiae)	miR-203	+++
SYNC1	Syncoilin, intermediate filament 1	miR-203	++
TFDP2	Transcription factor Dp-2 (E2F dimerization partner 2)	miR-30b-3p and 203	+++
TIA1	TIA1 cytotoxic granule-associated RNA binding protein	miR-30b3p	+++
TNFRSF19	Tumor necrosis factor receptor superfamily, member 19	miR-125a-3p	++
TNRC6B	Trinucleotide repeat containing 6B	miR-503-5p, 586 and 203	++
TSEN2	tRNA splicing endonuclease 2 homolog (S. cerevisiae)	miR-197-3p	+
VAMP1	Vesicle-associated membrane protein 1 (synaptobrevin 1)	miR-203	++
VGLL3	Vestigial like 3 (Drosophila)	miR-373-5p	++
VPS13A	Vacuolar protein sorting 13 homolog A (S. cerevisiae)	miR-586	++
WDFY3	WD repeat and FYVE domain containing 3	miR-203	+
ZBTB44	Zinc finger and BTB domain containing 44	miR-203	++
ZNF169	Zinc finger protein 169	miR-125a-3p	++

Gene cluster defining a molecular signature of up- (highlighted in red in Figure 3) and down-regulated (highlighted in green in Figure 3) genes linked to up-regulated miRNAs in TIA-depleted HeLa cells. Gene symbol, gene title as description and associated miRNAs are indicated. Estimation of the density of binding sites of TIA proteins on up- and down-regulated target genes by iCLIP analysis is shown. The relative quantification estimated as low (+), medium (++) and high (+++) density is indicated.

Table 4 Biological processes and pathways linked to experimentally defined and differentially expressed genes in TIA-depleted HeLa cells

Categories enriched in	Description	Number of genes	Gene symbol	p-value
Up-regulated genes				
GO term				
GO:0070555	Response to interleukin-1	2	IL1R1, IRAK2	6.21E-04
GO:0045087	Innate immune response	2	IL1R1, IRAK2	6,88E-03
GO:0007165	Signal transduction	2	IL1R1, IRAK2	2,88E-02
KEGG term				
Kegg:04080	Neuroactive ligand-receptor interaction	2	CNR1, F2RL2	6.88E-03
Kegg:04210	Apoptosis	2	IL1R1, IRAK2	1.43E-03
Down-regulated genes				
GO term				
GO:0032836	Glomerular basement membrane development	2	COL4A4, NID1	1.09E-02
KEGG term				
Kegg:04144	Endocytosis	4	RAB22A, AP2B1, DAB2, PLD1	3.23E-02
Kegg:04142	Lysosome	3	CTSC, AP1S2, LAMP2	4.77E-02

Gene Ontology (GO) and Kyoto Encyclopedia of Genes and Genomes (KEGG) databases analysis were carried out using GeneCodis3 software.

the post-transcriptional levels (i.e. mRNA stability or translational activation), to dampen its expression in order to promote or counteract the cellular phenotypes associated to the absence of the TIA proteins.

During the last decade, much progress has been made in the understanding of network topology and the relevance and properties of its basic modules. In this study, we analyze and assess module networks inferred from both miRNAs and gene expression data using a bioinformatic tool as CytoScape [32]. By matching expressed miRNAs and experimentally defined up- and down-regulated genes in TIA-depleted HeLa cells, a putative regulatory network of TIA-associated genes and miRNAs was constructed. Based on the number of potential gene interactions with single miRNAs, both up- and down-expressed genes regulating TIA-mediated differential expression were connected (Figure 5). This network draws a cellular scenario where the reduction of TIA proteins could lead to molecular responses mediated by individual miRNAs. Such a computational approach, starting from expression data alone, can be helpful in the future process of identification of the function of these miRNAs by suggesting modules of co-expressed genes in which they could play a regulatory role.

Discussion

Current data suggest that a significant portion of the information containing the human genome is regulated by miRNAs. These small RNAs recognize and regulate target genes. In this regard, more than 60% of protein-coding genes are predicted *in silico* as targets, based on

conserved base-pairing between the 3'- and 5'-untranslated regions (UTR) of the mRNA and the seed sequence of the miRNAs [33] without considering putative target on coding sequences. At present, near to 2,000 miRNAs have been identified in the human genome and about 20-30% of human genes are controlled by one or more miRNAs [33-36]. Multiple lines of evidence indicate that they are key regulators of numerous critical functions in development and disease, including cancer. Many of them have been reported to have molecular features and act either as tumour suppressors or oncogenes [33-35]. The changes in miRNA signature identified in this study might directly and indirectly function as encouraging/counteracting mechanism of biological processes and cellular pathways to promote/attenuate the inflammatory, angiogenic and proliferative responses linked to TIA-depleted HeLa cells [10].

Given that many target genes associated with the identified up-regulated miRNAs are down-regulated (Table 3) in TIA-depleted HeLa cells [10], from a mechanistic viewpoint our results indicate that mRNA abundance in most targeted genes was somewhat affected by miRNAs, thus suggesting that, for a substantial number of genes regulated by TIA-protein absence, destabilization of mRNA may be the main mechanism of protein repression by these miRNA-mediated regulators. This observation agrees with a recent study, suggesting that mammalian miRNAs predominantly reduce target mRNA levels [34]. However, some miRNAs such as miR-744 enhances cyclin B1 mRNA expression through a novel mechanism. miRNA positively-regulates gene expression by targeting promoter elements; this phenomenon is known as RNA activation [37]. In this

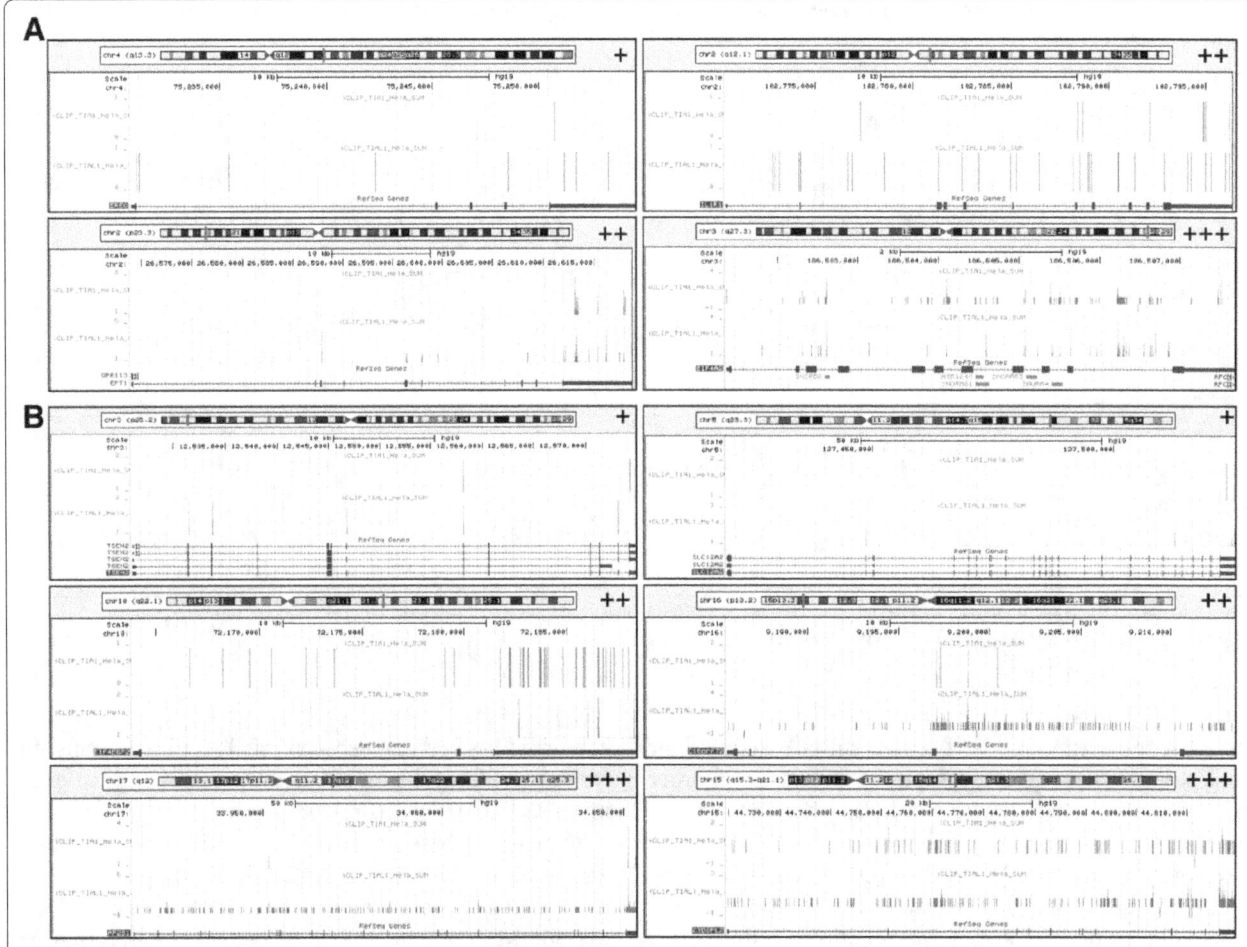

Figure 4 RNA map of TIA proteins in targeted pre-mRNAs related with miRNAs in TIA-deficient HeLa cells. (**A** and **B**) Profiles of experimental crosslink sites of TIA1 and TIAR proteins (adapted from TIA-iCLIP analysis [8]) on up- (**A**) and down-regulated (**B**) pre-mRNAs in TIA-depleted HeLa cells. Representative examples of up- (**A**) and down-regulated (**B**) genes with low (+), medium (++) and high (+++) density of TIA binding sites are shown. The bar graph in each panel indicates the number of cDNAs that identified each crosslink site.

regard, it is reasonable to think that this mechanism could be operating in the up-regulation observed for seven target genes associated with the up-regulated miRNAs (Table 3). Furthermore, miRNAs do not only regulate the expression of protein-encoding genes but also other miRNAs: for instance, let-7a controls the expression of important epigenetic regulators, including epigenetic miRNA regulatory circuits, and organizes the whole gene expression profile [38]. On the other hand, there are miRNA-target interactions that involve multiple sites for a given target and confer much stronger repression. More often, different miRNAs work together to co-target a given mRNA, therefore their combined repressive effect greatly exceeds the individual contributions [34-36]. Both regulatory situations are observed in the down-regulated genes associated to up-regulated miRNAs in TIA-depleted HeLa cells (Table 3). This suggests that clusters of miRNAs can play a more prominent role than only reinforcing

the expression patterns dictated by transcriptome dynamics. The existence of interactions among these regulators and the interactions between their regulatees suggests that these miRNAs generate networks that modulate antagonistic cellular responses, such as apoptosis or cell proliferation induction and/or repression. Our set of data indicates that the action of miRNAs could potentially be another important mechanism in the regulation of gene expression and some gene regulatory networks mediated by TIA proteins. These observations thus suggest the existence of feedback mechanisms that promote miRNA expression, which might therefore contribute to dampen the phenotypes observed in TIA-lacking HeLa cells. A type 2 incoherent feedforward loop [39] may contribute to the repression model between TIA proteins, target genes and associated miRNAs. This feedforward loop could reinforce the functional role of TIA proteins as repressors of inflammatory, angiogenic, and proliferative responses (Figure 6).

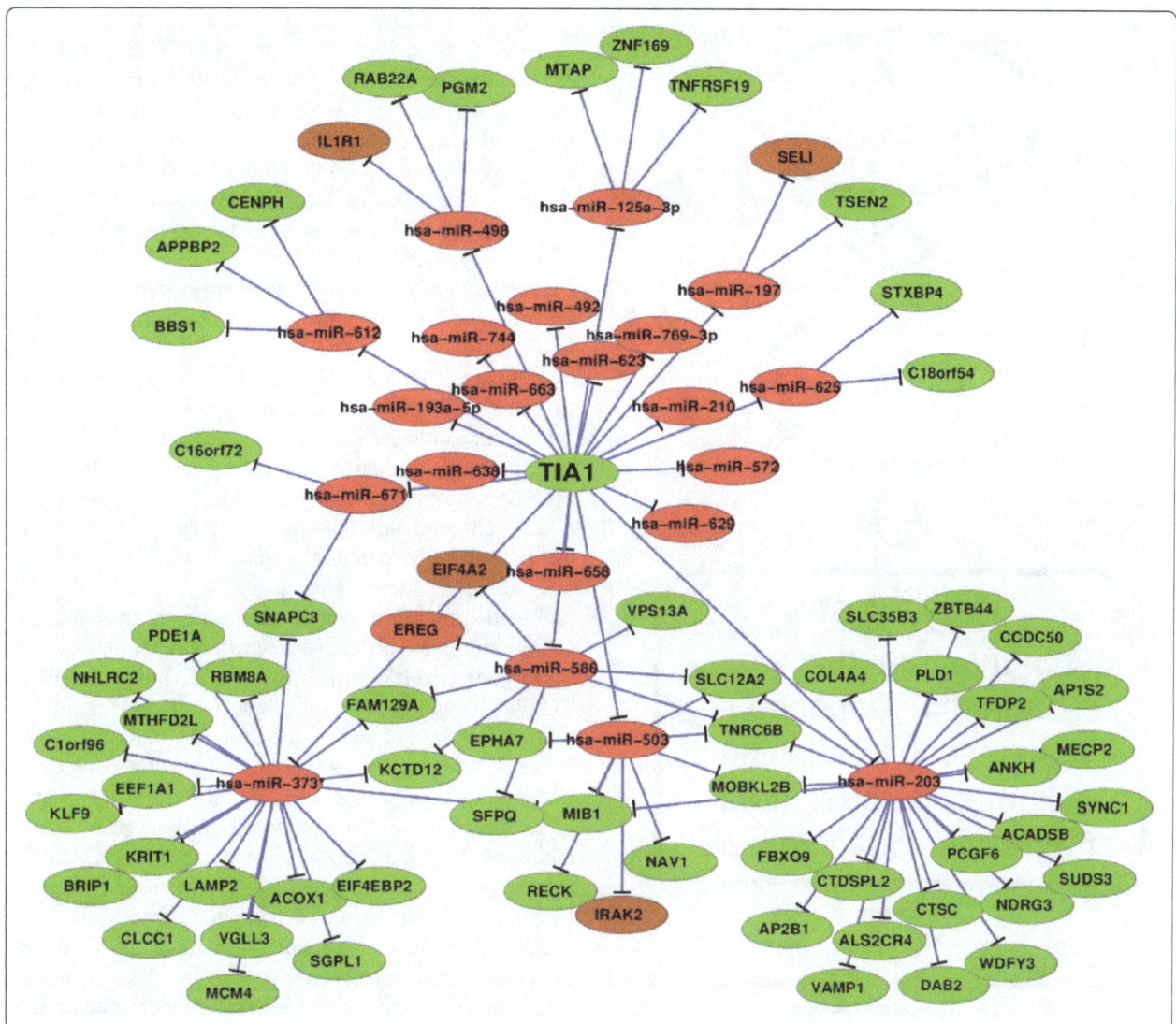

Figure 5 Regulatory network of up-regulated miRNAs and intersection with target genes associated with TIA-depleted HeLa cells. Elongated circles represent differentially expressed miRNAs and genes (red and brown- increased, green- decreased). Lines represent regulatory relations between differentially expressed miRNAs and genes.

Regarding some miRNAs identified in this study, these could be related to miRNA-mediated gene expression fluctuations and, more faithfully, to signal outcomes in the context of specific regulatory networks. This suggest that miRNAs can strenghten transcriptome patterns by buffering the deleterious effects of some network states linked to random fluctuations in gene expression program, in agreement with previous findings [25]. Since little experimental information is available on potential target genes and miRNA cluster identified in our study and we have not experimentally carried out analysis of gain- and loss-of function, we used experimental data on miRNA function to understand the potential implication and relevance of these miRNAs in the development and progression of (patho)-physiological conditions. For example, miRNA-

30b has been implicated in angiogenesis [40], TRAIL-induced apoptosis in glioma cells [41] and oral squamous cell cancers [42]. miR-193a-5p targets YY1-APC regulatory axis in human endometrioid endometrial adenocarcinoma [43]. miR-203 participates in a regulatory network that modulates epithelial to mesenchymal transition [44] as well as to promote suppression immune [45]. miR-210 is induced under hypoxia and works as a iron sensor to stimulate cell proliferation and promote cell survival in the hypoxic region within tumor [46-48]. Further, this miRNA modulates the mitochondrial functioning and metabolism [49,50], represses FGFRL1 and E2F3 expression, which inhibits cell apoptosis in hypoxia response [51,52], predicts poor survival in patients with breast cancer [53] and activates notch signalling pathway in

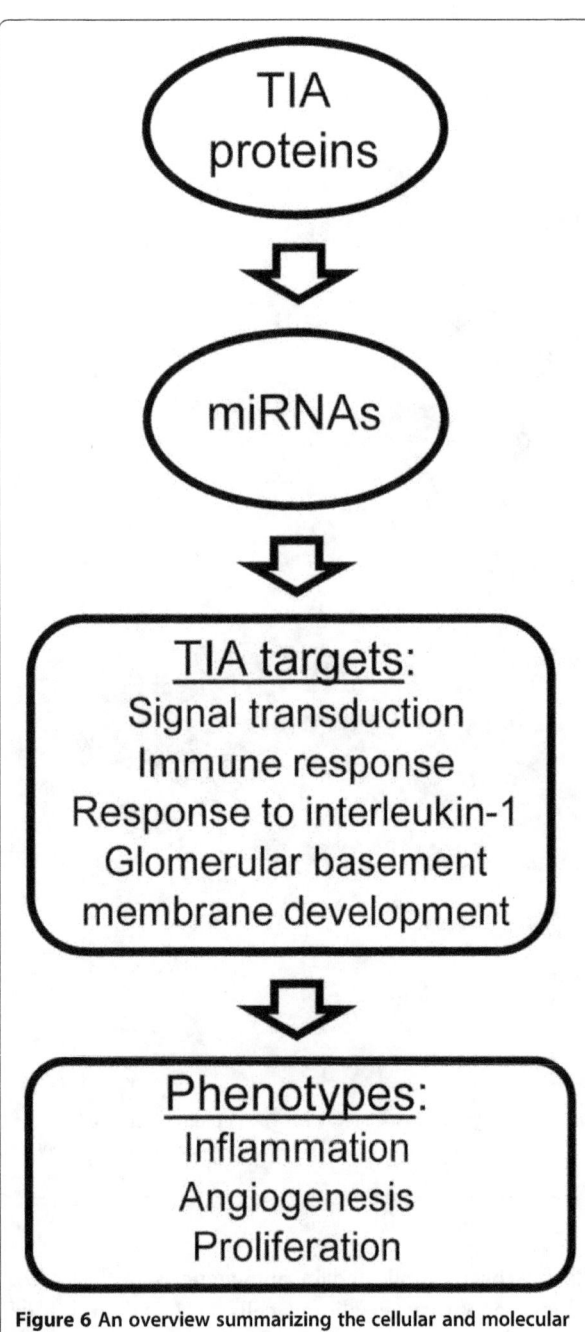

Figure 6 An overview summarizing the cellular and molecular events associated to the reduction of TIA proteins. The availability of TIA proteins could be contributing to a regulatory loop mediated by miRNAs to modulate target gene expression related to inflammatory, angiogenic and/or proliferative responses/phenotypes in HeLa cells.

tumor invasion and metastasis in testis cancer [58,59]. Further, both miRNA-210 and miRNA-373 participate in the gene expression control of DNA repair in hypoxic stress [60]. miRNA-492 and miRNA-498 are highly expressed and regulate metastatic functions in hepatoblastoma, rectal cancer, adenocarcinoma and retinoblastoma [61-64]. In addition, miR-503 is induced in angiogenesis, down-regulates CUGBP1 and modulates metastatic function in hepatocellular cancer cell [65-68]. miRNA-572 has been involved in chronic lymphocytic leukaemia via targeting anti-apoptotic genes [69,70]. miRNA-586 is up-regulated in colorectal cancer [71]. miRNA-615-3p dysregulates CDKN2A, NF2 and JUN in malignant mesothelioma [72] and enhances the phagocytic capacity of splenic macrophages [73]. miRNA-625 promotes invasion and metastasis of gastric cancer by targeting ILK [74]. miR-629 is associated to lung cancer by targeting NBS1 [75]. miRNA-638 and miR-658 are up-regulated in cell transformation and human gastric cancer [76,77]. In addition, other miRNAs show a great potential as activators of cell proliferation/transformation phenotypes; for example miRNA-663 functions as an oncogene promoting tumorigenesis by targeting p21(WAF1/CIP1), VEGF, JunB, JunD and TGFB1 genes [78-80] and plays also a role in inflammatory response of endothelial cells [81]. miRNA 671 regulates CD44 inducing metastasis by regulating extracellular matrix functions [82] as well as miRNA-dependent gene silencing involving Ago2-mediated cleavage of a circular antisense RNA [83]. miRNA-744 up-regulates Cyclin B1 expression [37] and down-regulates Transforming Growth Factor Beta-1 expression [84]. On the other hand, other miRNAs identified in our study have bipolar features and putatively antagonistic behaviours as opposite regulators acting either positively or negatively under specific biological programmes. In this regards, there are at least 2 miRNAs showing functions to repress proliferation, angiogenesis and transformation phenotypes. This is the case of miRNA-483-5p, which suppresses the proliferation of glioma cells via directly targeting ERK1 [85] and angiogenesis in vitro by targeting serum response factor [86]. However, microRNA-483-5p and miRNA-195 have been identified as predictors of poor prognosis in adrenocortical cancer [87]. miR-125a-3p is down-regulated in non-small cell lung cancer, having inverse effects on invasion and migration of lung cancer cells [88]. This miRNA is a potent prognostic marker in gastric cancer [89]. Collectively, the miRNA expression profiling identified in this study has important aftermath in tumorigenesis and could contribute/repress to the development of cell proliferation, angiogenesis, transformation phenotypes, because individual miRNAs are associated with diagnosis, prognosis and treatment efficacy linked to the biology of human tumors. Therefore, these miRNAs can target to oncogenes or tumor suppressor genes (also they can even

angiogenesis [54]. In addition, miR-371-5p is increased in gastric cancer [55] and miR-373 functions as an oncogene in hepatocellular carcinoma given that is a new regulator of protein phosphatase 6 [56], an repressor of the large tumor suppressor homolog 2 [57] and promotes

function themselves as such) and take part in the promotion or inhibition of tumorigenesis and cancer progression, thus being able to confer robustness to these pathological phenotypes.

In summary, our findings suggest that expression changes in individual miRNAs in TIA-depleted HeLa cells could directly or indirectly impact on biological processes and signaling pathways to favour cell phenotypes associated to the down-regulation of TIA proteins (Figure 6). Further, our observations suggest that some of the identified miRNAs are consistent with an adaptive response aimed at attenuating the inflammatory, angiogenic, and proliferative responses developed in TIA-protein absence. Therefore, cross talk between regulatory mechanisms that promote increased/reduced expression of miRNA-mediated genes may contribute to modify the relative expression levels of the mRNAs associated with TIA-lacking HeLa cells. Based on this integrative analysis, our results provide a prominent stand for future approaches aimed at characterizing the role of specific miRNAs in TIA-mediated gene expression regulation.

Conclusions

Gene expression profiling approaches have improved our understanding on how the human transcriptome dynamics orchestrates global responses at molecular level to answer to changing environmental challenges. Our study identifies a human miRNA collection that displays significant changes in the transient absence of TIA proteins in HeLa cells. The most prominent changes are linked to the up-regulation of miR-30b-3p, miR125a-3p, miR-193a-5p, miR-197-3p, miR-203a, miR-210, miR-371-5p, miR-373-5p, miR-483-5p, miR-492, miR-498, miR-503-5p, miR-572, miR-586, miR-612, miR-615-3p, miR-623, miR-625-5p, miR-629-5p, miR-638, miR-658, miR-663a, miR-671-5p, miR-769-3p and miR-744-5p. The integration of the identified miRNAs with the potential target genes and previous gene expression data under the same experimental conditions revealed enrichment of biological processes and signaling pathways controlling relevant members of the pathways associated to the oncogenesis. Down-regulation of these cellular components may contribute to establish the inflammatory, angiogenic, and proliferative responses previously described in TIA protein-lacking HeLa cells. Therefore, our results are consistent with the existence of regulatory networks that generate correlated expressions, commonly involving miRNAs [33,90]. This regulatory architecture may increase the fidelity of inhibition of the downstream components by acting on them redundantly. In other words, a transient loss of TIA proteins can be partially compensated for by the adaptive presence of specific miRNAs. Collectively, these findings suggest that TIA proteins can act as multifunctional regulators to provide a novel meeting point between the mechanisms for cross-talking among

concerted post-transcriptional regulatory layers that co-ordinate complex cellular responses.

Additional files

Additional file 1: Summary of the miRNA array analyses. The following additional data are available in this file: 1) Correlation of Hy3 and Hy5 signals for the spike-in controls between slides. 2) Hy5 vs Hy3 scatter, MA and ratio distribution plots (before and after normalization) and Hy5 vs Hy3 scatter plot for the spike-in controls, for each independent experiments. 3) Statistical test implemented in the limma R.

Additional file 2: Update of miRNAs sequences previously identified as miRPlus. The miRPlus sequences are licensed human sequences (Exiqon, Denmark). Many of them are already annotated in the miRBase database version 18.

Additional file 3: List of predicted potential target genes associated to up-regulated miRNAs in transiently TIA-depleted HeLa cells. Potential target genes of up-regulated miRNAs identified in TIA-deficient HeLa cells were predicted by TargetScan 5.2, PicTar-Vert and miRDB database downloaded using web-app miRBase (http://www.mirbase.org). The Gene Ontology (GO) and Kyoto Encyclopedia of Genes and Genomes (KEGG) database analyses were conducted using software programmes provided by GenCodis3 (http://genecodis.cnb.csic.es).

Abbreviations

iCLIP: In vivo ultraviolet (UV)-crosslinking and immunoprecipitation to identify the RNA crosslinking sites of TIA1 and TIAR proteins; GO: Gene ontology; KEGG: Kyoto and encyclopedia of genes and genomes; siRNA: Small interfering RNA; TIA1: T-cell intracellular antigen 1; TIAR/TIAL1: TIA1 related/like protein; UTR: Untranslated region of the eukaryotic mRNAs.

Competing interests
The authors declare that they have no competing interests.

Authors' contributions
JMI conceived the research and designed all experiments. CSJ, IC and JB carried out the experiments and analyzed the data presented in this paper. CSJ, IC, JB and JMI wrote the paper. All authors read and approved the final manuscript.

Acknowledgements
We are grateful to Jernej Ule for providing iCLIP data analysis of TIA proteins and Juan Carlos Oliveros for helping us with bioinformatics tools. This work was supported by grants from the Spanish Ministry of Innovation and Science through FEDER funds (BFU2008-00354 and BFU2011-29653). The CBMSO receives an institutional grant from Fundación Ramón Areces.

Author details
[1]Centro de Biología Molecular Severo Ochoa, Consejo Superior de Investigaciones Científicas, Universidad Autónoma de Madrid (CSIC/UAM), C/ Nicolás Cabrera 1, Cantoblanco, Madrid 28049, Spain. [2]Current address: Programa de Biología de Sistemas, Centro Nacional de Biotecnología, Consejo Superior de Investigaciones Científicas, C/Darwin 3, Cantoblanco, Madrid 28049, Spain.

References
1. McAlinden A, Liang L, Mukudai Y, Imamura T, Sandell LJ: **Nuclear protein TIA1 regulates COL2A1 alternative splicing and interacts with precursor mRNA and genomic DNA.** *J Biol Chem* 2007, **282:**24444–24454.
2. Suswam EA, Li YY, Mahtani H, King PH: **Novel DNA-binding properties of the RNA-binding protein TIAR.** *Nucleic Acids Res* 2005, **33:**4507–4518.
3. Kim HS, Wilce MC, Yoga YM, Pendini NR, Gunzburg MJ, Cowieson NP, Wilson GM, Williams BR, Gorospe M, Wilce JA: **Different modes of

interaction by TIAR and HuR with target RNA and DNA. *Nucleic Acids Res* 2011, **39**:1117–1130.

4. Das R, Yu J, Zhang Z, Gygi MP, Krainer AR, Gygi SP, Reed R: **SR proteins function in coupling RNAP II transcription to pre-mRNA splicing.** *Mol Cell* 2007, **26**:867–881.

5. Del Gatto-Konczak F, Bourgeois CF, Le Guiner C, Kister L, Gesnel MC, Stevenin J, Breathnach R: **The RNA-binding protein TIA1 is a novel mammalian splicing regulator acting through intron sequences adjacent to a 5 splice site.** *Mol Cell Biol* 2000, **20**:6287–6299.

6. Förch P, Puig O, Kedersha N, Martínez C, Granneman S, Séraphin B, Anderson P, Valcárcel J: **The apoptosis-promoting factor TIA1 is a regulator of alternative pre-mRNA splicing.** *Mol Cell* 2000, **6**:1089–1098.

7. Izquierdo JM, Majós N, Bonnal S, Martínez C, Castelo R, Guigó R, Bilbao D, Valcárcel J: **Regulation of Fas alternative splicing by antagonistic effects of TIA1 and PTB on exon definition.** *Mol Cell* 2005, **19**:475–484.

8. Wang Z, Kayikci M, Briese M, Zarnack K, Luscombe NM, Rot G, Zupan B, Curk T, Ule J: **iCLIP predicts the dual splicing effects of TIA-RNA interactions.** *PLoS Biol* 2010, **8**:e1000530.

9. Yamasaki S, Stoecklin G, Kedersha N, Simarro M, Anderson P: **T-cell intracellular antigen-1 (TIA1)-induced translational silencing promotes the decay of selected mRNAs.** *J Biol Chem* 2007, **282**:30070–30077.

10. Reyes R, Alcalde J, Izquierdo JM: **Depletion of T-cell intracellular antigen (TIA)-proteins promotes cell proliferation.** *Genome Biol* 2009, **10**:R87.

11. López De Silanes I, Galbán S, Martindale JL, Yang X, Mazan-Mamczarz K, Indig FE, Falco G, Zhan M, Gorospe M: **Identification and functional outcome of mRNAs associated with RNA-binding protein TIA1.** *Mol Cell Biol* 2005, **25**:9520–9531.

12. Mazan-Mamczarz K, Lal A, Martindale JL, Kawai T, Gorospe M: **Translational repression by RNA-binding protein TIAR.** *Mol Cell Biol* 2006, **26**:2716–2727.

13. Kim HS, Kuwano Y, Zhan M, Pullmann R Jr, Mazan-Mamczarz K, Li H, Kedersha N, Anderson P, Wilce MC, Gorospe M, Wilce JA: **Elucidation of a C-rich signature motif in target mRNAs of RNA-binding protein TIAR.** *Mol Cell Biol* 2007, **27**:6806–6817.

14. Liao B, Hu Y, Brewer G: **Competitive binding of AUF1 and TIAR to MYC mRNA controls its translation.** *Nat Struct Mol Biol* 2007, **14**:511–518.

15. Damgaard CK, Lykke-Andersen J: **Translational coregulation of 5/TOP mRNAs by TIA-1 and TIAR.** *Genes Dev* 2011, **25**:2057–2068.

16. Piecyk M, Wax S, Beck AR, Kedersha N, Gupta M, Maritim B, Chen S, Gueydan C, Kruys V, Streuli M, Anderson P: **TIA1 is a translational silencer that selectively regulates the expression of TNF-alpha.** *EMBO J* 2000, **19**:4154–4163.

17. Beck AR, Miller JJ, Anderson P, Streuli M: **RNA-binding protein TIAR is essential for primordial germ cell development.** *Proc Natl Acad Sci* 1998, **95**:2331–2336.

18. Kharraz Y, Salmand PA, Camus A, Auriol J, Gueydan C, Kruys V, Morello D: **Impaired embryonic development in mice overexpressing the RNA-binding protein TIAR.** *PLoS One* 2010, **5**:e11352.

19. Bartel DP: **MicroRNAs: target recognition and regulatory functions.** *Cell* 2009, **136**:215–233.

20. Smyth GK, Speed TP: **Normalization of cDNA microarray data.** *Methods* 2003, **31**:265–273.

21. Bolstad BM, Irizarry RA, Astrand M, Speed TP: **A comparison of normalization methods for high density oligonucleotide array data based on bias and variance.** *Bioinformatics* 2003, **19**:185–193.

22. Smyth GK: *Linear models and empirical bayes methods for assessing differential expression in microarray experiments. In statistical applications in genetics and molecular biology 2004. Volume 3. Number 1. Article 3.* http://www.bepress.com/sagmb/vol3/iss1/art3.

23. Benjamini Y, Hochberg Y: **Controlling the false discovery rate: a practical and powerful approach to multiple testing.** *J R Statist Soc B* 1995, **57**:289.

24. Oliveros JC: *FIESTA at BioinfoGP. An interactive server for analyzing DNA microarray experiments with replicates.* http://bioinfogp.cnb.csic.es/tools/FIESTA.

25. Lewis BP, Burge CB, Bartel DP: **Conserved Seed Pairing, Often Flanked by Adenosines, Indicates that Thousands of Human Genes are MicroRNA Targets.** *Cell* 2005, **120**:15–20.

26. Krek A, Grün D, Poy MN, Wolf R, Rosenberg L, Epstein EJ, MacMenamin P, da Piedade I, Gunsalus KC, Stoffel M, Rajewsky N: **Combinatorial microRNA target predictions.** *Nat Genet* 2005, **37**:495–500.

27. Wang X, El Naqa IM: **Prediction of both conserved and nonconserved microRNA targets in animals.** *Bioinformatics* 2008, **24**:325–332.

28. Wang X: **miRDB: a microRNA target prediction and functional annotation database with a wiki interface.** *RNA* 2008, **14**:1012–1017.

29. Kozomara A, Griffiths-Jones S: **miRBase: integrating microRNA annotation and deep-sequencing data.** *Nucleic Acids Res* 2011, **39**:D152–D157.

30. Carmona-Saez P, Chagoyen M, Tirado F, Carazo JM, Pascual-Montano A: **GENECODIS: a web-based tool for finding significant concurrent annotations in gene lists.** *Genome Biol* 2007, **8**:R3.

31. Nogales-Cadenas R, Carmona-Saez P, Vazquez M, Vicente C, Yang X, Tirado F, Carazo JM, Pascual-Montano A: **GeneCoDis: interpreting gene lists through enrichment analysis and integration of diverse biological information.** *Nucleic Acids Res* 2009, **37**:W317–W322.

32. Smoot M, Ono K, Ruscheinski J, Wang P-L, Ideker T: **Cytoscape 2.8: new features for data integration and network visualization.** *Bioinformatics* 2011, **27**:431–432.

33. Friedman RC, Farh KK, Burge CB, Bartel DP: **Most mammalian mRNAs are conserved targets of microRNAs.** *Genome Res* 2009, **19**:92–105.

34. Guo H, Ingolia NT, Weissman JS, Bartel DP: **Mammalian microRNAs predominantly act to decrease target mRNA levels.** *Nature* 2010, **466**:835–840.

35. Ebert MS, Sharp PA: **Roles for microRNAs in conferring robustness to biological processes.** *Cell* 2012, **149**:515–524.

36. Krol J, Loedige I, Fillipowicz W: **The widespread regulation of microRNA biogenesis, function and decay.** *Nat Rev Genet* 2010, **11**:597–610.

37. Huang V, Place RF, Portnoy V, Wang J, Qi Z, Jia Z, Yu A, Shuman M, Yu J, Li LC: **Upregulation of Cyclin B1 by miRNA and its implications in cancer.** *Nucleic Acids Res* 2012, **40**:1695–1707.

38. Iliopoulos D, Hirsch HA, Struhl K: **An epigenetic switch involving NF-kappaB, Lin28, Let-3 MicroRNA, and IL6 links inflammation to cell transformation.** *Cell* 2009, **139**:693–706.

39. Mangan S, Alon U: **Structure and function of the feed-forward loop network motif.** *Proc Natl Acad Sci USA* 2003, **10**:11980–11985.

40. Bridge G, Monteiro R, Henderson S, Emuss V, Lagos D, Georgopoulou D, Patient R, Boshoff C: **The microRNA-30 family targets DLL4 to modulate endothelial cell behavior during angiogenesis.** *Blood* 2012, in press.

41. Quintavalle C, Donnarumma E, Iaboni M, Roscigno G, Garofalo M, Romano G, Fiore D, De Marinis P, Croce CM, Condorelli G: **Effect of miR-21 and miR-30b/c on TRAIL-induced apoptosis in glioma cells.** *Oncogene* 2012, doi:10.1038/onc.2012.410.

42. Shao C, Yu Y, Yu L, Pei Y, Feng Q, Chu F, Fang Z, Zhou Y: **Amplification and up-regulation of microRNA-30b in oral squamous cell cancers.** *Arch Oral Biol* 2012, **57**:1012–1017.

43. Yang Y, Zhou L, Lu L, Wang L, Li X, Jiang P, Chan LK, Zhang T, Yu J, Kwong J, Cheung TH, Chung T, Mak K, Sun H, Wang H: **A novel miR-193a-5p-YY1-APC regulatory axis in human endometrioid endometrial adenocarcinoma.** *Oncogene* 2012, doi:10.1038/onc.2012.360. Aug 20.

44. Moes M, Le Béchec A, Crespo I, Laurini C, Halavatyi A, Vetter G, Del Sol A, Friederich E: **A novel network integrating a miRNA-203/SNAI1 feedback loop which regulates epithelial to mesenchymal transition.** *PLoS One* 2012, **7**:e35440.

45. Moffatt CE, Lamont RJ: **Porphyromonas gingivalis induction of microRNA-203 expression controls suppressor of cytokine signaling 3 in gingival epithelial cells.** *Infect Immun* 2011, **79**:2632–2637.

46. Kelly TJ, Souza AL, Clish CB, Puigserver P: **A hypoxia-induced positive feedback loop promotes hypoxia-inducible factor 1alpha stability through miR-210 suppression of glycerol-3-phosphate dehydrogenase 1-like.** *Mol Cell Biol* 2011, **31**:2696–2706.

47. Noman MZ, Buart S, Romero P, Ketari S, Janji B, Mari B, Mami-Chouaib F, Chouaib S: **Hypoxia-inducible miR-210 regulates the susceptibility of tumor cells to lysis by cytotoxic T cells.** *Cancer Res* 2012, **72**:4629–4641.

48. Yoshioka Y, Kosaka N, Ochiya T, Kato T: **Micromanaging iron homeostasis: hypoxia-inducible miR-210 suppresses iron homeostasis-related proteins.** *J Biol Chem* 2012, **287**:34110–34119.

49. Favaro E, Ramachandran A, McCormick R, Gee H, Blancher C, Crosby M, Devlin C, Blick C, Buffa F, Li JL, Vojnovic B, Pires Das Neves R, Glazer P, Iborra F, Ivan M, Ragoussis J, Harris AL: **MicroRNA-210 regulates mitochondrial free radical response to hypoxia and krebs cycle in cancer cells by targeting iron sulfur cluster protein ISCU.** *PLoS One* 2010, **26**:e10345.

50. Chen Z, Li Y, Zhang H, Huang P, Luthra R: **Hypoxia-regulated microRNA-210 modulates mitochondrial function and decreases ISCU and COX10 expression.** *Oncogene* 2010, **29**:4362–4368.

51. Tsuchiya S, Fujiwara T, Sato F, Shimada Y, Tanaka E, Sakai Y, Shimizu K, Tsujimoto G: **MicroRNA-210 regulates cancer cell proliferation through targeting fibroblast growth factor receptor-like 1 (FGFRL1).** *J Biol Chem* 2011, **286**:420–428.

52. Gou D, Ramchandran R, Peng X, Yao L, Kang K, Sarkar J, Wang Z, Zhou G, Raj JU: miR-210 has an anti-apoptotic effect in pulmonary artery smooth muscle cells during hypoxia. *Am J Physiol Lung Cell Mol Physiol*, in press.

53. Hong L, Yang J, Han Y, Lu Q, Cao J, Syed L: High expression of miR-210 predicts poor survival in patients with breast cancer: a meta-analysis. *Gene* 2012, **507**:135–138.

54. Lou YL, Guo F, Liu F, Gao FL, Zhang PQ, Niu X, Guo SC, Yin JH, Wang Y, Deng ZF: miR-210 activates notch signaling pathway in angiogenesis induced by cerebral ischemia. *Mol Cell Biochem*, in press.

55. Liu P, Wilson MJ: miR-520c and miR-373 upregulate MMP9 expression by targeting mTOR and SIRT1, and activate the Ras/Raf/MEK/Erk signaling pathway and NF-κB factor in human fibrosarcoma cells. *J Cell Physiol* 2012, **227**:867–876.

56. Wu N, Liu X, Xu X, Fan X, Liu M, Li X, Zhong Q, Tang H: MicroRNA-373, a new regulator of protein phosphatase 6, functions as an oncogene in hepatocellular carcinoma. *FEBS J* 2011, **278**:2044–2054.

57. Lee KH, Goan YG, Hsiao M, Lee CH, Jian SH, Lin JT, Chen YL, Lu PJ: MicroRNA-373 (miR-373) post-transcriptionally regulates large tumor suppressor, homolog 2 (LATS2) and stimulates proliferation in human esophageal cancer. *Exp Cell Res* 2009, **315**:2529–2538.

58. Voorhoeve PM, le Sage C, Schrier M, Gillis AJ, Stoop H, Nagel R, Liu YP, van Duijse J, Drost J, Griekspoor A, Zlotorynski E, Yabuta N, De Vita G, Nojima H, Looijenga LH, Agami R: A genetic screen implicates miRNA-372 and miRNA-373 as oncogenes in testicular germ cell tumors. *Cell* 2006, **124**:1169–1181.

59. Huang Q, Gumireddy K, Schrier M, le Sage C, Nagel R, Nair S, Egan DA, Li A, Huang G, Klein-Szanto AJ, Gimotty PA, Katsaros D, Coukos G, Zhang L, Puré E, Agami R: The microRNAs miR-373 and miR-520c promote tumour invasion and metastasis. *Nat Cell Biol* 2008, **10**:202–210.

60. Crosby ME, Kulshreshtha R, Ivan M, Glazer PM: MicroRNA regulation of DNA repair gene expression in hypoxic stress. *Cancer Res* 2009, **69**:1221–1229.

61. von Frowein J, Pagel P, Kappler R, von Schweinitz D, Roscher A, Schmid I: MicroRNA-492 is processed from the keratin 19 gene and up-regulated in metastatic hepatoblastoma. *Hepatology* 2011, **53**:833–842.

62. Gaedcke J, Grade M, Camps J, Søkilde R, Kaczkowski B, Schetter AJ, Difilippantonio MJ, Harris CC, Ghadimi BM, Møller S, Beissbarth T, Ried T, Litman T: The rectal cancer microRNAome - microRNA expression in rectal cancer and matched normal mucosa. *Clin Cancer Res* 2012, **18**:4919–4930.

63. Schultz NA, Werner J, Willenbrock H, Roslind A, Giese N, Horn T, Wøjdemann M, Johansen JS: MicroRNA expression profiles associated with pancreatic adenocarcinoma and ampullary adenocarcinoma. *Mod Pathol*, in press.

64. Schepeler T, Reinert JT, Ostenfeld MS, Christensen LL, Silahtaroglu AN, Dyrskjøt L, Wiuf C, Sørensen FJ, Kruhøffer M, Laurberg S, Kauppinen S, Ørntoft TF, Andersen CL: Diagnostic and prognostic microRNAs in stage II colon cancer. *Cancer Res* 2008, **68**:6416–6424.

65. Caporali A, Meloni M, Völlenkle C, Bonci D, Sala-Newby GB, Addis R, Spinetti G, Losa S, Masson R, Baker AH, Agami R, le Sage C, Condorelli G, Madeddu P, Martelli F, Emanueli C: Deregulation of microRNA-503 contributes to diabetes mellitus-induced impairment of endothelial function and reparative angiogenesis after limb ischemia. *Circulation* 2011, **123**:282–291.

66. Zhou J, Wang W: Analysis of microRNA expression profiling identifies microRNA-503 regulates metastatic function in hepatocellular cancer cell. *J Surg Oncol* 2011, **104**:278–283.

67. Cui YH, Xiao L, Rao JN, Zou T, Liu L, Chen Y, Turner DJ, Gorospe M, Wang JY: miR-503 represses CUG-binding protein 1 translation by recruiting CUGBP1 mRNA to processing bodies. *Mol Biol Cell* 2012, **23**:151–162.

68. Caporali A, Emanueli C: MicroRNA-503 and the extended microRNA-16 family in angiogenesis. *Trends Cardiovasc Med* 2011, **21**:162–166.

69. Zhu DX, Zhu W, Fang C, Fan L, Zou ZJ, Wang YH, Liu P, Hong M, Miao KR, Liu P, Xu W, Li JY: miR-181a/b significantly enhances drug sensitivity in chronic lymphocytic leukemia cells via targeting multiple anti-apoptosis genes. *Carcinogenesis* 2012, **33**:1294–1301.

70. Xiao W, Bao ZX, Zhang CY, Zhang XY, Shi LJ, Zhou ZT, Jiang WW: Upregulation of miR-31* is negatively associated with recurrent/newly formed oral leukoplakia. *PLoS One* 2012, **7**:e38648.

71. Cummins JM, He Y, Leary RJ, Pagliarini R, Diaz LA Jr, Sjoblom T, Barad O, Bentwich Z, Szafranska AE, Labourier E, Raymond CK, Roberts BS, Juhl H, Kinzler KW, Vogelstein B, Velculescu VE: The colorectal microRNAome. *Proc Natl Acad Sci* 2006, **103**:3687–3692.

72. Guled M, Lahti L, Lindholm PM, Salmenkivi K, Bagwan I, Nicholson AG, Knuutila S: CDKN2A, NF2, And JUN are dysregulated among other genes by miRNAs in malignant mesothelioma. A miRNA microarray analysis. *Genes Chromosomes Cancer* 2009, **48**:615–623.

73. Jiang A, Zhang S, Li Z, Liang R, Ren S, Li J, Pu Y, Yang J: miR-615-3p promotes the phagocytic capacity of splenic macrophages by targeting ligand-dependent nuclear receptor corepressor in cirrhosis-related portal hypertension. *Exp Biol Med* 2011, **236**:672–680.

74. Wang M, Li C, Nie H, Lv X, Qu Y, Yu B, Su L, Li J, Chen X, Ju J, Yu Y, Yan M, Gu Q, Zhu Z, Liu B: Down-regulated miR-625 suppresses invasion and metastasis of gastric cancer by targeting ILK. *FEBS Lett* 2012, **586**:2382–2388.

75. Yang L, Li Y, Cheng M, Huang D, Zheng J, Liu B, Ling X, Li Q, Zhang X, Ji W, Zhou Y, Lu J: A functional polymorphism at microRNA-629-binding site in the 3′-untranslated region of NBS1 gene confers an increased risk of lung cancer in Southern and Eastern Chinese population. *Carcinogenesis* 2012, **33**:338–347.

76. Li D, Wang Q, Liu C, Duan H, Zeng X, Zhang B, Li X, Zhao J, Tang S, Li Z, Xing X, Yang P, Chen L, Zeng J, Zhu X, Zhang S, Zhang Z, Ma L, He Z, Wang E, Xiao Y, Zheng Y, Chen W: Aberrant expression of miR-638 contributes to benzo(a)pyrene-induced human cell transformation. *Toxicol Sci* 2012, **125**:382–391.

77. Guo J, Miao Y, Xiao B, Huan R, Jiang Z, Meng D, Wang Y: Differential expression of microRNA species in human gastric cancer versus non-tumorous tissues. *J Gastroenterol Hepatol* 2009, **24**:652–657.

78. Liu ZY, Zhang GL, Wang MM, Xiong YN, Cui HQ: MicroRNA-663 targets TGFB1 and regulates lung cancer proliferation. *Asian Pac J Cancer Prev* 2011, **12**:2819–2823.

79. Tili E, Michaille JJ, Adair B, Alder H, Limagne E, Taccioli C, Ferracin M, Delmas D, Latruffe N, Croce CM: Resveratrol decreases the levels of miR-155 by upregulating miR-663, a microRNA targeting JunB and JunD. *Carcinogenesis* 2010, **31**:1561–1566.

80. Yi C, Wang Q, Wang L, Huang Y, Li L, Liu L, Zhou X, Xie G, Kang T, Wang H, Zeng M, Ma J, Zeng Y, Yun JP: MiR-663, a microRNA targeting p21(WAF1/CIP1), promotes the proliferation and tumorigenesis of nasopharyngeal carcinoma. *Oncogene* 2012, **1**:1–13.

81. Ni CW, Qiu H, Jo H: MicroRNA-663 upregulated by oscillatory shear stress plays a role in inflammatory response of endothelial cells. *Am J Physiol Heart Circ Physiol* 2011, **300**:H1762–H1769.

82. Rutnam ZJ, Yang BB: The non-coding 3′ UTR of CD44 induces metastasis by regulating extracellular matrix functions. *J Cell Sci* 2012, **125**:2075–2085.

83. Hansen TB, Wiklund ED, Bramsen JB, Villadsen SB, Statham AL, Clark SJ, Kjems J: miRNA-dependent gene silencing involving Ago2-mediated cleavage of a circular antisense RNA. *EMBO J* 2011, **30**:4414–4422.

84. Martin J, Jenkins RH, Bennagi R, Krupa A, Phillips AO, Bowen T, Fraser DJ: Post-transcriptional regulation of transforming growth factor beta-1 by microRNA-744. *PLoS One* 2011, **6**:e25044.

85. Wang L, Shi M, Hou S, Ding B, Liu L, Ji X, Zhang J, Deng Y: MiR-483-5p suppresses the proliferation of glioma cells via directly targeting ERK1. *FEBS Lett* 2012, **586**:1312–1317.

86. Qiao Y, Ma N, Wang X, Hui Y, Li F, Xiang Y, Zhou J, Zou C, Jin J, Lv G, Jin H, Gao X: MiR-483-5p controls angiogenesis in vitro and targets serum response factor. *FEBS Lett* 2011, **585**:3095–3100.

87. Soon PS, Tacon LJ, Gill AJ, Bambach CP, Sywak MS, Campbell PR, Yeh MW, Wong SG, Clifton-Bligh RJ, Robinson BG, Sidhu SB: miR-195 and miR-483-5p identified as predictors of poor prognosis in adrenocortical cancer. *Clin Cancer Res* 2009, **15**:7684–7692.

88. Jiang L, Huang Q, Zhang S, Zhang Q, Chang J, Qiu X, Wang E: Hsa-miR-125a-3p and hsa-miR-125a-5p are downregulated in non-small cell lung cancer and have inverse effects on invasion and migration of lung cancer cells. *BMC Cancer* 2010, **10**:318.

89. Hashiguchi Y, Nishida N, Mimori K, Sudo T, Tanaka F, Shibata K, Ishii H, Mochizuki H, Hase K, Doki Y, Mori M: Down-regulation of miR-125a-3p in human gastric cancer and its clinicopathological significance. *Int J Oncol* 2012, **40**:1477–1482.

90. Friard O, Re A, Taverna D, De Bortoli M, Corà D, Circuits DB: a database of mixed microRNA/transcription factor feed-forward regulatory circuits in human and mouse. *BMC Bioinforma* 2010, **11**:435.

Efficient 5'-3' DNA end resection by HerA and NurA is essential for cell viability in the crenarchaeon *Sulfolobus islandicus*

Qihong Huang[1,2], Linlin Liu[1], Junfeng Liu[1], Jinfeng Ni[1], Qunxin She[2*] and Yulong Shen[1*]

Abstract

Background: ATPase/Helicases and nucleases play important roles in homologous recombination repair (HRR). Many of the mechanistic details relating to these enzymes and their function in this fundamental and complicated DNA repair process remain poorly understood in archaea. Here we employed *Sulfolobus islandicus*, a hyperthermophilic archaeon, as a model to investigate the *in vivo* functions of the ATPase/helicase HerA, the nuclease NurA, and their associated proteins Mre11 and Rad50.

Results: We revealed that each of the four genes in the same operon, *mre11*, *rad50*, *herA*, and *nurA*, are essential for cell viability by a mutant propagation assay. A genetic complementation assay with mutant proteins was combined with biochemical characterization demonstrating that the ATPase activity of HerA, the interaction between HerA and NurA, and the efficient 5'-3' DNA end resection activity of the HerA-NurA complex are essential for cell viability. NurA and two other putative HRR proteins: a PIN (PilT N-terminal)-domain containing ATPase and the Holliday junction resolvase Hjc, were co-purified with a chromosomally encoded N-His-HerA *in vivo*. The interactions of HerA with the ATPase and Hjc were further confirmed by *in vitro* pull down.

Conclusion: Efficient 5'-3' DNA end resection activity of the HerA-NurA complex contributes to necessity of HerA and NurA in *Sulfolobus*, which is crucial to yield a 3'-overhang in HRR. HerA may have additional binding partners in cells besides NurA.

Keywords: Homologous recombination repair, ATPase, Helicase, Nuclease, HerA, NurA, Archaea

Background

Of the various types of DNA lesions, double-strand breaks (DSBs) are one of the most detrimental, capable of causing chromosomal rearrangements and eventually cell death if not repaired appropriately [1]. In eukaryotes, two major DSB repair pathways are known: non-homologous end joining (NHEJ) and homologous recombination (HR). The former is an error-prone process, while the latter mechanism exhibits high fidelity [2]. It has been suggested that in eukaryotes, DSBs that occur in the G1 phase of the cell cycle are most likely to be repaired via NHEJ, while those occurring in the S/G2 phase are preferentially processed via HRR [2,3].

HRR has been investigated extensively in bacteria and eukaryotes. Bacteria encode multiple pathways for DSB repair, including RecBCD, the primary HRR pathway, SbcC-SbcD, and one backup system, RecFOR [4,5]. The HRR pathway can be divided into five general steps: (1) recognition of the break sites and formation of a repair center (RC), (2) end-processing of the broken ends, (3) loading of RecA onto single-strand DNA, homology search, and strand invasion, (4) branch migration and resolution, or dissolution of the recombination intermediates and replication restart, and (5) disassembly of the recombination apparatus and segregation of sister chromosomes [5]. The eukaryotic HRR machinery is comprised of a core protein complex containing Mre11-Rad50-Xrs2/Mre11-Rad50-NBS1(MRX/N); the nucleases/helicases exodeoxyribonuclease 1 (Exo1/EXO1), Dna2/DNA2, and Sgs1/BLM; the recombinase Rad51/RAD51; and several other accessory and regulatory proteins [6]. The HRR

* Correspondence: yulgshen@sdu.edu.cn; qunxin@bio.ku.dk
[1]State Key Laboratory of Microbial Technology, Shandong University, 27 Shanda Nan Rd., Jinan 250100, P. R. China
[2]Archaea Centre, Department of Biology, University of Copenhagen, Ole MaaløesVej 5, Copenhagen N DK-2200, Denmark

pathway in eukaryotes proceed similarly with several steps of the bacterial pathway, but multiple layers of regulation exist to ensure these repair pathways are accurate and restricted to the appropriate cellular contexts [3]. In eukaryotes, DSBs are recognized by the MRX(N) complex which is involved in most DNA end-associated processes including damage checkpoint signaling, HR, NHEJ, telomere maintenance, and meiotic recombination [1]. DNA end processing is initiated by MRX/MRN in conjunction with Sae2/CtIP and proceeds along one of the two distinct pathways, Exo1/EXO1 or Dna2/DNA2-Sgs1/BLM, forming a long 3'-tail of single-stranded DNA (ssDNA) that is then utilized in Rad51-dependent strand exchange in HR [7-11].

Archaea appear to primarily encode the HRR pathway for DSB repair, since homologs of several eukaryotic HRR components have been identified, including Mre11 and Rad50 of the MR complex and the recombinase RadA [12,13], and homologs of the NHEJ proteins Ku70/80 are only present in a limited number of archaeal genomes [14,15]. The RecQ-like helicase Hjm (**H**olliday **j**unction **m**igration) and the 5'-flap endonuclease which has both endonuclease and 5'-3' exonuclease activities have been identified in archaea [16,17]; however, it is unclear whether they are involved in dsDNA end resection. Intriguingly, two archaeal genes, *herA* and *nurA*, are implicated in HRR by their genetic association with *mre11* and *rad50* in thermophilic archaea [18]. This has been supported by biochemical characterization of the encoded proteins: HerA exhibits ATPase activity and some exhibit dipolar helicase activities, while NurA is a 5'-3' ssDNA/double-stranded (ds) DNA exonuclease and ssDNA endonuclease [18-22]. Several studies have demonstrated that Mre11, Rad50, HerA, and NurA are capable of working in concert to process dsDNA *in vitro* [22-26]. Thus, HerA and NurA are regarded as the functional homologs of the eukaryotic Exo1/EXO1, Dna2/DNA2, and Sgs1/BLM proteins, and the archaeal Mre11-Rad50-HerA-NurA system can serve as a simple model system for studying HRR.

The functional role of these putative HRR proteins in archaea has been investigated only in a few reports [27-31]. However, gene function is still hypothesized based on the negative results of genetic analyses, such as the inability to isolate null mutants for *radA*, *mre11*, *rad50*, *herA*, and *nurA* in *Thermococcus kodakaraensis* or *radA*, *herA*, and *hjm* in *Sulfolobus islandicus* [29,32,33]. This is due in large part to the fact that suitable tools for conducting sophisticated analyses of gene function in archaea are still lacking. Recently, we reported a genetic complementation test for *S. islandicus* based on simvastatin selection [34]. This method can be utilized to analyze protein function by rescuing an essential gene deletion with expression of a mutant derivative from plasmids.

In this study, we analyzed the necessity of four genes putatively involved in DSB repair in *S. islandicus* REY15A, a genetic model for which the genome has been sequenced [35] and versatile genetic tools have been developed [36]. We revealed that all the four genes, *mre11*, *rad50*, *herA*, and *nurA* are essential for cell viability. Furthermore, we demonstrated that the ATPase activity of HerA, the interaction between HerA and NurA, and the high 5'-3' exonuclease activities of the HerA-NurA complex are essential for cell viability. We provide further evidence that HerA and NurA form a complex *in vivo*. The co-purification of HerA with a PIN (PilT N-terminal)-domain containing ATPase and the Holliday junction resolvase Hjc implies that HerA may also be involved in the HJ processing.

Methods

Strains and growth conditions

Sulfolobus islandicus strain REY15A(E233S) (△*pyrEF* △*lacS*) (hereafter E233S) (Additional file 1: Table S1) [37] was grown at 75°C in STVyU medium containing mineral salts, 0.2% (wt/vol) sucrose (S), 0.2% (wt/vol) tryptone (T), a mixed vitamin solution, 0.005% (wt/vol) yeast extract (y) and 0.01% (wt/vol) uracil (U), as described previously [37]. SCVy medium where tryptone was replaced by casamino acid (C) was used for cultivating uracil prototrophic transformants. ATVy medium where sucrose was replaced with 0.2% (wt/vol) arabinose (A) was used for protein expression (Additional file 1: Table S1). The STVyU medium supplemented with 5-fluorotic acid (5-FOA) (STVyUF) was used for counter-selection of the *pyrEF* auxotroph. Phytagel (0.7% [wt/vol]) was used for making the culture plates. The strains carrying the simvastatin-resistant selection marker (Additional file 1: Table S1) were grown in a medium supplemented with 12 µM simvastatin (Hangzhou Deli Chemical, Hangzhou, China) as described previously [34].

Construction of knockout plasmids

All plasmids used in this study are listed in Additional file 2: Table S2. The plasmids for gene knockout, pMID-*mre11*, pMID-*rad50*, and pMID-*nurA*, were constructed in the similar way as for pMID-*herA* [34]. All of the plasmids contained three homologous DNA arms: the integration (IN) arm, looping-out (OUT) arm, and target gene (TG) arm. The fragments were amplified using *S. islandicus* genomic DNA and their corresponding primers (Additional file 3: Table S3). After digestion with the corresponding restriction enzymes, these fragments were cloned into the knockout vector pMID which contains *pyrEF-lacS* selection markers [38].

Construction of plasmids for protein overexpression in *Escherichia coli* and for the genetic complementation assay

The construction of plasmids for protein expression and for the genetic complementation assay are described in Additional file 4: Supplementary methods. Briefly, the vector pET29a and pSSR carrying a simvastatin-resistant marker *hmg* [34] were used as the vectors.

Construction of a plasmid for the addition of a His-tag-coding sequence to 5′ end of chromosomal *herA* (*in situ* His-tagged)

The method for the construction of the plasmid pMIDHis-*herA* for HerA *in situ* His-tagging used the pUC19 as the original vector similar to that for the knockout plasmids. The plasmids contained three arms: two copies of L-arm (a 213 bp fragment at the upstream of the *herA* start codon), L-arm-1 and L-arm-2, and G-arm (a fragment of 5′ *herA* gene). The sequence of L-arm-1 was exactly the same as L-arm-2. L-arm-1 was cloned into pUC19 at the restriction sites of *Sal*I and *Mlu*I yielding pL-arm-1. The *herA* gene with N-terminal histidine coding sequence was amplified from pSeSDA-N-His-HerA (Additional file 2: Table S2), an expression plasmid carrying *herA* inserted at the restriction sites of *Cla*I and *Sal*I in pSeSD [39], with the primers 5′-6His-HerA-G-arm F/5′HerA-G-arm-*Sph*I R. L-arm-2, amplified with the primers 5′HerA L-arm-1-*Nco*I F/5′HerA L-arm-2-6His R, was ligated to the G-arm by SOE PCR. The fragment containing L-arm-2 and G-arm was cloned into pL-arm-1 at the *Nco*I and *Sph*I sites yielding the plasmid pMIDHis-*herA*.

Transformation of *S. islandicus* strains

The expression plasmids or linearized knockout plasmids were transformed into *S. islandicus* cells by electroporation as previously described [37].

X-gal assay

To detect the presence of the *lacS* marker in cells, X-Gal (5-bromo-4-chloro-3-indolyl-β-D-galactopyranoside) staining was performed as previously described [34].

Mutant propagation assay

The gene necessity was determined by a mutant propagation assay as described with minor modifications [38]. Briefly, 5-FOA was added into the culture of the purified pMID transformant (pMID-T) at $OD_{600} \sim 0.4$ and the culture continued for counter-selection, resulting in enrichment culture 1 (En1). Cells of En1 were diluted with the same fresh media to $OD_{600} \sim 0.1$ after they grew to the stationary phase. The OD_{600} values were measured at indicated times until the culture grew to the stationary phase or died.

Complementation assay of HerA and NurA mutants

pSSR vectors harboring genes encoding wild type HerA, NurA, or the site-directed mutants were used to complement the chromosomal *herA* or *nurA* deletion. The pSSR vector was transformed into pMID-*herA*-T or pMID-*nurA*-T and the transformants were selected on STVy plates supplemented with 12 μM simvastatin (STVy + sim). Single colonies were subsequently transferred into a 25 ml tube containing 10 ml of STVy + sim medium. After the culture grew, the cells were spread onto STVyUF + sim plates for counter selection. The colonies were picked and individually cultured in a tube containing 10 ml of STVyUF + sim medium and subsequently inoculated in a flask containing 50 ml of the same medium.

Genotype verification of the complementation strains by PCR and sequencing

Sulfolobus genomic DNA of the complementation strain was extracted from 3 ml of culture using Bacterial DNA Kit from Omega Bio-Tek (Norcross, GA, USA). The genotype at the loci was determined by PCR with the locus flanking primers.

For verification by sequencing, plasmids were isolated from 20–30 ml of the *Sulfolobus* cell cultures with Plasmid Mini Kit I (Norcross, GA, USA). The plasmids were re-transformed to *E. coli* DH5α for amplification. More than three *E. coli* single colonies were picked for plasmid extraction. The target genes in the plasmids were sequenced by BGI (Shenzhen, China).

Protein purification

The pET29a plasmids carrying *herA*, *nurA*, or their mutant genes were transformed into *E. coli* BL21 (DE3)-Codon-Plus-RIL for expression. The procedures for induction and purification of the proteins from *E. coli* cells were described in the Additional file 4: Supplementary methods.

To purify HerA from an *in situ* N-His-tagged HerA E233S strain, 9 L of the cells cultivated in the STV medium were collected by centrifugation. The cells were disrupted by sonication and the soluble fraction was precipitated by ammonium sulfate (0.6 g/ml). The precipitate was re-suspended in buffer A (50 mM Tris-HCl pH 8.0, 100 mM NaCl) and purified by Ni-NTA column described in the Additional file 4: Supplementary methods. The eluted protein was pooled, concentrated and diluted in buffer A. The subsequent sample was loaded onto a Superdex™ 200 10/300 GL column which was pre-equilibrated with buffer A. The proteins fractions were analyzed by SDS-PAGE and Western blot.

Analysis of HerA-NurA interaction by gel filtration chromatography

Physical interactions between HerA and NurA (wild type and I295L, I295E, F300Y, and F300E mutants) were detected by gel filtration. HerA (500 μg) and NurA (500 μg) were mixed and incubated at 60°C for 20 min in a total 500 μl of gel filtration buffer (20 mM Tris pH 8.0, 300 mM NaCl, 5% glycerol, 1 mM DTT). The mixtures were then spun down to remove any precipitated protein and loaded onto a Superdex™ 200 10/300 GL column pre-equilibrated by gel filtration buffer. The fractions (0.5 ml each) were collected and analyzed by 15% SDS-PAGE.

DNA substrates for the activity assays

Three oligonucleotides were synthesized for preparation of substrates for the nuclease assays (Additional file 3: Table S3). Strand E (34-mer in length) was 5′ end labeled with [γ-^{32}P-ATP] as previously described [22]. The labeled oligonucleotides were purified with Illustra™ Microspin™ G-25 columns (GE Healthcare, UK). The annealing was performed as previously described [22]. Various substrates were constructed by combinations of different oligonucleotides (Additional file 5: Table S4). The DNA substrates were stored at 4°C.

ATPase activity assay

The ATPase activity assay was performed as previously described. SsDNA, dsDNA, or NurA was added to a final concentration of 20 nM in the reaction [22].

The nuclease assay of HerA-NurA

The nuclease assay was performed in 20 μl reaction mixtures consisting of indicated amounts of wild-type HerA (or its mutants) and wild-type NurA (or its mutant D58A), 50 mM Tris-HCl, pH 8.0, 50 mM NaCl, 10 mM MgCl$_2$, 1 mM DTT, 1 mM ATP, 0.001% BSA, and 1 nM [γ-^{32}P]-labeled dsDNA. The mixture was incubated at 65°C for 30 min and then stopped by addition of a 5× loading buffer (50 mM EDTA, 0.5% SDS, 25% glycerol, and 0.025% bromophenol blue). The products were separated by electrophoresis in a native polyacrylamide gel at 120 V for 90 min in 1 × TBE. To examine the degradation products of the labeled ssDNA, the sample was analyzed by a 15% denatured polyacrylamide gel as previously described [21]. The electrophoresis was run at 300 V for 90 min in 1 × TBE. The gels were exposed to a phosphorimager and scanned with Typhoon 9410.

Western blot analysis

Aliquots (50 μl for each) of the gel filtration fractions were mixed with 5 × SDS-PAGE loading buffer and loaded onto the gel for SDS-PAGE analysis. The proteins in the PAGE gel were transferred onto a PDVF membrane at 30 mA for 15 hrs at 4°C. The membrane was washed and incubated with a primary antibody and a secondary antibody anti-rabbit HRP-conjugate (HuaAn Biotechnology limited company, Hangzhou, China) following the standard protocol for Western blot. The protein-specific rabbit

polyclonal antibodies were made by HuaAn Biotechnology limited company with a 15 aa synthetic peptide specific to each target protein as immunogens (Additional file 6: Table S5). The band was visualized with Immobilon™ Western Chemiluminescent HRP Substrate (Millipore Corporation, Billerica, MA, USA) and the images were obtained by Imagequant™ 400 (GE Healthcare, UK).

For assessment of HerA and NurA levels in cells by Western blot, cells from 0.5 ml E233S culture (OD$_{600}$ ~ 0.4) were collected and lysed by mixing with 5 × SDS-PAGE loading buffer and boiled for 10 min. Total proteins were separated by SDS-PAGE for Western blot analysis as described above. Purified proteins were loaded as standards for quantification.

Results

herA, mre11, rad50, and nurA are all essential for cell viability

It has been reported that each of the four genes in the *mre11* operon, namely *herA*, *mre11*, *rad50*, and *nurA* are possibly essential since deletion mutants were not obtainable (reviewed in Zhang *et al.* [33]). In these experiments, gene knockouts were tested by construction of a target gene-specific strain pMID-T and two-step selections (Figure 1A). A vector pMID (**m**arker **i**nsertion and target gene **d**eletion) containing a marker cassette (*pyrEF-lacS*) and three homologous arms (TG-arm, IN-arm and OUT-arm) specific to the target gene was constructed and a marker-integrated strain pMID-T was obtained by selection of a uracil prototroph (Figure 1A). The resulting strain pMID-T was subjected to the second selection of a uracil auxotroph by the addition of 5-FOA and uracil into the culture medium. If the targeted gene is successfully knocked out, cells are able to grow up easily under the second selection (Figure 1A). However, no deletion mutant for any of *herA*, *mre11*, *rad50*, or *nurA* was obtained after 5-FOA and uracil (Figure 1A).

To affirm the essentiality of the HRR genes, these recombinant pMID-T strains were employed in a mutant propagation assay [38]. If a gene is essential, no deletion mutant can be obtained, but spontaneous mutation at *pyrEF* would occur at a low frequency close to the corresponding rates reported for the protein-encoding genes of *E. coli* [40]. The mutant may then accumulate and the culture will grow, albeit at a slow rate (Figure 1A, bottom scheme). The culture will stop growing when the original growth rate is too low for the spontaneous mutants to accumulate (Figure 1A).

In the current study, two reference strains were used in the mutant propagation assay: the wild-type strain, REY15A [41], was used to investigate the growth of spontaneous *pyrEF* mutants, whereas pMID-*orc1-3*-T was used to investigate the growth of mutant cells of a nonessential target gene deletion [42] (Additional file 1: Table S1). It

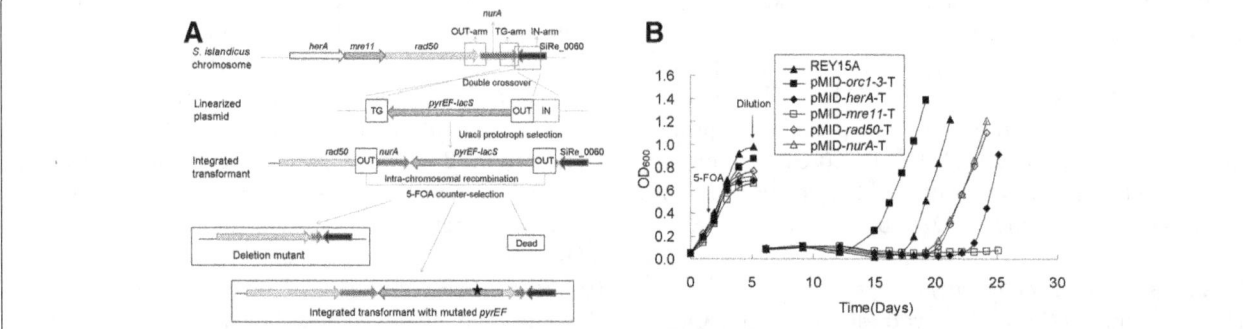

Figure 1 The four HRR genes, *herA*, *mre11*, *rad50*, and *nurA*, which are encoded in a single operon, are all essential for cell viability.
(A) Schematic diagram depicting the construction of the strain used in the gene knockout analysis. Marker insertion and target gene deletion (MID) and a subsequent mutant propagation assay were utilized. The star indicates a putative spontaneous mutation site in *pyrEF*. **(B)** Mutant propagation assay for *pyrEF-lacS*-integrated transformants (pMID-*herA*-T, pMID-*mre11*-T, pMID-*rad50*-T, and pMID-*nurA*-T). *orc1-3*, a nonessential gene coding for one of the three replication initiation proteins, was used as a positive control, whereas REY15A is a reference for *pyrEF* mutation. Strains were cultivated in STVy medium supplemented with 2 mg/ml uracil. The counter selection chemical, 5-FOA, was added when OD_{600} reached 0.4. The culture was diluted with fresh medium to $OD_{600} \sim 0.1$ when the cells had grown to early stationary phase. Absorbance of the culture was measured.

has been shown that *orc1-3*, which encodes a replication initiator protein, can be deleted in *S. islandicus* [42]. As shown in Figure 1B, pMID-*orc1-3*-T culture grew most rapidly, followed by the wild-type strain REY15A. Cultures of pMID-*rad50*-T, pMID-*nurA*-T, and pMID-*mre11*-T grew much slower than REY15A or not at all (Figure 1B). As for the wild-type strain REY15A, the growth of these pMID transformants was not due to deletion of the respective gene, but due to propagation of cells with a mutation in *pyrEF* (Figure 1A). The strain pMID-*mre11*-T never grew. As shown in Figure 1B, up to day 5, the enrichment of pMID-*mre11*-T caused the cells to grow more slowly than the others. We also noticed that colonies of this strain appeared later than others on plates. The integration of the marker cassette and OUT arm at the upstream of *mre11* locus may have affected its gene expression, resulting in growth retardation. The results demonstrate that all four putative HRR genes are essential for the viability of *S. islandicus*.

ATPase activity is essential but not sufficient for the *in vivo* function of HerA

Recently, we developed a genetic complementation method for *S. islandicus* in which an essential gene deficiency in the chromosome is able to be rescued by ectopic expression of the wild-type protein from a plasmid [34]. In this method, a pMID-T strain (pMID-*herA*-T) of the target gene *herA* as described above was transformed with a pSSR plasmid (harboring a simvastatin resistance marker and expressing the wild-type gene) under selection by simvastatin. Upon selection of the transformed strain with uracil, 5-FOA, and simvastatin, the essential chromosomal gene together with the *pyrEF-lacS* maker was looped out by intra-chromosomal recombination, and the plasmid was maintained in the absence of antibiotics.

Here we employed that system to test the possibility of rescuing *herA* deficiency with different HerA mutant proteins, and investigated the functions of the residues critical for various properties, including ATPase activity of HerA, nuclease activity of the HerA-NurA complex, and the interaction between HerA and NurA.

In vitro site-directed mutagenesis of archaeal HerA proteins has revealed conserved amino acid residues critical for its ATPase and helicase activities [18,22,23]. These amino acid residues are K154, D176, E356, and R381 of *S. islandicus* HerA, among which K154 and E356 are located in the Walker A and Walker B motifs, respectively. Point mutations at the four conserved sites were introduced by site-directed DNA mutagenesis and mutant genes encoding K154R, D176E, D176N, E356D, E356Q, or R381K substitutions were cloned into pSSR to create genetic complementation plasmids (Additional file 2: Table S2). The genetic host used for the experiment was pMID-*herA*-T, the recombinant strain carrying a marker-target gene cassette (*pyrEF-lacS-herA*) at the locus of *herA* gene (Figure 2A). A complementation plasmid was introduced into the host via transformation, and transformants were obtained by simvastatin selection.

Strikingly, only the transformants harboring pSSR, pSSRA-HerA or pSSRA-D176E were obtained, while no transformants with pSSR carrying any of other five mutants grew successfully after transformation. We have shown that ectopic wild type HerA was expressed in higher amounts in the cell than chromosomal HerA [34]. It is likely that the mutant protein was also in much higher abundance than the chromosome-coded HerA in the cells if the cells were transformed with the plasmid and the gene expression was induced. Loss of the ATPase and helicase activity in the HerA mutant was highly lethal to the cells. In a recent report, an *in vitro* mutant-doping

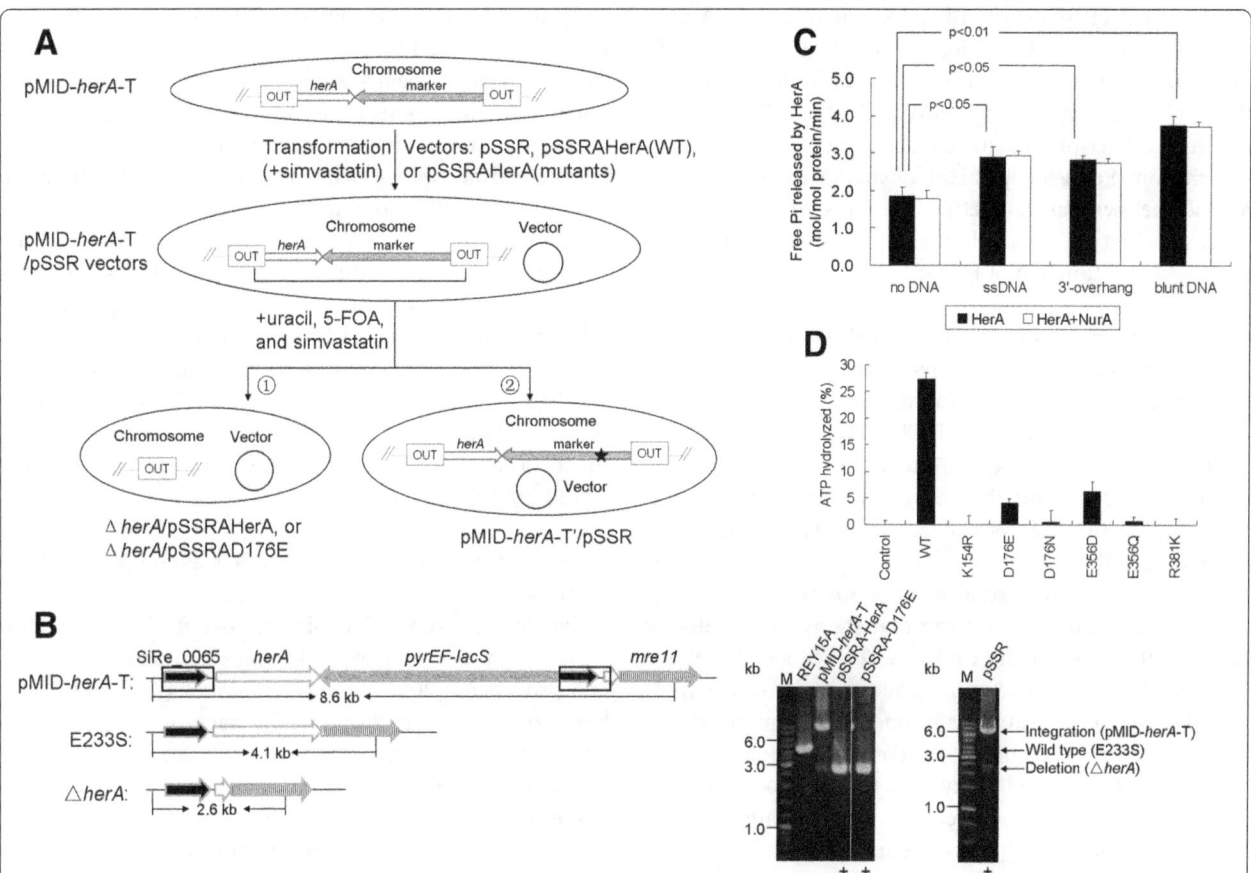

Figure 2 A HerA(D176E) mutant with reduced ATPase activity is able to complement deletion of the wild-type *herA* gene. (A) A schematic map depicting the procedures of the complementation analysis. pMID-*herA*-T cells were transformed with pSSR vectors. The cells were selected in the presence of simvastatin. Counter-selection was performed with solid medium containing 5-FOA, uracil, and simvastatin. Two genotypes of the cell grown up were indicated as ① and ②. The star indicates a putative spontaneous mutation in *pyrEF*. **(B)** Genotype analysis of cells by PCR using *herA* flanking primers. The genomic DNA of each strain was isolated and used as the template. The pSSR plasmids in cells were isolated and amplified in *E. coli* DH5α for sequencing. Left panel, a schematic map showing the lengths of PCR products of different strains. Right panel, agarose gel electrophoresis of the PCR products. "+", the designed plasmid maintained. **(C)** ATPase activity of HerA alone or with NurA in the absence or presence of DNA. The assay was performed as described in the Methods. Reaction products were separated using thin layer chromatography and analyzed by Phosphorimager. The means ± standard errors of three independent experiments and p values (n = 3) are shown. **(D)** ATPase activity of wild-type and mutant HerA proteins. The means ± standard deviations of three independent experiments are shown.

assay showed that increasing the ratio of *S. solfataricus* K154A to WT HerA resulted in an exponential drop in the proportion of unwound DNA substrate, where the ratio of 1:1 almost inactivated the unwinding activity [43]. The result is consistent with our *in vivo* data. Our results are also in good agreement with the above result that *herA* is essential for cell viability.

The three transformants above were then subjected to counter-selection against *pyrEF* on plates supplemented with uracil, 5-FOA, and simvastatin (Figure 2A). As shown in Figure 2B, two genotypes survived the selection. One genotype represents a Δ*herA* carrying plasmid harboring pSSRA-HerA or pSSRA-D176E and the other represents pMID-*herA*-T cells carrying a *pyrEF* mutation with the empty vector pSSR (Figure 2A). PCR and sequencing confirmed that the culture with pSSRA-HerA(D176E)

contained a deleted chromosomal *herA* allele as that of pSSRA-HerA, and the plasmid carried the D176E *herA* substitution (Figure 2B). Thus, these results demonstrate that only wild type HerA and HerA(D176E) could functionally complement *herA* deficiency *in vivo* under the assay conditions.

To gain insight into why only D176E could rescue *S. islandicus* cells with a chromosomal *herA* deletion, all six HerA mutant derivatives were expressed in and purified from *E. coli*. We showed that the wild type HerA was able to hydrolyze approximately 1.8 mole of ATP per mole of enzyme per min. The ATPase activity of HerA was enhanced in the presence of either ssDNA, 3'-overhang, or blunt-ended dsDNA (Figure 2C). In addition, the ATPase activity of HerA was not enhanced in the presence of NurA (Figure 2C). Except for D176E and

E356D, the ATPase activity of all the remaining HerA mutants was very low or undetectable. D176E and E356D maintained about one-seventh and one-fifth the activity of the wild-type enzyme, respectively (Figure 2D). Notably, the ATPase-active mutant E356D could not rescue the chromosomal deficiency of HerA, which was further studied (see below). Taken together, these results indicate that ATPase activity is essential, but not sufficient on its own, for the *in vivo* function of HerA.

Nuclease activity is essential for the *in vivo* function of NurA
Previous works established that a few amino acid residues in NurA are critical for its enzyme activity. The corresponding residues are D58, E116, and D133 of *S. islandicus* NurA [26,44]. Another residue (K202 in *S. islandicus*) critical for the enzyme activity was revealed by crystallization analysis of *S. solfataricus* NurA. It is located near the active site of the NurA and substitution of the corresponding residue in *S. solfataricus* yielded an inactive enzyme, indicating that the lysine residue is possibly involved in the catalysis [26]. Since the NurA proteins from *S. solfataricus* and *S. islandicus* exhibit high identity (90%) in their amino acid sequences, the conserved residues should function in the same way. Therefore, we focused on two of the residues D58 and K202 and addressed the necessity of NurA nuclease activity using the same approach employed for HerA. The empty vector and pSSR carrying genes encoding the wild type NurA or its mutants D58E, D58A, K202R, and K202A were constructed and transformed into pMID-*nurA*-T cells. Surprisingly, all of the transformants grew under simvastatin selection, not only for the empty vector and pSSR carrying the wild type gene, but also for all the four mutants. After subsequent counter-selection by 5-FOA, colonies were selected from the plates and cultivated for further analyses. PCR analyses showed that the chromosomal *nurA* had been deleted in all strains supplemented with NurA (wild type or mutants), while that supplemented with the empty vector had not been deleted (Figure 3A and 3B).

Subsequently, *nurA* on the complementation plasmid was amplified and sequenced from each *nurA*-rescued strain. Intriguingly, all plasmids carried the wild-type *nurA* gene rather than any of the various mutant *nurA* genes cloned into the pSSR vector. This process was repeated three times, and the same results were obtained each time. Therefore, we conclude that the *nurA* mutants reverted to the wild-type during the experiment. Since reverse mutation occurs at a very low rate, we reasoned that the *nurA* mutant complementation plasmids could not rescue *nurA* deficiency in this archaeon. Furthermore, NurA and HerA work as a complex for dsDNA end processing in *S. solfataricus* and the NurA (K202A) mutant did not affect the interaction [26],

suggesting that the inability to rescue *nurA* deficiency was not due to failure in the complex formation. Our results provide evidence supporting the hypothesis that nuclease activity is essential for the *in vivo* function of NurA. The reason why the plasmid carrying lethal NurA mutant existed in the cell but not for HerA mutants is not quite clear. We assume that the cells carrying plasmid with lethal NurA mutant have the possibility to assemble a functional NurA as a dimer, allowing the presence of the plasmid and the reversion to occur. While for the cells carrying plasmid with lethal HerA mutant, incorporation of a lethal deficient subunit into HerA hexamer could result in a non-functional HerA. As a result, the probability of assembling a functional HerA as a hexamer should be much lower than that for NurA as a dimer, and the cells would not likely survive.

Interaction between NurA and HerA is essential to cell viability
A biochemical study has shown that in *S. solfataricus*, NurA interacts with HerA, forming a complex for dsDNA end processing [26]. Two NurA residues (I295 and F300), both found on a hydrophobic surface according to the protein's crystal structure, have been implicated in the HerA-NurA interaction, as mutation at either residue to glutamic acid abolishes formation of the complex [26].

To address the importance of the HerA-NurA interaction *in vivo*, pSSR plasmids carrying four *S. islandicus* NurA mutants, I295L, I295E, F300Y, and F300E, were individually transformed into pMID-*nurA*-T cells to test for complementation. I295L and F300Y were able to achieve complementation (Figure 3B), whereas I295E and F300E failed to do so. Plasmid sequencing confirmed that each plasmid carried the original mutant *nurA* substitution. It is notable that the changes from isoleucine and phenylalanine to leucine and tyrosine (I295L and F300Y) do not affect the hydrophobic character at the interaction surface between NurA and HerA, whereas the I295E and F300E substitutions do. We speculated that NurA I295L and F300Y mutants should maintain the interaction, while I295E and F300E should not. To test this, the interaction between the HerA and NurA mutants I295L, I295E, F300Y, and F300E of *S. islandicus* were experimentally evaluated by gel filtration, confirming that I295L and F300Y maintained the interaction and complex formation while I295E and F300E did not (Figure 3C). It is interesting that the pSSR plasmid carrying NurA(I295E) or F300E was maintained in cells. We assume that I295E or F300E are not as toxic as the nuclease-dead NurA mutants (D58E, D58A, K202R, and K202A) since they were unable to interact with HerA and did not interfere with the DNA end resection process. The plasmid with either NurA mutant could exist in the cell and there is no selection for generation of the reversion genotype. Taken

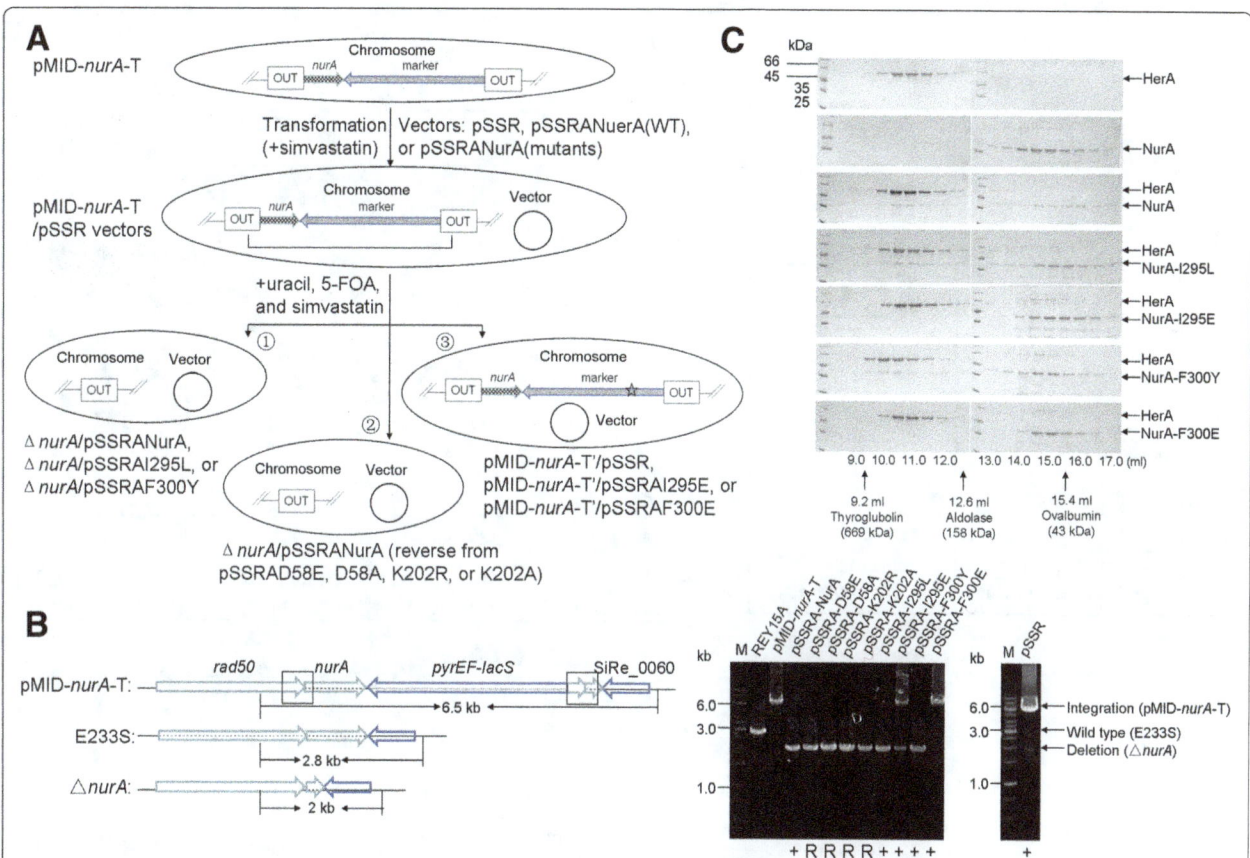

Figure 3 The nuclease activity of NurA and its interaction with HerA are essential for cell viability. (A) A schematic map depicting the procedures of the complementation analysis. pMID-*nurA*-T cells were transformed with pSSRA-NurA-C-His plasmids or empty vector. Three genotypes of the cells grown up after counter-selection were indicated as ①, ②, and ③. The star indicates a putative spontaneous mutation in *pyrEF*. **(B)** Genotype analysis of the cells grown up by PCR using *nurA* flanking primers and sequencing of the *nurA* on the isolated plasmid. "+", the designed plasmid maintained; "R", plasmid maintained, but *nurA* gene had reverted to wild-type. **(C)** Gel filtration analysis of the interaction between wild type HerA and NurA mutants I295L, I295E, F300Y, and F300E by a Superdex™ 200 10/300 GL column. A total of 500 µg HerA, 500 µg NurA, and the mixture were heated incubated at 60°C for 20 min before gel filtration. The sample fractions were analyzed by SDS-PAGE. A small amount of NurA formed dimers in gels when NurA concentration was high. The elution peaks of the molecular markers are indicated by arrows at the bottom.

together, we speculate that the interaction between NurA and HerA is essential for cell viability.

The HerA(D176E)-NurA complex, but not the HerA (E356D)-NurA complex, retains 5'-3' DNA resection activity *in vitro*

As described above, although HerA(E356D) showed a higher level of ATPase activity than D176E, only HerA (D176E) rescued *herA* deficiency. Here, we further characterized the two mutant proteins. The activities of HerA, NurA, and their complex were examined using a 3'-overhang or blunt-ended DNA as the substrate (Additional file 3: Table S3 and Additional file 5: Table S4). The substrates were labeled, as illustrated in Figure 4A and Additional file 5: Table S4, and products and substrates were analyzed by native as well as denaturing polyacrylamide gel electrophoresis. The native gels were used to investigate both endonuclease and exonuclease activities whereas

denaturing gels were used to investigate exonuclease activity of the substrates.

We first investigated the DNA degradation activity of the wild-type HerA and NurA. In the absence of ATP, the HerA-NurA complex failed to degrade any of the DNA substrates, indicating that the nuclease activity of NurA was dependent on the ATP hydrolysis by HerA (Additional file 7: Figure S1B). Furthermore, HerA also lost its helicase activity when it was alone or combined with nuclease-dead NurA protein (Additional file 7: Figure S1D). This indicates that the DNA degradation activity of HerA and NurA relies on both ATPase (and/or helicase) activity of HerA and the nuclease activity of NurA, in agreement with a previous report [26]. HerA-NurA complex degraded 3'-overhang and blunt-ended DNA in the 5'-3' direction, although it also exhibited partial endonuclease activity on 3'-overhang DNA (Additional file 7: Figure S1B and S1C). Subsequently, since we were unable to detect the helicase

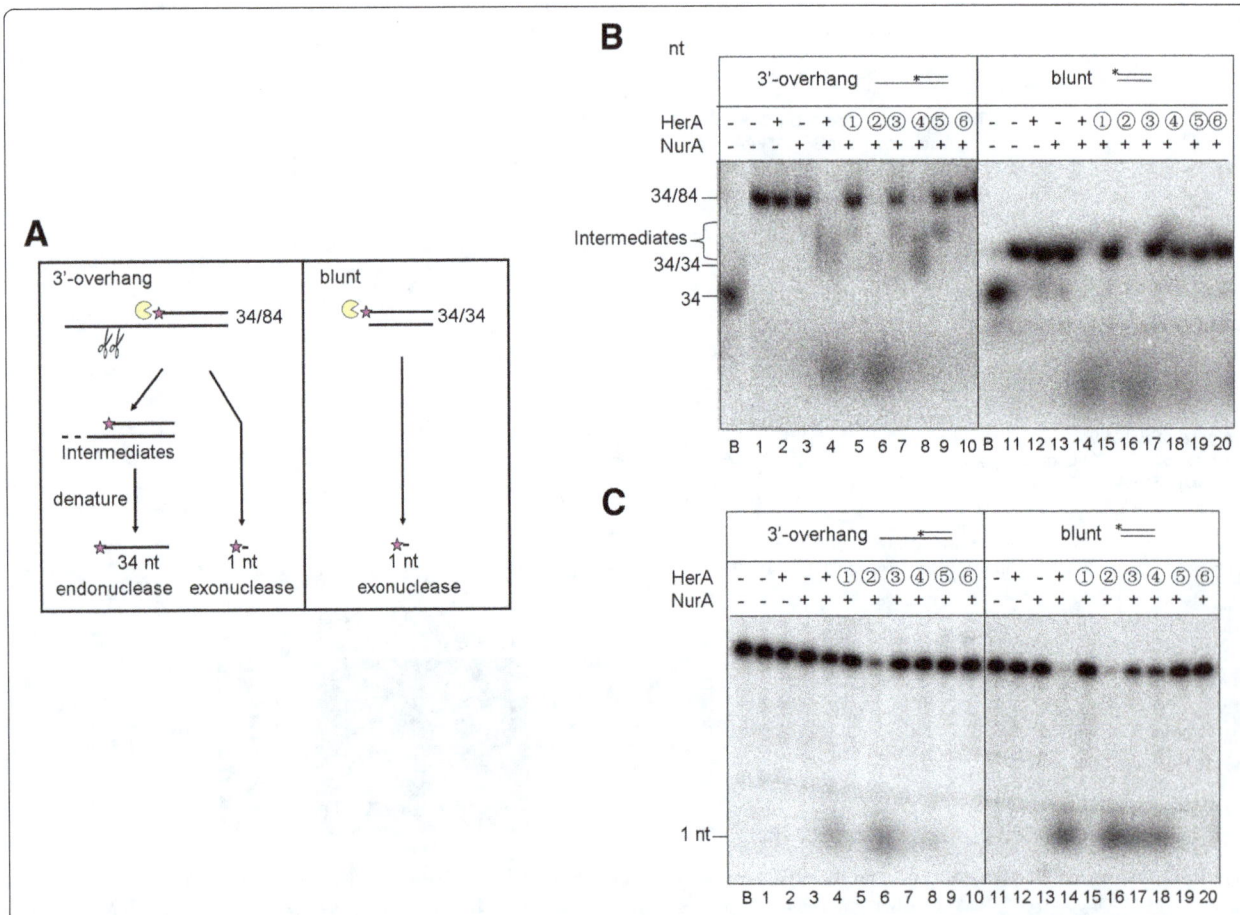

Figure 4 HerA(D176E)-NurA maintains strong 5'-3' exonuclease activity. (A) A schematic diagram of the nuclease activity of the HerA-NurA complex using two different substrates. **(B)** Represent gel profiles showing the nuclease activity of HerA-NurA complexes. Wild-type or mutant HerA (27.8 nM hexamer) was mixed with NurA (83.4 nM dimer) for 30 min at 65°C. After the reactions, the products were analyzed on a 10% native polyacrylamide gel. **(C)** Same as in **(B)**, but the products were analyzed on a 15% denaturing polyacrylamide gel. "+", wild-type HerA or NurA. ①-⑥,site-directed HerA mutants: K154R, D176E, D176N, E356D, E356Q, and R381K. **B**, boiled sample.

activity of HerA in the absence of NurA, the DNA degradation activity of various HerA mutants complexed with wild-type NurA was analyzed (Figure 4). For blunt-ended substrates, the 5'-3' exonuclease activity of D176E-NurA was similar to wild-type HerA-NurA (Figure 4B and 4C, lanes 14 and 16; Figure 5B). For 3'-overhang substrate, D176E-NurA exhibited higher exonuclease activity (Figure 4C, lanes 4 and 6; Figure 5A) and endonuclease activity (Figure 5A) than the wild-type HerA-NurA, while for both substrates, R381K-NurA were unable to degrade DNA. This was thought to be due to loss of ATPase activity (Figure 4B and 4C, lanes 10 and 20). K154R-NurA, D176N-NurA, and E356Q-NurA exhibited much lower activity than wild-type HerA-NurA or were completely inactive (Figure 4B and 4C, lanes 5, 7, 9, 15, 17, and 19; Figure 5). Importantly, we found E356D-NurA exhibited nuclease activities distinct from both wild-type HerA-NurA and HerA(D176E)-NurA. E356D-NurA maintained endonuclease activity on the 3'-overhang substrate but with less than half of the

exonuclease activity of the wild type on any of the substrate (Figure 4B and 4C, lanes 8 and 18; Figure 5). Since we have shown that only HerA(D176E) rescued the deficiency of the chromosomal *herA* but HerA(E356D) did not, these results reveal, for the first time, that high 5'-3' exonuclease activity of HerA-NurA is essential for cell viability.

HerA interacts with NurA *in vivo*

To further investigate the *in vivo* role of HerA, we conducted *in situ* poly-histidine (His)-tagged protein purification. This method facilitated the isolation of HerA-interacting proteins in *S. islandicus* cells under physiological growth conditions. Using a recombination scheme similar to marker insertion and target gene deletion (MID), the chromosomal version of *herA* was replaced with an N-His-tagged version of *herA* (Figure 6A). The resulting strain was confirmed by PCR using *herA* locus-specific flanking primers and X-gal staining (Figure 6B). His-tagged HerA and associated proteins were co-purified

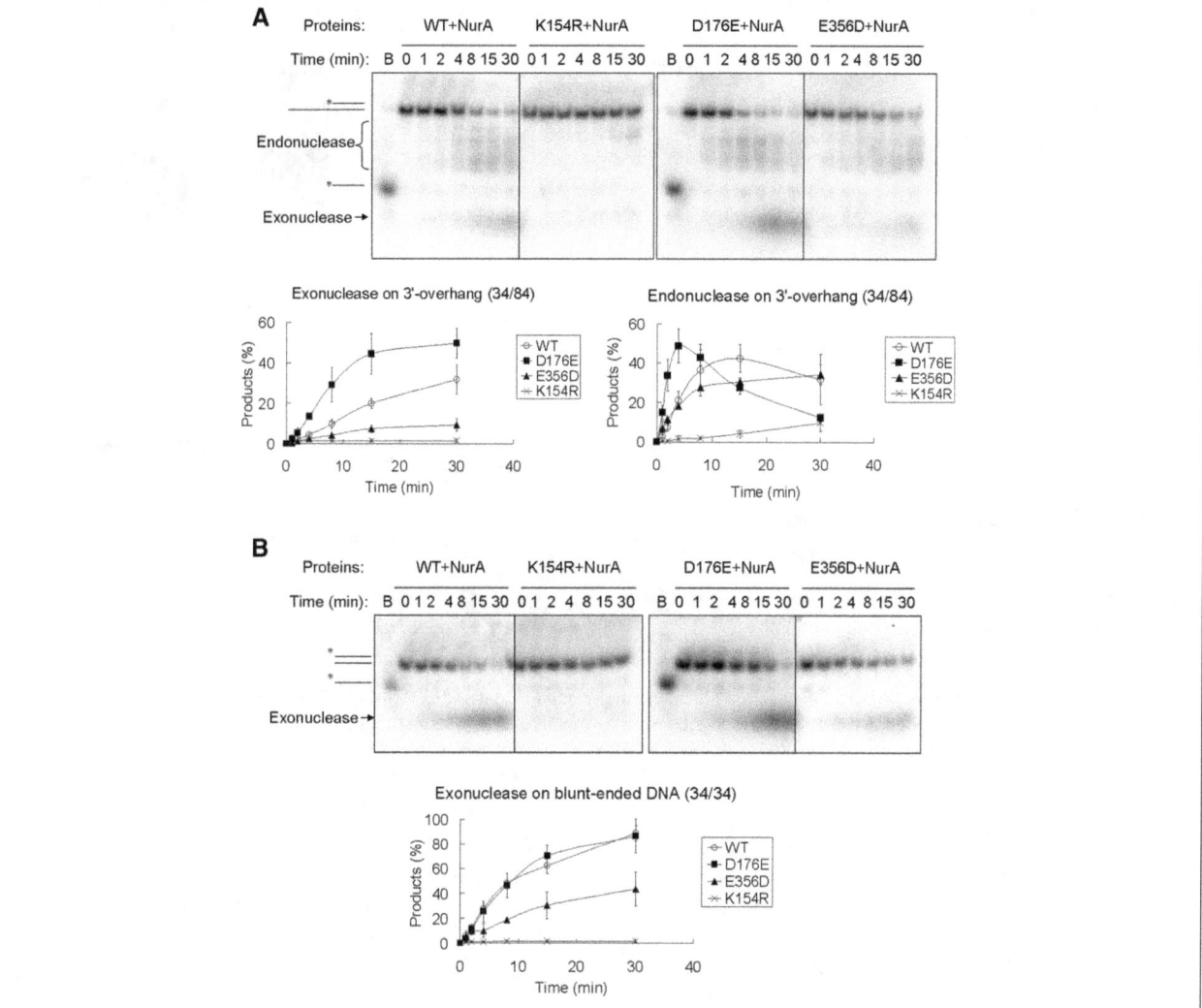

Figure 5 The time courses of the exonuclease and endonuclease activities of HerA-NurA complexes on dsDNA. The assay was performed in 20 µl reaction mixture as described in the Methods. The reactions were stopped at the indicating times by the addition of 5× loading buffer and putting the tubes on ice. The representative gels and the means ± standard deviations of three independent experiments are shown. **(A)** 3'-overhang. **(B)** blunt-ended DNA.

from cell extract of a 9-L culture by ammonium sulfate precipitation, nickel-nitrilotriacetic acid (Ni-NTA) affinity purification, and gel filtration (Figure 6C). Purified proteins were analyzed by SDS-PAGE and Western blotting. Western blot analysis using protein-specific antibodies revealed that NurA eluted in same fractions with HerA (Figure 6D). The fractions containing HerA were pooled, concentrated, and analyzed further by liquid chromatography-mass spectrometry. Strikingly, we identified two other proteins co-purified with HerA, Hjc (Holliday junction cleavage, SiRe_1431) and a PIN-domain ATPase (SiRe_1432). Western blot analysis showed that HerA, Hjc, and the PIN-domain ATPase were present in earlier fractions of the co-purified sample than that of HerA and NurA (Figure 6D). However, the HRR proteins, Mre11, Rad50, and RadA, were not found in either MS

(data not shown) or Western blot analyses (Figure 6D). The interaction between HerA and the PIN-domain ATPase or Hjc may be weak in the absence of other factors as shown in an *in vitro* pull-down assay (Figure 6E and 6F). The results support that HerA forms a complex with NurA *in vivo* and suggest that HerA has additional partners in the cell.

Discussion

Essentiality of the *mre11* operon and the 5'-3' exonuclease activity of the HerA-NurA complex

Previous works suggested that each gene in the *mre11* operon could be essential in *S. islandicus* and *T. kodakaraensis*. However, the gene essentiality in *T. kodakaraensis*, an archaeon belonging to euryarchaeota, one of the two main branches of archaea, was deduced from

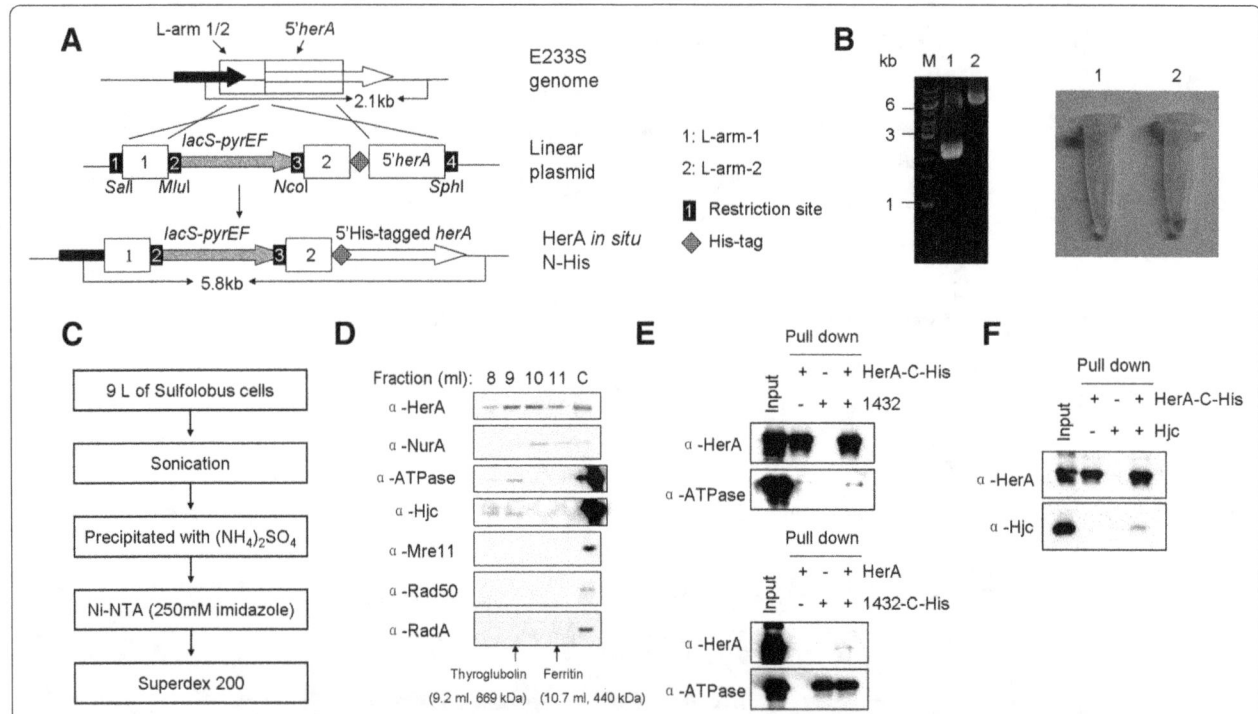

Figure 6 Identification of co-purified proteins using a chromosomally-coded His-tagged HerA. (A) Schematic diagram showing the construction of a strain that can encode a N-terminal His-tagged HerA by modification of the chromosomal *herA*. The MID strategy was used for the construction. **(B)** Confirmation of the constructed strain using PCR and X-gal staining. *herA* locus-specific flanking primers were used for PCR amplification. 1, the control strain E233S; 2, the constructed strain pMIDHis-*herA*-T. **(C)** Schematic map showing the purification of the proteins. **(D)** Western blot analyses of *in situ* N-His-tagged HerA and its co-purified proteins with specific antibodies. The fractions of 8, 9, 10, and 11 ml collected from the gel filtration were analyzed. C, purified protein. The standard molecular mass markers are indicated with arrows. **(E)** Pull-down assay confirming the interaction between HerA and the ATPase. A total of 50 μg HerA-C-His (or HerA) was mixed with 50 μg ATPase (or ATPase-C-His) at 70°C for 30 min in a buffer containing 25 mM Tris-HCl, pH 9.0, and 200 mM NaCl. The mixture was incubated with nickel-agarose beads. Unbound proteins were washed with the buffer above supplemented with 25 mM imidazole and target proteins were eluted by 250 mM imidazole. The fractions were detected by western blot with protein-specific antibodies. **(F)** Detection of the interaction between HerA and Hjc by pull-down assay. The procedure was as those used in **(E)**. A total of 42 μg HerA-C-His and 12 μg Hjc were used in the assay.

the inability of isolating mutants of respective genes whereas only an overview has been presented for the genetic study for the HRR genes in *S. islandicus*, an archaeon belonging to crenarchaeota, another main branch of archaea [29,32,33]. Here, we have evaluated the gene essentiality of the four genes in *S. islandicus* individually, using a mutant propagation assay previously developed in our laboratory [38]. We found that mutant cells generated from the pMID transformants (pMID-T) lost their propagation capability, strongly demonstrating that each of the genes in this operon is essential for cell viability of *S. islandicus*. Our previous studies have revealed that a RecQ-like DNA helicase Hjm and RadA, the archaeal homolog of Rad51, are essential in this archaeon [32,33]. These results indicate that all these HRR genes are essential for the viability of *S. islandicus*. Strikingly, hyperthermophilic archaea are the only known organisms among microorganisms, in which the *mre11*, *rad50*, and *radA* are absolutely required for cell growth.

We attempted to quantify the levels of HerA and NurA in the cell by quantitative Western blot analysis and found that the amounts were about 956 ± 30 and 848 ± 40 molecules per cell for HerA and NurA, respectively (Additional file 8: Figure S2). These are much higher than that for RecBCD, a DNA end resection complex in *E. coli*, which is only about 10 molecules per cell [45]. This may indicate that a large number of DSBs exist in *Sulfolobus* which need to be repaired constantly and efficiently. In response to the essential requirement of DSB processing, HerA and NurA may be constitutively expressed in high levels in the cells.

The essentiality of the activities of the HRR proteins in this organism was further characterized using a genetic complementation assay recently developed in our laboratory [34]. We have found that the ATPase activity of HerA and the efficient 5′-3′ degradation activity by the HerA and NurA complex is essential for cell viability in *S. islandicus* and this should reflect the *in vivo* activity of the enzyme complex. We have observed that HerA(D176E) have high exonuclease activity and could maintain the essential function of HerA *in vivo*. Another mutant (E356D) of HerA and NurA wild-type protein complex maintain the ssDNA endonuclease activity of the wild type enzyme complex, but with much reduced exonuclease activity. We have shown that HerA(E356D) did not complement the essential function of the protein complex and therefore the endonuclease activity apparently does not reflect the *in vivo* essentiality of the helicase/nuclease complex; however, the exonuclease to generate ssDNA does. The ssDNA generated during DSB end resection is a critical intermediate for strand invasion in eukaryote. It could also be utilized in a synthesis-dependent strand-annealing pathway (SDSA) [46]. Therefore, although the ssDNA is mostly likely employed in HRR in archaea, the possibility that it is a substrate for other processes such as a putative SDSA can not be ruled out.

Roles of the conserved residues of HerA

We revealed that out of the six HerA mutants, only HerA(D176E) could maintain the essential function of HerA *in vivo*. Both HerA(K154R) and HerA(R381K) lost the ATPase activity of HerA, confirming the importance of the Walker A and arginine finger motif in ATP hydrolysis for DNA helicase [47-49]. It was recently shown that E356 of the Walker B motif in *S. solfataricus* HerA forms a salt bridge with trans-acting R381. This interaction is thought to be necessary for both stable interface formation and water activation prior to catalysis [43]. The change of glutamate acid (E) to glutamine (Q) may impair the formation of the salt bridge leading to loss of the ATPase activity. Furthermore, HerA(E356D)-NurA maintained the same level of the endonuclease activity as the wild type, but had much lower 5'-3' exonuclease activity on dsDNA. E365 may be also involved in fine coupling of ATP hydrolysis and DNA unwinding. Slight change of the residue from glutamic acid to aspartic acid could lead to coupling impairment resulting in the reduction of the 5'-3' resection of HerA(E356D)-NurA. In agreement with this, *S. solfataricus* HerA structure showed that ATP binding, hydrolysis, and release led to the local changes in the positioning of R381, which resulted in DNA-binding loop movement and substrate translocation [43]. The E356D mutant may maintain the salt bridge with R381 for ATP binding and hydrolysis, but affect the changes of R381 positioning, which would impair its DNA unwinding activity. Finally, we showed that HerA(D176N) exhibited as low ATPase activity as that of E356Q. The conserved residue D176 in *S. solfataricus* HerA locates close to the ATP-binding site. Our results suggest that mutation to either glutamic acid or asparagine affects the ATP activity, and is in good agreement with the structural prediction of HerA.

The end resection activity requires both HerA and NurA

It is surprising that we could not detect the helicase activity of HerA or the nuclease activity of NurA, even though various reaction conditions were applied in our assay. The DNA end resection activity on dsDNA was only observed in the presence of both HerA and NurA (Additional file 7: Figure S1B and S1D). The result was consistent with that in *S. solfataricus* and *P. furiosus* [23,26]. However, the helicase activity of HerA and the nuclease activity of NurA could be detected using the HerA and NurA proteins from *S. tokodaii* and *S. acidocaldarius* [18,19,22,44]. Due to the closer phylogenetic relationship of *S. islandicus* to *S. solfataricus* than that to either *S. tokodaii* or *S. acidocaldarius*, it may be reasonable that the properties of *S. islandicus* HerA and NurA proteins resemble to their counterparts in *S. solfataricus*. It may indicate that the mode of end resection by HerA and NurA in *S. islandicus* and *S. solfataricus* is different from that of *S. tokodaii* and *S. acidocaldarius*. Further study on how HerA interacts with NurA for dsDNA end resection and how the length of the resected ssDNA is controlled will reveal the detailed mechanisms of DNA end processing in archaea.

Conclusion

Our results revealed that the high 5'-3' DNA end resection activity of HerA-NurA complex is essential for cell viability. This may indicate that a large number of DSBs occurs in *Sulfolobus* cells under high temperature conditions need to be efficiently resected. Two other HerA-interacting proteins, an ATPase and Hjc, were identified, suggesting that HerA have other roles in addition of that in DNA end resection. These results will help better understand HRR in thermophilic archaea.

Additional files

Additional file 1: Table S1. Strains used in this study.

Additional file 2: Table S2. Plasmids used in this study.

Additional file 3: Table S3. Oligonucleotides used in this study.

Additional file 4: Supplementary methods.

Additional file 5: Table S4. Substrates used in this study.

Additional file 6: Table S5. Peptides for making protein-specific antibodies used in this study.

Additional file 7: Figure S1. Both HerA and NurA are required for DNA degradation. (A) A schematic illustrating nuclease activity analysis. (B) Analysis of DNA degradation of 3'-overhangs and blunt-ends in the presence and absence of ATP by a HerA-NurA mixture. HerA (27.8 nM hexamer) was mixed with NurA (83.4 nM dimer) and incubated at 65°C for 30 min. Samples were analyzed on a 10% native polyacrylamide gel. B, boiled samples. (C) Same as in (B), but the samples were analyzed on a 15% denaturing polyacrylamide, and only samples with ATP were analyzed. M, size marker. (D) A HerA-NurA complex containing a nuclease inactive form of NurA (D58A) loses DNA degradation activity. B, boiled samples. The concentrations of HerA and NurA(D58A) used for 3'-overhang substrates were 0.0625, 0.125, 0.25, 0.5, 1, and 2 μM.

The concentrations of HerA and NurA(D58A) used for blunt-ended DNA substrates were 0.125, 0.25, 0.5, and 1 μM.

Additional file 8: Figure S2. Determination of HerA and NurA amounts in *S. islandicus* cells. Total proteins of wild type cells were separated by SDS-PAGE for quantitative Western blot. Purified HerA (19 ng) and NurA (45 ng) proteins from *E. coli* were loaded as standards for quantification.

Abbreviations
DSBs: Double-strand breaks; HRR: Homologous recombination repair; NHEJ: Non-homologous end joining; SDSA: Synthesis-dependent strand-annealing pathway; 5-FOA: 5-fluorotic acid; MID: Marker insertion and target gene deletion.

Competing interests
The authors declare that they have no competing interest.

Authors' contributions
QHH participated in the study design and data collection, carried out most experiments and drafted the manuscript. LLL and JFL carried out a part of plasmids construction and protein expression in *E. coli*. JFN helped revise the manuscript. QXS and YLS participated in the experiment design, manuscript revision and supervised the project. All authors read and approved the final manuscript.

Acknowledgements
We would like to thank Tong Wu for help in plasmids construction and NurA complementation assays, Changyi Zhang for providing the marker integrated transformants, pMID-*mre11*-T, pMID-*rad50*-T, and pMID-*nurA*-T, and our lab members for helpful discussions. We would like to thank Prof. Li Huang and Prof. Guihua Hou for providing the facilities to perform the enzymatic characterization of HerA-NurA complex. This work was supported by National Natural Science Foundation of China (3093002, 31170072, 31470184 to Y.S., 31470200 to J.N.); Danish Council of Independent Research (FTP/11-106683, DFF–1323-00330) and the Carlsberg Foundation (to Q.S.). Funding for open access charge: National Natural Science Foundation of China.

References
1. Lammens K, Bemeleit DJ, Mockel C, Clausing E, Schele A, Hartung S, et al. The Mre11:Rad50 structure shows an ATP-dependent molecular clamp in DNA double-strand break repair. Cell. 2011;145(1):54–66.
2. Bernstein KA, Rothstein R. At loose ends: resecting a double-strand break. Cell. 2009;137(5):807–10.
3. Chapman JR, Taylor MR, Boulton SJ. Playing the end game: DNA double-strand break repair pathway choice. Mol Cell. 2012;47(4):497–510.
4. Kuzminov A. Recombinational repair of DNA damage in Escherichia coli and bacteriophage lambda. Microbiol Mol Biol Rev. 1999;63(4):751–813.
5. Ayora S, Carrasco B, Cardenas PP, Cesar CE, Canas C, Yadav T, et al. Double-strand break repair in bacteria: a view from Bacillus subtilis. FEMS Microbiol Rev. 2011;35(6):1055–81.
6. Krejci L, Altmannova V, Spirek M, Zhao X. Homologous recombination and its regulation. Nucleic Acids Res. 2012;40(13):5795–818.
7. Gravel S, Chapman JR, Magill C, Jackson SP. DNA helicases Sgs1 and BLM promote DNA double-strand break resection. Genes Dev. 2008;22(20):2767–72.
8. Mimitou EP, Symington LS. Sae2, Exo1 and Sgs1 collaborate in DNA double-strand break processing. Nature. 2008;455(7214):770–4.
9. Zhu Z, Chung WH, Shim EY, Lee SE, Ira G. Sgs1 helicase and two nucleases Dna2 and Exo1 resect DNA double-strand break ends. Cell. 2008;134(6):981–94.
10. Mimitou EP, Symington LS. DNA end resection: many nucleases make light work. DNA Repair (Amst). 2009;8(9):983–95.
11. Cannavo E, Cejka P, Kowalczykowski SC. Relationship of DNA degradation by Saccharomyces cerevisiae exonuclease 1 and its stimulation by RPA and Mre11-Rad50-Xrs2 to DNA end resection. Proc Natl Acad Sci U S A. 2013;110(18):E1661–1668.
12. Haldenby S, White MF, Allers T. RecA family proteins in archaea: RadA and its cousins. Biochem Soc Trans. 2009;37(Pt 1):102–7.
13. White MF. Homologous recombination in the archaea: the means justify the ends. Biochem Soc Trans. 2011;39(1):15–9.
14. Bartlett EJ, Brissett NC, Doherty AJ. Ribonucleolytic resection is required for repair of strand displaced nonhomologous end-joining intermediates. Proc Natl Acad Sci U S A. 2013;110(22):E1984–1991.
15. Blackwood JK, Rzechorzek NJ, Bray SM, Maman JD, Pellegrini L, Robinson NP. End-resection at DNA double-strand breaks in the three domains of life. Biochem Soc Trans. 2013;41(1):314–20.
16. Matsui E, Kawasaki S, Ishida H, Ishikawa K, Kosugi Y, Kikuchi H, et al. Thermostable flap endonuclease from the archaeon, Pyrococcus horikoshii, cleaves the replication fork-like structure endo/exonucleolytically. J Biol Chem. 1999;274(26):18297–309.
17. Fujikane R, Komori K, Shinagawa H, Ishino Y. Identification of a novel helicase activity unwinding branched DNAs from the hyperthermophilic archaeon. Pyrococcus furiosus J Biol Chem. 2005;280(13):12351–8.
18. Constantinesco F, Forterre P, Koonin EV, Aravind L, Elie C. A bipolar DNA helicase gene, herA, clusters with rad50, mre11 and nurA genes in thermophilic archaea. Nucleic Acids Res. 2004;32(4):1439–47.
19. Constantinesco F, Forterre P, Elie C. NurA, a novel 5'-3' nuclease gene linked to rad50 and mre11 homologs of thermophilic Archaea. EMBO Rep. 2002;3(6):537–42.
20. Manzan A, Pfeiffer G, Hefferin ML, Lang CE, Carney JP, Hopfner KP. MlaA, a hexameric ATPase linked to the Mre11 complex in archaeal genomes. EMBO Rep. 2004;5(1):54–9.
21. Wei T, Zhang S, Zhu S, Sheng D, Ni J, Shen Y. Physical and functional interaction between archaeal single-stranded DNA-binding protein and the 5'-3' nuclease NurA. Biochem Biophys Res Commun. 2008;367(3):523–9.
22. Zhang S, Wei T, Hou G, Zhang C, Liang P, Ni J, et al. Archaeal DNA helicase HerA interacts with Mre11 homologue and unwinds blunt-ended double-stranded DNA and recombination intermediates. DNA Repair (Amst). 2008;7(3):380–91.
23. Hopkins BB, Paull TT. The P. furiosus mre11/rad50 complex promotes 5' strand resection at a DNA double-strand break. Cell. 2008;135(2):250–60.
24. Quaiser A, Constantinesco F, White MF, Forterre P, Elie C. The Mre11 protein interacts with both Rad50 and the HerA bipolar helicase and is recruited to DNA following gamma irradiation in the archaeon Sulfolobus acidocaldarius. BMC Mol Biol. 2008;9:25.
25. Chae J, Kim YC, Cho Y. Crystal structure of the NurA-dAMP-Mn2+ complex. Nucleic Acids Res. 2011;40(5):2258–70.
26. Blackwood JK, Rzechorzek NJ, Abrams AS, Maman JD, Pellegrini L, Robinson NP. Structural and functional insights into DNA-end processing by the archaeal HerA helicase-NurA nuclease complex. Nucleic Acids Res. 2012;40(7):3183–96.
27. Kish A, DiRuggiero J. Rad50 is not essential for the Mre11-dependent repair of DNA double-strand breaks in Halobacterium sp. strain NRC-1. J Bacteriol. 2008;190(15):5210–6.
28. Delmas S, Shunburne L, Ngo HP, Allers T. Mre11-Rad50 promotes rapid repair of DNA damage in the polyploid archaeon Haloferax volcanii by restraining homologous recombination. PLoS Genet. 2009;5(7):e1000552.
29. Fujikane R, Ishino S, Ishino Y, Forterre P. Genetic analysis of DNA repair in the hyperthermophilic archaeon, Thermococcus kodakaraensis. Genes Genet Syst. 2010;85(4):243–57.
30. Lestini R, Duan Z, Allers T. The archaeal Xpf/Mus81/FANCM homolog Hef and the Holliday junction resolvase Hjc define alternative pathways that are essential for cell viability in Haloferax volcanii. DNA Repair (Amst). 2010;9(9):994–1002.
31. Liang PJ, Han WY, Huang QH, Li YZ, Ni JF, She QX, et al. Knockouts of RecA-Like Proteins RadC1 and RadC2 Have Distinct Responses to DNA Damage Agents in Sulfolobus islandicus. J Genet Genomics. 2013;40(10):533–42.
32. Hong Y, Chu M, Li Y, Ni J, Sheng D, Hou G, et al. Dissection of the functional domains of an archaeal Holliday junction helicase. DNA Repair (Amst). 2012;11(2):102–11.
33. Zhang C, Tian B, Li S, Ao X, Dalgaard K, Gokce S, et al. Genetic manipulation in Sulfolobus islandicus and functional analysis of DNA repair genes. Biochem Soc Trans. 2013;41(1):405–10.
34. Zheng T, Huang Q, Zhang C, Ni J, She Q, Shen Y. Development of a simvastatin selection marker for a hyperthermophilic acidophile, Sulfolobus islandicus. Appl Environ Microbiol. 2012;78(2):568–74.
35. Guo L, Brugger K, Liu C, Shah SA, Zheng H, Zhu Y, et al. Genome analyses of Icelandic strains of Sulfolobus islandicus, model organisms for genetic and virus-host interaction studies. J Bacteriol. 2011;193(7):1672–80.

36. She Q, Zhang C, Deng L, Peng N, Chen Z, Liang YX. Genetic analyses in the hyperthermophilic archaeon Sulfolobus islandicus. Biochem Soc Trans. 2009;37(Pt 1):92–6.

37. Deng L, Zhu H, Chen Z, Liang YX, She Q. Unmarked gene deletion and host-vector system for the hyperthermophilic crenarchaeon Sulfolobus islandicus. Extremophiles. 2009;13(4):735–46.

38. Zhang C, Guo L, Deng L, Wu Y, Liang Y, Huang L, et al. Revealing the essentiality of multiple archaeal pcna genes using a mutant propagation assay based on an improved knockout method. Microbiology. 2010;156(Pt 11):3386–97.

39. Peng N, Deng L, Mei Y, Jiang D, Hu Y, Awayez M, et al. A synthetic arabinose-inducible promoter confers high levels of recombinant protein expression in hyperthermophilic archaeon Sulfolobus islandicus. Appl Environ Microbiol. 2012;78(16):5630–7.

40. Jacobs KL, Grogan DW. Rates of spontaneous mutation in an archaeon from geothermal environments. J Bacteriol. 1997;179(10):3298–303.

41. Contursi P, Jensen S, Aucelli T, Rossi M, Bartolucci S, She Q. Characterization of the Sulfolobus host-SSV2 virus interaction. Extremophiles. 2006;10(6):615–27.

42. Samson RY, Xu Y, Gadelha C, Stone TA, Faqiri JN, Li D, et al. Specificity and function of archaeal DNA replication initiator proteins. Cell Rep. 2013;3(2):485–96.

43. Rzechorzek NJ, Blackwood JK, Bray SM, Maman JD, Pellegrini L, Robinson NP. Structure of the hexameric HerA ATPase reveals a mechanism of translocation-coupled DNA-end processing in archaea. Nat Commun. 2014;5:5506.

44. Wei T, Zhang S, Hou L, Ni J, Sheng D, Shen Y. The carboxyl terminal of the archaeal nuclease NurA is involved in the interaction with single-stranded DNA-binding protein and dimer formation. Extremophiles. 2011;15(2):227–34.

45. Taylor A, Smith GR. Unwinding and rewinding of DNA by the RecBC enzyme. Cell. 1980;22(2 Pt 2):447–57.

46. Symington LS, Gautier J. Double-strand break end resection and repair pathway choice. Annu Rev Genet. 2011;45:247–71.

47. Neuwald AF, Aravind L, Spouge JL, Koonin EV. AAA+: A class of chaperone-like ATPases associated with the assembly, operation, and disassembly of protein complexes. Genome Res. 1999;9(1):27–43.

48. James JA, Escalante CR, Yoon-Robarts M, Edwards TA, Linden RM, Aggarwal AK. Crystal structure of the SF3 helicase from adeno-associated virus type 2. Structure. 2003;11(8):1025–35.

49. Tuteja N, Tuteja R. Unraveling DNA helicases. Motif, structure, mechanism and function. Eur J Biochem. 2004;271(10):1849–63.

Downregulation of microRNA-100 protects apoptosis and promotes neuronal growth in retinal ganglion cells

Ning Kong[1,2], Xiaohe Lu[1*] and Bin Li[2]

Abstract

Background: Retinal ganglion cells (RGCs) are preferentially lost in glaucoma or optic neuritis. In the present study, we investigated the protective effect of mircoRNA 100 (miR-100) against oxidative stress induced apoptosis in RGC-5 cells.

Results: Rat RGC-5 cells were cultured in plates and H_2O_2 was added to induce oxidative stress. TUNEL assay and qRT-PCR showed H_2O_2 induced apoptosis and up-regulated miR-100 in a dose-dependent manner in RGC-5 cells. Conversely, lentiviral-mediated miR-100 down-regulation protected H_2O_2 induced apoptosis, promoted neurite growth and activated AKT/ERK and TrkB pathways through phosphorylation. Luciferase assay confirmed that IGF1R was directly regulated by miR-100 in RGC-5 cells, and siRNA-mediated IGF1R knockdown activated AKT protein through phosphorylation, down-regulated miR-100, therefore exerted a protective effect on RGC-5 apoptosis.

Conclusion: Down-regulating miR-100 is an effective method to protect H_2O_2 induced apoptosis in RGC-5 cells, possible associated with IGF1R regulation.

Keywords: Retinal ganglion, miR-100, Oxidative stress, Apoptosis, IGF-1

Background

Retinal ganglion cells (RGCs) are a group of specialized sensory neurons in the central nervous system, rested in the inner layer of the retina. RGCs capture visual signals from bipolar and amacrine cells through the photoreceptors on the outer retina, and electrically transmit that information to the brain. In glaucoma patients, increased intraocular pressure damage the cell bodies as well as the axons of RGCs, resulting in permanent and irreversible blindness [1]. The degeneration of retina in glaucoma patients is often accompanied by apoptosis of RGCs, which may be caused by excitotoxicity, hypoxia, or oxidative stress [2-4]. Many of the molecular pathways, including Akt, ERK, insulin-like growth factor 1 (IGF-1) and its receptor IGF1R, brain-derived neurotrophic factor (BDNF) and its tyrosine kinase receptor TrkB, repulsive guidance molecule, RGMa and its receptor neogenin, are actively involved in the development or the protection/

regeneration of sensory neurons including RGC [5-13]. However, though many of the molecular pathways have been identified to protect RGCs from degeneration and induce optical nerve regeneration [14], the exact mechanism of RGC damage, or feasible clinical strategy for RGC regeneration are largely unknown.

MicroRNAs (miRNAs) are a group of noncoding RNAs with ~20 nucleotides, that inhibit endogenous gene expression through translational cleavage [15]. In both animal and human visual systems, miRNAs have been demonstrated to be involved in various aspects of retina development, including retinogenesis, retinal homeostasis, or retinal damage. For example, studies in mutant mice have shown that, miR-183/96/182 clusters and miR132/212 were essential for the synaptic development in retina [16-18], and miR-124 was critical for the maturation of cone cells [19] in retina. Study also showed that miR-218 inhibits Robo/Slit pathway to maintain normal vascular function in retina [20]. In retinal ganglion cells, a recent study specifically demonstrated that miR-132 was involved in BDNF-mediated retinal ganglion axonal branching and maturation [21]. Specifically, emerging evidences

* Correspondence: libin_bin@aol.com
[1]Department of Ophthalmology, Zhujiang Hospital, Southern Medical University, Guangzhou 510280, Guangdong Province, China
Full list of author information is available at the end of the article

demonstrate that miRNAs are highly associated with many aspects of retinal degeneration [22,23], yet the exact mechanisms of miRNAs on regulating retinal apoptosis or degeneration are largely unknown. MicroRNA-100 (miR-100), a member of miR-99 family (including miR-99a, miR-99b, miR-100), is a key apoptotic regulator in various cell types [24,25]. In retina, miR-100 was found to be associated with diabetic retinopathy, a common retinal disease leading to total blindness [26]. However, the function role of miR-100 in mediating retinal development or pathology remains elusive.

In the present study, we used an *in vitro* culture system to induce oxidative stress by H_2O_2 application in rat ganglion cell line RGC-5 cells. We found that microRNA 100 (miR-100) was specifically upregulated in RGC-5 cells upon apoptosis. We then investigated the anti-apoptosis effect of down-regulating miR-100, as well as the association between miR-100 down-regulation and apoptosis-related pathways, Akt/ERK/TrkB and IGF1R, in retina ganglion cells.

Methods
Ethics statement
All experimental procedures are approved by the Ethics Committee at Guangzhou Panyu Central Hospital, Guangzhou, China.

RGC-5 cell culture
The RGC-5 cell line, a transformed retinal ganglion progenitor cell line originally generated by Dr. N. Agarwal at the University of North Texas Health Science in the United States, was obtained from the Zhonghan Ophthalmic Center in Guangzhou, China. RGC-5 cells were cultured in low-glucose Dulbecco's modified Eagle's medium (DMEM, Gibco, USA) containing 10% fetal calf serum (FBS, Sigma Aldrich, USA), 100 U/ml penicillin and 100 μg/ml streptomycin in a 37°C tissue culture chamber supplied with 95% air and 5% CO_2. The RGC-5 cells were passaged by trypsinization every 2 to 3 days.

Apoptosis induction
To induce apoptosis in RGC-5 cell, an oxidative stress model was used. Briefly, RGC-5 cells were seeded into 24-well plates at concentration of $1 \times 10^5/cm^2$. Various concentrations of hydrogen peroxide (H_2O_2) (100 μm ~1000 μm) were added into the culture for 24 hours.

TUNEL assay
RGC-5 cells were fixed with 4% paraformaldehyde (PFA) in phosphate-buffered saline (PBS) for 20 min, and permeabilized with 1% Triton X-100 for 10 min. TUNEL staining was conducted with an In-Situ Apoptosis Detection Kit (Chemicon, USA) according to the manufacturer's protocol. A primary antibody against neural progenitor

marker Brn3a (1:500, Santa Cruz, USA) was also used to identify RGC-5 cells. Imaging was done on an Axiovert 200 fluorescence microscope (Carl Zeiss, Germany). For each condition, 5 ~ 8 regions (~100 cells per region) were randomly taken for analysis. The percentage of TUNEL-positive RGC-5 cells were calculated against all brn3a positive cells and normalized to the control condition. Each experiment was repeated at least 3 times.

RNA isolation and quantitative real-time reverse transcription-PCR (qRT-PCR)
Cultured RGC-5 cells were trypsinized and collected. RNA and miRNA fractions were then extracted using a Trizol RNA purification Kit according to manufacturer's instruction (Qiagen, USA). Total RNA concentrations were confirmed with a NanoDrop ND-1000 spectrophotometer (NanoDrop Technologies, USA) at 260 and 280 nm (A260/280), and analyzed with an Agilent 2100 Bioanalyzer (Agilent Technologies, USA). Quantitative real-time reverse transcription-PCR (qRT-PCR) was performed by a TaqMan miRNA Assay according to manufacturer's instruction (Applied Biosystems, USA). The amplification conditions were 35 cycles of 20 s at 97°C and 2 min at 56°C. The internal housing keeping genes were U6 for miR-100, and GAPDH for insulin-like growth factor 1 receptor (IGF1R) respectively.

Lentivirus transfection
The oligonucleotides of rno-miR-100 inhibitor, rno-miR-100 mimics and non-specific control were puchased from Ribo-Bio (Ribo-Bio, Shanghai, China). The coding sequences were then amplified and cloned into pCDH-CMV-MCS-EF1-coGFP constructs (System Biosciences, USA) to construct miR-100 inhibitor vector (miR100-Inhibitor), miR-100 mimics vector (miR100-mimics) and non-specific miRNA vector (miR-NC). The lentiviral vectors were co-transfected with pPACK packaging plasmid into 293 T cells. The viral products were then collected and titered. The transfection of miR-100-Inhibitor or miR-NC into RGC-5 cells was conducted by a Lipofectamine 2000 reagent according to manufacturer's recommendation (Invitrogen, USA). The culture medium was changed 24 hours after transfection.

Immunocytochemistry
RGC-5 cells were fixed with 10% neutral-buffered formalin containing 5% methanol for 10 minutes, followed by washing in PBS (3 × 10 mins). The cells were then treated with permeabilization solution containing 0.3% trition X-100 and 3% horse serum in PBS for 30 minutes, followed by washing PBS (3 × 10 mins). The cells were incubated with primary antibody of Thy-1 (1:250 in permeabilization solution, Sigma Aldrich, USA) at 4°C overnight in a humidified box. On second day, cells were

incubated with secondary antibody of Alexa-Fluor 488 (1:500 in PBS) for 12 hours. RGC-5 cells were also stained with the fluorescent nuclear binding label 4',6-diamidino-2-phenylindole (DAPI; 500 ng/mL).

Western blot analysis

RGC-5 lysates of were prepared with a lysis buffer containing 50 mM Tris (pH 7.6), 150 mM NaCl, 1 mM EDTA, 10% glycerol, and 0.5% NP-40 and protease inhibitor cocktail (Invitrogen, USA). The total protein were then separated on 10% SDS-PAGE gel and transferred to the nitrocellulose membranes. We used primary antibodies against, Thy-1 (1:2000, Santa Cruz Biotechnology, USA), ERK1/2 (1:1000, Cell Signaling Technology, USA), phospho-Erk1/2 (pERK1/2, 1:2000, Cell Signaling Technology, USA), PI3k3 /serine-threonine kinase (AKT, 1:1000, Cell Signaling Technology, USA), phosphor-AKT (pAKT, 1:1000, Cell Signaling Technology, USA), TrkB (1:1000, Cell Signaling Technology, USA), phosphorylated TrkB (1:200, Cell Signaling Technology, USA) and IGF1R (1:200, Santa Cruz Biotechnology, USA). Horseradish peroxidase-conjugated secondary antibodies (Bio-Rad, USA) were used and Actin was used as internal control. The western blots were visualized with an enhanced chemiluminesence system (Amersham Biosciences, USA) according to the manufacturer's protocol. Band intensities normalized to the total protein immuno-precipitation under each control condition and quantified with an ImageJ software (NIH, USA).

Luciferase reporter assay

RGC-5 cells were collected and regular PCR was performed to amplify the cDNAs of wild-type 3'-UTR of IGF1R. A mutant (MT) 3'-UTR of IGF1R (with modified binding site to rno-miR-100, Figure 1A) was also created by Site-Directed Mutagenesis Kit (SBS Genetech, China). The amplified cDNAs were then inserted into a pmiR-REPORT luciferase reporter vector (Ambion, USA) to construct Luc- IGF1R and Luc- IGF1R -mu vectors. The sequences of the vectors were verified by DNA sequencing. A non-specific control luciferase vector, Luc-Ctrl was also constructed. All three vectors were co-transfected with β-galactosidase and miR-100 mimics (Ribo-Bio, Shanghai, China) into HEK293 cells in 6-well plates by using Lipofectamine 2000 according to the manufacturer's protocol. Twenty-four hours after transfection, the activity was assessed by a luciferase reporter assay system (Promega, USA) according to the manufacturer's recommendtaion. All luciferase activities were normalized to the β-galactosidase signal under luc-Ctrl condition.

siRNA transfection

The gene silencing IGF1R siRNA (IGF1R-siRNA) and non-specific control siRNA (NC-siRNA) were purchased

from Stanta Cruz (Santa Cruz Biotechnology, USA). The transfection of siRNAs in RGC-5 cells was conducted with a Lipofectamine 2000 reagent according to manufacturer's recommendation. After reaching 50% ~60% confluence, RGC-5 cells were transfected with IGF1R-siRNA (100 nM) or NC-siRNA (100 nM). The effect of siRNA on knocking down IGF1R protein, as well as other related proteins were examined by western blotting 48 hours after transfection.

Statistical analysis

All data were shown as the mean ± S.E.M. For statistical analysis, a windows-based SPSS software was used. Comparison was made with student's t-test and the difference was termed to be significant if $P < 0.05$. All experiments were conducted in triplicates.

Results

H_2O_2 induced apoptosis and upregulated miR-100 in RGC-5 cells in a dose-dependent manner

First, we investigated the effect of oxidative stress on RGC-5 cell apoptosis. In our study, we cultured RGC-5 cells in 6-well plate for 24 hours, then included different concentration of H_2O_2 (100, 200, 400, 500 and 1000 µM) in RGC-5 cell culture for another 24 hours, followed by immunocytochemistry. The apoptosis was assessed by TUNEL assay. RGC-5 cells were identified by an antibody against Brn3a, a RGC-5 specific marker. Our results showed that H_2O_2 at the concentrations of 400 and 1000 µM induced significant apoptosis among RGC-5 cells (Figure 2A). Further analysis by quantifying the percentage of TUNEL-positive cells among all Brn3a-positive RGC-5 cells revealed that the apoptosis in RGC-5 cells was induced by H_2O_2 in a dose dependent manner (Figure 2B, *: $P < 0.05$). We then looked at the effect of H_2O_2-induced apoptosis on the expression level of miR-100 in RGC-5 cells. Our results demonstrated that miR-100 was significantly up-regulated, and also in a dose-dependent manner by the application of H_2O_2 in RGC-5 cells (Figure 2C, *: $P < 0.05$).

Downregulation of miR-100 reduces H_2O_2 induced apoptosis and promoted neurite growth in RGC-5 cells

We then investigated whether miR-100 played a functional role in modulating H_2O_2-induced apoptosis in RGC-5 cells. For that purpose, we constructed lentiviral vector expressing the inhibitory oligonucleotides of miR-100 (miR100-inhibitor) to specifically knock down the endogenous expression of miR-100. The efficiency of down-regulating miR-100 by lentivirus was verified by qRT-PCR. In that experiment, RGC-5 cells were transfected with either miR100-inhibitor, or its non-specific miRNA (miR-NC) for 24 hours. The results of qRT-PCR showed that the endogenous expression level of miR-100 was

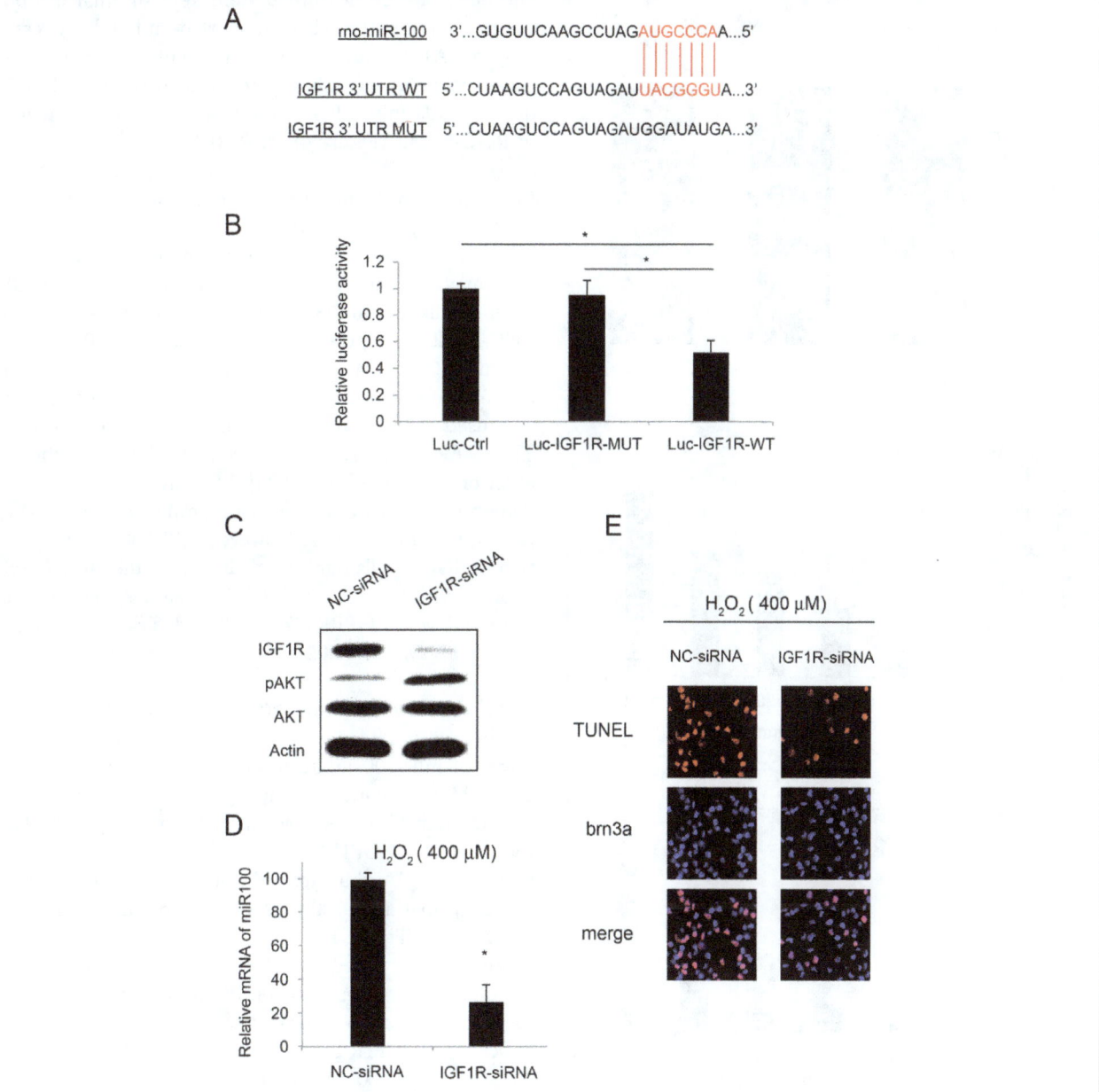

Figure 1 miR-100 interacted with IGF1R in RGC-5 cells. (A) Schematic diagram was shown for the predicted binding between rno-miR-100 and IGF1R 3'-UTR. The mutated 3'-UTR of IGF1R (IGF1R-mu) was also demonstrated. **(B)** In a luciferase report assay, HEK 293 T cells were transfected with pmiR-REPORT control vector (Luc-Ctrl), IGF1R vector with mutated 3'-UTR (Luc- IGF1R -mu) or IGF1R with wild-type 3'-UTR (Luc- IGF1R), along with β-galactosidase and miR-100 mimics for 24 hours. Luciferase signals were measured and normalized to the signal of control vector. (*: $P <0.05$). **(C)** RGC-5 cells were treated with IGF1R-siRNA or its non-specific siRNA (NC-siRNA), followed by western blotting on AKT/pAKT and IGF1R in 48 hours. RGC-5 cells were pre-treated with IGF1R-siRNA (100 nM) or its non-specific siRNA (NC-siRNA, 100 nM) for 24 hours, followed by H_2O_2 (400 μM) treatment for another 24 hours. **(D)** The mRNA expression level of miR-100 was assessed by qRT-PCR (*: $P <0.05$). **(E)** RGC-5 apoptosis was assessed by TUNEL staining.

significantly and specifically down-regulated by miR100-inhibitor (Figure 3A, *: $P <0.05$).

To evaluate the effect of down-regulating miR-100 on RGC-5 apoptosis, we pre-treated RGC-5 cells with either miR100-inhibitor or miR-NC for 24 hours, followed by H_2O_2 (400 μM) treatment for another 24 hours. Our results of TUNEL staining showed that significantly less apoptosis

were observed in miR-100 down-regulated RGC-5 cells than in control RGC-5 cells (Figure 3B). Quantification results demonstrated that the percentage of apoptotic cells reduced from 55.5% in control RGC-5 cells to around 30% in cells with miR-100 down-regulation (Figure 3C, *: $P <0.05$).

Interestingly, once the culture period was extended to 14 days after lentiviral transfection, we discovered

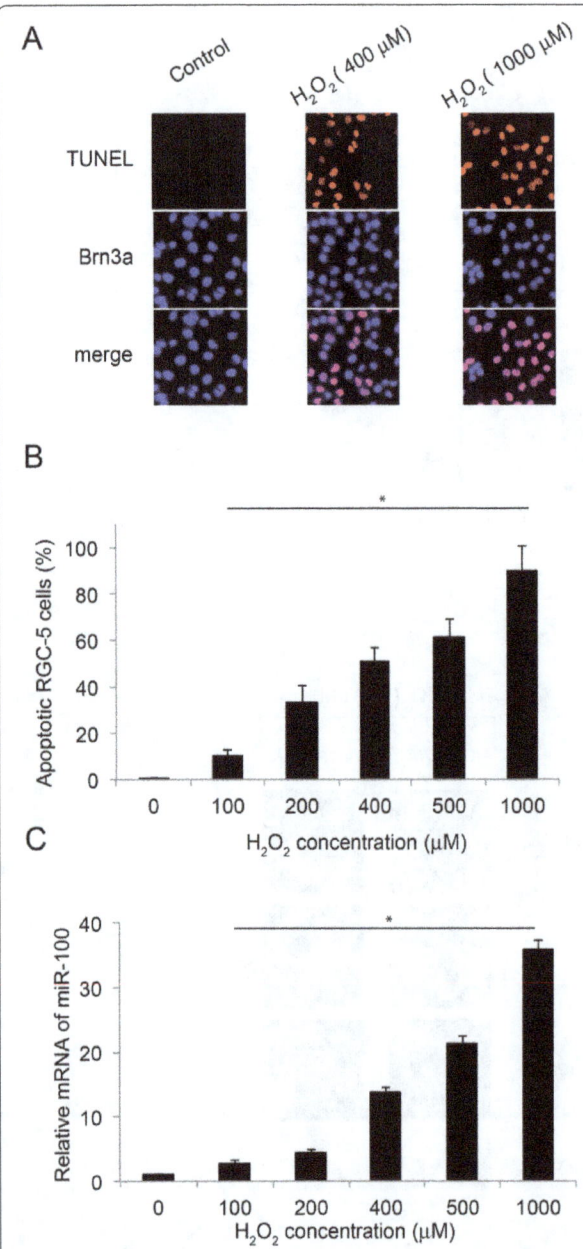

Figure 2 H$_2$O$_2$ induced apoptosis and upregulated miR-100 in RGC-5 cells. RGC-5 cells were treated with various concentrations of H$_2$O$_2$ for 24 hours to induce apoptosis. **(A)** Representative fluorescent images were shown for the untreated RGC-5 cells (control), and RGC-5 cells treated with 400 µM or 1000 µM H$_2$O$_2$. An antibody against Brn3a (blue) was used to identify RGC-5 cells, and TUNEL staining was applied to identify apoptotic cells among them. **(B)** Quantification of the percentage of apoptotic RGC-5 cells (*: P <0.05, as compared to control). **(C)** The expression levels of miR-100, corresponding to the application of various concentrations H$_2$O$_2$, were measured by qRT-PCR (*: P <0.05, as compared to control).

without miR-100 downregulation kept original morphology of progenitor cell with short peripheral processes (Figure 4A). Quantitative measurement with ImagJ software (http://imagej.nih.gov/ij/) confirmed the effect of down-regulating miR-100 on promoting neurite length in RGC-5 cells (Figure 4B, *: P <0.05).

Downregulation of miR-100 phosphorylated AKT/ERK/TrkB pathways in RGC-5 cells

We then investigated the intracellular signaling pathways that might be involved in the protection of miR-100 down-regulation on RGC-5 apoptosis. In order to do that, RGC-5 cells were pre-treated with either miR100-inhibitor or miR-NC for 24 hours, followed by H$_2$O$_2$ (400 µM) treatment for another 24 hours. Western blotting analysis was used to measure the protein expression levels of Thy-1, ERK1/2, AKT and TrkB, as well as phosphorylation of ERK1/2, AKT and TrkB (Figure 5A). Our results demonstrated that down-regulating miR-100 did not alter the expression of Thy-1. However, miR-100 downregulation in RGC-5 cells significantly increased the phosphorylation in ERK1/2, AKT and TrkB, while keeping the total protein levels of ERK1/2, AKT and TrkB unchanged (Figure 5B-D, *: P <0.05).

IGF1R was involved in miR-100 regulation in RGC-5

Finally, we intended to explore the possible molecular target of miR-100 modulation in RGC-5 cells. Through online bioinformatic screening (Target Scan, http://www.targetscan.org/), we noticed that, insulin-like growth factor 1 receptor (IGF1R) was a likely target of rat miR-100 (Figure 1A). We thus applied a luciferase reporter assay to verify that miR-100 was directly binding IGF1R in RGC-5 cells (Figure 1B).

We then speculated IGF1R might be directly involved in the modulation of miR-100 on RGC-5 apoptosis. In order to examine that, we constructed siRNA to specifically knock down IGF1R gene in RGC-5 cells (IGF1R-siRNA). After applying 100 nM IGF1R-siRNA or its non-specific control siRNA (NC-siRNA) in RGC-5 cells for 48 hours, we examined the knocking down efficiency by western blotting analysis. Our results showed that IGF1R protein was significantly reduced by IGF1R-siRNA (Figure 1C). The results also showed that down-regulation of IGF1R induced phosphorylation of AKT, whereas the total protein of AKT was unchanged (Figure 1C).

Finally, we examined the effect of knocking down IGF1R on miR-100 regulation and RGC-5 apoptosis. We pre-treated RGC-5 cells with either IGF1R-siRNA (100 nM) or NC-siRNA (100 nM) for 24 hours, followed by H$_2$O$_2$ (400 µM) treatment for another 24 hours. The results showed that the mRNA expression level of miR-100, presumably up-regulated upon apoptosis, was significantly down-regulated by knocking down IGF1R in RGC-5 cells

that the morphology of RGC-5 cells was dramatically modified by miR-100 down-regulation. The RGC-5 cells with down-regulated miR-100 extended the peripherals to be long and neurite-like processes, whereas the cells

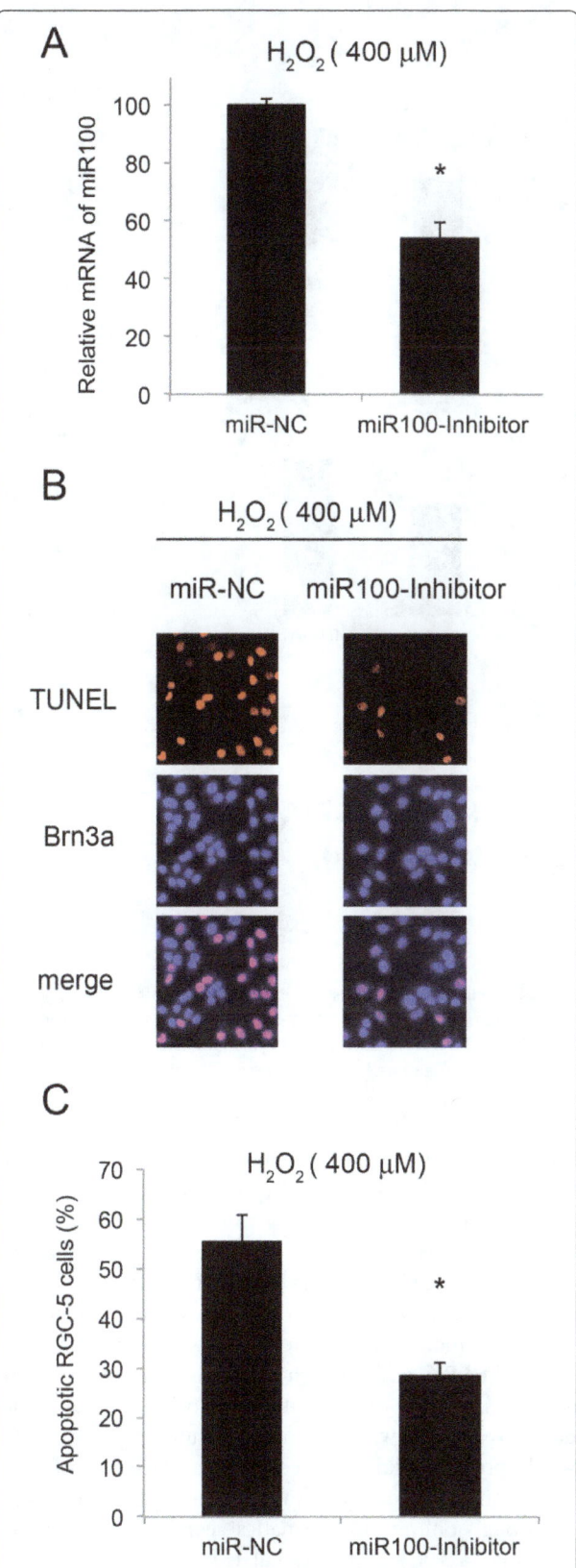

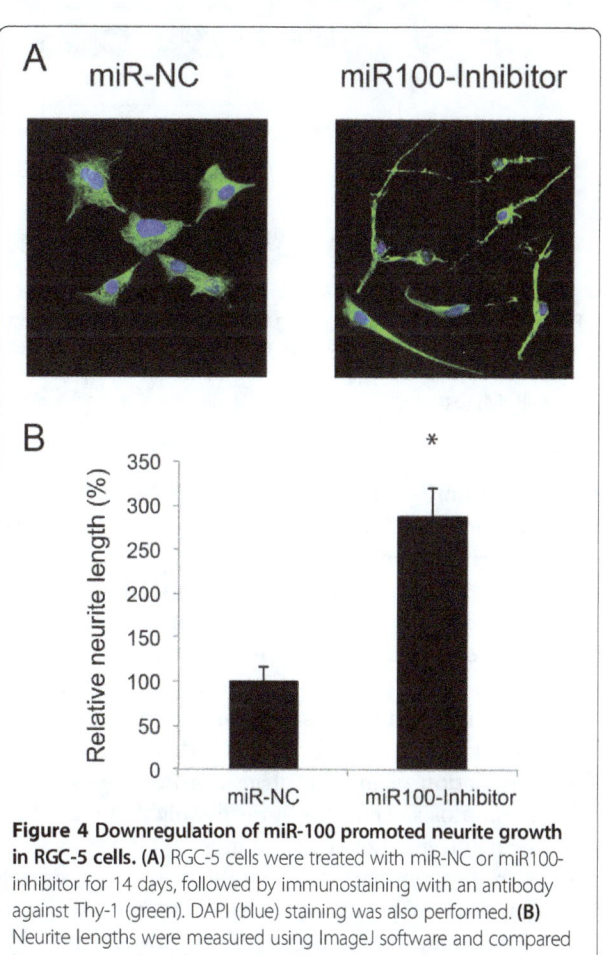

Figure 3 Downregulation of miR-100 protected apoptosis in RGC-5 cells. RGC-5 cells were pre-treated with lentiviral vector of miR100-inhibitor, or its non-specific miRNA vector (miR-NC) for 24 hours, followed by H_2O_2 (400 µM) treatment for another 24 hours. **(A)** The efficiency of down-regulating miR-100 by lentiviral vectors was evaluated by qRT-PCR (*: P <0.05). The effect of pre-treating RGC-5 cells with miR-100 down-regulation on oxidative stress induced apoptosis was then examined by TUNEL assay. **(B)** TUNEL staining was performed and representative fluorescent images were shown for the RGC-5 cells pre-treated with miR-NC, or miR100-inhibitor, and then H_2O_2 400 (µM). **(C)** The percentages of apoptotic RGC-5 cells were also quantified (*: P <0.05).

(Figure 1D). Moreover, the percentage of apoptotic RGC-5 was markedly reduced by knocking down IGF1R, from 78% to 33% (Figure 1E).

Discussion

Oxidative stress has been suggested to be a major mechanism causing RGC degeneration [1], and the application of H_2O_2 is known to induce apoptosis among cultured RGCs. In the present study, we utilized an *in vitro* culture system to induce RGC-5 apoptosis through H_2O_2 and

Figure 4 Downregulation of miR-100 promoted neurite growth in RGC-5 cells. (A) RGC-5 cells were treated with miR-NC or miR100-inhibitor for 14 days, followed by immunostaining with an antibody against Thy-1 (green). DAPI (blue) staining was also performed. **(B)** Neurite lengths were measured using ImageJ software and compared between miR100-inhibitor treated RGC-5 cells and miR-NC treated RGC-5 cells (*, P <0.05).

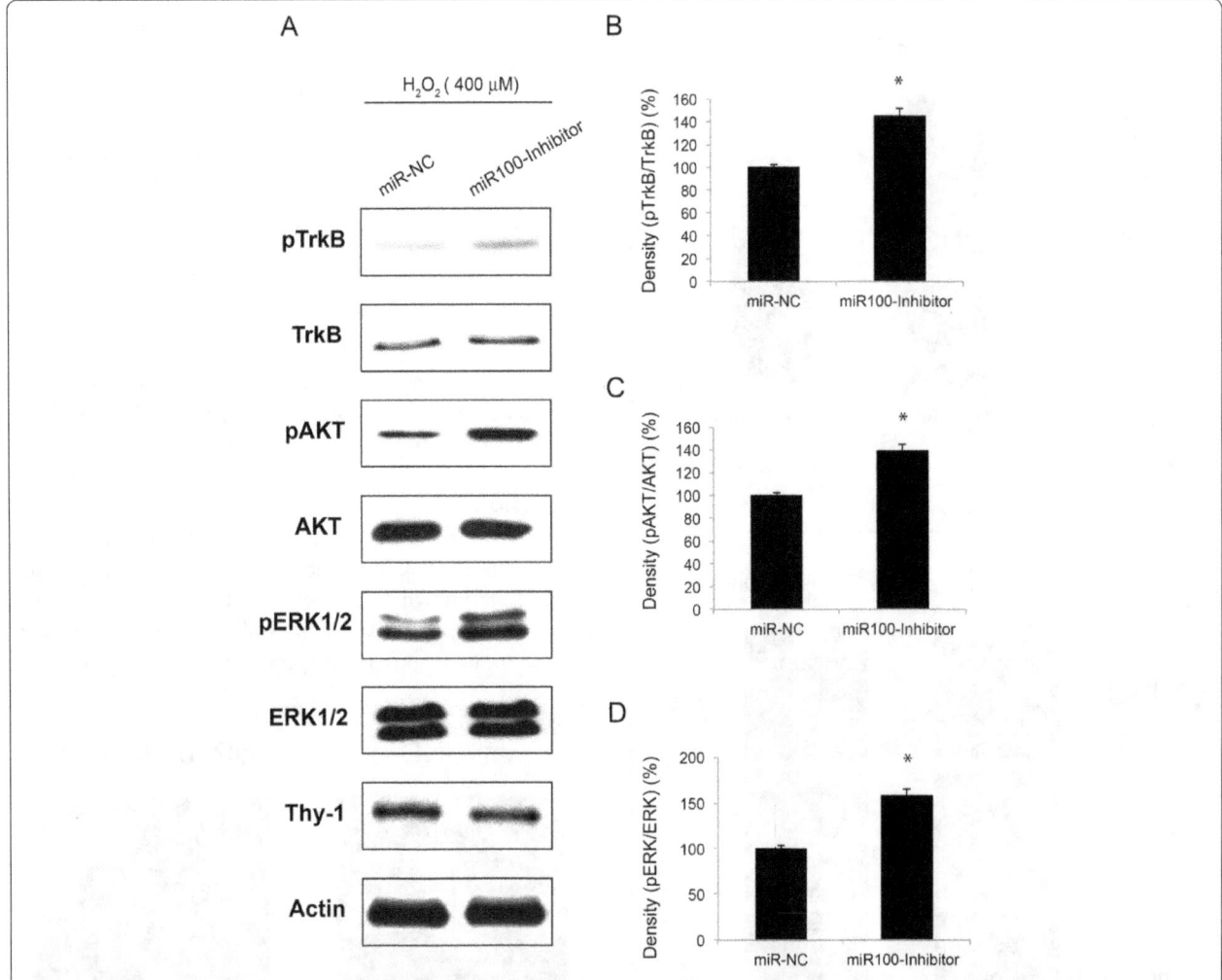

Figure 5 Effects of down-regulating miR-100 on AKT/ERK/TrkB pathways in RGC-5 cells. RGC-5 cells were pre-treated with miR100-inhibitor, or miR-NC for 24 hours, followed by H_2O_2 (400 μM) treatment for another 24 hours. **(A)** Western blotting analysis was performed to examine the protein expressions of Thy-1, ERK1/2, phspho-ERK1/2 (pERK1/2), AKT, phospho-AKT (pAKT), TrkB and phosphorylated-TrkB (pTrkB), in miR100-inhibitor pre-treated and miR-NC pre-treated RGC-5 cells. Semi-quantitative examination was performed to compare the phosphorylation levels of TrkB **(B)**, AKT **(C)** and ERK **(D)**. (*, $P <0.05$)

found that miR-100 was up-regulated by H_2O_2 in a dose-dependent manner. It has been commonly shown in the literature that miR-100 is an active factor in cancer regulation, modulating carcinoma development, metastasis or apoptosis [25,27]. In retina, though miR-100 is expressed, little is known of its molecular mechanism [28]. In the present study, not only did we show an active change in expression profile of miR-100 upon H_2O_2-induced apoptosis, but we also demonstrated that down-regulation of miR-100 had a protective effect on RGC-5 apoptosis. Thus, these results highlighted a first ever report on functional role of miR-100 in retina.

Also in the present study we showed that, during H_2O_2-induced apoptosis, down-regulating miR-100 activated TrkB signaling pathways and its downstream Akt/ERK pathways, through phosphorylation. Thus, it is likely that miR-100 downregulation may serve a role of TrkB

agonist to exert a protective effect on H_2O_2-induced apoptosis in RGC-5 cells. This hypothesis was further supported by our results showing miR-100 down-regulation promoted neurite growth, possibly induced neural differentiation in RGC-5 cells, as TrkB activation is known to induce maturation in retina [29]. However, to elucidate the exact molecular interaction between miR-100 and TrkB/Akt/ERK pathways in protecting retinal ganglion apoptosis, future experiments of blocking TrkB or its downstream pathways upon the inhibition of miR-100 would provide solid evidence on direct targeting of miR-100 on anti-apoptotic pathways in retinal ganglion cells.

Through online bioinformatics search, as well as our luciferase assay and functional experiments (Figure 1), we further revealed that IGF1R was very likely to be directly involved in the modulation of miR-100 on H_2O_2-induced apoptosis in RGC-5 cells. Furthermore, our results demonstrated that siRNA-mediated IGF1R knockdown activated Akt pathway and rescued H_2O_2-induced apoptosis. This is consistent with previous study showing that IGF1R mutant mice had increased lifespan and significant resistance to oxidative stress in retinal ganglion cells [8]. Interestingly, our experiment also showed that knocking down IGF1R was able to down-regulate miR-100. This result suggests that, instead of being downstream target of miR-100, IGF1R might be mutually interacted with miR-100 in regulating RGC apoptosis, acting through miR-100 itself or its upstream pathways. Along with our early results demonstrating the functional association between miR-100 and Akt/ERK/TrkB pathways, a much more complex molecular network might be involved in the regulation of miR-100 in retina.

Conclusions

Overall, our study identified a novel regulator, miR-100 in modulating apoptosis in retinal ganglion cells. The method of down-regulating miR-100 might help to further our understanding on the mechanisms of degeneration and regeneration in retina tissues.

Competing interest
The authors declare that they have no competing interest.

Authors' contribution
NK carried out the experiments and drafted manuscript. BL performed analysis. XL designed the study approved the final version of manuscript. All authors read and approved the final manuscript.

Acknowledgements
We thank Drs. Luli Zhang and Mingjie Song for their critical reviews of the manuscript.

Author details
[1]Department of Ophthalmology, Zhujiang Hospital, Southern Medical University, Guangzhou 510280, Guangdong Province, China. [2]Department of Ophthalmology, Guangzhou Panyu Central Hospital, Guangzhou 510280, Guangdong Province, China.

References
1. Almasieh M, Wilson AM, Morquette B, Cueva Vargas JL, Di Polo A: The molecular basis of retinal ganglion cell death in glaucoma. Prog Retin Eye Res 2012, 31(2):152–181.
2. Tezel G: Oxidative stress in glaucomatous neurodegeneration: mechanisms and consequences. Prog Retin Eye Res 2006, 25(5):490–513.
3. Arjamaa O, Nikinmaa M: Oxygen-dependent diseases in the retina: role of hypoxia-inducible factors. Exp Eye Res 2006, 83(3):473–483.
4. Casson RJ: Possible role of excitotoxicity in the pathogenesis of glaucoma. Clin Experiment Ophthalmol 2006, 34(1):54–63.
5. Kanamori A, Nakamura M, Nakanishi Y, Nagai A, Mukuno H, Yamada Y, Negi A: Akt is activated via insulin/IGF-1 receptor in rat retina with episcleral vein cauterization. Brain Res 2004, 1022(1–2):195–204.
6. Kermer P, Klocker N, Labes M, Bahr M: Insulin-like growth factor-I protects axotomized rat retinal ganglion cells from secondary death via PI3-K-dependent Akt phosphorylation and inhibition of caspase-3 In vivo. J Neurosci 2000, 20(2):2–8.
7. Averbukh E, Weiss O, Halpert M, Yanko R, Moshe R, Nephesh I, Flyvbjerg A, Yanko L, Raz I: Gene expression of insulin-like growth factor-I, its receptor

and binding proteins in retina under hypoxic conditions. Metab Clin Exp 1998, 47(11):1331–1336.
8. Holzenberger M, Dupont J, Ducos B, Leneuve P, Geloen A, Even PC, Cervera P, Le Bouc Y: IGF-1 receptor regulates lifespan and resistance to oxidative stress in mice. Nature 2003, 421(6919):182–187.
9. Klocker N, Kermer P, Weishaupt JH, Labes M, Ankerhold R, Bahr M: Brain-derived neurotrophic factor-mediated neuroprotection of adult rat retinal ganglion cells in vivo does not exclusively depend on phosphatidyl-inositol-3'-kinase/protein kinase B signaling. J Neurosci 2000, 20(18):6962–6967.
10. Kilic U, Kilic E, Jarve A, Guo Z, Spudich A, Bieber K, Barzena U, Bassetti CL, Marti HH, Hermann DM: Human vascular endothelial growth factor protects axotomized retinal ganglion cells in vivo by activating ERK-1/2 and Akt pathways. J Neurosci 2006, 26(48):12439–12446.
11. Tong M, Brugeaud A, Edge AS: Regenerated synapses between postnatal hair cells and auditory neurons. J Assoc Res Otolaryngol 2013, 14(3):321–329.
12. Koeberle PD, Tura A, Tassew NG, Schlichter LC, Monnier PP: The repulsive guidance molecule, RGMa, promotes retinal ganglion cell survival in vitro and in vivo. Neuroscience 2010, 169(1):495–504.
13. Brugeaud A, Tong M, Luo L, Edge AS: Inhibition of repulsive guidance molecule, RGMa, increases afferent synapse formation with auditory hair cells. Dev Neurobiol 2014, 74(4):457–466.
14. Pernet V, Schwab ME: Lost in the jungle: new hurdles for optic nerve axon regeneration. Trends Neurosci 2014, 37(7):381–387.
15. Pillai RS: MicroRNA function: multiple mechanisms for a tiny RNA? RNA 2005, 11(12):1753–1761.
16. Lumayag S, Haldin CE, Corbett NJ, Wahlin KJ, Cowan C, Turturro S, Larsen PE, Kovacs B, Witmer PD, Valle D, Zack DJ, Nicholson DA, Xu S: Inactivation of the microRNA-183/96/182 cluster results in syndromic retinal degeneration. Proc Natl Acad Sci U S A 2013, 110(6):E507–E516.
17. Zhu Q, Sun W, Okano K, Chen Y, Zhang N, Maeda T, Palczewski K: Sponge transgenic mouse model reveals important roles for the microRNA-183 (miR-183)/96/182 cluster in postmitotic photoreceptors of the retina. J Biol Chem 2011, 286(36):31749–31760.
18. Remenyi J, van den Bosch MW, Palygin O, Mistry RB, McKenzie C, Macdonald A, Hutvagner G, Arthur JS, Frenguelli BG, Pankratov Y: miR-132/212 knockout mice reveal roles for these miRNAs in regulating cortical synaptic transmission and plasticity. PLoS One 2013, 8(4):e62509.
19. Sanuki R, Onishi A, Koike C, Muramatsu R, Watanabe S, Muranishi Y, Irie S, Uneo S, Koyasu T, Matsui R, Cherasse Y, Urade Y, Watanabe D, Kondo M, Yamashita T, Furukawa T: miR-124a is required for hippocampal axogenesis and retinal cone survival through Lhx2 suppression. Nat Neurosci 2011, 14(9):1125–1134.
20. Small EM, Sutherland LB, Rajagopalan KN, Wang S, Olson EN: MicroRNA-218 regulates vascular patterning by modulation of Slit-Robo signaling. Circ Res 2010, 107(11):1336–1344.
21. Marler KJ, Suetterlin P, Dopplapudi A, Rubikaite A, Adnan J, Maiorano NA, Lowe AS, Thompson ID, Pathania M, Bordey A, Fulga T, Van Vactor DL, Hindges R, Drescher U: BDNF promotes axon branching of retinal ganglion cells via miRNA-132 and p250GAP. J Neurosci 2014, 34(3):969–979.
22. Loscher CJ, Hokamp K, Wilson JH, Li T, Humphries P, Farrar GJ, Palfi A: A common microRNA signature in mouse models of retinal degeneration. Exp Eye Res 2008, 87(6):529–534.
23. Loscher CJ, Hokamp K, Kenna PF, Ivens AC, Humphries P, Palfi A, Farrar GJ: Altered retinal microRNA expression profile in a mouse model of retinitis pigmentosa. Genome Biol 2007, 8(11):R248.
24. Nagaraja AK, Creighton CJ, Yu Z, Zhu H, Gunaratne PH, Reid JG, Olokpa E, Itamochi H, Ueno NT, Hawkins SM, Anderson ML, Matzuk MM: A link between mir-100 and FRAP1/mTOR in clear cell ovarian cancer. Mol Endocrinol 2010, 24(2):447–463.
25. Shi W, Alajez NM, Bastianutto C, Hui AB, Mocanu JD, Ito E, Busson P, Lo KW, Ng R, Waldron J, O'Sullivan B, Liu FF: Significance of Plk1 regulation by miR-100 in human nasopharyngeal cancer. Int J Cancer 2010, 126(9):2036–2048.
26. Kovacs B, Lumayag S, Cowan C, Xu S: MicroRNAs in early diabetic retinopathy in streptozotocin-induced diabetic rats. Invest Ophthalmol Vis Sci 2011, 52(7):4402–4409.
27. Leite KR, Sousa-Canavez JM, Reis ST, Tomiyama AH, Camara-Lopes LH, Sanudo A, Antunes AA, Srougi M: Change in expression of miR-let7c, miR-100, and miR-218 from high grade localized prostate cancer to metastasis. Urol Oncol 2011, 29(3):265–269.

Pre-amplification in the context of high-throughput qPCR gene expression experiment

Vlasta Korenková[1*], Justin Scott[2], Vendula Novosadová[1], Marie Jindřichová[1], Lucie Langerová[1], David Švec[1], Monika Šídová[1] and Robert Sjöback[3]

Abstract

Background: With the introduction of the first high-throughput qPCR instrument on the market it became possible to perform thousands of reactions in a single run compared to the previous hundreds. In the high-throughput reaction, only limited volumes of highly concentrated cDNA or DNA samples can be added. This necessity can be solved by pre-amplification, which became a part of the high-throughput experimental workflow. Here, we focused our attention on the limits of the specific target pre-amplification reaction and propose the optimal, general setup for gene expression experiment using BioMark instrument (Fluidigm).

Results: For evaluating different pre-amplification factors following conditions were combined: four human blood samples from healthy donors and five transcripts having high to low expression levels; each cDNA sample was pre-amplified at four cycles (15, 18, 21, and 24) and five concentrations (equivalent to 0.078 ng, 0.32 ng, 1.25 ng, 5 ng, and 20 ng of total RNA). Factors identified as critical for a success of cDNA pre-amplification were cycle of pre-amplification, total RNA concentration, and type of gene. The selected pre-amplification reactions were further tested for optimal Cq distribution in a BioMark Array. The following concentrations combined with pre-amplification cycles were optimal for good quality samples: 20 ng of total RNA with 15 cycles of pre-amplification, 20x and 40x diluted; and 5 ng and 20 ng of total RNA with 18 cycles of pre-amplification, both 20x and 40x diluted.

Conclusions: We set up upper limits for the bulk gene expression experiment using gene expression Dynamic Array and provided an easy-to-obtain tool for measuring of pre-amplification success. We also showed that variability of the pre-amplification, introduced into the experimental workflow of reverse transcription-qPCR, is lower than variability caused by the reverse transcription step.

Keywords: High-throughput qPCR, Exponential pre-amplification, Microfluidics, Gene expression, Fluidigm, BioMark, Degraded samples, FFPE

Background

The popularity of real time PCR steadily increases as well as the number of platforms, detection chemistries and multiple choices of analytical methods. Several years ago, the boom in high-throughput instruments changed the way of studying gene expression and enabled researchers to perform large scale studies based on the most sensitive and specific quantitative PCR method.

The first commercially available high-throughput qPCR instrument was the BioMark™ System from Fluidigm that was launched in 2006. Microfluidic Dynamic Arrays provided by Fluidigm are able to combine either 48 samples with 48 assays or 96 samples with 96 assays in a combinatorial manner inside the integrated fluidic circuit (IFC) [1]. The BioMark System is able to process a high number of reactions (9,216) in a single run, each reaction taking place in volume of 6.7 nl [2]. With this number of reactions in a single run and its versatility and the freedom of the custom designed assays, BioMark System outperforms other high-throughput qPCR systems. There

---

* Correspondence: vlasta.korenkova@ibt.cas.cz
[1]Laboratory of Gene Expression, Institute of Biotechnology, Academy of Sciences of the Czech Republic, Prague, Czech Republic
Full list of author information is available at the end of the article

are only a few high-throughput qPCR instruments on the market that can be compared with BioMark System: OpenArray using a chip with 3,072 reactions, each for 33-nanolitre reaction volumes (Life Technologies) [3] and SmartChip with 5,184 reactions, each for 100-nanolitre reaction volumes (Wafergen) [4]. All these systems are designed to significantly simplify experimental workflow, increase throughput and reduce costs, while providing excellent data quality. Even though these instruments are built on different platforms, one attribute is common for all of them and that is a need for highly concentrated starting sample material.

The problem with an insufficient number of copies of the target in the reaction can be overcome with the help of pre-amplification. For the purposes of BioMark System a specific target amplification (also known as STA) is used, which is a multiplex PCR run with cDNA template and with a limited number of cycles, which is an exponential type of pre-amplification enabling simultaneous gene expression measuring of multiple targets in a single sample [5-7]. This kind of pre-amplification increases the amount of the initial cDNA or DNA template molecules several-fold, quantitatively amplifies just the target genes to be measured, and preserves the relationships between the transcripts. Even though pre-amplification has been used for many years [8,9] and it has been incorporated in high-throughput qPCR instruments workflows [10-13], it is still the least studied part of qPCR workflow that might introduce an additional bias if it is used without caution and appropriate controls.

In last few years, we witnessed that along with new techniques and new bioinformatic approaches come praise-worthy effort for proper standardization and control of the whole experimental process to eliminate widespread publication of poor data, resulting in inappropriate conclusions [14]. Because of the initiator of the whole process, MIQE guidelines [15], the quality and transparency of the laboratory results has been improved considerably. Pre-amplification process should not be omitted from this effort and it should be thoroughly validated and correctly reported as well as other parts of reverse transcription-qPCR workflow. It means that controls of pre-amplification should include at least paired non-preamplified and pre-amplified samples and each assay should be tested independently before the main experiment as described by Rusnakova [16]. For unbiased pre-amplification, the same difference between Cq values of non-preamplified and pre-amplified cDNA samples is expected for all assays; only reproducible small deviations are acceptable. Reproducibility is critical. Other controls as pre-amplified no template control (NTC) and pre-amplified control of reverse transcription without reverse transcriptase (RT-) should also be included. The reason is to ensure that quantification will not be influenced by eventual primer-

dimer formation or by assays that would amplify gDNA. RT- control could be successfully replaced by a valid prime assay, which accurately corrects all reactions in BioMark Array for signals derived from gDNA using only one extra valid prime assay and pre-amplified genomic DNA (gDNA) [17]. As the pre-amplification reaction is a highly complex multiplex system (it is possible to pre-amplify almost limitless number of measured genes), simultaneous amplifications of the large number of targets may interfere; therefore it is necessary to use highly optimized qPCR assays with high efficiency and high precision and to run only a limited number of cycles and avoiding high-abundant targets if possible [18]. Even though it is possible to use fewer cycles of pre-amplification (10–14) for qPCR experiments with conventional qPCR instruments, high-throughput qPCR experiments require more than 14 pre-amplification cycles. Fluidigm advanced protocols recommend 14 cycles for conventional profiling [19] and 18 cycles for single-cell profiling [20]. These numbers of pre-amplification cycles are calculated for highly optimized assays but in practice pre-amplification PCR efficiencies are not close to 100% that is why these numbers are minimal and often suboptimal [18].

Here, we focus on identifying factors which influence the pre-amplification reaction and the pre-amplification limits, especially a limiting higher number of cycles for pre-amplification, which has not been studied systematically yet. Our aim is to find out the optimal conditions for BioMark Array that would give us an optimal distribution of quantifiable Cq values across the Array by using the proper amount of mRNA transferred into a reverse transcription reaction; the proper fraction of the cDNA used for pre-amplification and the proper fraction of the pre-amplified and correctly diluted cDNA, transferred into each sample well in BioMark Array.

Results and discussion
Evaluating variables in pre-amplification reaction using regular qPCR instrument

The primary purpose of pre-amplification is to enhance amount of input material, which can be, in some instances, very low even for conventional qPCR: single cell analysis [16,21], microRNA analysis [22], analysis of formalin-fixed, paraffine embedded tissues [23] or to enhance initial amount of material to be sufficient for high-throughput instrument [1]. The amount of pre-amplified transcripts correlates with the initial cDNA target copy numbers as has been shown previously for both good quality samples [24] and bad quality samples, e.g. formalin-fixed paraffin-embedded samples [23]. The exponential pre-amplification should not be affected by the quality of original RNA because the product of reverse transcription, cDNA molecule, is pre-amplified. That is why the quality of RNA will influence only the reverse transcription step.

Even though the pre-amplification reaction itself is quite simple, there are several factors that can influence the final result. To identify and evaluate these factors we performed pre-amplification experiment combining different conditions. We evaluated four donors and five genes having high, medium and low expression levels. The genes were FKBP, STK10, EIF3M, CD83, and RND1 and were selected as representative from 24 well-characterized assays (Additional file 1) that were used later on for the summarizing BioMark experiment. Their mean Cq values for four non-pre-amplified samples were 18.7, 21.5, 23.7, 26.7, 34.0, respectively, which expression is spanning four orders of magnitude of dynamic range. Each sample was pre-amplified using four different cycles (15, 18, 21, and 24) and at five different concentrations (equivalent to 0.078 ng, 0.32 ng, 1.25 ng, 5 ng, and 20 ng of total RNA). The copy number of each transcript and sample was estimated for each assay. The estimated copy number for the low expressed gene RND1 was confirmed by dPCR. The limit of detection (LOD), the limit of quantification (LOQ) and the efficiency were determined for all 5 assays (Additional file 1).

Obtained non-pre-amplified Cq data and pre-amplified Cq data were subtracted to calculate an 'experimental difference' of pre-amplification: $\Delta Cq_{experimental} = Cq_{non-preamp} - Cq_{preamp}$. A 'theoretical difference' of pre-amplification was calculated as: $\Delta Cq_{theoretical}$ = number of pre-amplification cycles − $\log_2$ (all dilutions during the processing of the sample). The final formula was $\Delta\Delta Cq = \Delta Cq_{theoretical} - \Delta Cq_{experimental}$. An obtained $\Delta\Delta Cq$ value, 'expression differential', close to zero indicates pre-amplification uniformity (example of calculation in Additional file 2). We set $\Delta\Delta Cq = 1.5$ as a quality threshold for an acceptable pre-amplification. This threshold value is in agreement with the threshold value recommended by Applied Biosystems in TaqMan PreAmp Master Mix Kit guide [11]. The values lower than the quality threshold ($\leq \pm 1.5$) were named a 'success'. The values higher than a quality threshold and the missing values, caused by missing copies in the reaction, were categorized as a 'failure' (16 or 4% of cases) (Additional file 2).

In order to evaluate which factors affect the 'success' of pre-amplification, we tested these data variables: Cycles (number of pre-amplification cycles), Log_copy ($\log_2$ copy number of cDNA used for pre-amplification), Log_concentn ($\log_2$ concentration of cDNA, presented as total RNA equivalent, used for pre-amplification), Donor, GeneNo (gene number = different transcripts) that were used in explanatory binomial candidate model. The optimal model was then derived in SPSS using the backward stepwise method to eliminate non-significant terms, which were Donor and Log_copy. Because all terms are known beforehand and controllable, the model

could serve also as a predictive model with sensitivity of 81% and specifity 67% (Additional file 3).

Individual statistical tests uncovered important details of the pre-amplification process. Concentration of cDNA sample used for pre-amplification had significant effect on the overall likelihood of 'success' when tested for all Genes and Cycles together (p = 0.012); the higher Concentration, the higher 'success' (Additional file 4A). When individual Genes were taken in account and all Cycles were together, Concentration had significant effect only on low copy genes, RND1 (p < 0.001) and CD83 (p = 0.001) (Additional file 4B). Both genes show high failure rates in the low concentrated pre-amplification reactions (up to 5 ng) because the low template concentration corresponds to the low number of copies in pre-amplification (<10 copies of cDNA). These findings are in agreement with Bengtsson [25], who claims that when amplifying less than 20 cDNA copies the level of technical noise of PCR amplification increases dramatically, technical reproducibility decreases, thus the accurate quantification is reached if >20 target molecules per PCR are amplified.

Copy number of cDNA used for pre-amplification was not significant in the predictive model because Copy number (Log_copy) did not have a significant effect on the overall likelihood of 'success' when all Genes and all Cycles were combined together (p = 0.322) (Additional file 5A). However, if each Gene was tested independently with all Cycles together, the same results as for variable Concentration were obtained. Copy number had significant effect on low copy genes RND1 (p = 0.0001) and CD83 (p = 0.0004) (Additional file 5B). Additional information was derived if Copy number was compared for all Genes and each Cycle independently. Whereas the likelihood of 'success' increased with increasing Copy number for cycles 15 (p = 0.0006) and 18 (p = 0.0002), it decreases for cycle 24 (p = 0.0007). The contradictory directions for individual Cycles can explain why there was no overall significant effect above (Additional file 5C). The increasing 'success'of pre-amplification with higher Copy numbers has been described before, for example, using different copy numbers of ERCC RNA-42 standard with 14 cycles of pre-amplification [26]. However, the effect of high copy number transcripts combined with higher pre-amplification cycles (>18 cycles) has not been systematically investigated for bulk experiments before.

Finally, effect of number of Cycles on pre-amplification 'success' was tested. We show that the number of Cycles had a highly significant effect on overall likelihood of 'success' (p < 0.001) if tested for all Genes and Concentration together. Increasing Cycle numbers decreased the likelihood of 'success' (Additional file 6A). If both Genes and Cycles were tested independently, Cycle number had significant effect only on high copy genes EIF3M (p = 0.001), STK10 (p < 0.001) and FKBP (p < 0.001). Increasing Cycle numbers drastically decreased the likelihood of 'success' (Additional file 6B). The presence of highly abundant transcripts has also effect on pre-amplification process, this effect was combined with number of Cycles. While pre-amplifiying 21 and 24 cycles, the quality of pre-amplification steeply dropped, which is shown in summary figure (Figure 1). The percentage of affected genes displayed in this figure can be found in Additional file 7. This would probably be caused by getting under optimal concentration of primers in the multiplex pre-amplification reaction. The possibility of exhausting reagents during qPCR reaction was ruled out by testing limiting dilutions of PCR product of FKBP (data not shown).

Applying the results, we can speculate why 18 s rRNA, which is often used as a reference gene using conventional qPCR would not be suitable transcript for pre-amplification as was also suggested by Stahlberg [18]. The previously published data demonstrated that the highest correlation observed for samples pre-amplified with 18 s rRNA measured with microfluidic BioMark Array and non-preamplified samples measured by conventional qPCR cycler was 0.801 [27]. The expression of 18 s rRNA is so abundant that we recommend to exclude it from pre-amplification reaction completely. 18 s rRNA would not be detected reliably because of the very high concentration of transcripts

present after pre-amplification. This reason would cause the inability of any instrument to set the correct baseline. On the other hand, for the same reason, it is possible to quantify 18 s rRNA in BioMark array without pre-amplification (Additional file 8). The simple clue for identifying possible unsuitable targets for pre-amplification is their measured Cq value. The Cq value of the non-preamplified high-abundant transcript should not be lower than the number of cycles being used for its pre-amplification.

After summarizing all results together, combination of significant variables Cycle and Concentration reveals that a cycle 15 or 18 combined with a concentration of 20 ng is the best pre-amplification option using good quality samples, although any concentration higher than 1.25 ng is likely to be sufficient if the cycle is 18 or 15 (Table 1). In other words, the solution is to minimize number of Cycles and maximize Concentration of the sample. Presented model (Table 1) can also be applied for degraded samples, e.g. formalin-fixed paraffin-embedded samples. If RNA samples are degraded, less cDNA could be formed during reverse transcription, thus less target copies of cDNA can be pre-amplified. Using our outcomes (recommended combinations of concentrations and cycles), the highly expressed transcripts will never be over-preamplificated.

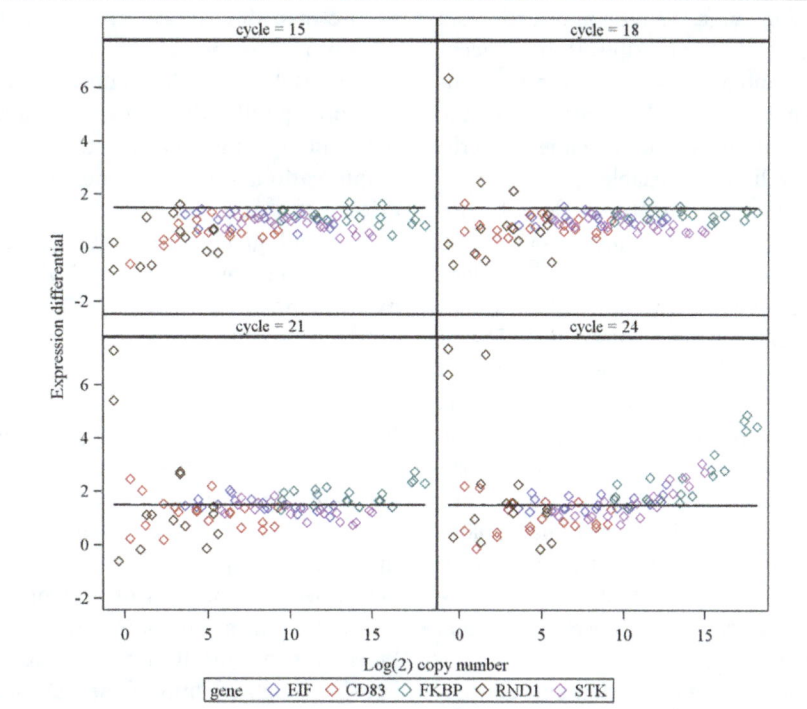

Figure 1 A plot showing the quality of pre-amplification. Successfully pre-amplified samples lie bellow the quality threshold, which corresponds to ΔΔCq = 1.5 (expression differential). The quality of pre-amplification gets worse with increasing number of pre-amplification cycles. During cycles 15 and 18 only small number of samples amplified with low copy gene RND1 (lowest concentrations) and high copy gene FKBP are affected. At cycles 21 and 24, the quality of pre-amplification is affected in all genes to some extent. The least affected gene is CD83, the most affected are high copy genes.

Table 1 A pivot table showing the success rate as a percentage for the possible combinations of Cycles (pre-amplification cycles) and concentrations for all five genes

Average of success	Concentration (equivalent of total RNA in ng used in pre-amplification)					
Cycles	0.078	0.32	1.25	5	20	Grand total
15	75%	90%	90%	90%	100%	89%
18	80%	80%	90%	95%	100%	89%
21	40%	30%	60%	75%	80%	57%
24	45%	50%	55%	40%	40%	46%
Grand total	60%	63%	74%	75%	80%	70%

Cycle 18 combined with a Concentration of 5 and 20 ng, and Cycle 15 combined with a concentration of 20 ng is optimal for successful BioMark experiment.

However, the right concentration and dilution of samples need to be tested for the low expressed genes.

Optimal dilution for BioMark system

The regular BioMark high-throughput gene expression experiment consists of high number of assays (up to 48 or up to 96) [1] that are amplified at the same time, resulting in the big spread of Cq values from highly expressed to lowly expressed genes. If either the concentration of the sample, the number of pre-amplification cycles or dilution after pre-amplification are not set correctly, the final result will not be optimal. Some transcripts could be under-amplified, which can result in a loss of detection sensitivity and generation of missing values. We should also avoid over-cycling of highly expressed transcripts. If the concentration of copies for a certain assay is too high in the sample, the instrument might not be able to distinguish the background of the reaction and set the baseline properly. The obtained Cq values will not be reliable.

For a successful BioMark experiment, it is desirable for the majority of the data to fall in the range about Cq = 6 to Cq = 23 [28,29]. In contrast to the regular qPCR cyclers, the optimal range of quantifiable Cq values in BioMark instrument is approximately 10 cycles lower [26]. This is caused by the fundamental difference between the size of the surface that comes into contact with the sample mix during the qPCR reaction in the Dynamic Array and in the conventional micro-titer plate. In contrast to the polypropylene surface in conventional micro-titer plates, the percentage of surface of polydimethylsiloxane nanochamber [30] that is connected with the reaction volume is much larger. It leads to a higher sensitivity of the microfluidic system, earlier detection of the fluorescence signal and thus shorter cycling time.

In order to identify the best concentration and dilution of pre-amplified samples that would be suitable for the BioMark experiment, we performed an experiment with 48.48 GE Dynamic Array using already pre-amplified samples from previous experiment with cycles 15, 18 and 21 respectively and with the concentrations 1.25 ng, 5 ng and

20 ng, respectively. The samples were diluted 20 and 40 fold, respectively to determine the best conditions for BioMark instrument. All obtained Cq data (from 21 assays excluding 3 reference genes) was normalized with GAPDH, PPIB and GUSB reference genes, which should cancel out the differences among the different concentrations and different amplification cycles but not the natural variability among donors. We set these criteria: missing data were not acceptable, the lowest Cq in Dynamic Array should be 6 and three samples should be distinguished and fall into respective groups. That is why, the two lowest concentrations (1.25 ng and 5 ng) for 15 cycle pre-amplification were removed from the analysis immediately. These criteria helped us to set up the principal component analysis that was used to reduce the dimensionality of a data set, which consisted of the 21 normalized gene assays, 3 pre-amplification cycles, 2 dilutions and 3 concentrations of 3 samples. PCA data was auto-scaled to reduce the effect of variation in the overall expression levels of the different genes. Only samples that created three separated compact groups were selected for further analysis. After removal of all samples pre-amplified 21 cycles and samples pre-amplified 18 cycles of concentration 1.25 ng, the data set was reanalyzed and the first 3 components of PCA explained 78.4% variability of auto-scaled data set. The right choice of selected samples from PCA was validated with another method, Kohonen's self-organizing feature map (SOM) that confirmed separation of samples into 3 distinct groups (Figure 2).

As a result, the highest concentration, 20 ng, for 15 cycles pre-amplification, both 20x and 40x diluted; and 5 ng and 20 ng concentrations for 18 cycles of pre-amplification, both 20x and 40x diluted fulfilled our criteria for successful and reliable pre-amplification and would give the best unbiased result with maximum detectable data for BioMark gene expression experiment.

Pre-amplification variability

In order to demonstrate how the combination of optimal conditions for success of pre-amplification would affect variability, the yield and standard deviations (SD) of pre-amplified FKBP, STK10, EIF3M, CD83, and RND1 were measured by qPCR. Pre-amplification reactions were performed in replicates of four on one cDNA that was synthetized from the same RNA pool. SD_{PRE} of combined pre-amplification and qPCR (Table 2) was calculated as weighted sum of the SDs of the qPCR (SD_{qPCR}) and SD of the pre-amplification reaction (SD_{pre}). SD_{PRE} was in the range of 0.14 – 0.24, which corresponds to variability 10% - 17% in estimated number of cDNA molecules (averaged variability for all genes is 13%). Variability increases towards the low expressed genes with higher Cq

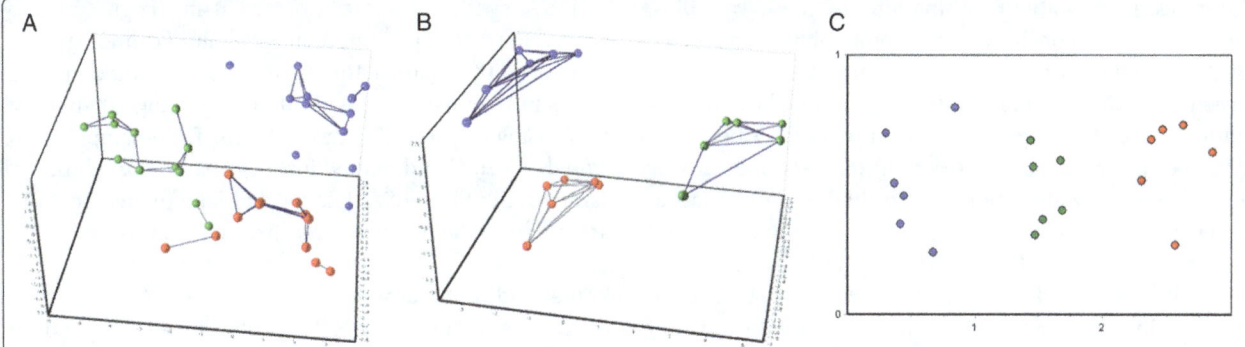

Figure 2 Identification of the best concentration and dilution of pre-amplified samples for the BioMark experiment. A. 3-D principal component analysis. PCA is based on 21 differentially expressed, auto-scaled genes, which classified pre-amplified samples into three groups according to donors (blue = donor 1, red = donor 2, green = donor 3). Samples pre-amplified 15, 18 and 21 cycles are shown, diluted both 20x and 40x. It is difficult to distinguish clearly 3 groups. **B**. Only acceptable pre-amplifications useful for BioMark GE Dynamic Array are selected: 15 cycles - 20 ng, dilution 20x and 40x, respectively; 18 cycles – 5 and 20 ng, dilution 20x and 40x, respectively. **C**. SOM with samples selected for Figure 2B, confirms 3 distinct groups.

values (Table 2), which is caused by well described statistical effects [31].

Previously, it has been described that experimental variation in the reverse transcription-qPCR (without pre-amplification) is mainly attributable to the reverse transcription step [32]. To confirm this statement also for reverse transcription–qPCR with the additional pre-amplification step, we performed the experiment where the yield and standard deviations of cDNA synthesis of the FKBP, STK10, EIF3M, CD83, and RND1 were measured by qPCR. This time reverse transcription reactions were performed in replicates of four on material from the same RNA pool as in previous pre-amplification experiment (Table 2). SD_{RT} of combined reverse transcription and qPCR (Table 2) was calculated as weighted sum of the SDs of the qPCR (SD_{qPCR}) and SD of the RT reaction (SD_{rt}). SD_{RT} was in the range of 0.24 – 0.41, which corresponds to variability 17% - 33% in estimated number of cDNA molecules (averaged variability for all genes is

23.6%). After comparison, the pre-amplification variability within the reverse transcription-qPCR experiment is significantly lower (p = 0.015) than variability caused by cDNA synthesis step.

Conclusion

In order to perform a valid experiment that would lead to reliable results, it is necessary to know both the capabilities and limitations of the used method and instrument. Even though BioMark instrument performs the regular qPCR reaction, we need to take some special properties into account when setting high-throughput qPCR experiment. The most distinct deviation from the regular qPCR experiment workflow is the necessity of pre-amplification.

As has been demonstrated, pre-amplification success is based on several variables, the most important ones are number of pre-amplification cycles, concentration of the sample used for pre-amplification, and the gene itself.

Table 2 Comparison of reverse transcription (RT) and pre-amplification (PRE) variability

	RND1	CD83	EIF3M	STK10	FKBP	AVG var.
Efficiency (E)	91%	100%	97%	94%	100%	
Cq RT	31.7	25.6	24.6	22.1	20.2	
$SD_{RT} = \sqrt{SD(rt)^2 + SD(qPCR)^2}$	0.38	0.41	0.24	0.40	0.29	
Variability RT $= (1 + E)^{SD_{RT}}$	28%	33%	18%	17%	22%	23.6%
Cq PRE*	23.6	17.1	16.2	13.4	12.1	
$SD_{PRE} = \sqrt{SD(pre)^2 + SD(qPCR)^2}$	0.24	0.18	0.18	0.14	0.16	
Variability PRE $= (1 + E)^{SD_{PRE}}$	17%	13%	13%	10%	12%	13%
Expression Differential	1.2	0.7	1.0	0.7	1.3	

*Equivalent of 5 ng of total RNA was used in 18 cycle pre-amplification, pre-amplified cDNA diluted 40x.

After testing possible combinations of these variables, we came to the conclusion that pre-amplification for the BioMark System using good quality samples is optimal between 15–18 pre-amplification cycles and higher concentrations of cDNA samples (5–20 ng of transcribed total RNA per pre-amplification reaction) diluted either 20x or 40x after pre-amplification. Use of higher amplification cycles (21 or 24) in bulk experiments (not in single cell experiments) is very limited because high abundant targets will cause an exhaustion of primers and reagents from pre-amplification reaction, thus they will cause lowering of pre-amplification success.

The success of the pre-amplification can be tested by our improved, easy-to-obtain, universal formula called "expression differential". The algorithm, which is presented here, evaluates the "expression differential" based on a $\Delta\Delta Cq$ value obtained subtracting ΔCq experimental - ΔCq expected or "theoretical". Formula can be used universally, for pre-testing of the quality of pre-amplification assays in high-throughput gene expression experiment as well as in RT-qPCR experiments with FFPE-RNA.

And finally, we show that variability of the pre-amplification, introduced into the experimental workflow of reverse transcription-qPCR, is lower than variability caused by the reverse transcription step.

Methods
Sample collection and preparation
Blood was collected in BD K_2EDTA tubes (BD, cat. no. 367525), 10 ml draw volume, from healthy volunteers. After approval by Norwegian south east regional committee for medical and health research ethics (REC South East), all participants signed a written informed consent before participating in the study in accordance with the Helsinki declaration. As soon as possible after the first blood tube collection, EDTA blood from each volunteer was transferred to and PAXgene® Blood RNA Tubes (PAXgene) (PreAnalytiX) to maintain gene expression, incubated at room temperature for 2 hours, and then stored at –80°C.

Isolation of RNA, quality control and reverse transcription
RNA from blood collected in the PAXgene tubes was extracted according to the standard protocol: PAXgene Blood RNA Kit (Qiagen) and stored at –80°C.

RNA quantity and purity was measured using Nano-DropTM 1000 Spectrophotometer (Thermo Scientific). $OD_{260/280}$ ratios for all samples were between 1.8 and 2.0. RNA integrity number (RIN) was checked using capillary electrophoresis performed on Agilent Bioanalyzer 2100, with RNA 6000 Nano Assay (Agilent Technologies). Sample 1 RIN = 8, sample 2 RIN = 7.3, sample 3 RIN = 7.7, sample 4 RIN = 7.6. Pooled sample 5 for variability modeling had RIN = 7.5.

cDNA synthesis was performed using High Capacity cDNA Reverse Transcription Kit (Life Technology) according to the manufacturer's protocol with random hexamers in the final volume of 50 µl containing 500 ng total RNA using a cycler C1000 (Bio-Rad). cDNA samples were stored at –20°C and diluted just before use. For dilution of samples GenElute-LPA (Sigma Aldrich) diluted in 1xTE according to the manufacturer instructions was used.

Primer and probe design
qPCR assays and a RND1 probe were designed by TATAA Biocenter, Sweden (Additional file 1). To avoid the amplification of genomic DNA all assays were placed to span and/or have one primer covering an intron/exon boundary. Criteria for the assays were: good linearity (5 log dynamic range at LC480 error < 0.2), efficiency (≥80%, ≤105%), specificity (no amplification of gDNA or at least 5 cycle's difference between target and genomic Cq-value) and clear NTCs. All assays were initially evaluated with SYBR green chemistry to test the primers. After approval of the primers a hydrolysis probe for RND1 was designed and evaluated in the same way as described for the primers. PCR products were analyzed for specificity (single product) on a pre-made 2.2% agarose gel (Flash Gel system, Lonza). All primer designs were performed with Primer BLAST [33] followed by probe design with Beacon Designer® (PREMIER Biosoft International). Primers and the probe were ordered from Eurofins. Primers were HPSF purified. Probe was labelled with FAM as reporter and BHQ1 as quencher and HPLC purified.

Real time PCR, copy number estimation, efficiency and limit of detection
10 µl qPCR reactions using SYBR green were prepared from 5 µl 2x TATAA SYBR GrandMaster Mix (TATAA Biocenter), 0.4 µl primers (final concentration 400 nM), 2.6 µl MB water, 2 µl cDNA (or pre-amplified cDNA diluted 20x). The qPCR was run in CFX384 (Bio-Rad) using the standard program 95°C for 1 min followed by 40 cycles 95°C for 3 s, 60°C for 60 s, and 72°C for 10s plus melting curve. At least triplicate qPCR reactions were performed for each qPCR experiment. Cq data were obtained by regression using Bio-Rad CFX Manager Software 3.0 (Bio-Rad).

For determination of the number of copies, PCR products were purified using QIAquick PCR Purification Kit (Oiagen) according to the manufacturer instructions, concentration was measured using Qubit® 2.0 Fluorometer (Life Technologies) and number of copies were calculated. Standard curves using PCR product of known copy numbers were generated and the copy numbers of tested samples were interpolated. The cDNA RND1 copy number for four donors was confirmed also by dPCR using a probe.

Limit of detections (LOD) was determined from standard curves with 6 replicates for each dilution, 8 dilutions 1:3 for assays: EIF3M, CD83, FKBP, RND1, STK10. (Additional file 1). Dilutions were made with carrier TE-LPA (Sigma Aldrich). The efficiency of remaining assays was determined from the standard curves generated from PCR product diluted 1:10 000 in TE-LPA with 3 replicates and 5 dilution 1:9 (Additional file 1). All standard curve experiments were run in CFX384 (Bio-Rad) with TATAA SYBR GrandMaster Mix (TATAA Biocenter).

Gene specific pre-amplification for experiments using intercalating dye

A single aliquot of each cDNA sample (diluted in carrier TE-LPA), equivalent to 20 ng RNA, 5 ng RNA, 1.25 ng RNA, 0.32 ng RNA, 0.078 ng RNA, respectively, was used for pre-amplification with TATAA PreAmp GrandMaster® mix (TATAA Biocenter) at either 15 cycles, 18 cycles, 21 cycles or 24 cycles, respectively. The total volume of pre-amplification was 10 μl for each sample. The reaction contained 5 μl of pre-amplification mastermix, 2 μl of cDNA, 1 μl of pooled primers with a final concentration of each primer of 25 nM and 2 μl of MB water. The cDNA samples were subjected to pre-amplification. The following temperature protocol was used: 95°C for 30 s, followed by 15, 18, 21, 24 cycles, respectively at 95°C for 15 s and 60°C for 4 min. 24 assays were pre-amplified as multiplex and only 5 selected assays (see above) were tested in the experiment. A list of 24 assays used for pre-amplification is described in Additional file 1. As a control, water (NTC) was included in the pre-amplification reaction. The pre-amplified cDNA was immediately used or placed in freezer at −20°C. The pre-amplified cDNA was diluted prior to use at either 20x or 40x with MB water.

High-throughput real time PCR with Eva green

qPCR was performed using the high-throughput platform BioMark™ HD System and the 48.48 GE Dynamic Arrays (Fluidigm) in duplicates in assays. 5 μL of Fluidigm sample premix consisted of 1 μL of either 20x or 40x diluted pre-amplified cDNA, 0.25 μL of 20x SG loading reagent (Fluidigm), 2.5 μL of Sso Fast Eva green mastermix (Bio-Rad), 0.1 μL of 4x diluted ROX (Invitrogen) and 1.15 μL of RNase/DNase-free water. Each 5 μL assay premix consisted of 2 μL of 10 μM primers (final concentration 400 nM primers), 2.5 μL 2x Assay loading reagent (Fluidigm) and 0.5 μL of RNase/DNase-free water. The samples and assays were mixed inside the chip using Nanoflex IFC controller (Fluidigm). Thermal conditions for qPCR were: 98°C for 40 s, 35 cycles of 95°C for 10 s, and 60°C for 40 s plus melting curve analysis. Data was processed by automatic threshold for each assay, with derivative baseline correction using BioMark Real-Time

PCR Analysis Software 3.1.2 (Fluidigm). The quality threshold was set at the default setting of 0.65.

qPCR data pre-processing and statistical analysis

The Cq data obtained from conventional qPCR cycler CFX384 was analyzed using IBM SPSS Statistics (Version 21) and an Excel (Version 14.3.4) pivot table. Tested variables were: Cycles (number of pre-amplification cycles), Log_copy ($\log_2$ copy number of cDNA used for pre-amplification), Log_concentn ($\log_2$ concentration of cDNA, presented as total RNA equivalent, used for pre-amplification), Donor, GeneNo (gene number = different transcripts). Copy number was analyzed as both a categorical and continuous variable. As the distribution of Copy number and Concentration were not normally distributed these were also log transformed (base 2). Each experiment was classified as a'success' or'failure'. An experiment was classified as a'success' if the'expression differential' was less than ± 1.5 (Additional file 2). An experiment was classified as a'failure' if the'expression differential' was greater than 1.5 or missing.'Expression differential' consisted of the'theoretical expression' minus the'experimental expression' (detailed description in results).

Measures of experiment behavior and outcome were compared against the likelihood of success to detect any statistical relationships. Univariate categorical measures were compared against experimental 'success' under specified conditions using the Chi-squared test (expected values were so high that the Fisher's Exact test was not used). Univariate continuous measures were compared against experimental 'success' using Box plots and group summary tables. A full variable logistic regression model was pared back to an optimal model using the backward stepwise method by eliminating non-significant terms. A classification (or confusion) table was produced and the sensitivity and specificity calculated. A pivot table was produced showing the success rate as a percentage for the possible combinations of Cycles and Concentrations.

All BioMark data were pre-processed in the software GenEx Enterprise 5.4.0.520 (MultiD Analyses AB). PPIB, GAPDH and GUSB were selected for normalization using Normfinder software. Principal component analysis (PCA) [34] and Kohonen self-organizing maps (SOM) [35] was performed using 21 original independent variables (21 normalized genes). PCA and SOM were performed with data that were normalized, the lowest expression was recalculated to 1, log2 transformed and auto-scaled using GenEx Enterprise software. All expression values were auto-scaled in order to remove the influence of both the expression level and the magnitudes of the changes and gave rise to classification based on the relative changes in expression. The SOM of size 3 x 1 dividing the samples into 3 groups was trained using GenEx with the following parameters: 0.1 learning rate, 3 neighbors and 5,000 iterations. The SOM analysis was repeated five times with identical classification.

Difference between variability of reverse transcription step and pre-amplification step was tested by paired, two tailed t test using GenEx Enterprise 5.4.0.520 (MultiD Analyses AB).

Additional files

Additional file 1: The Excel sheet with information on 24 assays used for pre-amplification. Five assays highlighted in gray were used for experiment 3.1. Additional information are added for these assays: LOD, LOQ.

Additional file 2: Table of all samples and their Cq and calculated characteristics as Concentration, Copy numbers, Cycle, Success and the example of the pre-amplification algorithm (expression differential) application.

Additional file 3: Construction and results of explanatory binomial candidate model explaining which combination of factors will influence the 'success'.

Additional file 4: Tables showing how Concentration of RNA (an equivalent of mRNA transferred into pre-amplification reaction) influences 'success'. A. Tested for all Genes and all Cycles together. B. Tested for each Gene independently and all Cycles together.

Additional file 5: Figures showing how Copy number (copy number of cDNA used for pre-amplification) influences 'success'. A. Tested for all Genes and all Cycles together. B. Tested for each Gene independently and all Cycles together. C. Tested for all Genes and each Cycle independently.

Additional file 6: Tables showing how number of Cycles (number of pre-amplification cycles) influences 'success'. A. Tested for all Genes and Concentrations together. B. Tested for each Gene independently.

Additional file 7: A pivot table showing the success rate as a percentage for the possible combinations of Cycles and Concentrations for individual genes. The additional information for Figure 1.

Additional file 8: A standard curve of non-preamplified sample detected by 18S rRNA used in GE Dynamic Array 48.48.

Abbreviations

Cq: Cycle of quantification; EDTA: Ethylenediaminetetraacetic acid; gDNA: Genomic DNA; GeneNo: Gene number = different transcript; IFC: Integrated fluidic circuit; LOD: Limit of detection; Log_concentn: Log_2 concentration of cDNA, presented as total RNA equivalent, used for pre-amplification; Log_copy: Log_2 copy number of cDNA used for pre-amplification; LOQ: Limit of quantification; MB water: Molecular biology grade water; NTC: No template control; PCA: Principal component analysis; qPCR: Quantitative polymerase chain reaction; RIN: RNA integrity number; RT-: Negative control in reverse transcription, without reverse transcriptase; SOM: Kohonen self-organising map; SD: Standard deviation; STA: Specific target amplification; TE-LPA: Linear polyacrylamide carrier in TE buffer.

Competing interests

All authors have read and understood BMC Molecular Biology policy on declaration of interests and declare that we have no competing interests, only Robert Sjöback is employed by TATAA Biocenter, which is a producer of TATAA PreAmp GrandMaster Mix and TATAA SYBR GrandMaster Mix.

Authors' contributions

VK wrote the manuscript, elaborated design of the study and evaluated results. JS provided main part of statistical analysis. VN participated in statistical analysis. MJ carried out the pre-amplification experiments. LL carried out the BioMark experiments. DS and MS designed and validated primers and a probe. RS participated in coordination of the study and helped to draft the manuscript. All authors read and approved the final manuscript.

Acknowledgements

The authors are thankful to volunteers who participated in the study and DiaGenic ASA that provided us with the isolated samples and cooperated with us within the scope of SPIDIA project. We thank Prof. Mikael Kubista for valuable comments. This project was funded by BIOCEV CZ.1.05/1.1.00/02.0109 from ERDF, Go8 Fellowship Australia, CZ: GACR: P304/12/1585, CZ: GACR:GA15-08239S and CZ: GACR: P303/13/02154S.

Author details

[1]Laboratory of Gene Expression, Institute of Biotechnology, Academy of Sciences of the Czech Republic, Prague, Czech Republic. [2]QFAB Bioinformatics, University of Queensland - St Lucia QLD, Brisbane, Australia. [3]TATAA Biocenter, Göthenburg, Sweden.

References

1. Spurgeon SL, Jones RC, Ramakrishnan R. High throughput gene expression measurement with real time PCR in a microfluidic dynamic array. PLoS One. 2008;3:e1662.
2. BioMark™ HD System. [http://www.fluidigm.com/biomark-hd-system.html]
3. Real-Time PCR Using OpenArray® Technology. [http://www.lifetechnologies.com/au/en/home/life-science/pcr/real-time-pcr/real-time-openarray.html?icid=fr-openarray-main%20http://www.lifetechnologies.com/au/en/home/life-science/pcr/real-time-pcr/real-time-openarray.html?icid=fr-openarray-main]
4. SmartChip Real-Time PCR System. [http://www.wafergen.com/products/smartchip-realtime-pcr-system]
5. Mengual L, Burset M, Marin-Aguilera M, Ribal MJ, Alcaraz A. Multiplex preamplification of specific cDNA targets prior to gene expression analysis by TaqMan Arrays. BMC Res notes. 2008;1:21.
6. Blow N. PCR's next frontier. Nat Meth. 2007;4:869–75.
7. Iscove NN, Barbara M, Gu M, Gibson M, Modi C, Winegarden N. Representation is faithfully preserved in global cDNA amplified exponentially from sub-picogram quantities of mRNA. Nat Biotechnol. 2002;20:940–3.
8. Noutsias M, Rohde M, Block A, Klippert K, Lettau O, Blunert K, et al. Preamplification techniques for real-time RT-PCR analyses of endomyocardial biopsies. BMC Mol Biol. 2008;9:3.
9. Sindelka R, Sidova M, Svec D, Kubista M. Spatial expression profiles in the Xenopus laevis oocytes measured with qPCR tomography. Methods (San Diego, Calif). 2010;51:87–91.
10. Fluidigm. Real-Time PCR Analysis, Appendix B: Fast Gene Expression Analysis Using EvaGreen on the BioMark of BioMark HD System, part No. 68000088. [https://www.fluidigm.com/documents]
11. TaqMan PreAmp Master Mix Kit, Protocol. [http://tools.lifetechnologies.com/content/sfs/manuals/cms_039316.pdf]
12. Targeted Enrichment of Limited RNA Samples via Pre-Amplification Prior to Analysis in the WaferGen SmartChip Real-Time PCR System. [http://www.wafergen.com/wp-content/uploads/2013/01/TargetEnrchmnt_RNA_TNf.pdf]
13. OpenArray Plates for microRNA expression analysis. [http://tools.lifetechnologies.com/content/sfs/manuals/cms_092509.pdf]
14. Johnson G, Nour AA, Nolan T, Huggett J, Bustin S. Minimum information necessary for quantitative real-time PCR experiments. Methods Mol Biol (Clifton, NJ). 2014;1160:5–17.
15. Bustin SA, Benes V, Garson JA, Hellemans J, Huggett J, Kubista M, et al. The MIQE guidelines: minimum information for publication of quantitative real-time PCR experiments. Clin Chem. 2009;55:611–22.
16. Rusnakova V, Honsa P, Dzamba D, Stahlberg A, Kubista M, Anderova M. Heterogeneity of astrocytes: from development to injury - single cell gene expression. PLoS One. 2013;8:e69734.
17. Laurell H, Iacovoni JS, Abot A, Svec D, Maoret JJ, Arnal JF, et al. Correction of RT-qPCR data for genomic DNA-derived signals with ValidPrime. Nucleic Acids Res. 2012;40:e51.
18. Stahlberg A, Kubista M. The workflow of single-cell expression profiling using quantitative real-time PCR. Expert Rev Mol Diagn. 2014;14:323–31.
19. Fluidigm. Fluidigm Gene Expression Specific Target Amplification Quick Reference, part No. 68000133. [https://www.fluidigm.com/documents]

20. Fluidigm. BioMark Advanced Development Protocol Number 5: Single-Cell Gene Expression Protocol for the BioMark 48.48 Dynamic Array–Real-Time PCR, part No. 68000107. [https://www.fluidigm.com/documents]

21. Stahlberg A, Bengtsson M. Single-cell gene expression profiling using reverse transcription quantitative real-time PCR. Methods (San Diego, Calif). 2010;50:282–8.

22. Chen Y, Gelfond JA, McManus LM, Shireman PK. Reproducibility of quantitative RT-PCR array in miRNA expression profiling and comparison with microarray analysis. BMC Genomics. 2009;10:407.

23. Li J, Smyth P, Cahill S, Denning K, Flavin R, Aherne S, et al. Improved RNA quality and TaqMan Pre-amplification method (PreAmp) to enhance expression analysis from formalin fixed paraffin embedded (FFPE) materials. BMC Biotechnol. 2008;8:10.

24. Fox BC, Devonshire AS, Baradez MO, Marshall D, Foy CA. Comparison of reverse transcription-quantitative polymerase chain reaction methods and platforms for single cell gene expression analysis. Anal Biochem. 2012;427:178–86.

25. Bengtsson M, Hemberg M, Rorsman P, Stahlberg A. Quantification of mRNA in single cells and modelling of RT-qPCR induced noise. BMC Mol Biol. 2008;9:63.

26. Devonshire AS, Elaswarapu R, Foy CA. Applicability of RNA standards for evaluating RT-qPCR assays and platforms. BMC Genomics. 2011;12:118.

27. Jang JS, Kolbert C, Jen J. High throughput quantitative PCR using low-input samples for mRNA and MicroRNA gene expression analyses [abstract]. J Biomol Tech. 2013;24:S56.

28. Svec D, Rusnakova V, Korenkova V, Kubista M. Dye-Based High-Throughput qPCR in Microfluidic Platform BioMark™. In: Nolan T, Bustin SA, editors. PCR Technology: Current Innovations. 3rd ed. Boca Raton: CRC Press; 2013. p. 323–36.

29. Sorg D, Danowski K, Korenkova V, Rusnakova V, Kuffner R, Zimmer R, et al. Microfluidic high-throughput RT-qPCR measurements of the immune response of primary bovine mammary epithelial cells cultured from milk to mastitis pathogens. Animal. 2013;7:799–805.

30. Perkel JM. Microfluidics, macro-impacts. Biotechniques. 2012;52:131–4.

31. Morrison TB, Weis JJ, Wittwer CT. Quantification of low-copy transcripts by continuous SYBR Green I monitoring during amplification. Biotechniques. 1998;24:954–8. 960, 962.

32. Stahlberg A, Hakansson J, Xian X, Semb H, Kubista M. Properties of the reverse transcription reaction in mRNA quantification. Clin Chem. 2004;50:509–15.

33. Primer-BLAST. [http://www.ncbi.nlm.nih.gov/tools/primer-blast/index.cgi?LINK_LOC=BlastHome]

34. Jolliffe IT. Principal Component Analysis. 2nd ed. Springer-Verlag New York: Springer; 2002.

35. Kohonen Teuvo. Self-Organizing Maps. 3rd ed. Springer-Verlag Berlin Heidelberg: Springer; 2001.

Spdef deletion rescues the crypt cell proliferation defect in conditional *Gata6* null mouse small intestine

Boaz E Aronson[1,2], Kelly A Stapleton[1], Laurens ATM Vissers[3], Eva Stokhuijzen[2], Hanneke Bruijnzeel[2] and Stephen D Krasinski[1*]

Abstract

Background: GATA transcription factors are essential for self-renewal of the small intestinal epithelium. *Gata4* is expressed in the proximal 85% of small intestine while *Gata6* is expressed throughout the length of small intestine. Deletion of intestinal *Gata4* and *Gata6* results in an altered proliferation/differentiation phenotype, and an up-regulation of SAM pointed domain containing ETS transcription factor (*Spdef*), a transcription factor recently shown to act as a tumor suppressor. The goal of this study is to determine to what extent SPDEF mediates the downstream functions of GATA4/GATA6 in the small intestine. The hypothesis to be tested is that intestinal GATA4/GATA6 functions through SPDEF by repressing *Spdef* gene expression. To test this hypothesis, we defined the functions most likely regulated by the overlapping GATA6/SPDEF target gene set in mouse intestine, delineated the relationship between GATA6 chromatin occupancy and *Spdef* gene regulation in Caco-2 cells, and determined the extent to which prevention of *Spdef* up-regulation by *Spdef* knockout rescues the GATA6 phenotype in conditional *Gata6* knockout mouse ileum.

Results: Using publicly available profiling data, we found that 83% of GATA6-regulated genes are also regulated by SPDEF, and that proliferation/cancer is the function most likely to be modulated by this overlapping gene set. In human Caco-2 cells, GATA6 knockdown results in an up-regulation of *Spdef* gene expression, modeling our mouse *Gata6* knockout data. GATA6 occupies a genetic locus located 40 kb upstream of the *Spdef* transcription start site, consistent with direct regulation of *Spdef* gene expression by GATA6. Prevention of *Spdef* up-regulation in conditional *Gata6* knockout mouse ileum by the additional deletion of *Spdef* rescued the crypt cell proliferation defect, but had little effect on altered lineage differentiation or absorptive enterocytes gene expression.

Conclusion: SPDEF is a key, immediate downstream effecter of the crypt cell proliferation function of GATA4/GATA6 in the small intestine.

Keywords: GATA6, SPDEF, Crypt cell proliferation, Intestinal differentiation

Background

The mature mammalian small intestine is lined by a highly specialized epithelium that regenerates itself in a tightly controlled manner resulting in a lineage distribution and gene expression patterning that is perfectly suited for the absorption of nutrients. The epithelium is organized into crypt-villus structures in which the Crypts of Lieberkühn contain stem cells that produce proliferating, transit-amplifying (TA) cells that differentiate into five principal post-mitotic cell types comprised of one type of absorptive cell (absorptive enterocytes) and four types of secretory cells (enteroendocrine, goblet, Paneth, and tuft cells [1]). Absorptive enterocytes express digestive enzymes and transporters necessary for the terminal digestion and absorption of nutrients. Mucus-secreting goblet cells and defensin-secreting Paneth cells maintain a dynamic mucosal defensive barrier. Enteroendocrine cells secrete hormones that regulate gastrointestinal processes. Tuft cells, recently shown to be an independent secretory lineage [2], secrete opioids and produce enzymes that synthesize

* Correspondence: stephen.krasinski@childrens.harvard.edu
[1]Division of Gastroenterology and Nutrition, Department of Medicine, Children's Hospital Boston, and Harvard Medical School, 300 Longwood Avenue, Boston, MA 02115, USA
Full list of author information is available at the end of the article

prostaglandins, suggesting a role in inflammation. Absorptive enterocytes, goblet cells, enteroendocrine cells, and tuft cells migrate up the crypt to populate the villi, whereas Paneth cells migrate to the base of crypts. The differentiated cells eventually undergo apoptosis and are shed into the lumen. Cells of the villus epithelium turn over in 3–4 days, whereas Paneth cells at the base of crypts turn over at a slower rate of 3–6 weeks.

Current models of intestinal epithelial renewal suggest that long-lived, multipotent stem cells produce progenitors that undergo a series of transitions that ultimately give rise to the individual cell lineages [1,3]. The Wnt, hedgehog, and bone morphogenetic protein signaling pathways regulate intestinal proliferation and differentiation, while the Notch signaling pathway plays a central role in determining epithelial cell fate. The first decision selects absorptive vs. secretory progenitors. Activated Notch signaling results in the transcriptional activation of its principal intestinal target, hairy and enhancer of split 1 (*Hes1*), which encodes a transcription factor that selects the absorptive enterocyte lineage. Progenitor cells that escape Notch signaling and activation of *Hes1* gene transcription express atonal homolog 1 (*Atoh1*, formerly called *Math1*), which encodes a transcription factor that selects the secretory cells. Additional regulators that function in secretory cell differentiation include: growth factor independent 1 (*Gfi1*) that distinguishes enteroendocrine from goblet/Paneth progenitors; neurogenin 3 (*Neurog3*) that specifies the enteroendocrine lineage; SAM pointed domain-containing Ets transcription factor (*Spdef*), a GFI1 target, that promotes goblet differentiation; and SRY-box containing gene 9 (*Sox9*) and ephrin type B receptor 3 (*Ephb3*), Wnt targets that are necessary for the differentiation of Paneth cells and their localization to the crypt base, respectively.

Recently, we showed that members of the GATA family, an ancient family of transcription factors that bind WGATAR motifs in DNA, play essential roles in crypt cell proliferation, secretory cell differentiation, and absorptive enterocyte gene expression. *Gata4* and *Gata6* are expressed in the intestinal epithelium, but whereas *Gata6* is expressed throughout the length of the small intestine, *Gata4* is expressed in the proximal 85% of small intestine and is sharply down-regulated in the distal ileum [4-6]. Using conditional knockout technology, we [5,7] and others [8] have shown that GATA4 functions to promote a 'jejunal' pattern of absorptive enterocyte gene expression and function while repressing an 'ileal' pattern. Using single and double conditional knockout approaches for *Gata4* and *Gata6*, we found that in the ileum of single *Gata6* conditional knockout mice, where *Gata4* is not normally expressed, or throughout the small intestine of double *Gata4/Gata6* conditional knockout mice, crypt cell proliferation and enteroendocrine cell specification are decreased,

Paneth cells are replaced by a goblet-like cell type, and the expression of specific absorptive enterocyte genes is altered [4]. We also noted that *Spdef*, a transcription factor expressed in secretory progenitors, goblet cells and Paneth cells that functions in goblet and Paneth cell differentiation [9,10], was up-regulated [4]. Using a conditional *Spdef* over-expression model, Noah et al. [10] described an intestinal phenotype that, with the exception of the changes in absorptive enterocyte gene expression, essentially phenocopies that of our *Gata4/Gata6* conditional knockout mice: crypt cell proliferation is decreased, enteroendocrine and Paneth cells are decreased, and goblet-like cells are increased. GATA6 is co-expressed with SPDEF in the same lineages in the small intestine [4]. Based on these findings [4], we hypothesized that GATA4/GATA6 regulates crypt cell proliferation and secretory cell differentiation in the small intestine by repressing *Spdef* gene expression. To test this hypothesis, we defined the functions most likely regulated by the overlapping GATA6/SPDEF target gene set in mouse intestine, delineated the relationship between GATA6 chromatin occupancy and *Spdef* gene regulation in Caco-2 cells, and determined the extent to which prevention of *Spdef* up-regulation by *Spdef* knockout rescues the GATA6 phenotype in conditional *Gata6* knockout mouse ileum.

Results and discussion
GATA6 and SPDEF regulate similar subsets of genes
To gain insight on the relationship between GATA6 and SPDEF in the small intestine, we scanned the overlap of gene targets using publicly available gene profiling data from conditional *Gata6* and *Spdef* knockout mouse intestine [4,9]. Previously, we identified 2564 genes whose expression is altered in ileum by conditional *Gata6* deletion [4]. Network analysis of this gene set indicated an up-regulation of p53 targets and a down-regulation of c-MYC targets (Additional file 1: Figure S1), consistent with a decrease in cellular proliferation. Of the 2564 genes altered by conditional *Gata6* knockout, 83% (2119) were also altered by *Spdef* knockout (Figure 1A), a far greater overlap than would be expected from a similar-sized, randomized allocation of genes (P < 10^{-60}, Fisher's Exact Test). The changes in expression of this subset when *Gata6* is deleted were analyzed by gene set array analysis. Using Database for Annotation, Visualization and Integrated Discovery (DAVID), we conducted functional annotation clustering of all the major pathways (listed in the KEGG, Biocarta and BBID databases). We found that Wnt signaling was the function most likely to be effected (Figure 1B), consistent with regulation of crypt cell proliferation. Using Gene Set Enrichment Analysis (GSEA), a publically available bioinformatics tool that delineates gene expression data for enrichment of pre-defined gene-sets, APC target network was one of the top three networks affected

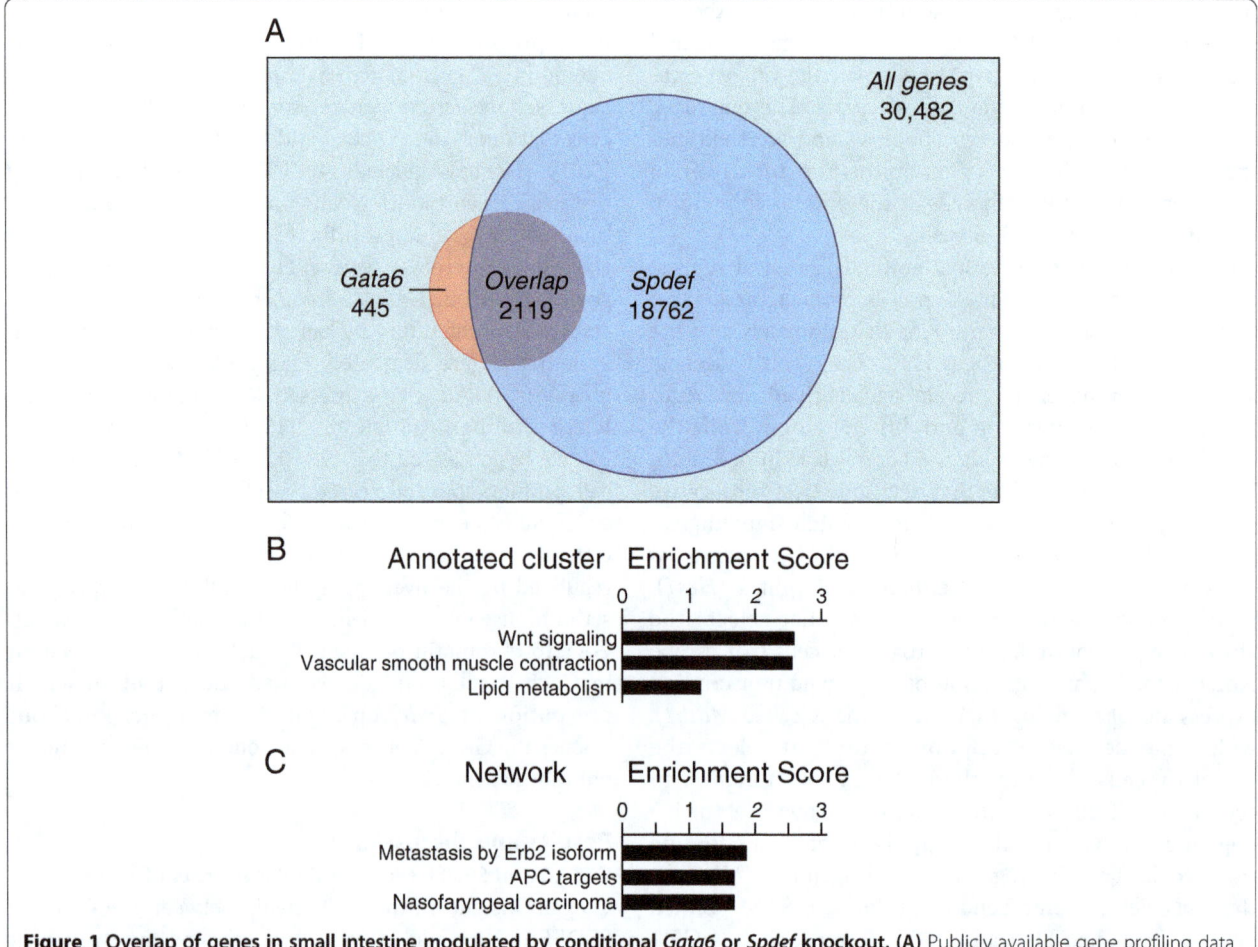

A

All genes
30,482

Gata6 ___
445

Overlap
2119

Spdef
18762

B   Annotated cluster   Enrichment Score

	0	1	2	3

Wnt signaling
Vascular smooth muscle contraction
Lipid metabolism

C   Network   Enrichment Score

	0	1	2	3

Metastasis by Erb2 isoform
APC targets
Nasofaryngeal carcinoma

Figure 1 Overlap of genes in small intestine modulated by conditional *Gata6* or *Spdef* knockout. (A) Publicly available gene profiling data from conditional *Gata6* (GSE22416) and *Spdef* (GSE14892) knockout mouse intestine show a significant overlap. **(B)** Functional annotation clustering of pathways shows that the top-enriched cluster, in overlapping *Gata6* and *Spdef* genes, contains the Wnt pathway as its main function (ES score 2.58). **(C)** GSEA analysis shows, in the overlapping *Gata6* and *Spdef* gene segment, APC targets as the second most highly enriched cluster (ES score 1.57).

by *Gata6* deletion with a Normalized Enrichment Score (NES) of −1.57 (Figure 1C). This analysis reveals a very strong overlap in gene targets between GATA6 and SPDEF, and suggests that the principal function of this overlap involves cellular proliferation.

GATA6 regulates *Spdef* gene expression and occupies a locus in the *Spdef* 5′-flanking region in human Caco-2 cells
Conditional *Gata6* deletion in mice produces an up-regulation of *Spdef* gene expression in ileum while knock-out or over-expression of *Spdef* has no effect on *Gata6* expression [4], suggesting that *Spdef* is regulated down-stream by GATA6. To determine the extent to which this process is conserved, and to further explore the role of GATA6 in regulating *Spdef* gene expression, we char-acterized the effect of *Gata6* knock-down on *Spdef* gene expression in Caco-2 cells. The Caco-2 cell line is a human colorectal adenocarcinoma-derived cell line that expresses GATA6, but very little GATA4 [11], similar to the ileum.

We screened five GATA6 short-hairpin RNA (shRNA) knockdown lentiviral constructs, and found that only one resulted in a statistically significant knockdown of *Gata6* mRNA (55%, P < 0.05, Figure 2A) and a concomitant decrease in GATA6 protein (Figure 2B). *Spdef* mRNA was up-regulated 2-fold in *Gata6* knock-down (*G6kd*) cells (P < 0.05, Figure 2A), demonstrating that the Caco-2 cells model the up-regulation of *Spdef* gene expression observed in conditional *Gata6* knockout mouse ileum [4].

To gain insight on the relationship between GATA6 and *Spdef* gene expression, we examined GATA6 chromatin occupancy at the *Spdef* gene locus. Because we were unable to immunoprecipitate bound chromatin from crosslinked mouse intestinal epithelial cells using existing antisera, we utilized the human Caco-2 intestinal cell culture model in which GATA6 ChIP assays have been performed previously [12]. Using a publicly available GATA6 chromatin immunoprecipitation-high throughput sequencing (ChIP-seq) database in Caco-2 cells [12], we

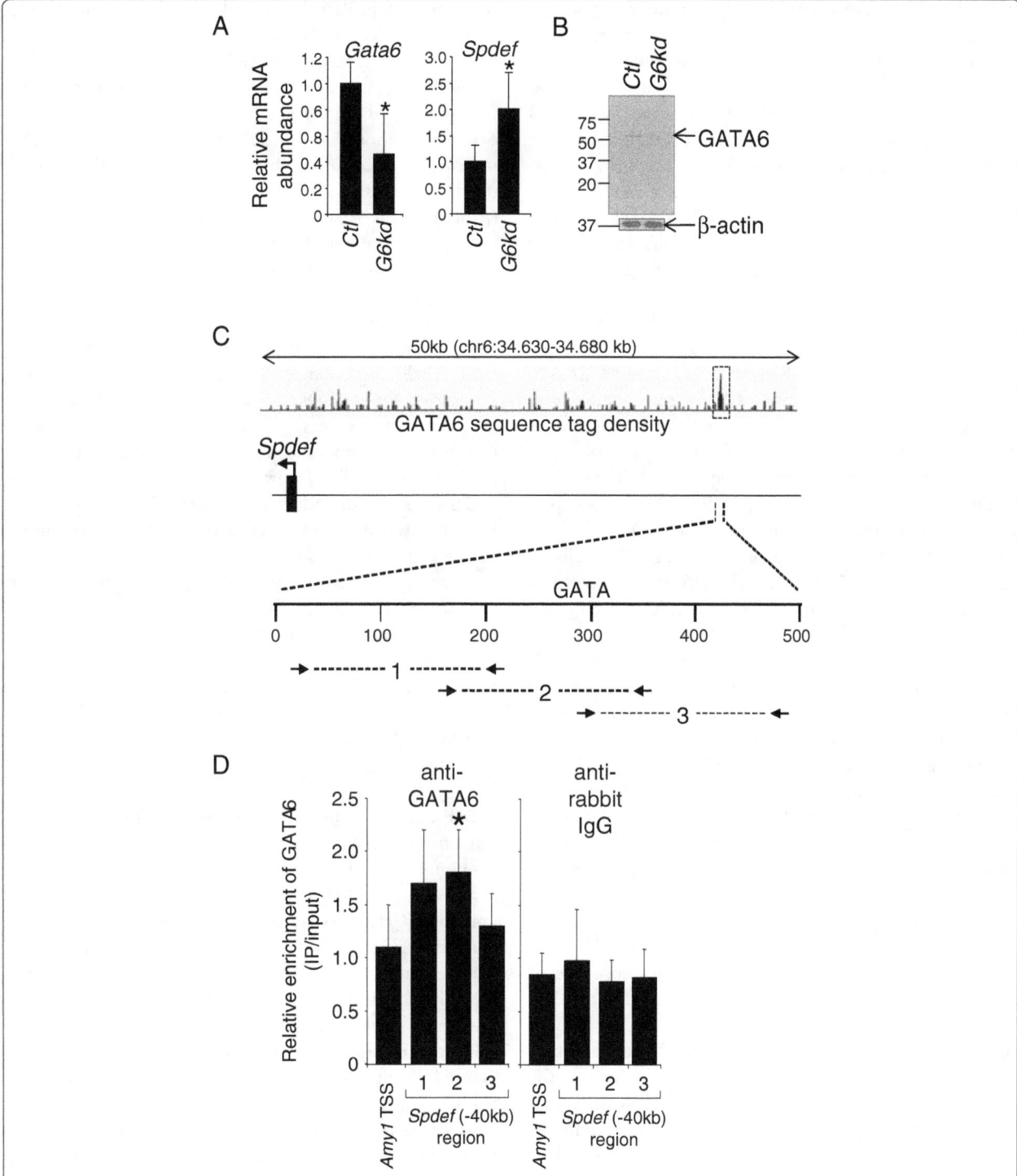

Figure 2 GATA6 regulates *Spdef* expression and occupies an enhancer in the *Spdef* 5'-flanking region in Caco-2 cells. (A) Quantitative RT-PCR analysis showing reduced expression of *Gata6* mRNA, and enhanced expression of *Spdef* mRNA in Caco-2 cells infected with a lentivirus vector expressing an shRNA for human *Gata6* mRNA (*G6kd*) (mean ± SD, n = 5, *P < 0.05). An shRNA vector for GFP was used as a control (*Ctl*). **(B)** Western analysis showing reduced abundance of GATA6 in *G6kd* Caco-2 cells. B-actin was used as an internal loading control. **(C)** Schematic representation of the human *Spdef* 5'-flanking region showing GATA6 occupancy at a locus ~40 kb upstream of the transcription start site (TSS). GATA6 sequence tag density is shown as a 'wiggle file', and a statistically significant GATA6-occupied locus was defined by MACS peak analysis [29] (*dotted box*). **(D)** ChIP assays on chromatin obtained from Caco-2 cells using a GATA6 antibody and three sets of overlapping primers centered on the GATA motif showing increased GATA6 occupancy (mean ± SD, n = 4, *P < 0.05). The salivary amylase-α1a (*Amy-1*) TSS was used as a negative control. ChIP assays using anti-rabbit IgG was used as a control for non-specific immunoprecipitation and primer efficiency.

identified a GATA6-occupied *cis*-regulatory region approximately 40 kb upstream of the *Spdef* transcription start site (TSS) that mapped to the *Spdef* gene based on closest distance to any TSS (Figure 2C). To confirm and localize GATA6 occupancy at this locus, we performed GATA6 ChIP assays using a multiple, overlapping primer strategy with amplicons of ~200 bp per primer pair (Figure 2C) centered at the only GATA motif within this region. We found increases of all three amplicons with amplicon from primer set 2 being significantly greater (~70%, P < 0.05, n = 4) than the *Amy1* TSS negative control (Figure 2D). Although our ChIP data reveal a modest increase in enrichment, it is nonetheless statistically significant and thus confirmatory of ChIP-seq data conducted by others [12] (see MACS peak data, Figure 2C). We also conducted an IgG control ChIP assay to show that enrichment with primer set 2 was not due to differences in primer efficiency from that of the *Amy1* TSS, or to non-specific antibody binding (Figure 2D). These data confirm previous ChIP-seq analysis that GATA6 occupies a *cis*-regulatory region that maps to the *Spdef* gene [12], and, together with *Spdef* up-regulation in *G6kd* cells, is consistent with the notion that GATA6 directly represses *Spdef* gene transcription. While mapping of occupancy sites to the nearest TSS is a well recognized method for defining transcription factor targets on a global basis [12,13], direct regulation of *Spdef* gene expression by GATA6 at this site will need to be confirmed using chromosome conformation capture or other techniques [14].

Transcriptional repression is highly complex, especially in eukaryotes, generally involving the recruitment of specific co-repressors and the local modification of histone tails and chromatin structure [15]. GATA factors are well known to mediate gene repression, and the mechanisms are beginning to be understood. GATA4 has been shown to directly repress cardiac genes [16] while GATA1 has been shown to directly repress hematopoietic genes [17,18]. Both GATA4 and GATA1 interact directly with corepressor complexes including the nucleosome remodeling and histone deacetylase (NuRD) complex and the polycomb repressive complex 2 (PRC2) [16], and are necessary for the multiple modifications of histone tails, but the extent to which GATA factors recruit co-repressors and/or modulate histone tails in the small intestine remains to be determined.

Spdef knockout rescues the proliferation defect in conditional *Gata6* knockout mice

We next asked to what extent *Spdef* deletion rescues the conditional *Gata6* knockout phenotype by analyzing single and double *Gata6/Spdef* knockout mice. In previous studies, conditional deletion of *Gata6* [4] or deletion of *Spdef* [9] resulted in greatly diminished levels of *Gata6* or *Spdef* mRNA, respectively, in the intestine that correlated

with reduced protein levels and altered intestinal phenotypes. Analysis of mRNA in the present study showed that *Gata6* and *Spdef* mRNAs were both expressed in *Ctl* mice, *Gata6* mRNA was nearly undetectable while *Spdef* mRNA was expressed in *Gata6ΔIE* mice, *Gata6* mRNA was expressed while *Spdef* mRNA was nearly undetectable in *SpdefKO* mice, and both *Gata6* and *Spdef* mRNAs were nearly undetectable in the double knockout (*DKO*) mice (Additional file 1: Figure S2), verifying our models.

Crypt cell proliferation is essential for the continuation of intestinal epithelial renewal. One of the principal phenotypic outcomes of conditional *Gata4/Gata6* deletion is a ~30% reduction in the number of Ki67- and BrdU-positive cells, and a concomitant decrease in villus height and villus epithelial cell number [4] resulting in a reduction in absorptive surface area. To define the role of SPDEF in mediating the decrease in crypt cell proliferation when *Gata6* is conditionally deleted, we stained ileal segments for Ki67 and BrdU (Figure 3A), markers of the non-G_0 and S-phases of the cell cycle, respectively. The number of Ki67-positive and BrdU-positive crypt cells was significantly reduced ~40% in *Gata6ΔIE* mice (Figure 3B), as previously reported [4], but unchanged in *SpdefKO* mice, as compared to controls. The number of Ki67- and BrdU-positive crypt cells in *DKO* mice was also similar to controls, indicating that the additional deletion of *Spdef* rescues the decrease in crypt cell proliferation observed in *Gata6ΔIE* mice. These data support the notion that GATA6 maintains crypt cell proliferation by down-regulating *Spdef* gene expression.

SPDEF does not regulate the secretory cell differentiation function of GATA6

Previously, conditional deletion of *Gata6* in the ileum resulted in a 25-40% reduction in the number of chromogranin A (CHGA)-positive cells, and in the mRNA abundances of *Chga* and *Neurog3* [4], consistent with a decrease in enteroendocrine lineage commitment. Though not statistically significant, we found a 15-50% reduction in the number of CHGA-positive cells (Figure 4A), and in the mRNA abundances of *Chga* and *Neurog3* (Figure 4B) in the *Gata6ΔIE* mice, in general agreement with our previous study [4]. We further found that although the pattern for these three measurements in the *SpdefKO* mice was similar to that in *Ctl* mice, consistent with previous data for distal intestine [9], the pattern in the *DKO* mice was similar to that in the *Gata6ΔIE* mice (Figure 4A and B). Together, these data show that enteroendocrine cell commitment is reduced in both *Gata6ΔIE* and *DKO* mice, indicating that GATA6 promotes enteroendocrine cell commitment independently of SPDEF.

Previously, we also showed that conditional *Gata6* deletion resulted in a transformation of Paneth cells into Mucin-2 (MUC2)-enriched goblet-like cells at the base of

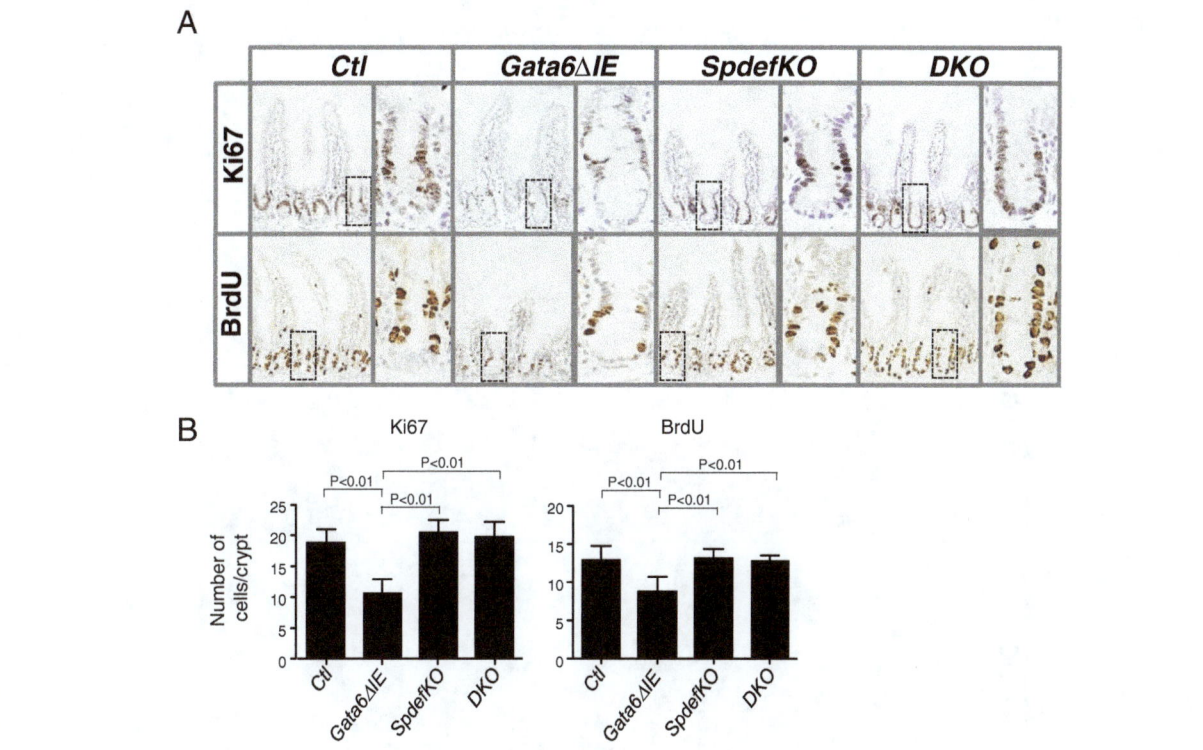

Figure 3 Crypt cell proliferation is decreased in *Gata6ΔIE* mice. (A) Immunostaining in ileum for the proliferation markers Ki67 (*top row*) and BrdU (*bottom row*). **(B)** Quantification of Ki67- and BrdU-positive cells/crypt in each group. Cells were counted as described in Methods. *Gata6ΔIE* mice had significantly fewer Ki67- or BrdU-positive cells than each of the other three groups (P < 0.01 in each case), consistent with a decrease in crypt cell proliferation.

crypts [4]. In the present study, immunostaining for CRS4C, a Paneth-specific marker, was slightly reduced in *SpdefKO* mice as compared to controls, consistent with previous data [9], but was greatly reduced in both *Gata6ΔIE* and *DKO* mice as compared to controls (Figure 5A). The number of crypt cross-sections with at least one CRS4C-positive cell was nearly 100% in *Ctl* and *SpdefKO* mice, but was less than 30% in both *Gata6ΔIE* and *DKO* mice (P < 0.01 for each, Figure 5B).

The mRNA abundance for lysozyme (*Lyz*), another Paneth marker, was not significantly different between *Ctl* and *SpdefKO* mice, but was significantly reduced in *Gata6ΔIE* and *DKO* mice as compared to controls (P < 0.01 for each, Figure 5C). PAS staining was greatly increased in crypts of *Gata6ΔIE* mice, as previously shown [4], noticeably reduced in *SpdefKO* mice, especially in villus goblet cells, as previously shown [9], and similar to controls in *DKO* mice (Figure 5A). MUC2 immunostaining was greatly

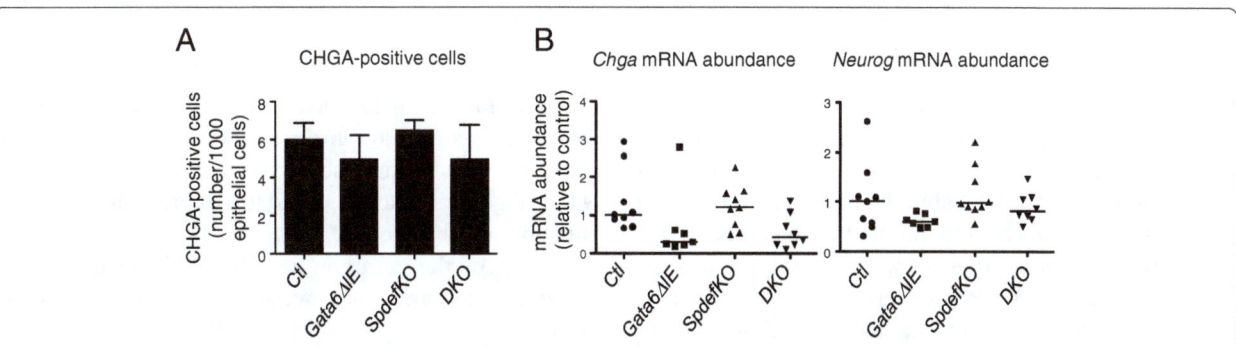

Figure 4 Enteroendocrine cell allocation is decreased in *Gata6ΔIE* and *DKO* mice. (A) Quantification of CHGA-positive cells in each group. Sections of ileum were stained for CHGA, and the number of positive cells/1000 epithelial cells was determined as described in Methods. **(B)** *Chga* and *Neurog3* mRNA abundance in each group. *Gata6ΔIE* and *DKO* mice had generally lower numbers of CHGA-positive cells and *Chga* and *Neurog3* mRNA abundances than *Ctl* mice, consistent with a decrease in enteroendocrine lineage allocation.

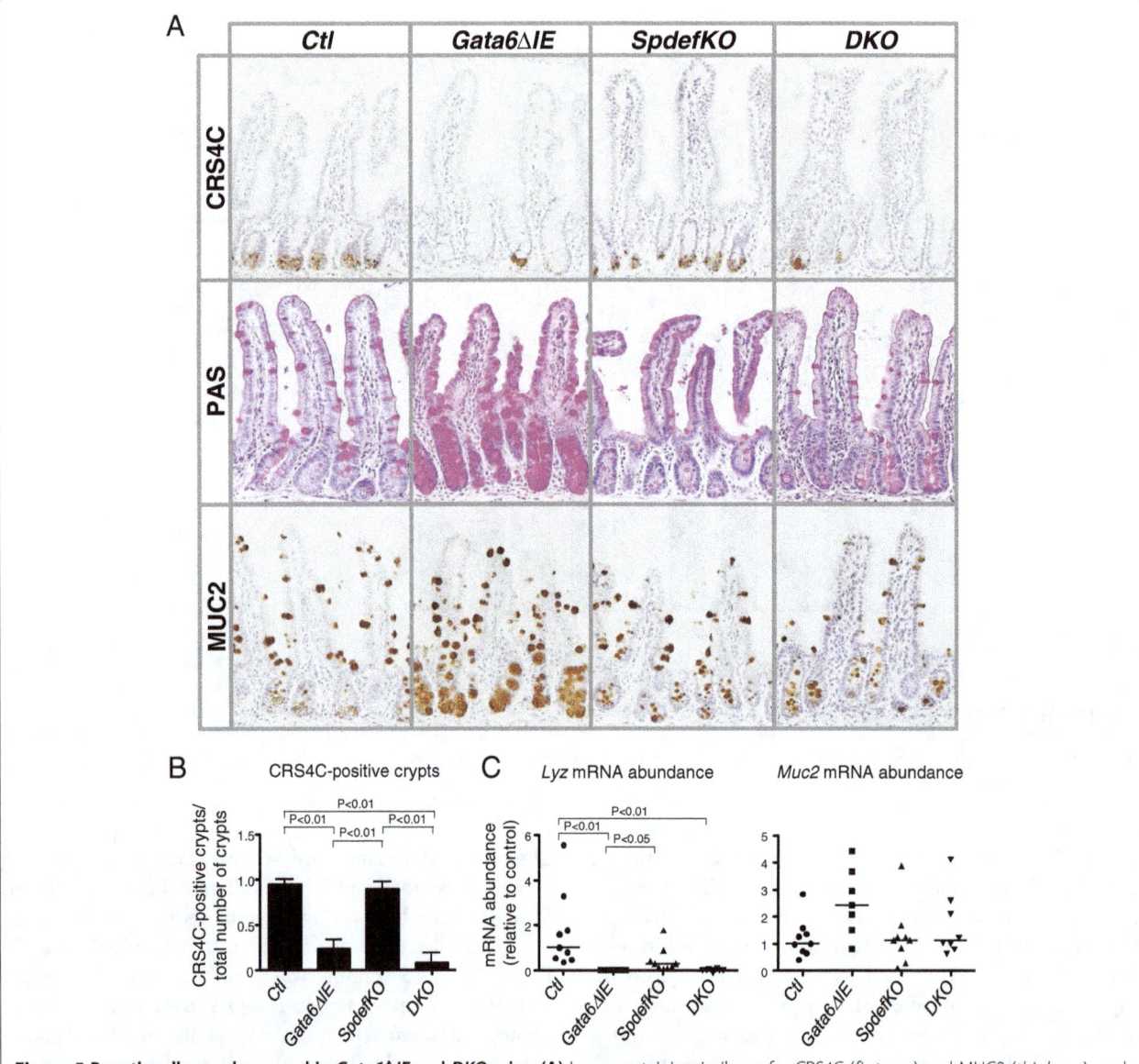

Figure 5 Paneth cells are decreased in *Gata6ΔIE* and *DKO* mice. (A) Immunostaining in ileum for CRS4C (*first row*) and MUC2 (*third row*), and chemical staining using the PAS reaction (*second row*). **(B)** Quantification of CRS4C-positive cells in each group. Sections of ileum were stained for CRS4C, and the number of positive cells/crypt cross section was determined as described in Methods. **(C)** *Lyz* and *Muc2* mRNA abundance in each group. *Gata6ΔIE* and *DKO* mice had significantly lower numbers of CRS4C-positive cells and *Lyz* mRNA abundance than *Ctl* mice, consistent with a decrease in mature Paneth cells.

enriched in crypts of *Gata6ΔIE* mice, consistent with our previous report [4], but similar to controls in *SpdefKO* and *DKO* mice (Figure 5A). Generally consistent with this observation, *Muc2* mRNA abundance in *Gata6ΔIE* mice was 2.5-fold higher as compared to controls, though not statistically significant, whereas *Muc2* mRNA abundances in *SpdefKO* and *DKO* mice were similar to controls (Figure 5C). These data indicate that the additional deletion of *Spdef* did not rescue the decrease in the terminal differentiation of Paneth cells observed in *Gata6ΔIE* mice, but did at least partially prevent their conversion to MUC2-enriched goblet-like cells. Thus, SPDEF is not

necessary for the GATA-mediated differentiation of Paneth cells, but does function in their default differentiation into goblet-like cells in the absence of *Gata6*. This is consistent with the general function of SPDEF to promote the differentiation of goblet cells [9,10].

In spite of the common secretory cell phenotypes in *Gata6* deletion and *Spdef* over-expression models [4,10], our data show that *Spdef* deletion did not rescue the *Gata6* knockout defects in enteroendocrine lineage commitment or in Paneth cell differentiation. Closer scrutiny suggests that the secretory cell phenotypes in the *Gata6* deletion and *Spdef* over-expression models are not identical. While the decline in enteroendocrine and Paneth cells, and accumulation of goblet cells, appear similar, the underlying alteration in Paneth and goblet cell differentiation is different. Conditional *Gata6* deletion has no effect on the normal commitment and differentiation of goblet cells on villi but reveals a conversion of Paneth cells at the base of crypts into a goblet-like cell type. These cells do not express defensins, but express abundant *Muc2;* they also express abundant *Sox9* and *Ephb3* [4], indicating that they are committed and targeted Paneth cells. On the other hand, conditional *Spdef* over-expression results in a generalized increase in goblet cells at the expense of all other epithelial cell types [10]. Hence, these mice show a decline in Paneth cell specification rather than a defect in the terminal differentiation of committed Paneth cells, as observed in our conditional *Gata6* knockout model.

SPDEF does not regulate the absorptive enterocyte gene expression function of GATA6

We next examined whether *Spdef* loss affected GATA6-dependent absorptive enterocyte gene expression. Conditional *Gata6* deletion resulted in a down-regulation of specific absorptive enterocyte genes in ileum that include lipid transporters and apolipoproteins, and an up-regulation of genes in absorptive enterocytes normally not expressed or expressed at low levels in small intestine, but expressed at high levels in colon [4]. Using marker genes for these two patterns, apolipoprotein A1 (*Apoa1*) and carbonic anyhdrase 1 (*Car1*), respectively, we found that their mRNA abundances were down-regulated and up-regulated, respectively, in *Gata6ΔIE* mice (Figure 6), consistent with our previous findings [4]. *Apoa1* and *Car1* mRNA abundance in *SpdefKO* mice was similar to controls, whereas that in *DKO* mice was similar to *Gata6ΔIE*, indicating that SPDEF does not play a role

in mediating the conditional *Gata6* knockout alteration in absorptive enterocyte gene expression.

GATA6 is expressed in crypts and in mature absorptive enterocytes [4,5], whereas SPDEF is expressed in secretory progenitors, and in goblet and Paneth cells, but is not expressed in mature absorptive enterocytes [9,10]. While it is possible that SPDEF could instruct gene expression in absorptive enterocytes through a process that originates in progenitors early in the differentiation process, we found that SPDEF did not regulate the GATA6 targets studied here. As in the GATA4-specific pathway, in which GATA4 regulates its targets in mature absorptive enterocytes on villi rather than in crypt progenitors [19], we believe that GATA6 also regulates absorptive enterocyte gene expression within mature absorptive enterocytes on villi.

Conclusion

Previously, we defined two fundamental pathways of GATA regulation in the small intestine, one mediated exclusively by GATA4 (GATA4-specific pathway), and one regulated by GATA4 or GATA6 (GATA4/GATA6-redundant pathway) (Figure 7). In the GATA4-specific pathway, GATA4, but not GATA6, activates and represses a subset of absorptive enterocyte genes, and by virtue of its expression in the proximal 85% of small intestine and lack of expression in distal ileum [7], distinguishes proximal intestinal from distal ileal gene expression and function [5,7,8]. For the GATA4/GATA6-redundant pathway, either GATA4 or GATA6 (which is expressed throughout the small intestine, including distal ileum) promote intestinal epithelial renewal by supporting crypt cell proliferation, enteroendocrine lineage commitment, Paneth cell differentiation, and absorptive enterocyte gene expression [4]. Here, we show that the maintenance of crypt cell proliferation function of the intestinal GATA4/GATA6-redundant pathway is dependent

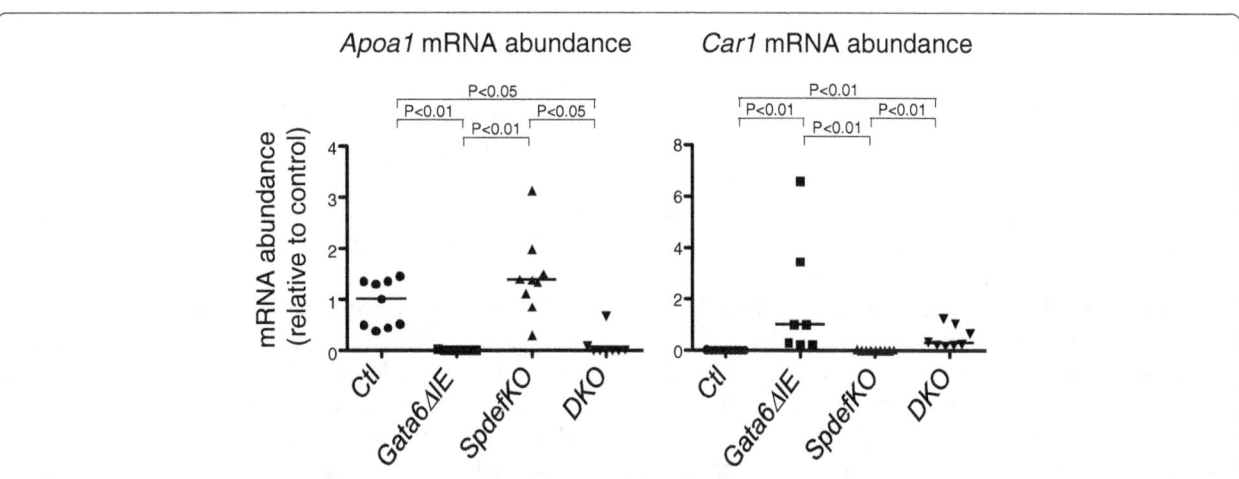

Figure 6 Absorptive enterocyte gene expression is altered in *Gata6ΔIE* and *DKO* mice. *Gata6ΔIE* and *DKO* mice had significantly lower *Apoa1*, and significantly higher *Car1* mRNA abundances than *Ctl* mice.

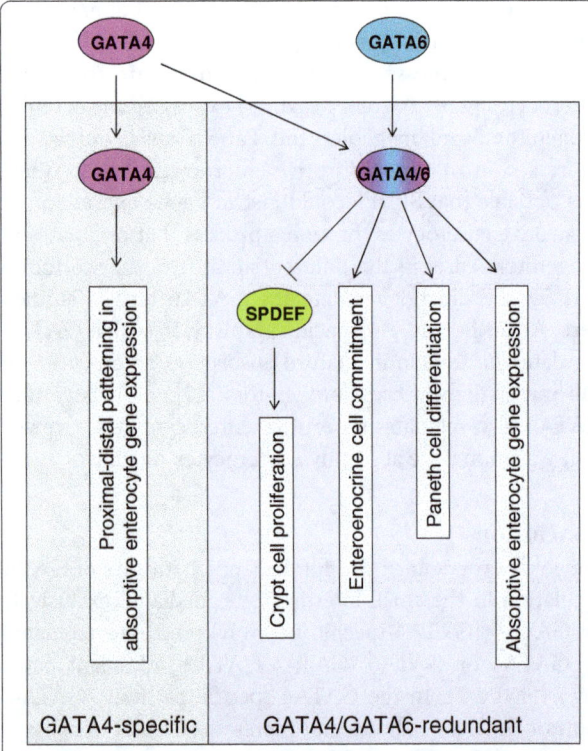

Figure 7 Model for the known GATA pathways in mature small intestine, and the placement of SPDEF within those pathways. GATA4 (*purple*), but not GATA6 (*blue*), activates and represses a subset of absorptive enterocyte genes (GATA4-specific), whereas either GATA4 or GATA6 (*purple/blue*) promote crypt cell proliferation, enteroendocrine lineage commitment, Paneth cell differentiation, and absorptive enterocyte gene expression (GATA4/GATA6-redundant). GATA4/GATA6 promotes the crypt cell proliferation function by repressing the gene encoding SPDEF (*green*).

on SPDEF (Figure 7). Specifically, our data support the notion that GATA4/GATA6 promotes crypt cell proliferation by directly repressing *Spdef* gene expression.

GATA6 has been suggested to be oncogenic in multiple cancers and pre-cancerous lesions, including Barrett's esophagus, gastric cancer, pancreatic cancer, and colon cancer [20]. *Gata6* expression is up-regulated in colon cancer epithelial cells [21-23], as well as in non-malignant cells along the stromal margins in human colorectal cancer [22], but it should be noted that it has not been determined whether this is a correlative, causative or protective event. Recently, Noah et al. [24] showed that SPDEF functions as a colorectal tumor suppressor. In colorectal tumors from patients, loss of SPDEF was observed in approximately 85% of tumors and correlated with progression from normal tissue, to adenoma, to adenocarcinoma. Further, SPDEF inhibited the expression of β-catenin-target genes in mouse colon tumors, and interacted with β-catenin to block its transcriptional activity in colorectal cancer cell lines, resulting in lower levels of cyclin D1 and c-MYC. Pathway analysis showed that *Gata6* deletion also modulated

c-MYC targets (Additional file 1: Figure S1) and that the GATA6/SPDEF overlapping gene set likely functions in Wnt signaling (Figure 1). Thus, our data here suggest that the possible tumorigenic effects of GATA6 could be mediated by its repression of *Spdef* gene expression.

Methods
Analysis of publicly available gene profiling data
From the National Center for Biotechnology Information (NCBI) website (http://www.ncbi.nlm.nih.gov/), we downloaded gene profiling data from the ileum of conditional *Gata6* knockout mice (GSE22416) [4], and from the small intestine of *Spdef−/−* mice (GSE14892) [9]. Data analysis was performed using Cistrome (www.cistrome.org), a flexible bioinformatics workbench with an analysis platform for ChIP-seq and gene expression microarray analysis [25]. Significant over-representation of overlapping target genes was determined by the Fisher's Exact Test. Gene ontology analysis and pathway analysis was conducted using Database for Annotation, Visualization and Integrated Discovery (DAVID) [26]. Gene set enrichment analysis was performed by using publicly available Gene Set Enrichment Analysis (GSEA) tools v2.3 from the Broad Institute [27].

Cell culture and lentiviral infection
Lentiviral infection was conducted as previously described [28]. 293 T and Caco-2 cells were cultured in Dulbecco's modified Eagle's medium (DMEM; Mediatech, Inc., Manassas, VA) containing glucose (4.5 g/liter), l-glutamine, and sodium pyruvate. The media was supplemented with 10% fetal bovine serum (Atlanta Biologicals, Lawrenceville, GA) and penicillin (100 U/ml)-streptomycin (1 mg/ml) (Sigma Chemical Company, St. Louis, MO). Both cell lines were maintained in 5% CO_2 at 95% relative humidity and 37°C. Media were replaced every 2 to 3 days. To produce lentivirus for the knockdown of *Gata6*, 293 T cells were plated at 10^5 cells/ml on 6-cm plates and transfected after 24 h with 1 μg of vesicular stomatitis virus glycoprotein G (VSV-G) envelope-expressing plasmid pMD. G, 1 μg of pCMV-dR8.91 (Delta 8.9) plasmid containing *gag*, *pol*, and *rev* genes, and 1 μg of shRNA-pLKO.1 plasmids expressing a knockdown short hairpin RNA (shRNA) (Sigma) for green fluorescent protein (GFP) (used as a control [*Ctl*]) (SHC005) or human *Gata6* (*G6kd*) using 6 μl of FuGENE 6 reagent (Roche Diagnostic Corporation, Indianapolis, IN). A total of five human *Gata6* knockdown constructs were screened, and only one (TRCN0000005392) was subsequently found to reduce *Gata6* mRNA significantly. Media from the transfected 293 T cells were changed 16 h after transfection. Culture media containing lentiviral particles were collected and filtered (0.45 μm pore size) 48 h after transfection and transferred immediately to Caco-2 cells plated at 30% confluence to infect for 2 h using 0.5 μl of polybrene/ml.

Infection was repeated the next day followed by selection of infected cells using 4 µg/ml puromycin. Infected cells were kept under conditions of selection until the day of harvesting. Trypsinized cells were homogenized using a QIA shredder (Qiagen, Inc., Valencia, CA), and RNA was isolated as described above for mouse tissue. Nuclear extracts were isolated as described previously [11]. Gata6 knockdown was determined by quantitative reverse transcriptase-(qRT-) PCR for human Gata6 (see Additional file 1: Figure S3A for primers) and by Western blot analysis as previously described [5] using rabbit anti-GATA6 (Cell Signaling Technology, Inc, Danvers, MA; Cat. No. 4253S).

Chromosomal immunoprecipitation (ChIP) assays

Using a publicly available GATA6 chromatin immunoprecipitation-high throughput sequencing (ChIP-seq) database in Caco-2 cells [12], sequences were mapped to reference genome *Homo Sapiens* build 18 (HG18) using ELAND tools, allowing 0 to 2 mismatches (Illumina), and binding peaks were identified by model-based analysis of ChIP-seq (MACS) [29] using default parameters and P value cutoffs of 10^{-10}.

Chromosomal immunoprecipitation (ChIP) assays were conducted as previously described [28]. Caco-2 cells were incubated in DMEM containing 1% formaldehyde (Fisher Scientific, Pittsburgh, PA) for 10 min at 37°C. The cells were washed 2 times with PBS, scraped, and resuspended in lysis buffer (50 mM Tris-Cl [pH 8.1], 10 mM EDTA, 1% sodium dodecyl sulfate [SDS]) containing protease inhibitor cocktail and PMSF (10 µl/ml). The samples were sonicated to obtain chromatin fragments of between 400 and 1,000 bp. Sonicated samples were resuspended in ChIP dilution buffer (1% Triton X-100, 2 mM EDTA, 150 mM NaCl, 20 mM Tris–HCl, pH 8.1) and incubated overnight at 4°C with Dynabeads beads (Life Technologies, Grand Island, NY) conjugated with rabbit anti-GATA6 (Cell Signaling; Cat. No. 4253) or rabbit IgG (negative control) (Millipore, Temecula, CA; Cat. No. PP64B). The IP samples were washed 6 times with radioimmunoprecipitation (RIPA) buffer (50 mM HEPES [pH 7.6], 0.5 M LiCl, 1 mM EDTA, 1% Nonidet P-40, 0.7% sodium deoxycholate), and the DNA was recovered by reverse cross-linking in 1% SDS–0.1 M NaHCO$_3$ for 7 h at 65°C. DNA was purified using a QIAquick PCR purification kit (Qiagen) and quantified by Picogreen (Life Technologies). One nanogram of DNA was used per qPCR reaction using primers shown in Additional file 1: Figure S3B.

Mice

Previously established and confirmed Gata6$^{loxP/loxP}$ [30], Spdef−/− [9], and transgenic VillinCreER^{T2} [31] mice were used in this study to produce four groups of mice, including controls (*Ctl*), conditional Gata6 deletion in the intestinal epithelium (*Gata6ΔIE*), germline Spdef knockout

(*SpdefKO*), and double Gata6/Spdef knockout (*DKO*). The genotypes of each group are as follows:

Ctl: Gata6$^{loxP/loxP}$, Spdef +/−, VillinCreER^{T2}-negative
Gata6ΔIE: Gata6$^{loxP/loxP}$, Spdef +/−, VillinCreER^{T2}-positive
SpdefKO: Gata6$^{loxP/loxP}$, Spdef −/−, VillinCreER^{T2}-negative
DKO: Gata6$^{loxP/loxP}$, Spdef −/−, VillinCreER^{T2}-positive

DNA was extracted from tail biopsies, and genotypes were determined by semiquantitive polymerase chain reaction (PCR) using previously validated primers (Additional file 1: Figure S3C). Male and female mice four weeks of age were treated with a single dose of tamoxifen (Sigma) (0.1 ml, 10 mg/ml; dissolved in ethanol/sunflower oil = 1:9 [vol/vol]) daily for five consecutive days as described [5], followed by a single dose two weeks later (see timeline, Additional file 1: Figure S4). Mice were killed and tissue was collected 28 days after the start of tamoxifen treatment. Bromodeoxyuridine (BrdU, 0.1 ml of 10 mg/ml) was injected two hours before dissection. Approval was obtained from the Institutional Animal Care and Use Committee.

Tissue isolation and processing

Mice were anesthetized for dissection as previously described [28]. The most distal 1.0 cm segment of small intestine adjacent to the ileocecal valve was snap frozen for RNA isolation, and the 6 cm segment proximal to that was removed and transferred to a glass plate on wet ice and prepared for sectioning. This segment was flushed with ice cold 4% paraformaldehyde (PFA) in PBS, cut longitudinally, pinned open onto paraffin wax in a petri dish filled with 4% PFA in PBS with the epithelium facing upward. After a 5 min incubation period, the pins were removed and the proximal end was grasped with forceps, rolled with the epithelium facing outwards, and a pin was inserted transversely to secure the roll in place. The roll was placed in 10 ml of 4% PFA in PBS and mixed gently on a tube rotator for 16–18 hr at 4°C. The PFA was decanted and the tissue was washed in PBS 3 times for 20 min each at 4°C, and then dehydrated in 70% ethanol. Tissue was processed at the Rodent Histopathology Core at the Dana Farber/Harvard Cancer Center (Boston, MA) for paraffin embedding and sectioning. Selected slides were stained using the periodic acid Schiff (PAS) reaction.

RNA isolation and gene expression analysis

RNA was isolated using the RNeasy kit (Qiagen), and mRNA abundances were determined by qRT-PCR as previously described [4], using validated primer pairs (Additional file 1: Figure S3A). Glyceraldehyde-3-phosphate dehydrogenase (GAPDH) mRNA abundance was measured for each sample and used to normalize the data. Data were expressed relative to the median value of control ileum. A minimum of five mice in each group was analyzed.

Immunohistochemistry

Tissue sections were immunostained as previously described [28]. Primary antibodies included rabbit anti-Ki67 (Thermo Fisher Scientific, Inc., Fremont, CA; Cat. No. RM-9106-S1) (1:200), mouse anti-BrdU (Thermo; Cat. No. MS-1058-PO) (1:250), goat anti-cryptdin related sequence 4C (CRS4C) (gift from Dr. A. J. Ouellette, University of Southern California, Los Angeles, CA) [32] (1:2000), and rabbit anti-MUC2 (Santa Cruz Biotechnology, Inc, Santa Cruz, CA; Cat. No. sc15334) (1:100). Secondary antibodies included biotinylated donkey anti-rabbit IgG, donkey anti-goat IgG, and donkey anti-mouse IgG (all from Vector Labs, Burlingame, CA). Biotinylated antibodies were linked to avidin-horseradish peroxidase conjugates (Vector Labs), visualized using 3,3′-diamino benzidine (Sigma) for 2 to 5 min, and lightly counterstained with hematoxylin.

Cell counting

The total number of Ki67-, BrdU-, or CRS4C-positive cells in crypts was determined as the total number per crypt. Only well oriented crypts with the epithelial layer on at least one side continuous with the villus epithelial layer were counted, and a minimum of 6 crypts per slide were analyzed. The average number of CHGA−positive cells was expressed as a fraction of total epithelial cells (villi and crypts) from a minimum of 5000 epithelial cells per slide, with equal representation of crypts and villi. All determinations were blinded and conducted on a minimum of 5 animals per group.

Statistical analyses

In Caco-2 cells, qRT-PCR data was compared by the student's t-test and ChIP data were compared by the analysis of variance (ANOVA) followed by the Tukey-Kramer multiple comparison test. In mice, mRNA measurements had unequal variances across groups requiring nonparametric statistics, and were thus compared by the Kruskal-Wallis test followed by the Dunn multiple comparison test, and presented as individual data points and medians. Cell count determinations had equal variances across groups allowing parametric statistics, and were thus compared by ANOVA followed by the Tukey-Kramer multiple comparison test, and presented as mean ± SD. Differences were considered statistically significant at a P-value of less than 0.05.

Additional file

Additional file 1: Figure S1. Intestinal *Gata6* deletion alters gene networks controlling cell proliferation in the mature ileum. Network analyses on microarray data of *Ctl* and *Gata6ΔIE* ileum (n=3 in each group) revealed (A) an increase in targets of the tumor surpressor gene

p53, and (B) a decrease in targets of the proto-oncogene c-MYC. (C) Legend defining symbols used in the networks. Arrows indicate the direction of the interaction. Red circles = up-regulated transcripts; Blue circles = down-regulated transcripts. Differentially expressed transcripts were determined at the 5% FDR level using Significance Analysis of Microarrays (SAM) and interaction networks were developed from the differentially expressed transcripts using Metacore. **Figure S2.** *Gata6* and *Spdef* mRNA abundance in ileum in each group of mice. *Gata6ΔIE* and *DKO* mice had significantly lower *Gata6* mRNA abundances than *Ctl* and *SpdefKO* mice. *SpdefKO* and *DKO* mice had significantly lower *Spdef* mRNA abundances than *Ctl* and *Gata6ΔIE* mice. **Figure S3.** Primers used for: (A) qRT-PCR, (B) ChIP assays, and (C) genotyping. **Figure S4.** Timeline for study. Mice 4 wks of age were given Tamoxifen as indicated (*black circle*) beginning on Day 0, and an injection of BrdU 2 hr before tissue collection at Day 28.

Competing interests
The authors declare no competing interest.

Authors' contributions
BEA, LATMV, ES and HB performed and interpreted the Caco-2 cell culture experiments. KAS performed and interpreted the in vivo mouse experiments. BEA, KAS and SDK contributed to the study design and interpretation of results, and the writing of the manuscript. All authors read and approved the final manuscript.

Acknowledgements
We would like to thank Drs. E. Beuling and M. P. Verzi for helpful suggestions; Dr. A. J. Ouellette for the CRS4C antibody; and Dr. H. Clevers for the *Spdef–/–* mice. This work was supported by National Institute of Diabetes and Digestive and Kidney Diseases grant RO1-DK-061382 (S.D.K.), the Harvard Digestive Disease Center (5P30-DK-34854), the Nutricia Research Foundation (B.E.A.), KWF Kankerbestrijding (B.E.A.), Prins Bernhard Cultuurfonds (B.E.A.) in The Netherlands, and the European Society for Pediatric Research (B.E.A.) in Switzerland.

Author details
[1]Division of Gastroenterology and Nutrition, Department of Medicine, Children's Hospital Boston, and Harvard Medical School, 300 Longwood Avenue, Boston, MA 02115, USA. [2]Academic Medical Center Amsterdam, Emma Children's Hospital, Amsterdam, the Netherlands. [3]University Medical Center Groningen, Groningen, the Netherlands.

References
1. Noah TK, Donahue B, Shroyer NF: **Intestinal development and differentiation.** *Exp Cell Res* 2011, **317**(19):2702–2710.
2. Gerbe F, van Es JH, Makrini L, Brulin B, Mellitzer G, Robine S, Romagnolo B, Shroyer NF, Bourgaux JF, Pignodel C, *et al*: **Distinct ATOH1 and Neurog3 requirements define tuft cells as a new secretory cell type in the intestinal epithelium.** *J Cell Biol* 2011, **192**(5):767–780.
3. Yeung TM, Chia LA, Kosinski CM, Kuo CJ: **Regulation of self-renewal and differentiation by the intestinal stem cell niche.** *Cell Mol Life Sci* 2011, **68**(15):2513–2523.
4. Beuling E, Baffour-Awuah NY, Stapleton KA, Aronson BE, Noah TK, Shroyer NF, Duncan SA, Fleet JC, Krasinski SD: **GATA factors regulate proliferation, differentiation, and gene expression in small intestine of mature mice.** *Gastroenterology* 2011, **140**(4):1219–1229. e1211-1212.
5. Bosse T, Piaseckyj CM, Burghard E, Fialkovich JJ, Rajagopal S, Pu WT, Krasinski SD: **Gata4 is essential for the maintenance of jejunal-ileal identities in the adult mouse small intestine.** *Mol Cell Biol* 2006, **26**(23):9060–9070.
6. van Wering HM, Bosse T, Musters A, de Jong E, de Jong N, Hogen Esch CE, Boudreau F, Swain GP, Dowling LN, Montgomery RK, *et al*: **Complex regulation of the lactase-phlorizin hydrolase promoter by GATA-4.** *Am J Physiol Gastrointest Liver Physiol* 2004, **287**(4):G899–G909.
7. Beuling E, Kerkhof IM, Nicksa GA, Giuffrida MJ, Haywood J, aan de Kerk DJ, Piaseckyj CM, Pu WT, Buchmiller TL, Dawson PA, *et al*: **Conditional Gata4 deletion in mice induces bile acid absorption in the proximal small intestine.** *Gut* 2010, **59**(7):888–895.

8. Battle MA, Bondow BJ, Iverson MA, Adams SJ, Jandacek RJ, Tso P, Duncan SA: **GATA4 is essential for jejunal function in mice.** *Gastroenterology* 2008, **135**(5):1676–1686. e1671.

9. Gregorieff A, Stange DE, Kujala P, Begthel H, van den Born M, Korving J, Peters PJ, Clevers H: **The ets-domain transcription factor Spdef promotes maturation of goblet and paneth cells in the intestinal epithelium.** *Gastroenterology* 2009, **137**(4):1333–1345. e1331-1333.

10. Noah TK, Kazanjian A, Whitsett J, Shroyer NF: **SAM pointed domain ETS factor (SPDEF) regulates terminal differentiation and maturation of intestinal goblet cells.** *Exp Cell Res* 2010, **316**(3):452–465.

11. Krasinski SD, Van Wering HM, Tannemaat MR, Grand RJ: **Differential activation of intestinal gene promoters: functional interactions between GATA-5 and HNF-1 alpha.** *Am J Physiol Gastrointest Liver Physiol* 2001, **281**(1):G69–G84.

12. Verzi MP, Shin H, He HH, Sulahian R, Meyer CA, Montgomery RK, Fleet JC, Brown M, Liu XS, Shivdasani RA: **Differentiation-specific histone modifications reveal dynamic chromatin interactions and partners for the intestinal transcription factor CDX2.** *Dev Cell* 2010, **19**(5):713–726.

13. Yang J, Mitra A, Dojer N, Fu S, Rowicka M, Brasier AR: **A probabilistic approach to learn chromatin architecture and accurate inference of the NF-kappaB/RelA regulatory network using ChIP-Seq.** *Nucleic Acids Res* 2013, **41**(15):7240–7259.

14. de Wit E, de Laat W: **A decade of 3C technologies: insights into nuclear organization.** *Genes Dev* 2012, **26**(1):11–24.

15. Payankaulam S, Li LM, Arnosti DN: **Transcriptional repression: conserved and evolved features.** *Curr Biol* 2010, **20**(17):R764–R771.

16. Zhou P, He A, Pu WT: **Regulation of GATA4 transcriptional activity in cardiovascular development and disease.** *Curr Top Dev Biol* 2012, **100**:143–169.

17. Yu M, Riva L, Xie H, Schindler Y, Moran TB, Cheng Y, Yu D, Hardison R, Weiss MJ, Orkin SH, et al: **Insights into GATA-1-mediated gene activation versus repression via genome-wide chromatin occupancy analysis.** *Mol Cell* 2009, **36**(4):682–695.

18. Fujiwara T, O'Geen H, Keles S, Blahnik K, Linnemann AK, Kang YA, Choi K, Farnham PJ, Bresnick EH: **Discovering hematopoietic mechanisms through genome-wide analysis of GATA factor chromatin occupancy.** *Mol Cell* 2009, **36**(4):667–681.

19. Beuling E, Bosse T, aan de Kerk DJ, Piaseckyj CM, Fujiwara Y, Katz SG, Orkin SH, Grand RJ, Krasinski SD: **GATA4 mediates gene repression in the mature mouse small intestine through interactions with friend of GATA (FOG) cofactors.** *Dev Biol* 2008, **322**(1):179–189.

20. Ayanbule F, Belaguli NS, Berger DH: **GATA factors in gastrointestinal malignancy.** *World J Surg* 2011, **35**(8):1757–1765.

21. Belaguli NS, Aftab M, Rigi M, Zhang M, Albo D, Berger DH: **GATA6 promotes colon cancer cell invasion by regulating urokinase plasminogen activator gene expression.** *Neoplasia* 2010, **12**(11):856–865.

22. Haveri H, Westerholm-Ormio M, Lindfors K, Maki M, Savilahti E, Andersson LC, Heikinheimo M: **Transcription factors GATA-4 and GATA-6 in normal and neoplastic human gastrointestinal mucosa.** *BMC Gastroenterol* 2008, **8**:9.

23. Shureiqi I, Zuo X, Broaddus R, Wu Y, Guan B, Morris JS, Lippman SM: **The transcription factor GATA-6 is overexpressed in vivo and contributes to silencing 15-LOX-1 in vitro in human colon cancer.** *FASEB J* 2007, **21**(3):743–753.

24. Noah TK, Lo YH, Price A, Chen G, King E, Washington MK, Aronow BJ, Shroyer NF: **SPDEF Functions as a Colorectal Tumor Suppressor by Inhibiting beta-Catenin Activity.** *Gastroenterology* 2013, **144**(5):1012–1023.

25. Liu T, Ortiz JA, Taing L, Meyer CA, Lee B, Zhang Y, Shin H, Wong SS, Ma J, Lei Y, et al: **Cistrome: an integrative platform for transcriptional regulation studies.** *Genome Biol* 2011, **12**(8):R83.

26. Huang da W, Sherman BT, Zheng X, Yang J, Imamichi T, Stephens R, Lempicki RA, Baxevanis AD, et al: **Extracting biological meaning from large gene lists with DAVID.** *Current Protoc Bioinformatics / editoral board* 2009, **Chapter 13**:Unit 13 11.

27. Subramanian A, Tamayo P, Mootha VK, Mukherjee S, Ebert BL, Gillette MA, Paulovich A, Pomeroy SL, Golub TR, Lander ES, et al: **Gene set enrichment analysis: a knowledge-based approach for interpreting genome-wide expression profiles.** *Proc Natl Acad Sci U S A* 2005, **102**(43):15545–15550.

28. Beuling E, Aronson BE, Tran LM, Stapleton KA, ter Horst EN, Vissers LA, Verzi MP, Krasinski SD: **GATA6 is required for proliferation, migration, secretory cell maturation, and gene expression in the mature mouse colon.** *Mol Cell Biol* 2012, **32**(17):3392–3402.

29. Zhang Y, Liu T, Meyer CA, Eeckhoute J, Johnson DS, Bernstein BE, Nusbaum C, Myers RM, Brown M, Li W, et al: **Model-based analysis of ChIP-Seq (MACS).** *Genome Biol* 2008, **9**(9):R137.

30. Sodhi CP, Li J, Duncan SA: **Generation of mice harbouring a conditional loss-of-function allele of Gata6.** *BMC Dev Biol* 2006, **6**:19.

31. el Marjou F, Janssen KP, Chang BH, Li M, Hindie V, Chan L, Louvard D, Chambon P, Metzger D, Robine S: **Tissue-specific and inducible Cre-mediated recombination in the gut epithelium.** *Genesis* 2004, **39**(3):186–193.

32. Ouellette AJ, Lualdi JC: **A novel mouse gene family coding for cationic, cysteine-rich peptides. Regulation in small intestine and cells of myeloid origin.** *J Biol Chem* 1990, **265**(17):9831–9837.

The ICP22 protein selectively modifies the transcription of different kinetic classes of pseudorabies virus genes

Irma F Takács, Dóra Tombácz, Beáta Berta, István Prazsák, Nándor Póka and Zsolt Boldogkői[*]

Abstract

Background: Pseudorabies virus (PRV), an alpha-herpesvirus of swine, is a widely used model organism in investigations of the molecular pathomechanisms of the herpesviruses. This work is the continuation of our earlier studies, in which we investigated the effect of the abrogation of gene function on the viral transcriptome by knocking out PRV genes playing roles in the coordination of global gene expression of the virus. In this study, we deleted the *us1* gene encoding the ICP22, an important viral regulatory protein, and analyzed the changes in the expression of other PRV genes.

Results: A multi-timepoint real-time RT-PCR technique was applied to evaluate the impact of deletion of the PRV *us1* gene on the overall transcription kinetics of viral genes. The mutation proved to exert a differential effect on the distinct kinetic classes of PRV genes at the various stages of lytic infection. In the *us1* gene-deleted virus, all the kinetic classes of the genes were significantly down-regulated in the first hour of infection. After 2 to 6 h of infection, the late genes were severely suppressed, whereas the early genes were unaffected. In the late stage of infection, the early genes were selectively up-regulated. In the mutant virus, the transcription of the *ie180* gene, the major coordinator of PRV gene expression, correlated closely with the transcription of other viral genes, a situation which was not found in the wild-type (*wt*) virus. A 4-h delay was observed in the commencement of DNA replication in the mutant virus as compared with the *wt* virus. The rate of transcription from a gene normalized to the relative copy number of the viral genome was observed to decline drastically following the initiation of DNA replication in both the *wt* and mutant backgrounds. Finally, the switch between the expressions of the early and late genes was demonstrated not to be controlled by DNA replication, as is widely believed, since the switch preceded the DNA replication.

Conclusions: Our results show a strong dependence of PRV gene expression on the presence of functional *us1* gene. ICP22 is shown to exert a differential effect on the distinct kinetic classes of PRV genes and to disrupt the close correlation between the transcription kinetics of *ie180* and other PRV transcripts. Furthermore, DNA replication exerts a severe constraint on the viral transcription.

Keywords: Herpesvirus, Pseudorabies virus, Real-time PCR, ICP22, *us1* gene

* Correspondence: boldogkoi.zsolt@med.u-szeged.hu
Department of Medical Biology, Faculty of Medicine, University of Szeged,
Somogyi B. st. 4, Szeged H-6720, Hungary

Background

The pseudorabies virus (PRV), an alpha-herpesvirus, is the etiological cause of Aujeszky's disease of swine [1]. PRV is related to the human pathogen varicella-zoster virus (VZV) and herpes simplex virus types 1 and 2 (HSV-1 and -2), and the animal herpesvirus bovine herpesvirus type 1 (BHV-1). PRV is widely used as a model organism in investigations of the molecular pathomechanisms of the herpesviruses [2], and is a useful tool for the mapping of neural circuits [3,4]. Attempts have additionally been made to utilize this virus as a gene delivery vector [5,6] and an oncolytic agent [7]. Besides the lytic phase, alpha-herpesviruses can enter a latent state, where they transcribe a limited set of *cis*-antisense RNAs [8]. Traditionally, the lytic herpesvirus genes are classified into three kinetic categories: immediate-early (IE) genes, early (E) genes and late (L) genes. On a finer scale, an intermediate category, the early/late (E/L = delayed early) genes can also be distinguished [9]. PRV encodes a single IE gene, the *ie180* gene. The IE180 protein, a transactivator, is the principal coordinator of the overall gene expression of the virus. E genes encode proteins required for the nucleotide metabolism and DNA replication. Other E genes such, as the early protein 0 (*ep0*) [10] and *ul54* genes [11], encode transcriptional regulators. Most of the L genes code for structural elements of the virus. ICP22 is one of the five IE proteins of HSV-1, which is encoded by the *us1* gene. Intriguingly, a large part of the HSV *us1* gene is located in the unique US region, whereas its promoter and a short 5' portion of the transcribed region are in the inverted repeat (IR) segment. In PRV, however, the entire *us1* gene (earlier called the *rsp40* gene) resides in the IR region; this gene is therefore represented in two copies in the PRV genome. There is no consensus as to whether the PRV *us1* gene is expressed in IE [12] or E kinetics [13]. We demonstrated in an earlier analysis that this gene is expressed in atypical kinetics, and that it is obviously not an IE gene [9]. The function of the ICP22 polypeptide has primarily been analyzed in HSV-1. The investigations have revealed that ICP22 is a multifunctional protein that plays roles in various aspects of HSV pathogenesis. It has not yet clearly established whether ICP22 acts to repress E genes [14] or to enhance the transcription of L genes [15]. It has been shown that not all L genes require this transactivator for their expressions [16]. The BICP22 protein of BHV-1, a homologue of ICP22, has been demonstrated to exert a general repressive effect on each kinetic class [17]. Rice and coworkers reported that ICP22 acts at the level of transcriptional regulation [18]. However, the level of ICP0 mRNA was also reduced in the *us1* knockout (KO) HSV [16], which raises the question of whether the direct cause of the reduced transcription is the lack of *us1*

gene activity or the low ICP0 mRNA level. ICP22 has to be phosphorylated by the viral UL13 protein kinase in order to accomplish the transcriptional activation of L genes [19]. An additional function of the ICP22 polypeptide is associated with the alteration of the activity of cyclin-dependent kinase cdc2, a regulator of the cell cycle, which results in a selective up-regulation of HSV L genes during lytic infection [20]. Furthermore, HSV ICP22 also acts to modify the phosphorylation of RNA polymerase II (RNAP II) [21], which carries out the transcription of viral genes. One of the major control regions of RNAP II is its carboxy terminal domain (CTD), residing on the large subunit of the molecule. The CTD, containing multiple repeats of a heptapeptide sequence, serves as a binding site for various cellular proteins involved in the regulation of transcription. ICP22 is presumed to trigger the loss of Ser-2 phosphorylation on the CTD, and thereby modify the activity of RNAP II [22]. A novel function of ICP22 was recently identified, involving alteration of the chaperon localization of the host cells [23]. It has been shown that ORF63, the ICP22 homolog of VZV, does not alter RNAP II phosphorylation and the host chaperon machinery [22], which might indicate that ICP22 acts in a species- or genus-specific manner. In the present study, we have investigated the effects of *us1* gene deletion on the overall transcription of PRV genes.

Results and discussion

Experimental design

An insertion mutant PRV strain was constructed which contains the mutation in both copies of the *us1* gene.

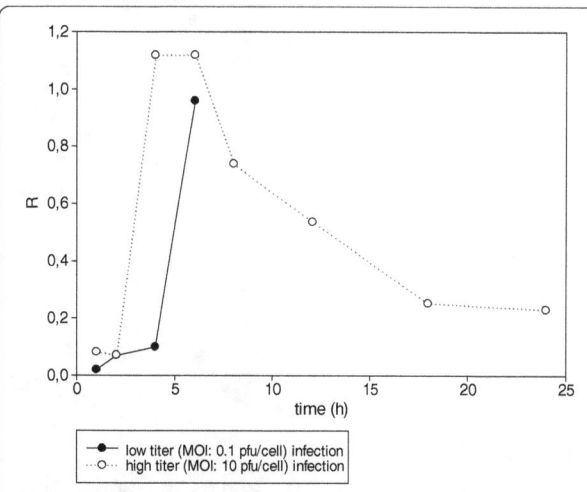

Figure 1 Transcription kinetics of the *us1* gene of the *wt* virus in low-MOI (0.1 pfu/cell) and high-MOI (10 pfu/cell) infection conditions. The low-MOI experiment was conducted only up to 6 h pi: after this time, the newly generated virions may be released from the infected cells and initiate new infections in the originally non-infected cells. The Figure shows a 2-h delay in the initiation of the steep rise in gene expression.

Mutation of a DNA sequence in the internal repeat region is copied to the terminal repeat by a mechanism called equalization [24]. From among the 70 PRV genes, 32 were selected for the transcription analysis, which reside at the upstream position of the tandem gene clusters and which represent each kinetic class of PRV genes. The reason for this choice was to exclude the distorting effect of the transcriptional read-through exerted by the upstream genes on the downstream genes. Furthermore, the genes selected for analysis play important roles in the regulation of the overall gene expression of the virus, including the *ie180*, the *ep0* (and their antisense transcripts, the LAT and the AST, respectively), the virion host shut-off (*vhs*), and the *ul54* genes. For each viral gene, a minimum of 3 parallel replicates were performed in order to achieve statistical reliability. Immortalized PK-15 cells were infected with either the wild-type (*wt*) or *us1* gene-deleted (*us1-KO*) PRV, using a high multiplicity of infection [MOI; 10 plaque-forming units (pfu)/cell]. The low-dose infections produce a much finer resolution of the cycle threshold

(Ct) values for the transcripts than in the case of high-dose infections; however, in the former case a large proportion of the cells remain uninfected, which allows the initiation of an additional infection cycle after 6 h post-infection (pi) by the newly generated virions [9], which would confuse the interpretation of the expression data obtained. The transcription of PRV genes was monitored at 9 different time points: 0.5, 1, 2, 4, 6, 8, 12, 18 and 24 h pi (multi-timepoint analysis). Strand-specific primers were used for the reverse transcription reaction so as to exclude the distorting effects of the potential *cis*-antisense transcripts that might be produced from the complementary DNA strands, which cannot be avoided on the use of other methods, such as oligo-dT- or random priming-based reverse transcription [25]. On the other hand, we found that the specificity of strand-specific primer-based RT is much higher than that of other methods [9]. In our work, we applied a modified version of the mathematical model described by Soong and colleagues for the relative quantification [26]. Specifically, we used the average of the 6-h ECt-sample values

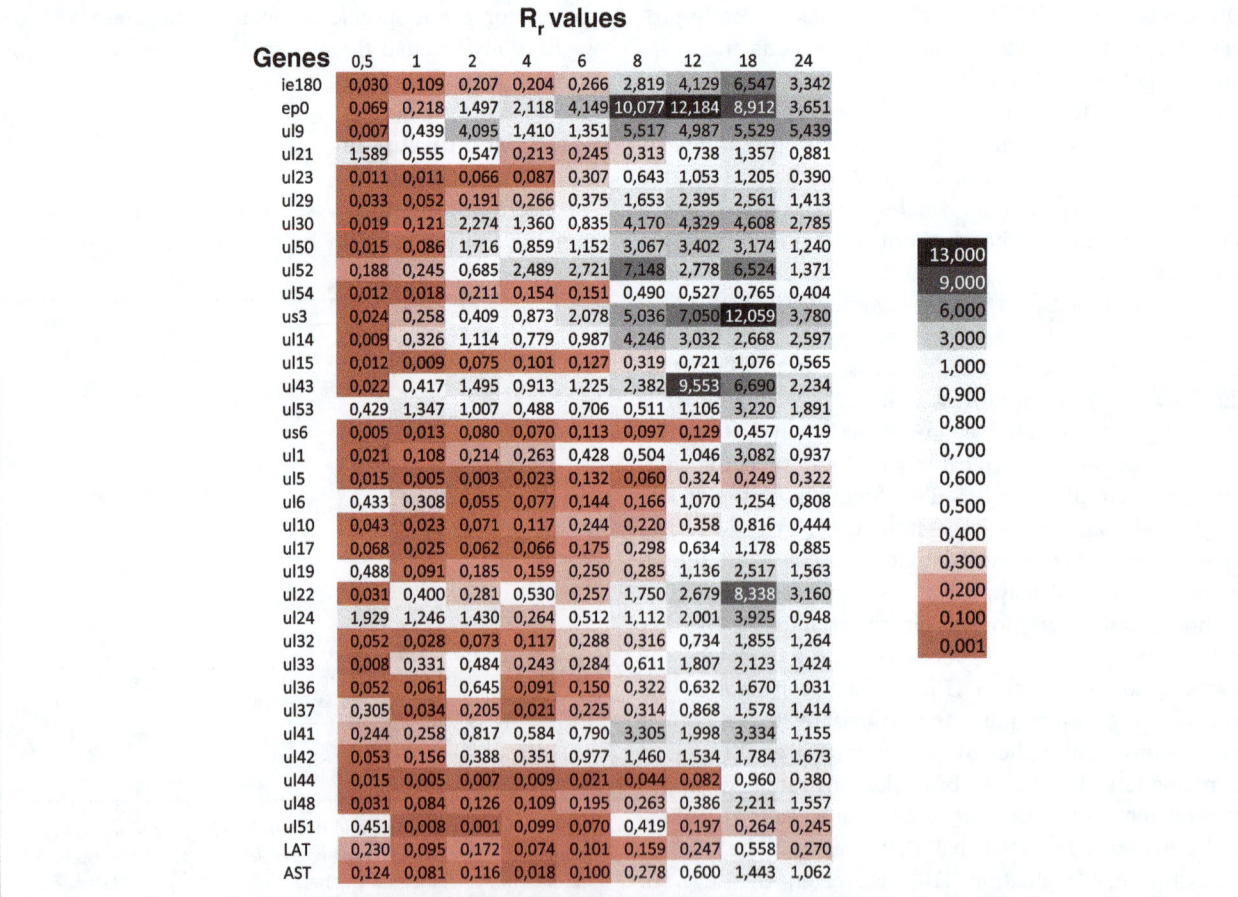

Figure 2 Heatmap visualization of the ratios of the transcripts of the *us1-KO* and the *wt* viruses ($R_r = R_{us1KO}/R_{wt}$). The red color ($R_r < 1$) indicates a transcript level that is lower in the mutant than in the *wt* virus at a certain period of infection, whereas the black ($R_r > 1$) indicates the opposite: the mutant virus produces a higher amount of mRNA than the wt virus.

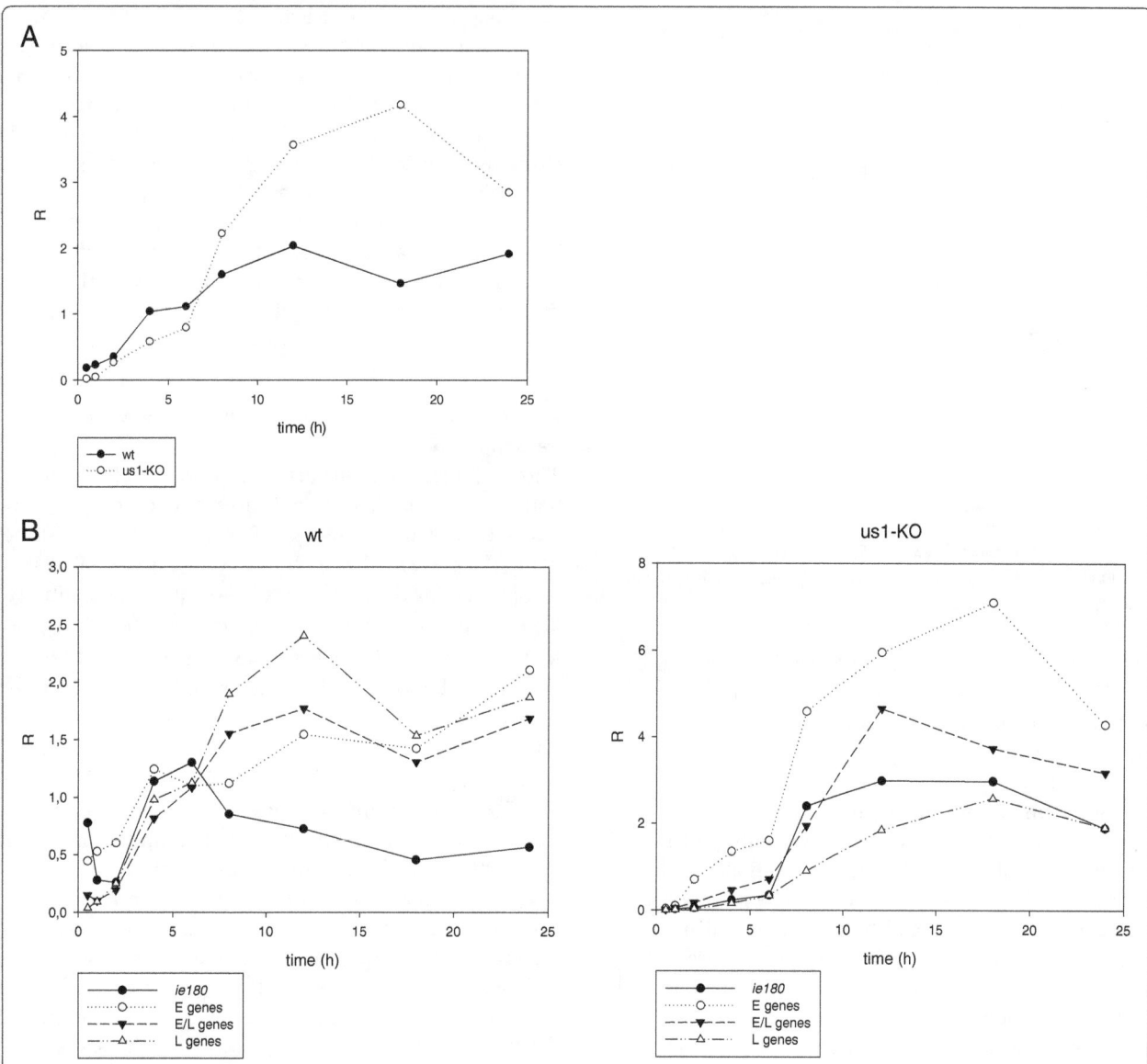

Figure 3 The average relative expression ratios ($\bar{R}$). (A) The $\bar{R}$ values of the total genes of the mutant and *wt* viruses plotted against time. The average level of the transcripts of the mutant virus is lower in the early stage, but higher in the late stage of infection. **(B)** The $\bar{R}$ values of the different kinetic classes of mutant and *wt* viruses. In the mutant virus, the L genes are repressed in the early stage of infection, whereas the E gene expressions are enhanced in the late stage of infection.

of the *wt* PRV for each gene in both the *wt* and the mutant backgrounds, as controls, which were normalized to the average of the corresponding 28S values (ECt-reference). The 28S RNA gene was used as a reference gene since the ribosomal RNAs are not substrates of VHS ribonuclease [27]. We used a selected Ct value (at 6 h in our system) as control for the comparability of the relative copy numbers of a transcript at different time points. The relative amounts of the transcripts of different genes cannot be compared directly due to the variation in the primer efficiencies in both RT and PCR. However, the use of multi-timepoint qPCR analysis allows a comparison of the transcription kinetics of

different viral genes throughout the whole period of infection. Furthermore, this method allows a comparison of the same genes in the two genetic backgrounds ($R_r = R_{us1KO)},/R_{wt}$), and of the mRNAs and the complementary antisense transcripts in the case of the *ep0*/LAT and *ie180*/AST pairs. A high R_r value indicates an inhibitory effect of the ICP22 protein on the transcript level of a particular gene in the *wt* virus. Conversely, a low R_r value indicates a stimulatory effect on the gene expression. We applied the same logic for the interpretation of the data obtained as is used in other knockout organisms, i.e. the normal role of the *us1* gene is considered to be the opposite of that of the phenotype

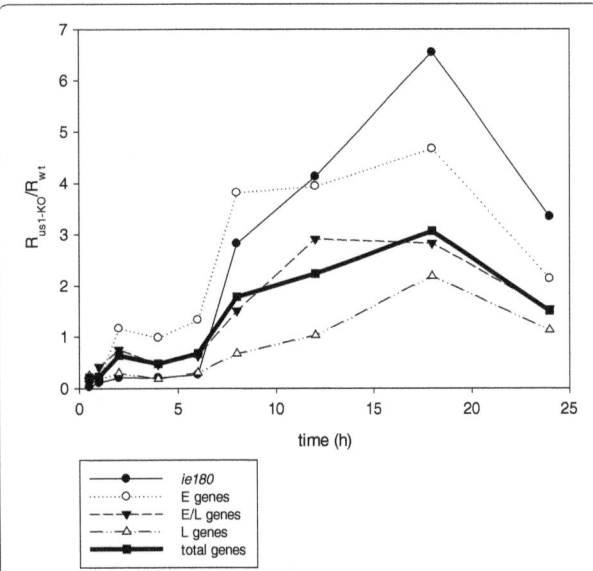

Figure 4 The average R_r values ($\bar{R}_r = \bar{R}_{us1-KO}/\bar{R}_{wt}$). The thick solid line depicts the $\bar{R}_r$ values of the total genes, and the thinner lines the $\bar{R}_r$ values of the different kinetic classes of viral genes.

caused by the mutation of this gene. In other words, an elevated expression of a gene in the *us1*-null mutant is indicative of a suppressive effect of the ICP22 protein on the expression of this gene.

Confirmation of the mutant genotype

The mutation in the *us1* gene was confirmed by PCR amplification of the DNA sequences containing the mutation, followed by pyrosequencing. The mutation of the *us1* gene was rescued, and this was followed by growth analysis in order to confirm that the altered kinetic properties can be solely explained by the abrogation of the *us1* function (data not shown). Besides these analyses, we also carried out more precise techniques for this purpose. Thus, we compared the rates of increase of viral DNA during the first twelve hours of viral infection. We observed similar dynamics in the growth rates of the DNA of the wild-type and rescued viruses, while both differed significantly from those of the us1-mutant virus (Additional file 1). In addition, we compared the transcription kinetics of the two most important transactivator genes, the ie180 and ep0 genes of pseudorabies virus. This revealed that the kinetics of the rescued virus resembled that of the wild-type virus, but differed significantly from that of the mutant virus (Additional file 2 and Additional file 3).

Transcription of the *us1* gene in the wild-type PRV

We compared the expression kinetics of the *us1* gene of the *wt* virus under the low- and high-dose infection conditions (Figure 1). Infection was analyzed for 6 h in the low-MOI infection, and for 24 h in the high-MOI

infection. The data revealed that in the low-MOI infection the *us1* gene behaved as an L gene [9], since the amounts of its transcripts started to rise rapidly from 4 h pi. This is in contrast with the situation of the high-dose experiment where this gene exhibited typical E characteristics in the early phase of infection, since its expression began to rise steeply from 2 h pi. Thus, there was a 2-h shift in the expression kinetics between the two experimental conditions up to 6 h pi. Another unusual feature of the *us1* mRNA kinetics was the rapid drop in the transcript level after 6 h pi, which is not typical in any of the kinetic classes of the viral genes.

The impact of the *us1* gene mutation on the viral transcriptome

The R values of individual genes were calculated for both *wt* and *us1-KO* mutant viruses, and compared by calculation of the ratios ($R_r = R_{us1KO}/R_{wt}$) for each time point (Figure 2). The average R values were calculated and plotted for the total genes ($\bar{R}$) (Figure 3A; Additional file 4A) and for each kinetic class of genes ($\bar{R}_E$, $\bar{R}_{E/L}$ and $\bar{R}_L$.) in both genetic backgrounds (Figure 3B; Additional file 4B). Additionally, the mutant and *wt* viruses were compared by calculating the average R_r values for the total genes ($\bar{R}_r$) and separately for each kinetic class ($\bar{R}_{r-E}, R_{r-E/L}$ and $\bar{R}_{r-L}$) (Figure 4B). Figures 3A and 4 reveal a significant decrease in the amounts of transcripts in each kinetic class of genes in the first hour of infection (Figure 2). The average levels of transcripts were not decreased significantly or at all in a few genes (e.g. *ul21*, *ul24*, *ul53* and *ul19*), but in most genes, and especially the E genes, the fold of the decrease was close to 2 orders of magnitude at this very early stage of infection. However, the average level of the E transcripts of the mutant virus reached approximately the same level as that of the *wt* virus in the interval 1–6 h pi. This was in contrast with the L genes, which on average were still expressed at a significantly lower level during the first 8 h of infection. The average amounts of the E/L transcripts falls between those of the E and L genes. The *ie180* gene was also expressed at a lower level in the first 6 h of infection in the mutant virus. The deletion of the *us1* gene led to a significant reduction in the average expression of the L genes and to a lesser reduction in the expression of the E/L genes in the first 6 h pi, but the E genes were expressed at a slightly higher level in the mutant than in the *wt* background in the period 2–6 h pi (Additional file 4B). The average rate of transcription of the IE, E and E/L genes from the mutant viral genome exceeded the rate of transcription from the *wt* genome after 6 h pi (Figure 4). This was in contrast with the transcript levels of the average L genes of the mutant virus, which remained below or equal to that of the *wt*

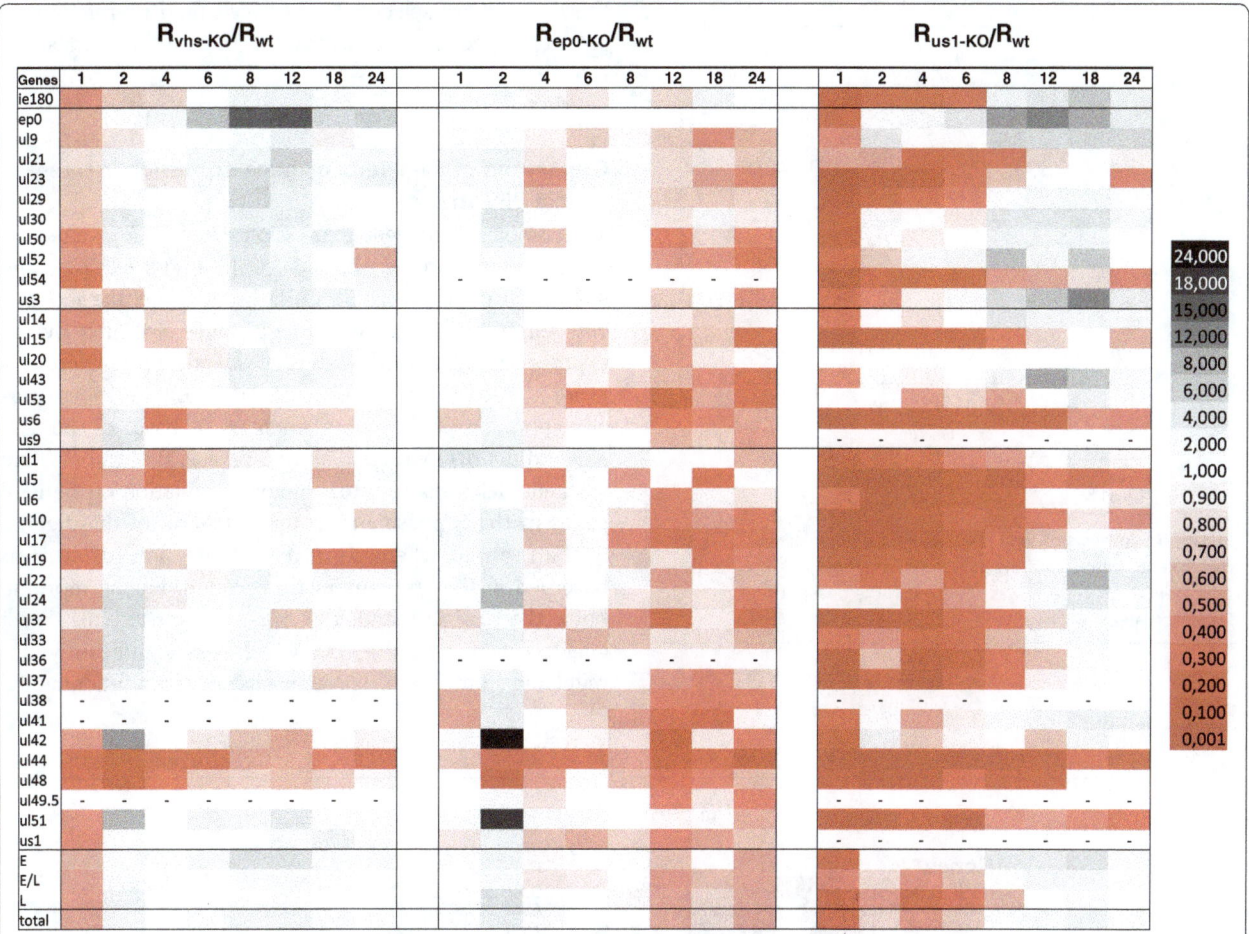

Figure 5 Comparison of the gene expression patterns (the $\bar{R}_r$ values) in three mutant viruses. The $\bar{R}_r$ values of the *us1*-, *ep0*- and *vhs*-null mutants were visualized in a heatmap presentation. The expression profile of the *us1-KO* virus is similar to that of the *vhs-KO*, i.e. complementary to that of the ep0-KO virus.

virus, except at 18 h pi. These data suggest that the ICP22 protein exerts an inhibitory effect especially on the transcription of the *ie180* and E genes following the initiation of viral DNA synthesis. The *us1* gene deletion exerted the highest impact on the expression of the *ep0* gene, an important regulator of PRV gene expression [28]: the amount of *ep0* transcripts was ~ 10 times higher in the cells infected with the mutant than that for the *wt* virus within the period 8–18 h pi. This result was the opposite of that described in the HSV, where the level of ICP0 mRNA was significantly lower in the KO virus [16]. The expressions of the *us3, ul22, ul43,* and *ul52* genes were also significantly elevated at certain time points in the late stage of infection (Figure 2). The *ul5, ul51* and LAT genes were the only examined PRV genes whose expressions were always lower in the *us1-KO* virus than in the *wt* virus. Overall, the above data suggest that the ICP22 protein exerts a selective effect on the expressions of PRV genes belonging in different

kinetic classes. ICP22 (possibly of tegument origin) appeared to have a significant stimulatory effect on the general gene expression of the virus at both 30 min and 1 h pi [PRV ICP22 has been shown to be localized in the viral tegument layer [29]. Though to a lower extent, its stimulatory effect continued up to 6 h in the E/L and L genes, but the effect became neutral or slightly inhibitory in the later stages of infection in these kinetic classes of genes. The expressions of the E genes exhibited different profiles on average in the mutant background: no discernible effect of the mutation within the 1–6 h pi period, followed by a profound inhibitory effect in the later phases of infection. Our results may resolve the debate as to whether ICP22 represses the E genes [14] or enhances the L genes [15]. The data demonstrate that the rates of transcription of the L genes in the mutant virus are repressed (and therefore they are selectively enhanced in the *wt* virus) in the very early phase of infection, whereas the rates of transcription of

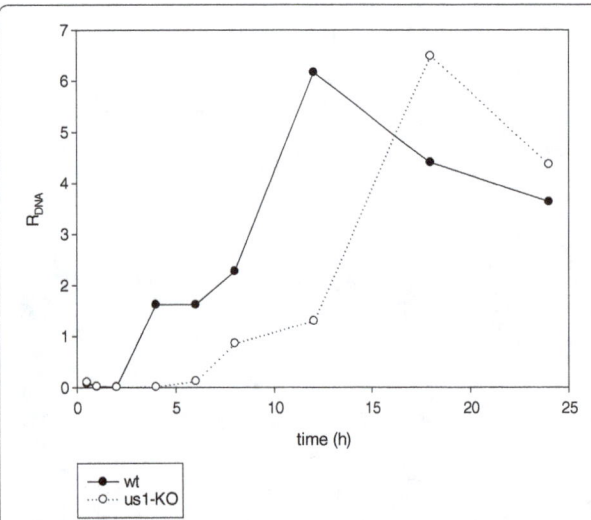

Figure 6 Synthesis of viral DNAs. The replication of viral DNA was monitored in the *wt* and the *us1-KO* viruses through the use of real-time RT-PCR. A 4-h delay was observed in the onset of DNA replication.

the E genes are selectively enhanced (and therefore they are repressed in the *wt* virus) in the late phase of infection.

Comparison of the effects on gene expression in three mutant PRV strains

We compared the gene expression profiles ($\bar{R}_r$ values) of three mutant viruses: the *vhs-KO* [30], *ep0-KO* [28] and *us1-KO* viruses (Figure 5). On a broad scale, the *us1* and *vhs* gene deletions appeared to produce a similar overall expression pattern (though the *us1* deletion exerted a more marked effect), whereas *ep0* deletion led to an overall expression profile complementary to those of the *us1-KO* and *ep0-KO* viruses. Figure 5 reveals that the *us1* and *vhs* gene deletions resulted in down-regulation of the PRV genes in the E phase and in up-regulation of these genes in the L phase, whereas a reverse expression pattern was observed in the *ep0* mutant background. Furthermore, in both the *us1-KO* and the *vhs-KO* viruses, the most affected gene was the *ep0*, which was significantly up-regulated from 2 h pi. These results suggest a fundamental

R$_{rn}$ values

Genes	0,5	1	2	4	6	8	12	18	24
ie180	0,022	0,076	0,254	19,072	3,377	7,423	19,501	4,451	2,778
ep0	0,050	0,152	1,837	197,640	52,775	26,534	57,543	6,059	3,034
ul9	0,005	0,305	5,025	131,548	17,190	14,526	23,555	3,759	4,520
ul21	1,142	0,385	0,672	19,865	3,111	0,823	3,485	0,923	0,732
ul23	0,008	0,007	0,082	8,128	3,900	1,694	4,975	0,819	0,324
ul29	0,024	0,036	0,235	24,844	4,772	4,353	11,313	1,741	1,174
ul30	0,014	0,084	2,791	126,895	10,615	10,980	20,447	3,133	2,315
ul50	0,011	0,059	2,106	80,191	14,653	8,077	16,067	2,158	1,030
ul52	0,135	0,170	0,841	232,261	34,617	18,823	13,118	4,436	1,140
ul54	0,009	0,013	0,259	14,409	1,918	1,290	2,490	0,520	0,336
us3	0,018	0,179	0,502	81,454	26,433	13,261	33,298	8,199	3,142
ul14	0,007	0,226	1,368	72,674	12,557	11,181	14,321	1,814	2,158
ul15	0,009	0,006	0,092	9,417	1,614	0,840	3,406	0,732	0,469
ul43	0,016	0,290	1,835	85,166	15,581	6,272	45,115	4,549	1,856
ul53	0,308	0,935	1,235	45,486	8,986	1,347	5,222	2,189	1,572
us6	0,004	0,009	0,098	6,574	1,438	0,255	0,608	0,311	0,348
ul1	0,015	0,075	0,262	24,556	5,444	1,328	4,941	2,096	0,779
ul5	0,011	0,004	0,003	2,133	1,676	0,159	1,530	0,170	0,267
ul6	0,311	0,214	0,068	7,188	1,825	0,438	5,056	0,853	0,671
ul10	0,031	0,016	0,087	10,959	3,100	0,579	1,689	0,555	0,369
ul17	0,049	0,018	0,077	6,181	2,227	0,785	2,995	0,801	0,736
ul19	0,351	0,063	0,227	14,821	3,182	0,750	5,366	1,711	1,299
ul22	0,022	0,278	0,345	49,404	3,272	4,609	12,653	5,669	2,626
ul24	1,386	0,865	1,755	24,638	6,506	2,928	9,451	2,669	0,788
ul32	0,037	0,020	0,090	10,961	3,662	0,832	3,465	1,261	1,051
ul33	0,005	0,230	0,593	22,626	3,614	1,609	8,532	1,444	1,183
ul36	0,037	0,042	0,791	8,484	1,904	0,848	2,983	1,135	0,857
ul37	0,219	0,024	0,251	1,981	2,857	0,828	4,101	1,073	1,175
ul41	0,175	0,179	1,003	54,503	10,046	8,702	9,438	2,267	0,960
ul42	0,038	0,109	0,476	32,725	12,425	3,844	7,245	1,213	1,391
ul44	0,011	0,004	0,009	0,876	0,272	0,116	0,390	0,653	0,316
ul48	0,022	0,058	0,154	10,206	2,480	0,692	1,824	1,503	1,294
ul51	0,324	0,006	0,002	9,196	0,889	1,104	0,933	0,180	0,204
LAT	0,165	0,066	0,211	6,924	1,281	0,419	1,165	0,379	0,225
AST	0,089	0,057	0,142	1,638	1,274	0,732	2,832	0,981	0,883

Scale
240,000
120,000
60,000
30,000
15,000
7,500
3,750
1,875
1,000
0,750
0,500
0,400
0,300
0,200
0,100
0,001

Figure 7 Heatmap visualization of the ratios of the transcripts of the *us1-KO* and the *wt* viruses normalized to the relative copy number of the DNAs (R$_{rn}$ = R$_{us1KO}$:R$_{DNA-us1KO}$/R$_{wt}$:R$_{DNA-wt}$). The R$_{rn}$ values were very low in the first 2 h pi. There was a drastic change in the period 4–6 h pi, when the R$_{rn}$ values became high. In the last stage of infection, the gene expressions became the same in the two viruses.

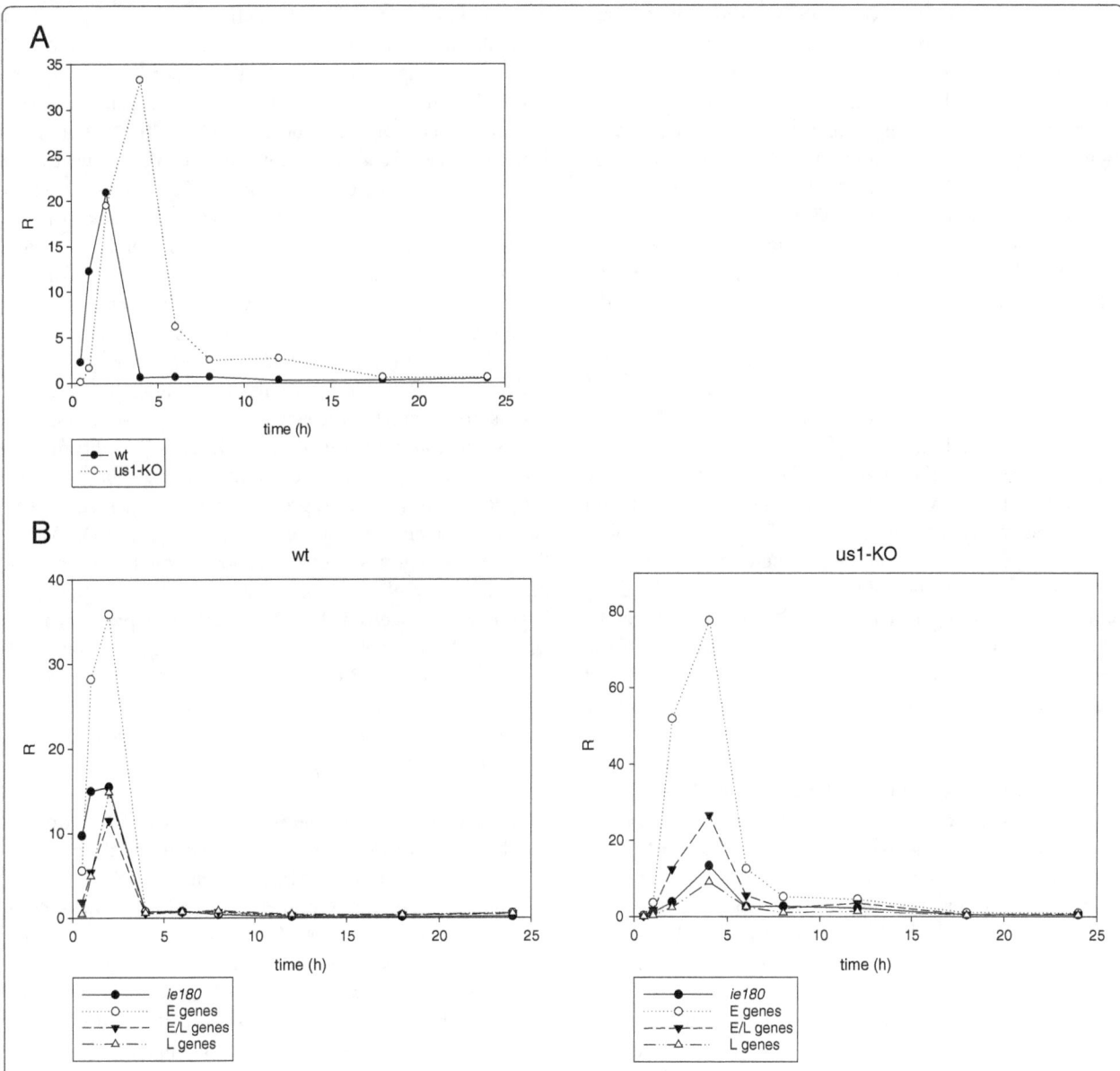

Figure 8 The average transcription rates are drastically decreased after the onset of DNA replication. The transcriptions from the genome of both genotypes are virtually shut down following the initiation of DNA synthesis. The transcription from the *us1-KO* genome is suppressed to a lower extent than that of wt virus.

role of the EP0 transactivator in the formation of the global gene expression profile of PRV.

The expression of the *ie180* gene is correlated with the expressions of the PRV genes in the *us1-KO* virus

We previously reported that the expression of the *ie180* gene, the major transactivator of the *wt* virus, was uncorrelated with the expressions of the remaining genes [30]. When we investigated this relationship in the *us1-KO* virus by using Pearson correlation analysis, very high correlations emerged between the *ie180* transcripts and the transcripts produced by all three kinetic classes of the PRV genes (Additional file 5A). The correlation was

especially high between the *ie180* and the E transcripts. Interestingly, we observed a similar effect in the *vhs*-null mutant as concerns of the correlation between the *ie180* and the other viral genes [30].

The initiation of DNA replication is delayed in the *us1-KO* as compared with the *wt* virus

Our real-time RT-PCR investigation of the kinetics of DNA replication of the two viral strains demonstrated that the amount of viral DNA was slightly less at 1 h pi than at 30 min pi in both viruses (Figure 6). We can only speculate about this phenomenon. It is possible that the cellular DNases digest some proportion of the infecting viral DNA molecules. DNA synthesis was initiated between 2 and 4 h pi in the *wt* virus. There was a decrease in the rate of multiplication of the DNA within the interval 4–6 h pi, which might be due to the switch between the theta-type and rolling circle-type of replications. A high rate of amplification of DNA molecules was observed between 6 and 12 h pi, followed by a slow decrease in the copy number of the *wt* viral DNA. The decrease in the amount of viral DNA in the late stage of viral infection is explained by the egress of mature virions from the infected cells. The onset of the DNA replication of the mutant virus exhibited a significant delay relative to that of the *wt* virus. The *us1-KO* DNA started replication 4 h later, and the rate of production of viral genomes was lower than that of the *wt* virus within the period 6–12 h pi. The copy numbers of the viral genomes were almost exactly the same at their peaks in the two viruses, but the maximum amount was reached 6 h later in the mutant virus. The mechanism whereby ICP22 affects the DNA replication remains to be determined, but it may be associated with the effect of the *us1* mutation on the delay of the onset of the general gene expression. Since the expressions of several key viral factors, such as the IE180 and the EP0, were profoundly changed in consequence of the mutation, the effect of ICP22 on the DNA replication, at least in part, must be indirect.

Investigation of the expression kinetics of PRV genes in the mutant and *wt* viruses through the use of R values normalized to the relative copy numbers of the DNA

We analyzed the viral gene expressions by normalizing the amounts of transcripts to the relative copy number of the DNA molecules, which allows a comparison of the transcriptional activity with the corresponding copy number of the DNA at different time points. The ratios of the normalized R values ($R_{rn} = R_{n-us1KO}/R_{n-wt}$) indicated that the level of transcripts in most genes was lower in the mutant than in the *wt* virus within the period 0.5-1 h pi with two exceptions (*ul21* and *ul24* at 30 min pi). This tendency was substantially reversed at 4 h pi, when the genes of the mutant virus were more active than those of the *wt* virus (Figure 7). Some early genes (*ep0, ul9, ul30,* and *ul52*) produced > 100 times more transcripts from a single copy of the genes in the mutant than in the *wt* virus. There was not such a large difference between the non-normalized genes of the two viruses. The switch from low to high amounts of transcripts occurred much later in the mutant virus, the exact time of the switch varying from gene to gene (Figure 2). The average expression from a single gene was very low at 30 min and 1 h pi in the *us1-KO* virus as compared with the *wt* virus, but it had become > 50 times higher (R_{us1KO}/R_{wt}) by 4 h pi (Figure 8A). This value was especially high for the average E genes (101.5), lower for the E/L genes (53), and much lower for

the L genes (15.2). This phenomenon obviously resulted from the delayed onset of DNA synthesis in the *us1-KO* virus, which suggests that DNA replication exerts a severe constraint on the transcription. Indeed, the gene transcription rate was profoundly decreased after the initiation of DNA synthesis (Figures 8A and 8B). In the *wt* virus, the average rate of gene transcription was 33 fold less at 4 h pi than at 2 h pi (R_{4h}/R_{2h}), and remained low or even became lower. In the mutant virus, the decrease in the rate of transcription started later (from 4 h pi), and occurred gradually, not suddenly as in the *wt* virus. The steepest drop in gene expression in the interval 4–6 h pi did not coincide with the highest rate of DNA synthesis, at 6 to 8 h pi, which suggests that the inhibition of overall gene expression is not directly controlled by DNA replication itself, as it was suggested by Huvet and colleagues [31]. From data obtained in the analysis of human cells, those authors proposed novel gene expression regulatory mechanism, based on the collision of the transcriptional and DNA replication machineries. However, those results could not be reproduced by Necsulea and colleagues [32]. Although, it is generally conceived that the switch in expression from the E to the L kinetic class of genes is triggered by DNA replication, Figure 8B clearly shows that the maximum expression as regards the individual L genes occurs at 2 h pi, i.e. before the onset of DNA replication, which casts severe doubt on the validity of this idea.

Normalization of the R values to the relative DNA copy number leads to the *ie180* gene correlating with the other viral genes in the *wt* background

Normalization of the R values of the cDNAs to the copy number of the viral DNAs leads to a very strong correlation between the *ie180* and the other genes in both genetic backgrounds (Additional file 5B). It is noteworthy that the strength of the correlation between the *ie180* and the different kinetic classes of genes is the opposite that in the non-normalized case. These data suggest that the synchronism in the transcription between the IE180 transactivator and the other genes is indicative of a real correlation. However, since the DNA replication imposes a severe constraint on the transcription of each gene, including the *ie180* gene, the high correlation between the normalized data could possibly be merely a statistical curiosity without any functional significance. Additional file 6 shows the R values of *ie180*, the average total viral genes and the average of various kinetic classes of genes normalized to the DNA copy number at different time points of infection.

Conclusions

Our kinetic data show that the abrogation of the *us1* function leads to a significant reduction of transcription in every kinetic class of genes in the first hour pi. In the period 1 to 6 h pi, the L genes are selectively down-regulated, and the E genes are later up-regulated in the *us1-KO* background relative to the *wt* virus. The questions as to whether the ICP22 protein exerts a direct effect upon the gene expression and, if so, at what levels remains to be answered. It is noteworthy that deletion of the *vhs* gene resulted in a similar overall expression pattern as that observed in the *us1-KO* virus, and in an expression profile complementary to those in the *ep0-KO* virus. Interestingly, *ep0* was the gene that was most affected in both the *us1* and *vhs* knockout viruses (highly elevated expressions), which might imply that EP0 could be a common link in the determination of the overall gene expression profile. We observed a strong inhibition of transcription during DNA replication. The question arises of whether the process of DNA replication itself could exert this inhibitory effect. In the mutant virus, the main inhibition in gene expression occurs between 4 and 6 h pi, while DNA synthesis exhibits the highest rate 2 h later. This suggests that the decrease in gene expression is not, or not only, a result of the potential collision of the transcription and DNA replication machineries. We have earlier shown that disruption of the function of the *vhs* gene of the virus results in synchronization of the gene expression profile of the *ie180* and the remaining PRV genes [30]. We obtained very similar results with the *us1* gene-deleted PRV: the expression of the *ie180* gene correlated with those of the other PRV genes in the mutant virus, which was not the case for the *wt* virus. The question may be posed of whether this correlation is directly associated with the disruption of the *us1* gene function, or is rather caused by the delay of DNA replication.

Methods

Cells and viruses
Monolayer cultures of porcine kidney epithelial cells (PK-15) were used for the propagation of the pseudorabies virus. Cells were grown in DMEM (Sigma Aldrich), supplemented with 5% fetal bovine serum (Gibco) and 80 μg of gentamicin per ml (Invitrogen) at 37°C in the presence of 5% CO_2. The Kaplan *wt* strain of PRV was used as the parental strain for the generation of the *us1*-null mutant virus (*us1-KO*).

Construction of the *us1* gene knockout virus
The *Ka-us1-KO* (*us1-KO*, in short) virus was generated as follows. As a first step the *Bam*HI-10 fragment of PRV was isolated from the gel, and was then subcloned to pRL525 [33], resulting in the generation of pRL525-B10. This plasmid was used as a template for the PCR amplification of the two arms of the flanking sequences providing homology with the target viral genomic region. A green-fluorescent protein (GFP) gene expression

cassette (Clontech) was inserted into the unique *Ecl136*II site of the targeting sequence, resulting in pUS1-gfp, which was used as the transfer plasmid for the generation of the knockout virus. The linearized transfer plasmid was transfected along with the purified *wt* viral DNA into PK-15 cells. The recombinant virus was generated by homologous recombination, then isolated and plaque-purified on the basis of the fluorescence. Rescued viruses were generated by using pUS1 as a transfer plasmid, which was co-transfected with the purified DNA of *us1-KO* to PK-15 cells. The revertant viruses were selected on the basis of the non-fluorescent plaque phenotype. Both mutant and rescued viruses were checked with DNA sequencing.

Infection
The virus stock used for the experiments was prepared by infecting PK-15 cells with low-dose viruses, followed by incubation of the cells until a complete cytopathic effect was observed. To assess the effect of the *us1* gene deletion on the transcription kinetics of PRV, rapidly-growing semi-confluent PK-15 cells were infected with the *wt* or *us1-KO* virus at a high multiplicity of infection [MOI; 10 plaque forming units (pfu)/cell], and incubated for 1 h, which was followed by removal of the virus suspension and washing of the cells with PBS. Subsequently, fresh culture medium was added to the cells, which were further cultivated for an additional 0.5, 1, 2, 4, 6, 8, 12, 18 or 24 h.

Isolation of RNAs
Infected or non-infected PK-15 cells were washed with PBS and harvested for RNA purification. Total RNA was isolated by using the NucleoSpin RNA II Kit (Macherey-Nagel GmbH and Co. KG) as recommended by the supplier. Briefly, harvested cells were collected by low-speed centrifugation, and lysed in the buffer included in the kit. Samples were treated with RNase-free rDNase solution (included in the Kit) to remove potential genomic DNA contamination. As the next step, the potential residual DNA contamination was removed by using Turbo DNase (Ambion Inc.). Subsequently, RNA samples were eluted in RNase-free water (supplied with the kit), resulting in a total volume of 60 μl of RNA solution. RNA concentrations were measured spectrophotometrically in a BioPhotometer Plus (Eppendorf). The RNA solution was stored at –80°C until use.

Reverse transcription
Total RNA samples were reverse-transcribed by using gene-specific primers, and SuperScript III reverse transcriptase (Invitrogen) as described in our earlier reports [9,34]. Briefly, RT mixtures containing total RNA, primer, SuperScript III enzyme, buffer, dNTP mix and

RNase inhibitor (RNAsin, Promega) were incubated at 55°C for 1 h. The amplification of the first-strand cDNA synthesis was terminated by keeping the samples at 70°C for 15 min. The cDNAs were diluted 10-fold with nuclease-free water (Promega Corp.) and the solutions were stored at −80°C until use.

Real-time PCR

SYBR Green-based (Absolute QPCR SYBR Green Mix, Thermo Scientific) quantitative real-time PCRs were performed on the first-strand cDNAs in a real-time PCR cycler (Rotor-Gene 6000, Corbett Life Sciences), as described in our previous studies [9,30]. The specificity of the reverse transcription and the PCR reactions was ensured by using no-RT, no-primer, and no-template controls. The accuracy of sampling was guaranteed by using 28S rRNA of swine as loading control. The specificity of the PCR products was confirmed by melting point analysis, PAGE and/or DNA sequencing.

DNA sequencing

We used pyrosequencing with a Pyromark Q24 pyrosequencer (Qiagen) to validate the mutation of the *us1* gene and the specificity of the PCR products obtained in the kinetic experiments in the event of doubt.

Data analysis

The relative expression ratios (R values) of the cDNAs of the PRV genes were calculated via the following formula:

$$R = \frac{\overline{\left(E_{sample6h}\right)^{Ct_{sample6h}}}}{\left(E_{sample}\right)^{Ct_{sample}}} : \frac{\overline{\left(E_{ref6h}\right)^{Ct_{ref6h}}}}{\left(E_{ref}\right)^{Ct_{ref}}}$$

The cDNAs were all normalized to the cDNAs of the 28S rRNAs of swine [9] by using the Comparative Quantitation module of the Rotor-Gene 6000 software (Version 1.7.28, Corbett Research), which automatically sets the thresholds and calculates the efficiency of PCR reactions. We used the average 6-h ECt values of the "samples", with those of the "references" as controls, as in our earlier works [9,30,34]. The R values of the viral DNAs were calculated similarly; the 6-h ECt values were taken as the control, and porcine 28S rRNA gene was used as the reference. The effect of deletion of the *us1* gene on the global gene expression was calculated by using R_r, the ratio of the R values of the *us1* mutant and the *wt* viruses ($R_r = R_{us1KO}/R_{wt}$), where R_{us1KO} and R_{wt} are the R values of a particular gene at a given time point in the *us1*-KO and *wt* genetic background, respectively. All data were analyzed by using the average and the standard deviance functions of Microsoft Excel. Pearson's correlation coefficient was calculated for the analysis of the

correlation between the expression kinetics of the genes, using the following formula:

$$r = \frac{\sum_{i=1}^{n}(X_i - \bar{X})(Y_i - \bar{Y})}{(n-1)S_x S_y}$$ [35]. The normalized R

values were calculated by dividing the appropriate R value of the cDNA by the R value of the viral DNA measured at the same time for the same sample. The Pearson correlation coefficient is a number between −1 and +1 that measures the linear relationship between two variables, denoted here as X and Y, which are the R values of two different genes or the average R values of genes belonging in the same kinetic class in the same time interval. X and Y are the average values, n is the sample number, and SX and SY are the standard errors of the mean values for X and Y, respectively. A positive value for the correlation indicates a positive association, while a negative value indicates an inverse association.

Additional files

Additional file 1: Comparison of the rates of increase of viral DNA during the first twelve hours of PRV infection. We observed similar dynamics in the growth rates of the DNA of the wild-type and rescued PRVs, while both differed significantly from those of the us1-mutant virus.

Additional file 2: Comparison of the transcription kinetics of the *ie180* gene in the wild-type, *us1*-KO and *us1*-rescued PRV. This revealed that the kinetics of the rescued virus resembled that of the wild-type PRV, but differed significantly from that of the mutant virus.

Additional file 3: Comparison of the transcription kinetics of the *ep0* gene in the wild-type, *us1*-KO and *us1*-rescued PRV. This revealed that the kinetics of the rescued virus resembled that of the wild-type PRV, but differed significantly from that of the mutant virus.

Additional file 4: The average relative expression ratios ($\bar{R}$). This table shows the $\bar{R}$ values for the total PRV genes (A) and for each kinetic class of viral genes (B) at different time points of infection.

Additional file 5: Correlations between the transcription of the *ie180* gene and other PRV genes in *wt* and *us1*-KO backgrounds A. The viral genes are expressed in synchronism with the *ie180* gene in the mutant virus, whereas the expressions are not correlated in the *wt* virus. B. Correlation between the transcription of the *ie180* gene and other PRV genes in the *wt* and *us1*-KO backgrounds with the use of normalized R values. The expression of viral genes becomes correlated with the expression of *ie180* genes in the *wt* virus, too.

Additional file 6: The normalized average relative expression ratios ($\bar{R}_n$) values This table shows the $\bar{R}_n$ values for the total PRV genes (A) and for each kinetic class of viral genes (B) at different time points of infection.

Competing interests
The authors declare that they have no competing interests.

Authors' contributions
IFT carried out the construction of the targeting plasmids, RNA purification, the reverse transcription reactions, the standard and real-time PCR and. DT carried out the DNA sequencing, and participated in the evaluation of the primary data. BB took part by performing the RT reactions and the real-time PCR. IP purified PRV RNA, propagated PK-15 cells and participated in the genotyping of the recombinant virus. NP carried out the agarose and polyacrylamide gel electrophoresis and participated in the propagation of cultured cells. ZB coordinated the study, propagated viruses, isolated the

viral DNAs and isolated the recombinant viruses. All authors have read and approved the final manuscript.

Acknowledgements
We would like to thank Katalin Révész, Csilla Papdi and Margit Kisapáti for the technical assistance. This study was supported by grant TÁMOP-4.2.1.B-09/1//KONV, TÁMOP-4.2.2/B-10/1-2010-0012, and Swiss-Hungarian Cooperation Programme grant SH/7/2/8 to ZB.

References

1. Aujeszky A: **A contagious disease, not readily distinguishable from rabies, with unknown origin.** *Veterinarius* 1902, **12**:387–396.

2. Pomeranz L, Reynolds A, Hengartner C: **Molecular biololgy of pseudorabies virus: Impact on neurovirology and veterinary medicine.** *Microbiology and Molecular Biology Reviews* 2005, **69**(3):462.

3. Boldogkoi Z, Balint K, Awatramani GB, Balya D, Busskamp V, Viney TJ, Lagali PS, Duebel J, Pasti E, Tombacz D, Toth JS, Takacs IF, Scherf BG, Roska B: **Genetically timed, activity-sensor and rainbow transsynaptic viral tools.** *Nature Methods* 2009, **6**(2):127–130.

4. Enquist LW: **Five Questions about Viral Trafficking in Neurons.** *Plos Pathogens* 2012, **8**(2):e1002472.

5. Boldogkoi Z, Nogradi A: **Gene and cancer therapy–pseudorabies virus: a novel research and therapeutic tool?** *Curr Gene Ther* 2003, **3**(2):155–182.

6. Prorok J, Kovacs PP, Kristof AA, Nagy N, Tombacz D, Toth JS, Ordog B, Jost N, Virag L, Papp JG, Varro A, Toth A, Boldogkoi Z: **Herpesvirus-mediated delivery of a genetically encoded fluorescent Ca(2+) sensor to canine cardiomyocytes.** *J Biomed Biotechnol* 2009, **2009**:361795.

7. Boldogkoi Z, Bratincsak A, Fodor I: **Evaluation of pseudorabies virus as a gene transfer vector and an oncolytic agent for human tumor cells.** *Anticancer Res* 2002, **22**(4):2153–2159.

8. Spivack J, Fraser N: **Detection of Herpes-Simplex Virus Type-1 Transcripts during Latent Infection in Mice.** *J Virol* 1987, **61**(12):3841–3847.

9. Tombacz D, Toth JS, Petrovszki P, Boldogkoi Z: **Whole-genome analysis of pseudorabies virus gene expression by real-time quantitative RT-PCR assay.** *BMC Genomics* 2009, **10**:491.

10. Boldogkoi Z, Braun A, Fodor I: **Replication and virulence of early protein 0 and long latency transcript deficient mutants of the Aujeszky's disease (pseudorabies) virus.** *Microb Infect* 2000, **2**(11):1321–1328.

11. Schwartz J, Brittle E, Reynolds A, Enquist L, Silverstein S: **UL54-null pseudorabies virus is attenuated in mice but productively infects cells in culture.** *J Virol* 2006, **80**(2):769–784.

12. Fuchs W, Ehrlich C, Klupp B, Mettenleiter T: **Characterization of the replication origin (Ori(S)) and adjoining parts of the inverted repeat sequences of the pseudorabies virus genome.** *J Gen Virol* 2000, **81**:1539–1543.

13. Zhang G, Leader D: **The Structure of the Pseudorabies Virus Genome at the End of the Inverted Repeat Sequences Proximal to the Junction with the Short Unique Region.** *J Gen Virol* 1990, **71**:2433–2441.

14. Bowman JJ, Orlando JS, Davido DJ, Kushnir AS, Schaffer PA: **Transient Expression of Herpes Simplex Virus Type 1 ICP22 Represses Viral Promoter Activity and Complements the Replication of an ICP22 Null Virus.** *J Virol* 2009, **83**(17):8733–8743.

15. Poffenberger K, Raichlen P, Herman R: **In-Vitro Characterization of a Herpes-Simplex Virus Type-1 Icp22 Deletion Mutant.** *Virus Genes* 1993, **7**(2):171–186.

16. Purves F, Ogle W, Roizman B: **Processing of the Herpes-Simplex Virus Regulatory Protein Alpha-22 Mediated by the U(I)13 Protein-Kinase Determines the Accumulation of a Subset of Alpha-Messenger Rnas and Gamma-Messenger Rnas and Proteins in Infected-Cells.** *Proc Natl Acad Sci U S A* 1993, **90**(14):6701–6705.

17. Koppel R, Vogt B, Schwyzer M: **Immediate-early protein BICP22 of bovine herpesvirus 1 trans-represses viral promoters of different kinetic classes and is itself regulated by BICP0 at transcriptional and posttranscriptional levels.** *Arch Virol* 1997, **142**(12):2447–2464.

18. Rice S, Long M, Lam V, Schaffer P, Spencer C: **Herpes-Simplex Virus Immediate-Early Protein Icp22 is Required for Viral Modification of Host Rna-Polymerase-Ii and Establishment of the Normal Viral Transcription Program.** *J Virol* 1995, **69**(9):5550–5559.

19. Long M, Leong V, Schaffer P, Spencer C, Rice S: **ICP22 and the UL13 protein kinase are both required for herpes simplex virus-induced modification of the large subunit of RNA polymerase II.** *J Virol* 1999, **73**(7):5593–5604.

20. Advani S, Weichselbaum R, Roizman B: **Herpes simplex virus 1 activates cdc2 to recruit topoisomerase II alpha for post-DNA synthesis expression of late genes.** *Proc Natl Acad Sci U S A* 2003, **100**(8):4825–4830.

21. Rice S, Long M, Lam V, Spencer C: **Rna-Polymerase-Ii is Aberrantly Phosphorylated and Localized to Viral Replication Compartments Following Herpes-Simplex Virus-Infection.** *J Virol* 1994, **68**(2):988–1001.

22. Fraser KA, Rice SA: **Herpes simplex virus immediate-early protein ICP22 triggers loss of serine 2-phosphorylated RNA polymerase II.** *J Virol* 2007, **81**(10):5091–5101.

23. Bastian TW, Livingston CM, Weller SK, Rice SA: **Herpes Simplex Virus Type 1 Immediate-Early Protein ICP22 Is Required for VICE Domain Formation during Productive Viral Infection.** *J Virol* 2010, **84**(5):2384–2394.

24. Boldogkoi Z, Braun A, Medveczky I, Glavits R, Gyuro B, Fodor I: **Analysis of the equalization of inverted repeats and neurovirulence using a pseudorabies virus mutant strain altered at the Ul/Ir junction.** *Virus Genes* 1998, **17**(1):89–98.

25. Boldogkoi Z: **Transcriptional interference networks coordinate the expression of functionally-related genes clustered in the same genomic loci.** *Frontiers in Genetics* 2012, **3**:00012.

26. Soong R, Tabiti K: **Detection of colorectal micrometastasis by quantitative RT-PCR of cytokeratin 20 mRNA.** *Proc Am Assoc Cancer Res* 2000, **41**:391. XP002149389.

27. Chambers J, Angulo A, Amaratunga D, Guo H, Jiang Y, Wan J, Bittner A, Frueh K, Jackson M, Peterson P, Erlander M, Ghazal P: **DNA microarrays of the complex human cytomegalovirus genome: Profiling kinetic class with drug sensitivity of viral gene expression.** *J Virol* 1999, **73**(7):5757–5766.

28. Tombacz D, Toth JS, Boldogkoi Z: **Effects of deletion of the early protein 0 gene of pseudorabies virus on the overall viral gene expression.** *Gene* 2012, **493**(2):235–242.

29. Kramer T, Greco TM, Enquist LW, Cristea IM: **Proteomic Characterization of Pseudorabies Virus Extracellular Virions.** *J Virol* 2011, **85**(13):6427–6441.

30. Tombacz D, Toth JS, Boldogkoi Z: **Deletion of the virion host shut-off gene of pseudorabies virus results in selective upregulation of the expression of early viral genes in the late stage of infection.** *Genomics* 2011, **98**(1):15–25.

31. Huvet M, Nicolay S, Touchon M, Audit B, D'Aubenton-Carafa Y, Arneodo A, Thermes C: **Human gene organization driven by the coordination of replication and transcription.** *Genome Res* 2007, **17**(9):1278–1285.

32. Necsulea A, Guillet C, Cadoret J, Prioleau M, Duret L: **The Relationship between DNA Replication and Human Genome Organization.** *Mol Biol Evol* 2009, **26**(4):729–741.

33. Elhai J, Wolk C: **A Versatile Class of Positive-Selection Vectors Based on the Nonviability of Palindrome-Containing Plasmids that Allows Cloning into Long Polylinkers.** *Gene* 1988, **68**(1):119–138.

34. Toth JS, Tombacz D, Takacs IF, Boldogkoi Z: **The effects of viral load on pseudorabies virus gene expression.** *Bmc Microbiology* 2010, **10**:311.

35. Campbell AM, Heyer LJ: **Basic research with DNA microarray.** In *Discovering Genomics Proteomics and Bioinformatics*. 2nd edition. Edited by Anonymous. San Francisco: CSHL Press; 2007:238–241.

MicroRNA-19a regulates lipopolysaccharide-induced endothelial cell apoptosis through modulation of apoptosis signal-regulating kinase 1 expression

Wei-Long Jiang[1][†], Yu-Feng Zhang[1][†], Qing-Qing Xia[1], Jian Zhu[2], Xin Yu[3], Tao Fan[2] and Feng Wang[4*]

Abstract

Background: MicroRNAs, small non-encoding RNAs that post-transcriptionally modulate expression of their target genes, have been implicated as critical regulatory molecules in endothelial cells.

Results: In the present study, we found that overexpression of miR-19a protects endothelial cells from lipopolysaccharide (LPS)-induced apoptosis through the apoptosis signal-regulating kinase 1 (ASK1)/p38 pathway. Quantitative real-time PCR demonstrated that the expression of miR-19a in endothelial cell was markedly down-regulated by LPS stimulation. Furthermore, LPS-induced apoptosis was significantly inhibited by over-expression of miR-19a. Finally, both a luciferase reporter assay and western blot analysis showed that ASK1 is a direct target of miR-19a.

Conclusions: MiR-19a regulates ASK1 expression by targeting specific binding sites in the 3' untranslated region of ASK1 mRNA. Overexpression of miR-19a is an effective method to protect against LPS-induced apoptosis of endothelial cells.

Keywords: miR-19a, ASK1, Apoptosis, Endothelial cells

Background

MicroRNAs (miRNAs) are endogenous, small non-coding RNA molecules consisting of about 22 nucleotides, which function in RNA silencing and post-transcriptional regulation of gene expression [1-4]. Many miRNAs are evolutionarily conserved and believed to play a role in controlling various biological process including developmental patterning, cell differentiation, and cell proliferation [5-7]. MiR-19a belongs to the MiR-17-92 cluster that encodes six single mature miRNAs (miR-17, miR-19a/b, miR-20, miR-92, and miR18) [8-10]. It is up-regulated in a variety of cancers including gliomas, medulloblastoma, gastric cancer, and thyroid cancer, and enhances proliferation, inhibits apoptosis, and induces tumor angiogenesis, indicating that miR-19a is an oncogene [11-17]. MiR-19a is also involved into controlling endothelial cell functions

and neovascularization [18,19]. It has been reported that miR-19a expression increases during induction of endothelial cell differentiation in embryonic stem cells [20].

Recently, Philippe et al. reported that lipopolysaccharide (LPS) down-regulates the expression of miR-19a and miR-19b, which is associated with toll-like receptor 2 up-regulation [21]. It is well known that LPS induces apoptosis in various types of endothelial cells including human umbilical vein endothelial cells (HUVECs) and lung-derived normal human microvascular endothelial cells [22-24]. Previous studies have also reported that LPS release into circulation induces endothelial cell apoptosis in vivo and thus causes microvascular injury in numerous tissues [25-27]. LPS induces the activity of apoptosis signal-regulating kinase 1 (ASK1) and activates the downstream mitogen-activated protein kinase (MAPK) pathways, leading to induction of JNK/p38 activity and resulting in apoptosis [28]. ASK1-deficient mice have been shown to be resistant to LPS-induced sepsis shock [29]. LPS-induced

* Correspondence: wfwangfeng@yahoo.com
[†]Equal contributors
[4]Department of Neurology, Shanghai First People's Hospital, Shanghai Jiaotong University School of Medicine, Shanghai 200080, China
Full list of author information is available at the end of the article

p38 activation and production of inflammatory cytokines are reduced in splenocytes and dendritic cells derived from ASK1-deficient mice [29]. As a member of the MiR-17-92 cluster, miR-20 has been also reported to target ASK1 [9]. Therefore, it might be interesting to determine whether miR-19a and miR-20 share a common mechanism in LPS-induced apoptosis.

In the present study, we identified miR-19a, whose expression was markedly down-regulated in LPS-stimulated HUVECs, as a novel modulator of ASK1 expression and LPS-induced endothelial cell apoptosis.

Methods
Cells and reagents
HUVECs and EAhy926 cells were purchased from the American Type Culture Collection (Manassas, VA, USA). A miRNA-19a inhibitor (Product Number: HSTUD0343) and control inhibitor (Product Number: NCSTUD001) were purchased from Sigma-Aldrich. The miRNA inhibitors were designed using the mature miRNA sequence information from miRBase and are 2'-O-methylated RNA duplexes with a miRNA-binding site on each strand.

Western blotting
To assess ASK1 expression, proteins from HUVECs were collected and analyzed by western blotting. Briefly, a protein sample (20 μg) was fractionated by SDS-polyacrylamide gel electrophoresis and then transferred to a polyvinylidene difluoride membrane (Immobilon-P; Millipore). The membrane was blocked with phosphate-buffered saline containing 0.3% Tween 20 and 5% dry milk, and then incubated with a primary antibody overnight at 4°C. The immune complexes were detected by chemiluminescence methods (ECL; Amersham International). Anti-ASK1 and anti-phospho-ASK1(Thr845) antibodies were purchased from abcam. Anti-p38, anti-phospho-p38, anti-cleaved caspase-3, and anti-glyceraldehyde-3-phosphate dehydrogenase (GAPDH) antibodies were purchase from Cell Signaling Technology. All antibodies were diluted at 1:1000. GAPDH was used as a loading control.

Generation of a miR-19a adenovirus
The miR-19a adenovirus used in this study contained the human miR-19a gene (NR_029489.1). The adenovirus was generated using the AdMax (Microbix) system according to the manufacturer's recommendations. Briefly, the pacAd5 9.2-100 Ad backbone vector was cotransfected with the pacAd5 K-NpA shuttle vector containing the miR-19a sequence into Ad293 cells using FuGene 6 Transfection Reagent (Roche, Indianapolis, IN). The viruses were propagated in Ad293 cells and purified using $CsCl_2$ banding followed by dialysis against 10 mmol/L Tris-buffered saline with 10% glycerol.

Titering was performed with HEK293 cells using an Adeno-X Rapid Titer kit (BD Biosciences Clontech, Palo Alto, CA) according to the manufacturer's instructions. An adenovirus bearing LacZ (Ad-LacZ) was obtained from Clontech.

Quantitative real-time PCR
Total RNAs were extracted from HUVECs using a miR-Neasy Mini Kit or RNeasy kit (QIAGEN). Quantitative real-time PCR (qRT-PCR) was performed with cDNA generated from 20 ng total RNA using a miRCURY LNATM Universal cDNA Synthesis kit and SYBR® Green Master Mix Kit (Exqion). MiR-19a primers were 5'-CCTCTG-TTAGTTTTGCATAGTTGC-3' and 5'-CAGGCCACCATCAGTTTTG-3'; miR-20a primers were 5'-ACACTCCAGCTGGGTAAAGTGCTTATAGTGC-3' and 5'-CTCAACTGGTGTCGTGGAGTCGGCAATTCAGTTGAGCTACCTGC-3' (stem-loop reverse primer). qRT-PCR analysis of ASK1 expression was performed with cDNA generated from 250 ng total RNA using HotStart-IT® SYBR® Green qPCR Master Mix with a UDG (2×) Tested User FriendlyTM kit (USB Corporation). ASK1 primers were 5'-AGACATCTGGTCTCTGGGCTGTAC-3' and 5'-AACATTCCCACCTTGAACATAGC-3'. The relative expression level was calculated by the 2-ΔΔCt method with the CT values normalized to 18S rRNA as the internal control for ASK1 and U6 snRNA as the internal control for miRNAs.

Luciferase reporter assay
Based on the human ASK1 mRNA sequence, firefly luciferase cDNA fused with the human ASK1 mRNA 3' untranslated region (UTR) containing the two seed sequences for miR-19a was amplified from the genomic DNA of HUVECs (two primers containing XbaI sites were used. Forward: 5'-TGTAGAGTTGAGAGTCTCTTTAATT-3'; Reverse: 5'-TGTAGACTGTTGCTCAATCTAATCTTC-3'). The ASK 3'UTR was cloned into pGL3-promoter luciferase reporter vector(Promega, Madison, USA) between luciferase coding sequence and SV40-poly(A) sequence using the XbaI site.Two miR-19a sites located in the ASK1 3'UTR (site-1: UUGCAC, starting at nt 304, and site-2: UUGCAC, starting at nt 620) were mutated to Luc-ASK1 3'UTR MU (UUCGTG) using a QuickChange II Site-Directed Mutagenesis Kit (Agilent Technologies, Santa Clara, CA). Specific primers were used for mutagenesis of miR-19a target site-1 and –2 in the human ASK1 3'UTR (site-1 forward: 5'-CAGCAGCTATTCGTGTTCAGCC-3', site-1 reverse: 5'-GGCTGAACACGAATAGCTGCTG-3'; site-2 forward: 5'-ACTGTACCAGTTCGTGATGCTTGA-3', site-2 reverse: 5'-CAAGCATCACGAACTGGTACAGT-3'). For the reporter gene assay, EAhy926 cells were cultured in 12-well plates, and transduced with Ad-LacZ and Ad-miR-19a for 24 h.

Then, the cells were transfected with 300 ng firefly luciferase reporter plasmid (pGL3-Luc-ASK1 3'UTR or pGL3-Luc-ASK1 3'UTR MU) and 20 ng Renilla luciferase reporter plasmid pRL-RSV (Promega) using Lipofectamine 2000 transfection reagent (Invitrogen). Luciferase assays were performed at 48 h after transfection using the Dual Luciferase Reporter Assay system (Promega Biotech Co., Ltd). Firefly luciferase activities were normalized to Renilla luciferase activities.

Sandwich enzyme-linked immunosorbent assay for histone-associated DNA fragments

Endothelial cell death was assessed by an enzyme-linked immunosorbent assay using a Death Detection Kit from Boehringer Mannheim (Indianapolis, IN). HUVECs were seeded at 2×10^4 cells per well in a 96-well plate and grown to 90% confluence. The cells were then treated with LPS (100 ng/ml) for the indicated times. The cells were harvested in lysis buffer, and the cytoplasmic and nuclear protein fractions were separated by centrifugation at $200 \times g$. The supernatant (cytoplasmic fraction) was used to measure histone-associated DNA fragments.

Statistical analyses

Data are expressed as means ± standard error. The statistical significance of differences was assessed by Student's t-tests or analysis of variance as appropriate. A value of $P < 0.05$ was considered statistically significant.

Results

LPS down-regulates miR-19a expression in endothelial cells

The human ASK1 3'-UTR was analyzed using a website tool (http://www.microrna.org/microrna/getGeneForm.do). As a result, we found ASK1 target miRNAs (Additional file 1: Figure S1). To investigate whether miR-19a expression is altered during LPS-induced endothelial cell apoptosis, HUVECs were treated with various doses of LPS for the indicated times. The expression level of miR-19a was detected by qRT-PCR. As shown in Figure 1, miR-19a expression was not affected by an LPS concentration of less than 10 ng/ml. However, miR-19a expression was significantly decreased by 100 ng/ml LPS (Figure 1A). In addition, we found a decrease in miR-19a expression by about 80% after treatment with 100 ng/ml LPS for 12 h (Figure 1B). Taken together, these data show that LPS down-regulates miR-19a expression in a dose and time-dependent manner.

Expression of ASK1 is up-regulated during LPS-induced endothelial cell apoptosis

To examine the possible link between miR-19a and ASK1, we also measured the expression of ASK1 at both

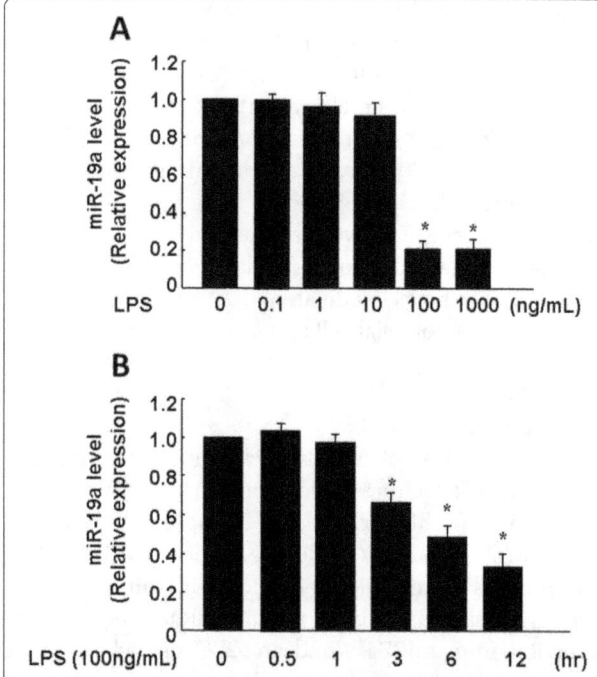

Figure 1 LPS down-regulates expression of miR-19a in endothelial cells. **(A)**, HUVECs were cultured in 6-well plate, after 80% confluence, cells were treated with different concentration of LPS for 12 h. **(B)**, HUVECs were treated with 100 ng/mL LPS for different times point as indicated. Quantitative real-time PCR showed that miR-19a expression was down-regulated, compared with that in the respective control group. Significance is indicated as *$P < 0.05$.

mRNA and protein levels. HUVECs were stimulated for 12 h with LPS, and then qRT-PCR was performed using RNA isolated from control and treated cells. We observed a 2–5-fold increase of Ask1 transcripts in response to LPS (Figure 2A). To determine whether the increase in Ask1 mRNA expression correlated with enhanced ASK1 protein expression, we performed western blotting and quantified ASK1 expression by densitometry. As shown in Figure 2B, treatment of HUVECs with LPS led to increased expression of ASK1 by up to 2.8-fold. ASK1 can be phosphorylated at several sites, and these phosphorylation sites regulate ASK1 activity in both positive and negative manners. Phosphorylation of ASK1 at Ser83 inhibits ASK1-induced apoptosis. On the other hand, phosphorylation of ASK1 at Thr845 promotes ASK1-induced apoptosis [30,31]. The activity of ASK1 induced by LPS was indicated by detecting phosphorylation of ASK1 at Thr845. As shown in Figure 2B, LPS significantly increased the phosphorylation of ASK1 at Thr845 by up to 5.2-fold. Finally, we detected LPS-induced HUVEC apoptosis by a TUNEL assay. As expected, LPS treatment resulted in an increase of HUVEC apoptosis, which is consistent with other studies [32,33].

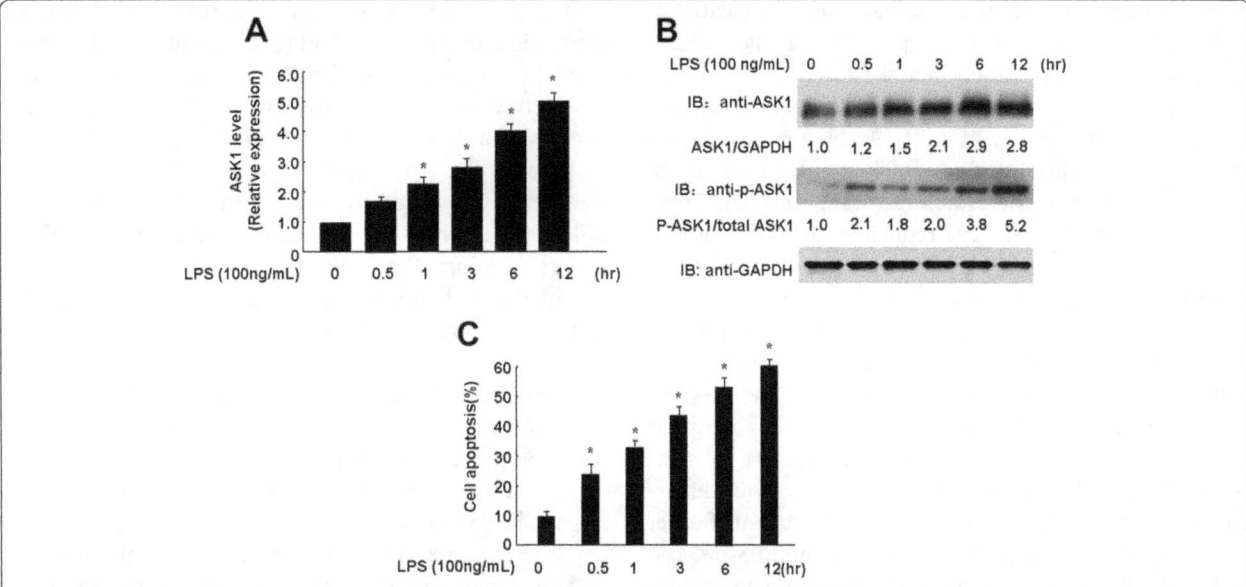

Figure 2 Expression of ASK1 was up-regulated during LPS-induced endothelial cell apoptosis. **(A)**, HUVECs were cultured in 6-well plate, after 80% confluence, cells were treated with 100 ng/mL LPS for different times point as indicated. Quantitative real time PCR revealed that the expression levels of ASK1 were increased after LPS stimuli. **(B)**, Cell lysates obtained from HUVECs were subjected to western blot. ASK1, phosphorylation of ASK1 at Thr845, and GAPDH were determined by indicated antibodies. **(C)**, the quantitative analysis of HUVECs apoptosis was determined by ELISA assay. Significance is indicated as *P <0.05.

Over-expression of miR-19a attenuates LPS-induced apoptosis in endothelial cells

We next tested whether overexpression of miR-19a affected ASK1 expression levels and apoptosis in endothelial cells. To this end, we generated Ad-miR19a. First, qRT-PCR was performed to evaluate miR-19a expression levels in HUVECs transduced with either Ad-miR-19a or transfected with the miR-19a inhibitor. Transduction of HUVECs with Ad-miR-19a led to a marked increase of miR-19a levels in a dose-dependent manner (Figure 3A). However, transfection of the miR-19a inhibitor significantly suppressed miR-19a expression (Figure 3B), but had no effect on miR-20a expression (Additional file 1: Figure S2). Second, we evaluated the expression of ASK1

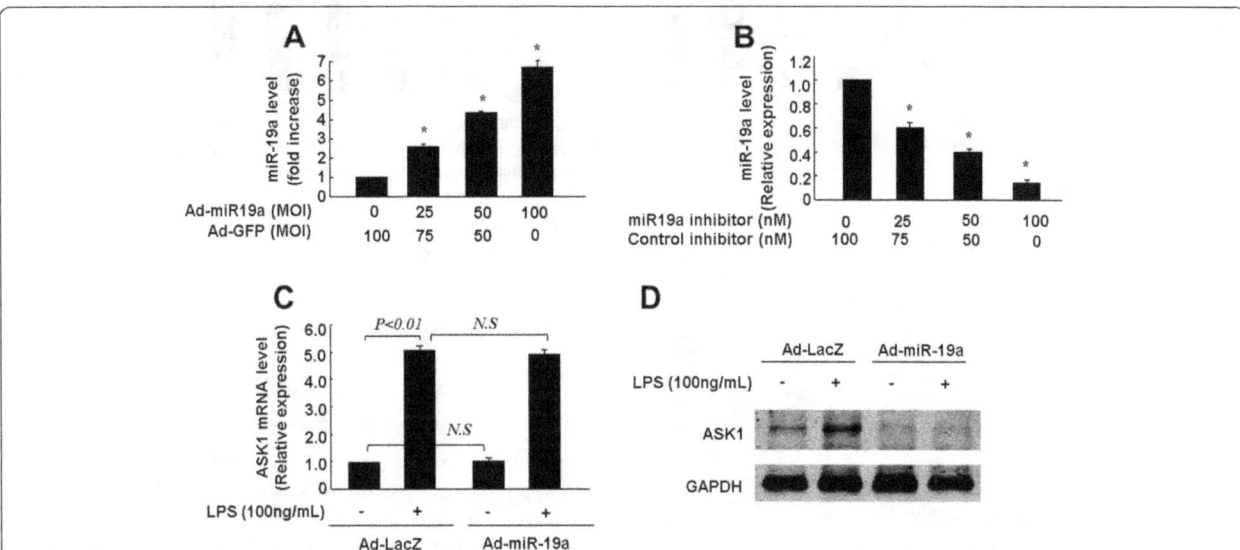

Figure 3 Expression of miR-19a in HUVECs either transduced with Ad-miR19a and transfected with miR-19a inhibitor. **(A)** and **(B)**, Quantitative real time PCR revealed that the expression levels of miR-19a were increased after Ad-miR-19a transduction **(A)**, but suppressed by miR-19a inhibitor in a dose-dependent manner **(B)**. Significance is indicated as *P < 0.05. **(C)** and **(D)**, MiR-19a regulates ASK1 expression. HUVECs were transduced with Ad-miR-19a or Ad-LacZ (50MOI) for 24 h, and then treated with 100 ng/mL LPS for another 24 h. The expression of ASK1 was determined by quantitative real time PCR **(C)** and western blot **(D)**.

in LPS-treated HUVECs. We found that LPS treatment significantly induced ASK1 expression at the mRNA level, but the up-regulation of Ask1 mRNA was not affected by overexpression of miR-19a (Figure 3C). Interestingly, overexpression of miR-19a by the adenovirus resulted in marked down-regulation of LPS-induced ASK1 expression at the protein level (Figure 3D). These results suggest that miR-19a regulates the expression of ASK1 at the translational level.

To investigate the effects of miR-19a on HUVEC apoptosis, we used Ad-miR-19a and the miR-19a inhibitor. We found that the miR-19a inhibitor increased HUVEC apoptosis in a dose-dependent manner. However, the miR-19a inhibitor had no effect on LPS-induced HUVEC apoptosis (Figure 4A and B). Over-expression of miR-19a substantially inhibited LPS-induced HUVEC apoptosis, but this effect was reversed by miR-19a inhibitor treatment, further indicating the involvement of miR-19a in LPS-induced endothelial cell apoptosis (Figure 4C). Because ASK1 is directly involved in the LPS-induced apoptotic pathway, we determined whether over-expression of miR-19a regulates the ASK1/p38 apoptotic pathway. We thus performed western blotting to detect the expression levels of key molecules in this apoptotic pathway, including phosphorylated p38 and cleaved caspase-3. The levels of ASK1 and phosphorylation of p38 were significantly increased in cells

treated with LPS, but markedly decreased by over-expression of miR-19a (Figure 4D). Indeed, LPS treatment markedly increased the levels of cleaved caspase-3, which were only decreased by miR-19a. However, the decreases in ASK1, phosphorylation of p38, and cleaved caspase-3 due to infection with Ad-miR-19a were significantly reversed by treating the HUVECs with the miR-19a inhibitor (Figure 4D). Taken together, these results support that miR-19a regulates LPS-induced endothelial cell apoptosis through modulation of the ASK1/p38 apoptotic pathway.

MiR-19a regulates ASK1 expression by targeting its 3′UTR

To examine whether ASK1 is a direct target of miR-19a, we employed a luciferase reporter assay. Sequence analysis of the human ASK1 3′UTR revealed two putative miR-19a-binding sites located at 287–309 nt and 620–625 nt (Figure 5A). Accordingly, we constructed a reporter plasmid by cloning the ASK1 3′UTR containing the two putative miR-19a-binding sites into the 3′UTR of a pGL3 vector. Two miR-19a-binding sites located in the ASK1 3′UTR (site-1: UUGCAC and site-2: UUGCAC) were mutated to Luc-ASK1 3′UTR MU (UUCGTG) (Additional file 1: Figure S3). As shown in Figure 5B, over-expression of miR-19a markedly down-regulated the activity of the luciferase gene fused with the wild-type ASK1 3′UTR. In

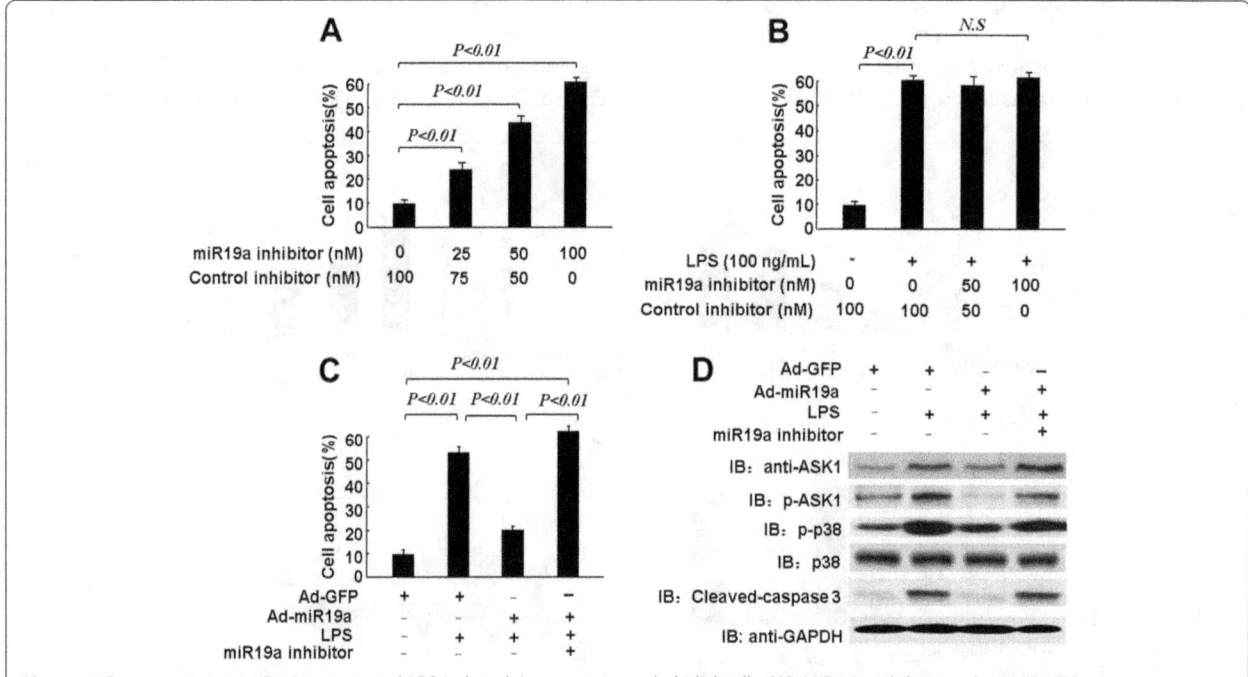

Figure 4 Over-expressing miR-19a attenuated LPS-induced Apoptosis in endothelial cells. **(A)**, MiR-19a inhibitor induces HUVECs apoptosis. HUVECs were transduced with miR19a inhibitor or Control inhibitor for 48 h, then cell apoptosis was determined by ELISA assay. **(B)**, HUVECs were transduced with miR19a inhibitor or Control inhibitor for 24 h, then treated with 100 ng/mL LPS for another 24 h. Cell apoptosis was determined by ELISA assay. **(C)**, HUVECs were transduced with Ad-miR-19a or miR19a inhibitor or Ad-GFP (50MOI) for 24 h, then treated with 100 ng/mL LPS for another 24 h. Cell apoptosis was determined by ELISA assay. **(D)**, Cell lysates obtained from HUVECs were subjected to western blot. ASK1, phosphorylation of ASK1 at Thr845, p38, phosphorylation of p38, cleaved caspase3 and GAPDH were determined by indicated antibodies.

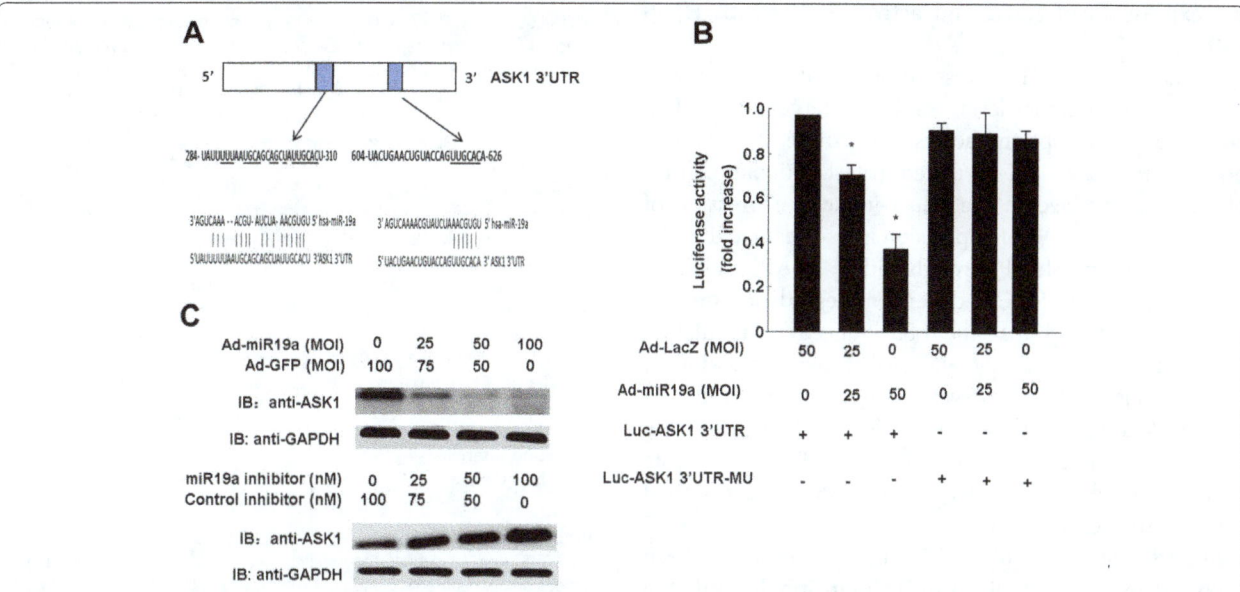

Figure 5 MiR-19a suppressed expression of ASK1 through binding to its 3'UTR. **(A)**, a representative illustration of the putative binding sites for miR-19a in human ASK1 3'UTR. **(B)**, Luciferase assay was performed in EAhy926 cells using pGL3 reporter vector fused with either ASK1 wild-type 3'UTR or ASK1 3'UTR mutant. Over-expression of miR-19a significantly decreased the activity of luciferase gene fused with ASK1 wild-type 3'UTR, but had no effect on the activity of luciferase fused with ASK1 3'UTR mutant. **(C)**, Detection of ASK1 expression by western blot in the whole lysates of HUVECs transduced with different dosages of AdmiR-19a. The result showed that the protein level of ASK1 was suppressed by miR-19a, however, increased by miR-19a inhibitor, in a dose-dependent manner. Significance is indicated as *P <0.05.

contrast, over-expression of miR-19a barely affected the activity of the luciferase gene fused with the mutant ASK1 3'UTR. The protein levels of ASK1 underwent marked dose-dependent down-regulation in cells transduced with Ad-miR-19a at various multiplicities of infection and significant up-regulation in cells treated with the miR-19a inhibitor (Figure 5C). Taken together, these results indicate that ASK1 is a direct target of miR-19a in endothelial cells. MiR-19a regulates ASK1 expression by targeting specific binding sites in the 3'UTR of ASK1.

Discussion

It is clear that miRNAs contribute to cell development and regulate many biological processes, including cell proliferation and apoptosis, by acting as oncogenes or tumor suppressor genes [5,34]. MiR-19a has been reported to be involved in control of endothelial cell functions and neovascularization [18,19]. LPS has been shown to contribute to damage observed in various types of endothelium [22,25,26]. The molecular pathways of apoptosis are only just being deciphered in endothelial cells. Choi et al. reported that LPS induces apoptosis in microdermal endothelial cells via recruitment of the adaptor Fas-associated death domain [35]. Luyendyk et al. reported that LPS induces the activity of ASK1 and activates downstream MAPK pathways [28]. LPS-induced p38 activation and apoptosis are reduced in splenocytes and dendritic cells from ASK1-deficient mice [29].

MiR-19a belongs to the miR-17-92 cluster [8,9]. Some members of this cluster are regulated by LPS [9,36]. For example, miR-19a and miR-20 are both involved in LPS-induced apoptosis of rheumatoid arthritis fibroblast-like synoviocytes [9,21]. Interestingly, miR-20 has been reported to target ASK1 [9], which prompted us to determine whether miR-19a regulates ASK1 expression. In the present study, we found that LPS down-regulated the expression of miR-19a and miR-20a (Additional file 1: Figure S4), but the decrease of miR-19a expression was more dramatic than that of miR-20a. These data indicate that miR-19a might play a major role in LPS-induced apoptosis. Indeed, ASK1, also known as mitogen-activated protein kinase kinase kinase 5 (MAP3K5), is a member of the MAPK kinase kinase family [37]. Under stress conditions such as oxidative stress and stimulation tumor necrosis factor-α (TNF-α) or LPS, p38 MAPK can be activated by MAP3K5, followed by activation of the downstream target caspase-3 [29,38]. In fact, over-expression of ASK1 has been reported to induce apoptotic cell death [37]. Under stress conditions, ASK1 is auto-phosphorylated at Thr845, resulting in activation of ASK1 and phosphorylation of p38 [39]. Expression of miR-19a was markedly inhibited by 100 ng/ml LPS. Moreover, 100 ng/ml LPS significantly increased the expression of ASK1 in HUVECs as well as activation of the ASK1/p38 pathway, leading to apoptosis. However, we found that over-expression of miR-19a by Ad-miR-19a inhibited LPS-induced HUVEC apoptosis by decreasing the expression of ASK1 and activity of the ASK1/p38 pathway.

Regulation of ASK1 expression occurs at various levels. At the transcriptional level, Ask1 is a target of the E2F family of transcription factors, but there is currently no evidence that E2F-mediated transcriptional control of Ask1 is involved in the injury-induced expression of Ask1 in vivo [40]. ASK1 expression is also regulated at the post-translational level through either c-IAP1-mediated ubiquitination or a SOCS1-dependent degradation process [41,42]. We found that the LPS-induced up-regulation of ASK1 mRNA was not affected by overexpression of miR-19a (Figure 3C). However, overexpression of miR-19a by the adenovirus impaired LPS-induced ASK1 expression at the protein level (Figures 3D and 4D). These data clearly indicate that miR-19a regulates ASK1 expression at the post-transcription level.

Indeed, regulation of miR-19 in apoptosis has been reported by other studies [8,17]. Interestingly, miR-19a target proteins, such as PTEN, p53, TNF-α and SMAD4, have been reported to regulate apoptosis [12,43-45]. In the present study, we found that miR-19a not only down-regulated the activity of ASK1, but also inhibited the expression of ASK1. These results indicate that miR-19a regulates LPS-induced endothelial cell apoptosis partially via regulating the expression of ASK1. Ad-miR-19a significantly decreased the activity of a luciferase reporter containing the 3'UTR of ASK1. These results further suggest that, in LPS-treated endothelial cells, miR-19a controls ASK1 expression by regulating mRNA translation. Therefore, our study suggests a novel mechanism for the regulation of ASK1 expression at the translational level in response to inflammatory stimuli.

Conclusions

In summary, our data suggest that miR-19a is expressed in HUVECs, and the expression of miR-19a is modulated by LPS. Moreover, we found that overexpression of miR-19a inhibits LPS-induced apoptosis of endothelial cells. Furthermore, we identified ASK1 as a direct target of miR-19a in HUVECs. MiR-19a regulates ASK1 expression by targeting specific binding sites in the 3'UTR of ASK1. Taken together, these results suggest that miR-19a may be a useful target to protect endothelial cells from LPS-induced apoptosis.

Additional file

Additional file 1: Figure S1. MicroRNAs target to human ASK13'-UTR. Human ASK1 3'-UTR was analyzed using a website tools (http://www.microrna.org/microrna/getGeneForm.do). **Figure S2.** MiR19a inhibitor has no effect on expression of miR-20a. HUVECs were cultured in 6-well plate, after 80% confluence, cells were treated with miR-19a inhibitor or control inhibitor for 24 h. The expression of miR-20a was determined by quantitative real time PCR. **Figure S3.** MiR-19a suppressed ASK1

3'UTR-Luc activity. The ASK 3'UTR was cloned into pGL3-promoter luciferase reporter vector between luciferase coding sequence and SV40-poly(A) sequence using the Xbal site. EAhy926 cells transfected with pGL3 reporter vector fused with either ASK1 wild-type 3'UTR or ASK1 3'UTR mutant for 12 h, and then the cells were infected with Ad-LacZ or Ad-miR19a (50MOI). After 36 h cells were harvested and luciferase assay was performed. Luc-ASK1 3'UTR-M1: site-1 "UUGCAC" was mutated; Luc-ASK1 3'UTR-M2: site-2 "UUGCAC" was mutated; Luc-ASK1 3'UTR-MU, both sites were mutated. (* indicates P < 0.01 compared with Ad-LacZ group). **Figure S4.** LPS down-regulates expression of miR-19a and miR-20a in endothelial cells. HUVECs were cultured in 6-well plate, after 80% confluence, cells were treated with 100 ng/mL LPS for different times point as indicated. Quantitative real-time PCR showed that miR-19a and miR-20a expression were down-regulated. (* indicates P < 0.05 compared with miR-19a/LPS 0 h; # indicates P < 0.05 compared with miR-20a/LPS 0 h; § indicates P < 0.05 compared with miR-19a).

Competing interests

The authors declare that they have no competing interests.

Authors' contributions

JWL, ZYF, FT and WF defined the research theme. JWL, ZYF and WF designed methods and experiments, carried out the experiments, analyzed the data, interpreted the results, and wrote the paper. XQQ, ZJ and YX co-designed experiments, and discussed analyses, interpretations, and presentation. ZYF and JWL carried out the additional experiments, analyzed the data, interpreted the results, and rewrote the paper. All authors have contributed to, seen and approved the manuscript.

Acknowledgements

This study was supported by a research grant from the Health Bureau of Jiangyin (M201405 to Tao Fan).

Author details

[1]Department of Respiration, Jiangyin Hospital of Traditional Chinese Medicine Affiliated to Nanjing University of Chinese Medicine, Jiangyin City, Jiangsu Province 214400, China. [2]Department of Neurology, Jiangyin Hospital of Traditional Chinese Medicine Affiliated to Nanjing University of Chinese Medicine, Jiangyin City, Jiangsu Province 214400, China. [3]Department of Internal Medicine, Jiangyin Hospital of Traditional Chinese Medicine Affiliated to Nanjing University of Chinese Medicine, Jiangyin City, Jiangsu Province 214400, China. [4]Department of Neurology, Shanghai First People's Hospital, Shanghai Jiaotong University School of Medicine, Shanghai 200080, China.

References

1. Kong Y, Han JH. MicroRNA: biological and computational perspective. Genomics Proteomics Bioinformatics. 2005;3(2):62–72.
2. McManus MT, Sharp PA. Gene silencing in mammals by small interfering RNAs. Nat Rev Genet. 2002;3(10):737–47.
3. Grabarek JB. RNA silencing. Adv Exp Med Biol. 2003;544:145–58.
4. He L, Hannon GJ. MicroRNAs: small RNAs with a big role in gene regulation. Nat Rev Genet. 2004;5(7):522–31.
5. Bukhari SI, Vasquez-Rifo A, Gagne D, Paquet ER, Zetka M, Robert C, et al. The microRNA pathway controls germ cell proliferation and differentiation in C. elegans. Cell Res. 2012;22(6):1034–45.
6. Chen Y, Gelfond J, McManus LM, Shireman PK. Temporal microRNA expression during in vitro myogenic progenitor cell proliferation and differentiation: regulation of proliferation by miR-682. Physiol Genomics. 2011;43(10):621–30.
7. Shenoy A, Blelloch RH. Regulation of microRNA function in somatic stem cell proliferation and differentiation. Nat Rev Mol Cell Biol. 2014;15(9):565–76.
8. Liu M, Wang Z, Yang S, Zhang W, He S, Hu C, et al. TNF-alpha is a novel target of miR-19a. Int J Oncol. 2011;38(4):1013–22.
9. Philippe L, Alsaleh G, Pichot A, Ostermann E, Zuber G, Frisch B, et al. MiR-20a regulates ASK1 expression and TLR4-dependent cytokine release in rheumatoid fibroblast-like synoviocytes. Ann Rheum Dis. 2013;72(6):1071–9.

10. He L, Thomson JM, Hemann MT, Hernando-Monge E, Mu D, Goodson S, et al. A microRNA polycistron as a potential human oncogene. Nature. 2005;435(7043):828–33.

11. Chen Q, Xia HW, Ge XJ, Zhang YC, Tang QL, Bi F. Serum miR-19a predicts resistance to FOLFOX chemotherapy in advanced colorectal cancer cases. Asian Pac J Cancer Prev. 2013;14(12):7421–6.

12. Feng Y, Liu J, Kang Y, He Y, Liang B, Yang P, et al. miR-19a acts as an oncogenic microRNA and is up-regulated in bladder cancer. J Exp Clin Cancer Res. 2014;33(1):67.

13. Jia Z, Wang K, Zhang A, Wang G, Kang C, Han L, et al. miR-19a and miR-19b overexpression in gliomas. Pathol Oncol Res. 2013;19(4):847–53.

14. Lin Q, Chen T, Lin Q, Lin G, Lin J, Chen G, et al. Serum miR-19a expression correlates with worse prognosis of patients with non-small cell lung cancer. J Surg Oncol. 2013;107(7):767–71.

15. Qin S, Ai F, Ji WF, Rao W, Zhang HC, Yao WJ. miR-19a promotes cell growth and tumorigenesis through targeting SOCS1 in gastric cancer. Asian Pac J Cancer Prev. 2013;14(2):835–40.

16. Wu Q, Yang Z, An Y, Hu H, Yin J, Zhang P, et al. MiR-19a/b modulate the metastasis of gastric cancer cells by targeting the tumour suppressor MXD1. Cell death Dis. 2014;5:e1144.

17. Wu TY, Zhang TH, Qu LM, Feng JP, Tian LL, Zhang BH, et al. MiR-19a is correlated with prognosis and apoptosis of laryngeal squamous cell carcinoma by regulating TIMP-2 expression. Int J Clin Exp Pathol. 2014;7(1):56–63.

18. He J, Li Y, Yang X, He X, Zhang H, He J, et al. The feedback regulation of PI3K-miR-19a, and MAPK-miR-23b/27b in endothelial cells under shear stress. Molecules. 2012;18(1):1–13.

19. Doebele C, Bonauer A, Fischer A, Scholz A, Reiss Y, Urbich C, et al. Members of the microRNA-17-92 cluster exhibit a cell-intrinsic antiangiogenic function in endothelial cells. Blood. 2010;115(23):4944–50.

20. Treguer K, Heinrich EM, Ohtani K, Bonauer A, Dimmeler S. Role of the microRNA-17-92 cluster in the endothelial differentiation of stem cells. J Vasc Res. 2012;49(5):447–60.

21. Philippe L, Alsaleh G, Suffert G, Meyer A, Georgel P, Sibilia J, et al. TLR2 expression is regulated by microRNA miR-19 in rheumatoid fibroblast-like synoviocytes. J Immunol. 2012;188(1):454–61.

22. Damico RL, Chesley A, Johnston L, Bind EP, Amaro E, Nijmeh J, et al. Macrophage migration inhibitory factor governs endothelial cell sensitivity to LPS-induced apoptosis. Am J Respir Cell Mol Biol. 2008;39(1):77–85.

23. Rafikov R, Dimitropoulou C, Aggarwal S, Kangath A, Gross C, Pardo D, et al. Lipopolysaccharide-induced lung injury involves the nitration-mediated activation of RhoA. J Biol Chem. 2014;289(8):4710–22.

24. Wang HL, Akinci IO, Baker CM, Urich D, Bellmeyer A, Jain M, et al. The intrinsic apoptotic pathway is required for lipopolysaccharide-induced lung endothelial cell death. J Immunol. 2007;179(3):1834–41.

25. Haimovitz-Friedman A, Cordon-Cardo C, Bayoumy S, Garzotto M, McLoughlin M, Gallily R, et al. Lipopolysaccharide induces disseminated endothelial apoptosis requiring ceramide generation. J Exp Med. 1997;186(11):1831–41.

26. Dauphinee SM, Karsan A. Lipopolysaccharide signaling in endothelial cells. Lab Investig. 2006;86(1):9–22.

27. Bannerman DD, Goldblum SE. Mechanisms of bacterial lipopolysaccharide-induced endothelial apoptosis. Am J Physiol Lung Cell Mol Physiol. 2003;284(6):L899–914.

28. Luyendyk JP, Piper JD, Tencati M, Reddy KV, Holscher T, Zhang R, et al. A novel class of antioxidants inhibit LPS induction of tissue factor by selective inhibition of the activation of ASK1 and MAP kinases. Arterioscler Thromb Vasc Biol. 2007;27(8):1857–63.

29. Matsuzawa A, Saegusa K, Noguchi T, Sadamitsu C, Nishitoh H, Nagai S, et al. ROS-dependent activation of the TRAF6-ASK1-p38 pathway is selectively required for TLR4-mediated innate immunity. Nat Immunol. 2005;6(6):587–92.

30. Jung H, Seong HA, Ha H. Murine protein serine/threonine kinase 38 activates apoptosis signal-regulating kinase 1 via Thr 838 phosphorylation. J Biol Chem. 2008;283(50):34541–53.

31. Zhang R, Luo D, Miao R, Bai L, Ge Q, Sessa WC, et al. Hsp90-Akt phosphorylates ASK1 and inhibits ASK1-mediated apoptosis. Oncogene. 2005;24(24):3954–63.

32. Munshi N, Fernandis AZ, Cherla RP, Park IW, Ganju RK. Lipopolysaccharide-induced apoptosis of endothelial cells and its inhibition by vascular endothelial growth factor. J Immunol. 2002;168(11):5860–6.

33. Xing YL, Zhou Z, Agula, Zhong ZY, Ma YJ, Zhao YL, et al. Protocatechuic aldehyde inhibits lipopolysaccharide-induced human umbilical vein endothelial cell apoptosis via regulation of caspase-3. Phytother Res. 2012;26(9):1334–41.

34. Yan L, Hao H, Elton TS, Liu Z, Ou H. Intronic microRNA suppresses endothelial nitric oxide synthase expression and endothelial cell proliferation via inhibition of STAT3 signaling. Mol Cell Biochem. 2011;357(1–2):9–19.

35. Choi KB, Wong F, Harlan JM, Chaudhary PM, Hood L, Karsan A. Lipopolysaccharide mediates endothelial apoptosis by a FADD-dependent pathway. J Biol Chem. 1998;273(32):20185–8.

36. Xu Z, Zhang C, Cheng L, Hu M, Tao H, Song L. The microRNA miR-17 regulates lung FoxA1 expression during lipopolysaccharide-induced acute lung injury. Biochem Biophys Res Commun. 2014;445(1):48–53.

37. Ichijo H, Nishida E, Irie K, ten Dijke P, Saitoh M, Moriguchi T, et al. Induction of apoptosis by ASK1, a mammalian MAPKKK that activates SAPK/JNK and p38 signaling pathways. Science. 1997;275(5296):90–4.

38. Liu H, Nishitoh H, Ichijo H, Kyriakis JM. Activation of apoptosis signal-regulating kinase 1 (ASK1) by tumor necrosis factor receptor-associated factor 2 requires prior dissociation of the ASK1 inhibitor thioredoxin. Mol Cell Biol. 2000;20(6):2198–208.

39. Hattori K, Naguro I, Runchel C, Ichijo H. The roles of ASK family proteins in stress responses and diseases. Cell Commun Signal. 2009;7:9.

40. Hershko T, Korotayev K, Polager S, Ginsberg D. E2F1 modulates p38 MAPK phosphorylation via transcriptional regulation of ASK1 and Wip1. J Biol Chem. 2006;281(42):31309–16.

41. Zhao Y, Conze DB, Hanover JA, Ashwell JD. Tumor necrosis factor receptor 2 signaling induces selective c-IAP1-dependent ASK1 ubiquitination and terminates mitogen-activated protein kinase signaling. J Biol Chem. 2007;282(11):7777–82.

42. He Y, Zhang W, Zhang R, Zhang H, Min W. SOCS1 inhibits tumor necrosis factor-induced activation of ASK1-JNK inflammatory signaling by mediating ASK1 degradation. J Biol Chem. 2006;281(9):5559–66.

43. Fan Y, Yin S, Hao Y, Yang J, Zhang H, Sun C, et al. miR-19b promotes tumor growth and metastasis via targeting TP53. RNA. 2014;20(6):765–72.

44. Chen B, She S, Li D, Liu Z, Yang X, Zeng Z, et al. Role of miR-19a targeting TNF-alpha in mediating ulcerative colitis. Scand J Gastroenterol. 2013;48(7):815–24.

45. Fuziwara CS, Kimura ET. High iodine blocks a Notch/miR-19 loop activated by the BRAF(V600E) oncoprotein and restores the response to TGFbeta in thyroid follicular cells. Thyroid. 2014;24(3):453–62.

Differential regulation of the *α-globin* locus by Krüppel-like factor 3 in erythroid and non-erythroid cells

Alister PW Funnell[1], Douglas Vernimmen[2], Wooi F Lim[1], Ka Sin Mak[1], Beeke Wienert[1], Gabriella E Martyn[1], Crisbel M Artuz[1], Jon Burdach[1], Kate GR Quinlan[1], Douglas R Higgs[3], Emma Whitelaw[4], Richard CM Pearson[1] and Merlin Crossley[1*]

Abstract

Background: Krüppel-like Factor 3 (KLF3) is a broadly expressed zinc-finger transcriptional repressor with diverse biological roles. During erythropoiesis, KLF3 acts as a feedback repressor of a set of genes that are activated by Krüppel-like Factor 1 (KLF1). Noting that KLF1 binds *α-globin* gene regulatory sequences during erythroid maturation, we sought to determine whether KLF3 also interacts with the *α-globin* locus to regulate transcription.

Results: We found that expression of a human transgenic *α-globin* reporter gene is markedly up-regulated in fetal and adult erythroid cells of *Klf3$^{-/-}$* mice. Inspection of the mouse and human *α-globin* promoters revealed a number of canonical KLF-binding sites, and indeed, KLF3 was shown to bind to these regions both *in vitro* and *in vivo*. Despite these observations, we did not detect an increase in endogenous murine *α-globin* expression in *Klf3$^{-/-}$* erythroid tissue. However, examination of murine embryonic fibroblasts lacking KLF3 revealed significant de-repression of *α-globin* gene expression. This suggests that KLF3 may contribute to the silencing of the *α-globin* locus in non-erythroid tissue. Moreover, ChIP-Seq analysis of murine fibroblasts demonstrated that across the locus, KLF3 does not occupy the promoter regions of the *α-globin* genes in these cells, but rather, binds to upstream, DNase hypersensitive regulatory regions.

Conclusions: These findings reveal that the occupancy profile of KLF3 at the *α-globin* locus differs in erythroid and non-erythroid cells. In erythroid cells, KLF3 primarily binds to the promoters of the adult *α-globin* genes, but appears dispensable for normal transcriptional regulation. In non-erythroid cells, KLF3 distinctly binds to the *HS-12* and *HS-26* elements and plays a non-redundant, albeit modest, role in the silencing of *α-globin* expression.

Keywords: KLF1, KLF3, Alpha globin, Globin gene regulation, Transcription factor

Background

Krüppel-like Factor 3 (KLF3/BKLF) belongs to the KLF family of transcription factors, of which there are 17 members with diverse biological roles in development and cellular differentiation [1,2]. KLFs are characterized by a highly homologous C-terminal DNA-binding domain, containing three C2H2 zinc fingers that direct binding to CACCC boxes and related GC-rich sequences in the control regions of target genes [3]. KLF3 is predominantly a transcriptional repressor which recruits a co-repressor complex containing C-terminal binding protein (CtBP) to facilitate silencing of its target genes [4]. KLF3 is broadly expressed and has been shown to have roles in several processes, including erythropoiesis [5,6], adipogenesis [7,8], muscle cell differentiation [9], and B cell development [10,11].

The *Klf3* gene is highly expressed in the red blood cell lineage due to the presence of an erythroid specific promoter, which is driven by a related KLF, Krüppel-like Factor 1 (KLF1) [12]. KLF1 is a master regulator of erythropoiesis, with functional roles in many facets of erythroid development, including red blood cell structure, heme biosynthesis and *globin* gene regulation [13,14]. Loss of KLF1 is embryonic lethal, with *Klf1$^{-/-}$* mice dying *in utero* from

* Correspondence: m.crossley@unsw.edu.au
[1]School of Biotechnology and Biomolecular Sciences, University of New South Wales, Sydney, NSW 2052, Australia
Full list of author information is available at the end of the article

lethal β-thalassemia, due to a failure of activation of β-globin gene expression [15,16]. In addition to regulating the β-globin gene, KLF1 has been shown to bind the α-globin locus [17-19], as a component of a complex of factors recruited when looping of enhancer elements to the proximal promoter occurs and initiates high level gene expression [17,20]. Loss of KLF1 leads to reduced α-globin gene expression and chromosome looping [21], although these effects are notably less severe than the down-regulation of β-globin expression, possibly due to functional redundancy between other KLF family members and related SP (specificity protein) factors [17]. In regulating both the α-globin and β-globin loci, it is probable that KLF1 contributes to the maintenance of globin chain balance, which is critical for red blood cell function and viability.

Given that KLF3 is required for normal erythropoiesis and is known to repress a subset of KLF1-driven target genes [5], we investigated whether KLF3 can also bind and repress the α-globin gene. In support of this, we found that expression of a GFP reporter transgene, driven by the human α-globin promoter and regulatory elements [22] is significantly up-regulated in Klf3$^{-/-}$ mice. Furthermore, inspection of the α-globin promoter revealed numerous KLF3 consensus recognition sites and we confirmed that KLF3 binds to this region both in vitro in electrophoretic mobility shift assays and in vivo by chromatin immunoprecipitation. However, despite demonstrating an in vivo interaction of KLF3 with the α-globin locus, we did not detect de-regulated endogenous α-globin expression in Klf3$^{-/-}$ erythroid tissue. In contrast, examination of α-globin mRNA levels in Klf3$^{-/-}$ murine embryonic fibroblasts revealed a significant increase in expression. In fibroblasts, KLF3 was found to bind not at α-globin promoter regions, but at the upstream HS-12 and HS-26 regulatory regions. Together, these results suggest that KLF3 may have a role in the silencing of the α-globin locus in non-erythroid tissue.

Methods

Mouse lines
The generation of GFP Line3 [22] and Klf3$^{-/-}$ [8] lines have been described previously. Mice were maintained on the FVBN/J background and animal work was carried out under the approval of the Animal Care and Ethics Committees of the University of Sydney (project numbers L02/1-2005/3/4048, L02/6-2006/3/4344 and L02/7-2009/3/5079) and the University of New South Wales (approval number 09/128A).

Cell sorting and flow cytometry
Flow cytometry was performed using a FACSCalibur Flow Cytometer (BD Biosciences, San Jose, CA) and data were analyzed using CellQuest Pro (BD Biosciences) or FlowJo v7.6.5 software (TreeStar, Ashland, OR). TER119 antibody was supplied by BD Biosciences and titrated to optimal concentration. TER119$^+$ cells were purified from embryonic day 14.5 fetal liver (Klf3$^{+/+}$, Klf3$^{+/-}$ and Klf3$^{-/-}$ littermates) using Magnetic Activated Cell Sorting with Anti-TER119 MicroBeads (Miltenyi Biotec Australia Pty Ltd, Macquarie Park, NSW, Australia) by positive selection using MS columns as per the supplier's instructions.

Cell culture
Mouse and human primary erythroblasts, murine erythro-leukemia (MEL) cells and interspecific MEL hybrids (containing a copy of human chromosome 16) were cultured and differentiated as previously described [17]. K562 cells were cultured at 37°C in RPMI medium and COS-7 cells were cultured in Dulbecco's Modified Eagle Medium (DMEM), each supplemented with 10% (v/v) fetal calf serum (FCS) and 1% (v/v) penicillin, streptomycin and glutamine solution (PSG) (Gibco-BRL Life Technologies, Grand Island NY). Murine embryonic fibroblasts (MEFs) were prepared from littermate E12.5 embryos (Klf3$^{+/+}$, Klf3$^{+/-}$ and Klf3$^{-/-}$). Briefly, heart, liver, intestinal, lung and brain tissue were removed and remaining embryonic tissue was homogenized in 3 mL trypsin/EDTA using an 18-gauge needle. MEFs were subsequently incubated for 2–3 minutes at 37°C and were then transferred to 100 mm plates containing 7 mL DMEM (10% FCS, 1% PSG). The cells were then left undisturbed for 48 h at 37°C and were passaged every 2–3 days. MEF cells (passage 2 or 3) were immortalized by transfecting with 5 μg pRSV-T [23] using the FuGENE6 transfection reagent protocol (Roche Diagnostics Australia Pty Ltd, Castle Hill, NSW, Australia). Immortalized Klf3$^{-/-}$ MEFs that have been stably rescued with KLF3-V5, or pMSCVpuro empty vector (Clontech Laboratories, Mountain View, CA) as a negative control, have been described previously [24].

RNA extraction and cDNA synthesis
RNA extraction was performed using TRI-Reagent, according to the manufacturer's guidelines (Sigma, St. Louis, MO). RNA samples were further purified using RNeasy columns (Qiagen, Victoria, Australia) and by treating with DNase I (Ambion, Austin, TX). Subsequently, cDNA was prepared using Superscript VILO cDNA synthesis kit (Invitrogen, Carlsbad, CA), according to the manufacturer's instructions.

Primers and real-time RT-PCR
Primer sequences for real-time RT-PCR were: mouse α-globin, 5'-GTCACGGCAAGAAGGTCGC-3' and 5'-GGGGTGAAATCGGCAGGGT-3'; mouse β-actin, 5'-GCTTCTTTGCAGCTCCTTCGT-3' and 5'-CCAGCGCAGCGATATCG-3'; mouse 18S, 5'-CACGGCCGGTACA

GTGAAAC-3′ and 5′-AGAGGAGCGAGCGACCAA-3′; mouse *Gapdh*, 5′-GTCTCCTGCGACTTCAGC-3′ and 5′-TCATTGTCATACCAGGAAATGAGC-3′; and as described previously for *Klf3*, *Klf8* and *Fam132a* [7,12,25]. Quantitative real-time PCR was performed using *Power SYBR Green PCR Master Mix* and the *7500 Fast Real-Time PCR System* (Applied Biosystems, Foster City, CA), as described previously [26]. Data were analyzed using *7500 Software v2.0.4* (Applied Biosystems).

Electrophoretic mobility shift assays (EMSAs)

EMSAs were carried out as described previously [27]. COS-7 cells in 100 mm plates were transfected with 5 μg vector (pMT3-empty or pMT3-Klf3 [28]) using FuGENE6 (Roche Diagnostics Australia Pty Ltd) as per the manufacturer's protocol. Nuclear extracts from COS-7, uninduced K562, uninduced MEL and MEF cell lines were harvested as previously described [28]. Oligonucleotides used in the synthesis of radiolabelled probes were: human *α-globin* promoter, 5′-CGCAGGCCCCGCCCGGGACTC-3′ and 5′-GAGTCCCGGGCGGGGCCTGCG-3′; mouse *α-globin* promoter, 5′-TGGAGGACACGCCCTTGGAGG-3′ and 5′-CCTCCAAGGGCGTGTCCTCCA-3′; mouse *HS-26* probe 1, 5′-AGGTGTACACACCCAGGCCAA-3′ and 5′-TTGGCCTGGGTGTGTACACCT-3′, and; *HS-26* probe 2, 5′-AGGCCAAGGGTGGAGCAGACCA-3′ and 5′-TGG TCTGCTCCACCCTTGGCCT-3′. Supershift recognition of KLF3 was achieved using specific antiserum that has been described previously [27]. Probe sequences were identified using CLC Main Workbench software version 6.6.2 (CLC Bio, Cambridge, MA).

Chromatin immunoprecipitation (ChIP)

ChIP assays were carried out as previously described [17,29], using the previously described anti-KLF3 antibody [27]. KLF3 ChIP-Seq analysis has previously been described [24] and enrichment tracks were visualized using Integrative Genomics Viewer [30].

Western blotting

Western blots of nuclear extracts from MEF, MEL and COS-7 cells were performed as previously described [31] using KLF3 anti-serum [27]. Full-Range Rainbow Molecular Weight Marker was supplied by GE Healthcare (Piscataway, NJ).

Results

KLF3 regulates expression of a human transgenic α-globin promoter *in vivo*

To begin our investigation into potential regulation of the *α-globin* gene by KLF3, we made use of an existing well-characterized transgenic mouse model, termed Line3, in which a GFP reporter gene is expressed under the control of the human *α-globin* proximal promoter and *HS-40*

enhancer region [22]. The red blood cells of Line3 mice express GFP and it is possible to accurately measure the level of expression by flow cytometry in either adult peripheral blood or erythroid cells purified from tissues, such as the fetal liver. To determine whether KLF3 has a role in regulating expression of the reporter gene, we introduced the homozygous transgene into $Klf3^{-/-}$ mice [8] by breeding and compared GFP expression in $Klf3^{+/+}$, $Klf3^{+/-}$ and $Klf3^{-/-}$ erythrocytes.

As previously reported [22], we found that GFP is expressed in Line3::$Klf3^{+/+}$ erythrocytes with a broad, but consistent and reproducible profile. These cells can be classified as expressing low, intermediate or high levels of GFP (Figure 1). Loss of a single allele of *Klf3* had no effect on transgene expression, as we did not find any notable difference between the GFP profiles of Line3:: $Klf3^{+/+}$ and Line3::$Klf3^{+/-}$ mice (Figure 1A and 1C). However, analysis of red blood cells from homozygous Line3:: $Klf3^{-/-}$ animals revealed a significant increase in GFP expression (Figure 1A). On average, we found that 46% of $Klf3^{-/-}$ cells express high levels of GFP, compared to 18% in $Klf3^{+/+}$ animals (Figure 1C). We also examined newly formed erythrocytes in the erythroid fetal liver. We purified TER119+ cells from the fetal livers of Line3::$Klf3^{+/+}$ and Line3::$Klf3^{-/-}$ mice and again observed a significant increase in transgenic promoter activity in the absence of KLF3 (Figure 1B and 1D). Together, these data suggest that KLF3 directly or indirectly represses the human transgenic *α-globin* promoter in this mouse model.

KLF3 binds the human and mouse α-globin promoters *in vitro* and *in vivo*

Having determined that KLF3 influences the expression of a transgene driven by *α-globin* gene regulatory sequences *in vivo*, we next investigated whether KLF3 interacts directly with the *α-globin* promoter. We inspected the human and mouse *α-globin* proximal promoters to identify potential high affinity KLF3 binding sites, which match the KLF consensus sequence, 5′-NCN CNC CCN-3′ [32]. This analysis revealed the presence of several sites, with the human promoter in particular containing 14 potential interaction motifs (Figure 2A and 2B). We then used our sequence analysis to design probes for electrophoretic mobility shift assays (EMSA) to investigate binding of KLF3 to the *α-globin* promoter *in vitro*. To assess binding to the human promoter, we based our probe on the most frequently seen consensus sequence, 5′-NCC CGC CCN-3′, which occurs four times (Figure 2A). In the case of the mouse promoter, where there are noticeably fewer potential KLF3 binding sites (Figure 2B), we used the sequence 5′-NCA CGC CCN-3′, which is found twice, to inform our probe design.

We began our investigation into *in vitro* binding by expressing KLF3 in COS-7 cells and assessing the ability

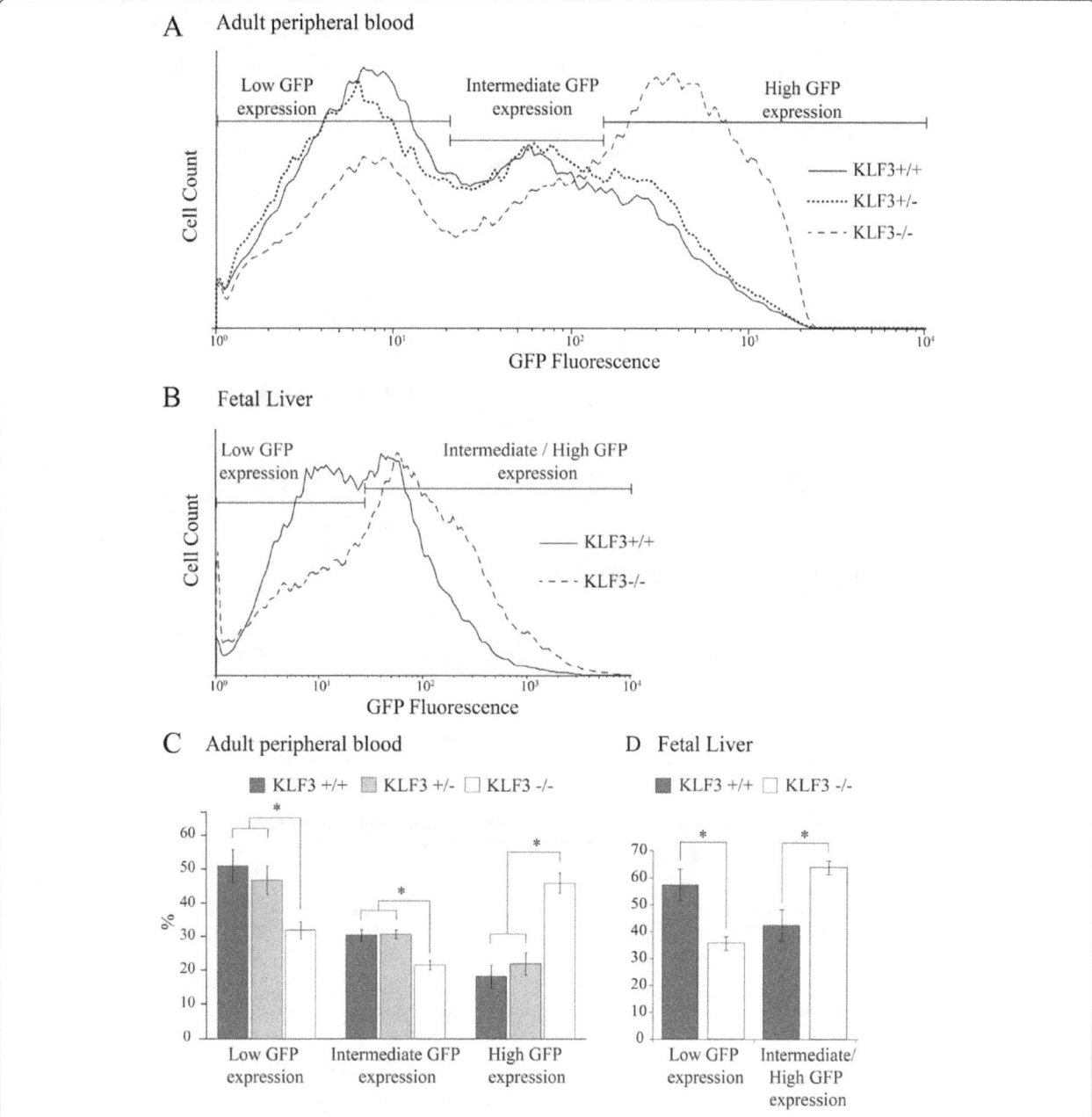

Figure 1 Loss of KLF3 results in up-regulation of the human *α-globin* gene in a transgenic mouse model. Line3 mice, containing a GFP transgene under the control of the human *α-globin* proximal promoter and *HS-40* enhancer [22], were crossed with *Klf3*[+/−] mice to generate Line3::*Klf3*[+/+], Line3::*Klf3*[+/−] and Line3::*Klf3*[−/−] mice, all homozygous for the transgene. Erythroid GFP fluorescence was then measured by flow cytometry. Shown are representative fluorescence profiles of **(A)** peripheral blood from mice at 3 weeks of age and **(B)** TER119[+] sorted erythrocytes from embryonic day E14.5 fetal liver. The populations were gated to identify cells expressing low, intermediate and high levels of GFP. Statistical analysis of these gated populations is shown for **(C)** erythrocytes from mice at 3 weeks of age and (D) TER119[+] fetal liver cells. For erythrocytes analyzed at 3 weeks of age, n = 32 for Line3::*Klf3*[+/+], n = 48 for Line3::*Klf3*[+/−] and n = 8 for Line3::*Klf3*[−/−]. For the analysis of fetal erythrocytes, n = 3 for Line3::*Klf3*[+/+] and n = 4 for Line3::*Klf3*[−/−]. Error bars represent standard deviation and * represents *P* < 0.05 (two tailed t-test).

of nuclear extracts purified from these cells to interact with the human and mouse *α-globin* promoter sequences by EMSA. We found that the nuclear extracts bound both human and mouse probes with high affinity and confirmed that this interaction was specific to KLF3 by supershift with anti-KLF3 antibody (Figure 3,

lanes 2–3 and 7–8). Minimal background binding was observed for nuclear extracts from mock transfected COS-7 cells (Figure 3, lanes 1 and 6). We next determined whether endogenous KLF3 present in erythroid cell lines also binds to the human and mouse *α-globin* promoter probes by preparing nuclear extracts from

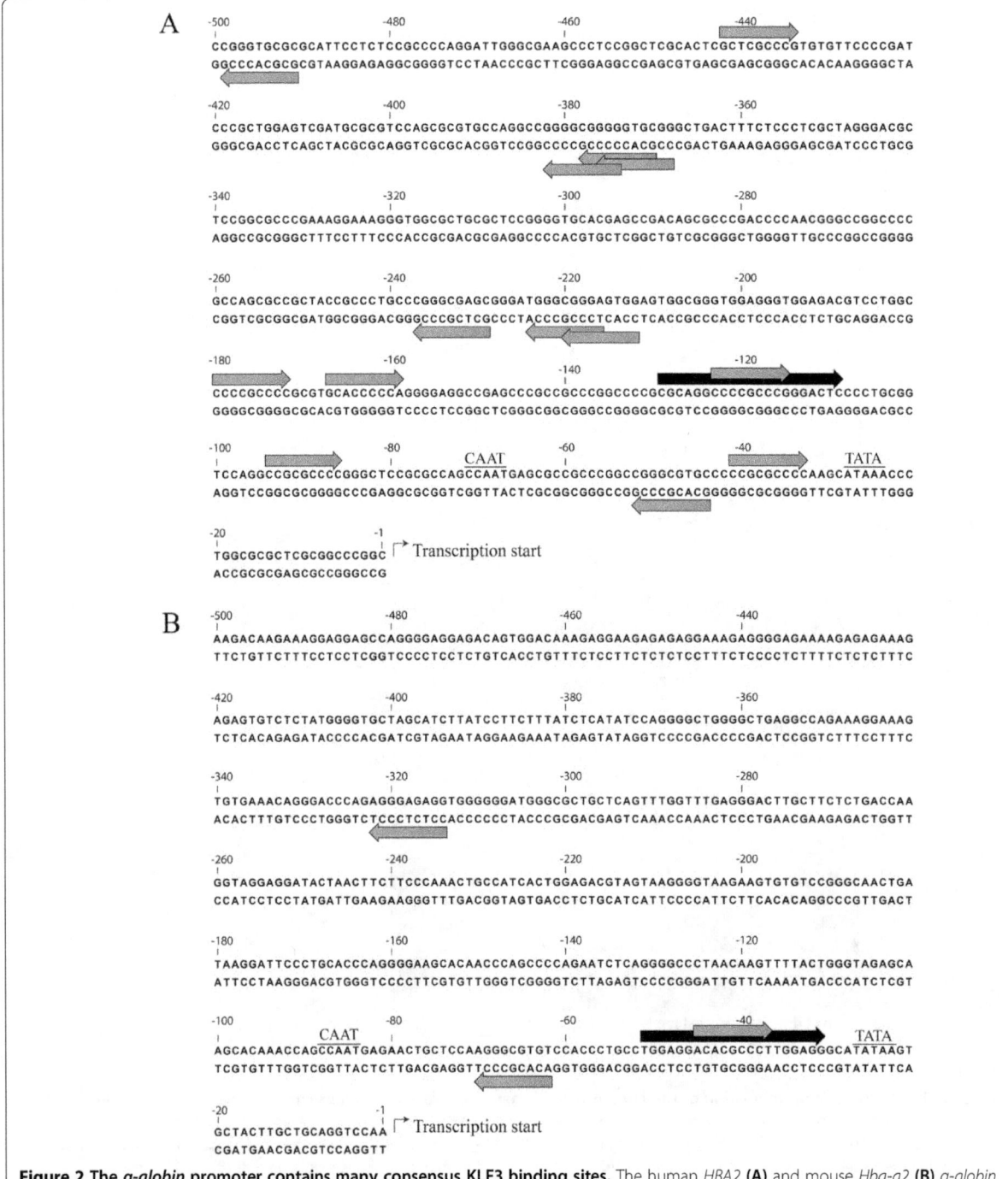

Figure 2 The *a-globin* promoter contains many consensus KLF3 binding sites. The human *HBA2* **(A)** and mouse *Hba-a2* **(B)** *a-globin* proximal promoter sequences, immediately 5′ to the transcriptional start site, were inspected for consensus binding sites, conforming to the sequence 5′-NCN CNC CCN-3′. The position and direction of binding sites are indicated by grey arrows. The sequences used in the design of probes for electrophoretic mobility shift assays are shown by black arrows. Also indicated are CAAT and TATA boxes. Sequences are numbered with respect to the transcription start site at +1.

human K562 and murine erythroleukemia (MEL) cells. We tested binding of the K562 nuclear extracts to the human *α-globin* probe and the MEL nuclear extracts to the mouse probe. Again, we found that proteins in both extracts bound to the promoter sequences and

confirmed the identity of a KLF3 complex by supershift with an anti-KLF3 antibody (Figure 3, lanes 4–5 and 9–10).

Having established that KLF3 can bind to both the human and mouse *α-globin* proximal promoters *in vitro*,

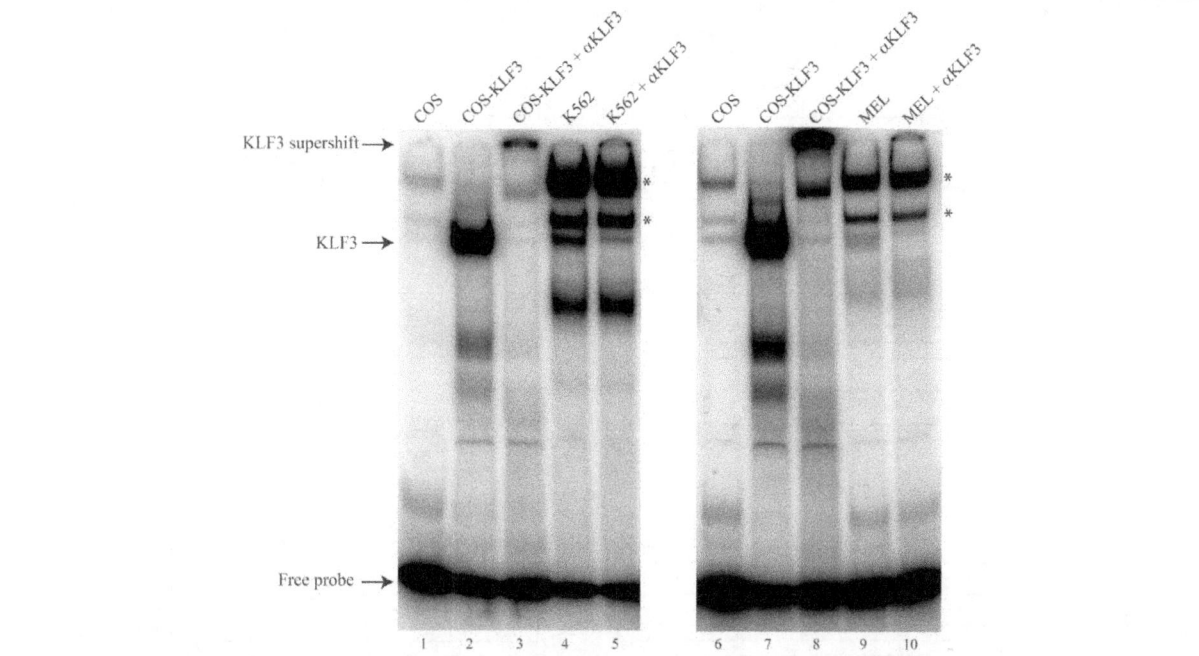

Figure 3 KLF3 binds the *a-globin* promoter *in vitro*. The binding of KLF3 to the *a-globin* promoter was assessed by EMSA, using radiolabeled probes designed from analysis of the human and mouse *a-globin* proximal promoter sequences (Figure 2). KLF3 was either expressed in COS-7 cells (lanes 2, 3, 7 and 8) or endogenous KLF3 was harvested in nuclear extracts from K562 (lanes 4 and 5) and MEL (lanes 9 and 10) erythroid cell lines. Nuclear extracts from mock transfected COS-7 cells have been included as a negative control (lanes 1 and 6). Binding to the human promoter sequence is shown in the left hand panel whilst binding to the mouse sequence is shown on the right. αKLF3 indicates an anti-KLF3 antibody used to validate KLF3 specific binding by supershift. Additional bands in lanes 4, 5, 9 and 10 (denoted by asterisks) most likely represent SP1 and SP3 as in [12].

we carried out chromatin immunoprecipitation (ChIP) assays on a number of erythroid cell types to determine whether KLF3 binds to the *α-globin* locus *in vivo*. Our approach was to conduct a primer walk across the locus, in which we used TaqMan real time RT-PCR probes to assess binding at the upstream HS (DNase hypersensitive) enhancers, the proximal promoter, the coding sequence, and at a number of control sites, including the *α-globin* intergenic region, and the *β-actin* and *β-globin* genes.

First, we investigated KLF3 binding to the *α-globin* locus in uninduced MEL cells and found only background binding at each of the sites we examined (Figure 4A). However, when we chemically induced erythroid maturation in these cells, we observed a marked enrichment of KLF3 at the *α-globin* proximal promoter (Figure 4B), consistent with what we have previously reported [6]. An examination of mouse primary erythroblasts confirmed that KLF3 also binds this site *in vivo* (Figure 4C). We then made use of an interspecies hybrid MEL cell line into which human chromosome 16, containing the *α-globin* locus, has been introduced [17,33]. Again, we saw only background binding of KLF3 across the locus in uninduced cells but observed noticeable enrichment at the human *α-globin* proximal promoter following erythroid maturation (Figure 4D and E). Finally, we assessed binding

in human primary erythroblasts and once again found high enrichment at the *α-globin* proximal promoter (Figure 4F).

KLF3 represses *a-globin* expression in non-erythroid tissue

Having confirmed that KLF3 can bind to the *α-globin* promoter *in vitro* and *in vivo*, we next asked whether loss of KLF3 results in de-regulation of endogenous *α-globin* gene expression. We first compared *α-globin* mRNA levels in red blood cells purified from the erythroid fetal liver of $Klf3^{+/+}$, $Klf3^{+/-}$ and $Klf3^{-/-}$ embryos (E14.5) by real time qRT-PCR. Despite our observation that KLF3 binds the *α-globin* gene promoter *in vivo*, we did not detect any up-regulation of *α-globin* expression in $Klf3^{-/-}$ erythroid cells (Figure 5A). In addition, we have previously analyzed the expression of multiple *globin* genes at an earlier stage of development (E13.5) and similarly observed no change in adult *α-globin* transcripts in the absence of KLF3 [6].

It is possible that in erythroid cells, loss of KLF3 has little effect because *α-globin* is expressed at maximal levels. We therefore turned our attention to non-erythroid cells, namely murine embryonic fibroblasts (MEFs), which express only low levels of *α-globin* transcripts. In both primary and immortalized MEFs lacking KLF3, we observed a modest de-repression of *α-globin* gene expression (by

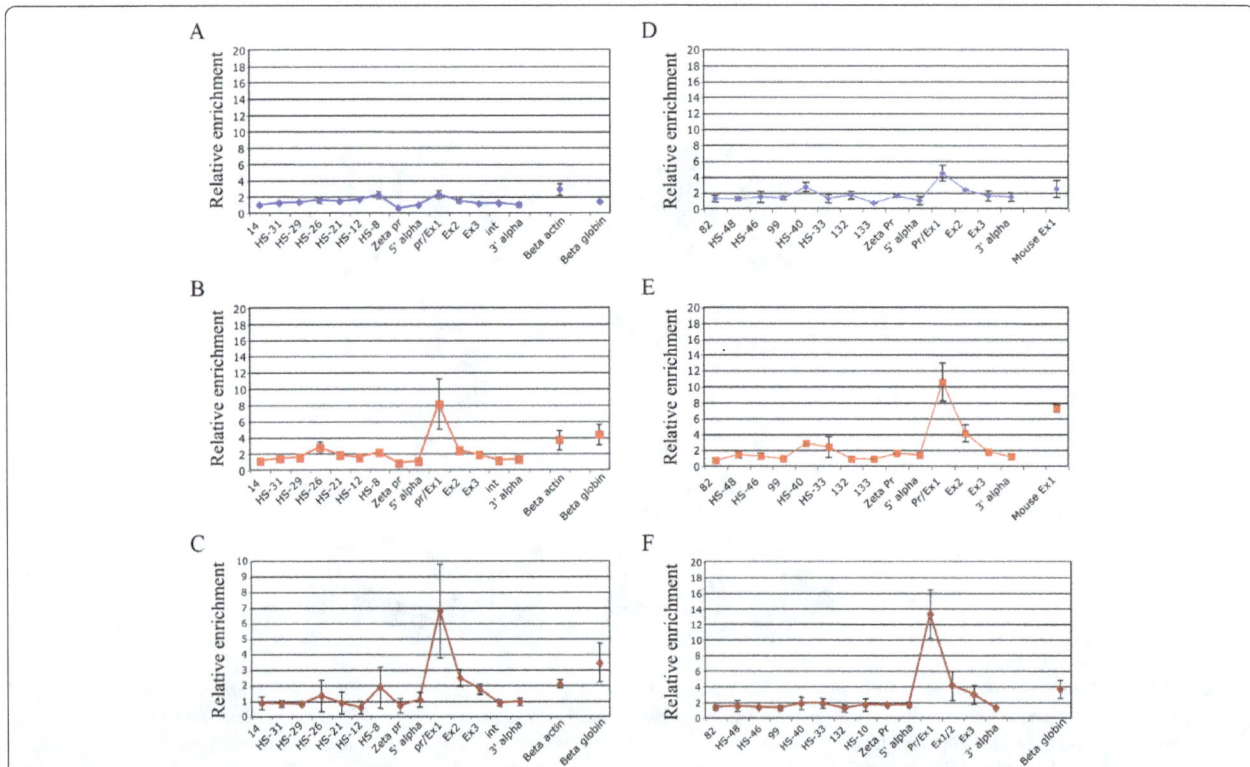

Figure 4 KLF3 binds the human and mouse *α-globin* promoters *in vivo* in chromatin immunoprecipitation assays. An anti-KLF3 antibody was used to immunoprecipitate chromatin from the following cell types: **(A)** uninduced MEL cells, **(B)** induced MEL cells, **(C)** mouse primary erythroblasts, **(D)** uninduced interspecific MEL hybrids containing a normal copy of human chromosome 16, **(E)** induced interspecific MEL hybrids, and **(F)** human primary erythroblasts. The y-axis represents enrichment over input DNA, normalized to a control sequence in the *Gapdh* gene (mouse) or *18S* (human). The x-axis represents the positions of the TaqMan probes used. The coding sequence is represented by the three exons (Promoter/Ex1, Ex2, and Ex3) of the *α-globin* genes. HS- primer sets refer to upstream DNase-hypersensitive regions. Zeta pr refers to the mouse and human embryonic *α-globin* promoters (*Hba-x* and *HBZ*). Inter, refers to the intergenic region (between mouse *Hba-a1* and *Hba-a2*). 5' and 3' are negative controls flanking the *α-globin* gene. β-actin and β-globin denote control sequences at the *β-actin* gene and *β-globin* promoter respectively. Error bars correspond to ±1 standard deviation from at least two independent ChIPs.

6.3-fold and 4.9-fold respectively compared to *Klf3*$^{+/+}$cells) (Figure 5B and 5C). Furthermore, stable rescue of *Klf3*$^{-/-}$ MEFs with V5-tagged KLF3 resulted in a significant diminution of *α-globin* mRNA expression (Figure 5D).

To explore KLF3's potential mode of regulation at the *α-globin* locus in non-erythroid cells, we analyzed recently generated KLF3 ChIP-Seq data from MEF cells [24]. We found that in these cells, KLF3 was not bound to the adult *α-globin* promoters (*Hba-a1* and *Hba-a2*), but showed significant occupancy at the upstream *HS-12* and *HS-26* regulatory regions (Figure 5E). This contrasted with our observation from a series of erythroid cells (Figure 4), in which KLF3 was primarily found at the *α-globin* promoter. Analysis of the *HS-26* region revealed two sites resembling the KLF binding consensus via which KLF3 might be recruited. Indeed, EMSA experiments confirmed that both of these sites are recognized by both KLF3 expressed in COS-7 cells and endogenous KLF3 in MEFs (Figure 5F). Taken together, these findings suggest that in non-erythroid cells, KLF3 binds the *HS-12* and *HS-26* regulatory regions and may be involved in

repressing and thereby maintain physiologically low levels of *α-globin* expression in these cells.

Lastly, we also analyzed the DNA-binding capacity of KLF3 extracted from erythroid (MEL) and non-erythroid (MEF) cells (Figure 6A and B). Equivalent levels of KLF3 from these two cellular sources exhibited comparable DNA-binding activity at sites in both the murine *α-globin* promoter and the *HS-26* regulatory element. This suggests that the differing *in vivo* occupancy of KLF3 across the *α-globin* locus in erythroid and non-erythroid cells (compare Figures 4B and 4C with Figure 5E) is not due to intrinsic differences in KLF3's ability to bind DNA.

Discussion

Our data show that KLF3 binds the adult mouse *α-globin* promoter in erythroid tissue *in vivo*. However, KLF3 does not appear to functionally repress the endogenous promoter in red blood cells. Similarly, we have previously observed KLF3 occupancy at the adult *β-globin* (*Hbb-b1*) promoter in erythroid cells and no associated perturbation

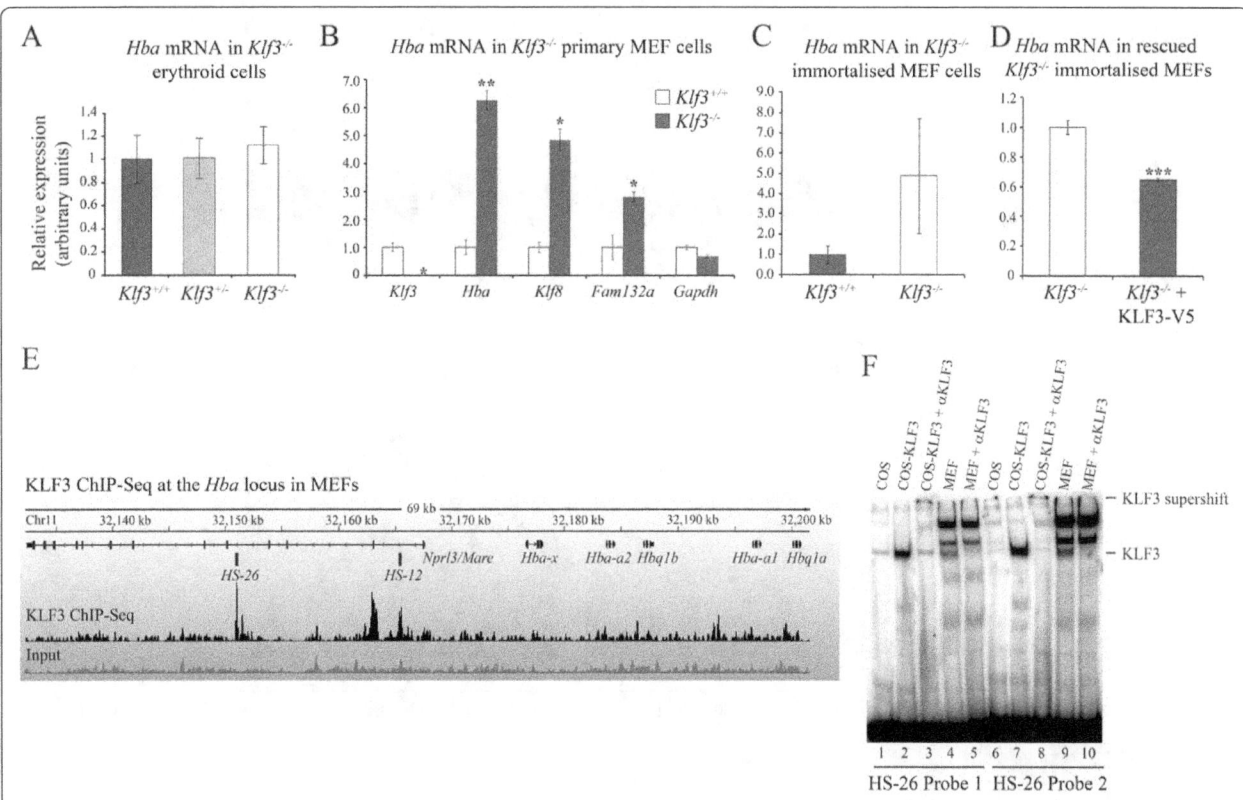

Figure 5 α-globin gene expression is de-repressed in murine embryonic fibroblasts lacking KLF3. α-globin mRNA expression levels were determined by real time qRT-PCR analysis of **(A)** TER119+ erythrocytes purified from embryonic day E14.5 fetal liver (Klf3+/+n = 3, Klf3+/− n = 5, Klf3−/− n = 6), **(B)** primary MEFs (Klf3+/+n = 2, Klf3−/− n = 2 or 3), **(C)** immortalized MEFs (Klf3+/+n = 2, Klf3−/− n = 2), and **(D)** immortalized Klf3−/− MEFs rescued with KLF3-V5 or empty vector (n = 2 for each). **(A-D)** In each case, relative expression of α-globin mRNA was normalized to 18S rRNA levels, and the expression levels of Klf3+/+ **(A-C)** or Klf3−/− **(D)** were set to 1.0. In **(B)**, mRNA levels of Klf3 and two known KLF3-repressed targets, Klf8 and Fam132a [7,25], have also been analyzed together with a negative control, Gapdh. In **(A-D)**, error bars shown represent standard error of the mean, *P < 0.05 (one-tailed t-test relative to Klf3+/+), **P < 0.002 (two-tailed t-test relative to Klf3+/+), ***P < 0.05 (two-tailed t-test relative to Klf3−/−). **(E)** KLF3 ChIP-Seq track across the murine α-globin locus in MEFs from [24]. The positions of the HS-12 and HS-26 regulatory regions are indicated. **(F)** EMSA showing the binding of KLF3 to two sites within the HS-26 region. Nuclear extracts were obtained from COS-7 cells that were mock-transfected (lanes 1 and 6) or transfected with pMT3-Klf3 (lanes 2, 3, 7 and 8). Nuclear extracts from MEFs are shown in lanes 4, 5, 9 and 10. Identification of KLF3:DNA complexes was achieved by addition of an antibody specific for KLF3 (αKLF3, lanes 3, 5, 8 and 10).

of Hbb-b1 transcription upon ablation of KLF3 [6]. It is notable that KLF3 binding is highest at the late stages of erythroid maturation (compare Figure 4A with 4B, and 4D with 4E) when the adult globin genes are expressed at very high levels and their promoters are presumably highly accessible. This is also when KLF3 levels peak [5] and it is possible that KLF3 gains access to these regions but is not sufficiently potent to limit KLF1 driven activation of the genes. This observation highlights the view that transcription factor binding sites discovered by ChIP may not always have functional relevance in the context in which they are identified, but may instead reflect the dynamic nature of transcription factor binding at permissive loci. Indeed, a number of recent ChIP-Seq experiments, performed in association with transcriptome analysis of gene knockout models have revealed that transcription factor binding is not always associated with changes in gene activity [34,35].

In contrast to the endogenous mouse α-globin promoter, we have shown that KLF3 does appear to regulate the expression of a human transgenic promoter in erythroid cells. The transgene is driven by a minimal human α-globin promoter and HS-40 and perhaps this subset of elements is more reliant on repression by KLF3 than the entire set of globin regulatory elements. In the case of the endogenous α-globin locus, chromatin conformation capture experiments suggest that gene expression is dependent upon chromosomal looping of distal enhancers to the proximal promoter, in a process that is dependent upon many regulatory factors [17]. The removal of such complexity in the transgene most likely offers a far greater opportunity for observing the contribution that single factors make to expression levels. Alternatively, it should be noted that the experiments presented here primarily analyzed KLF3 function in murine cells, and thus it remains possible that KLF3

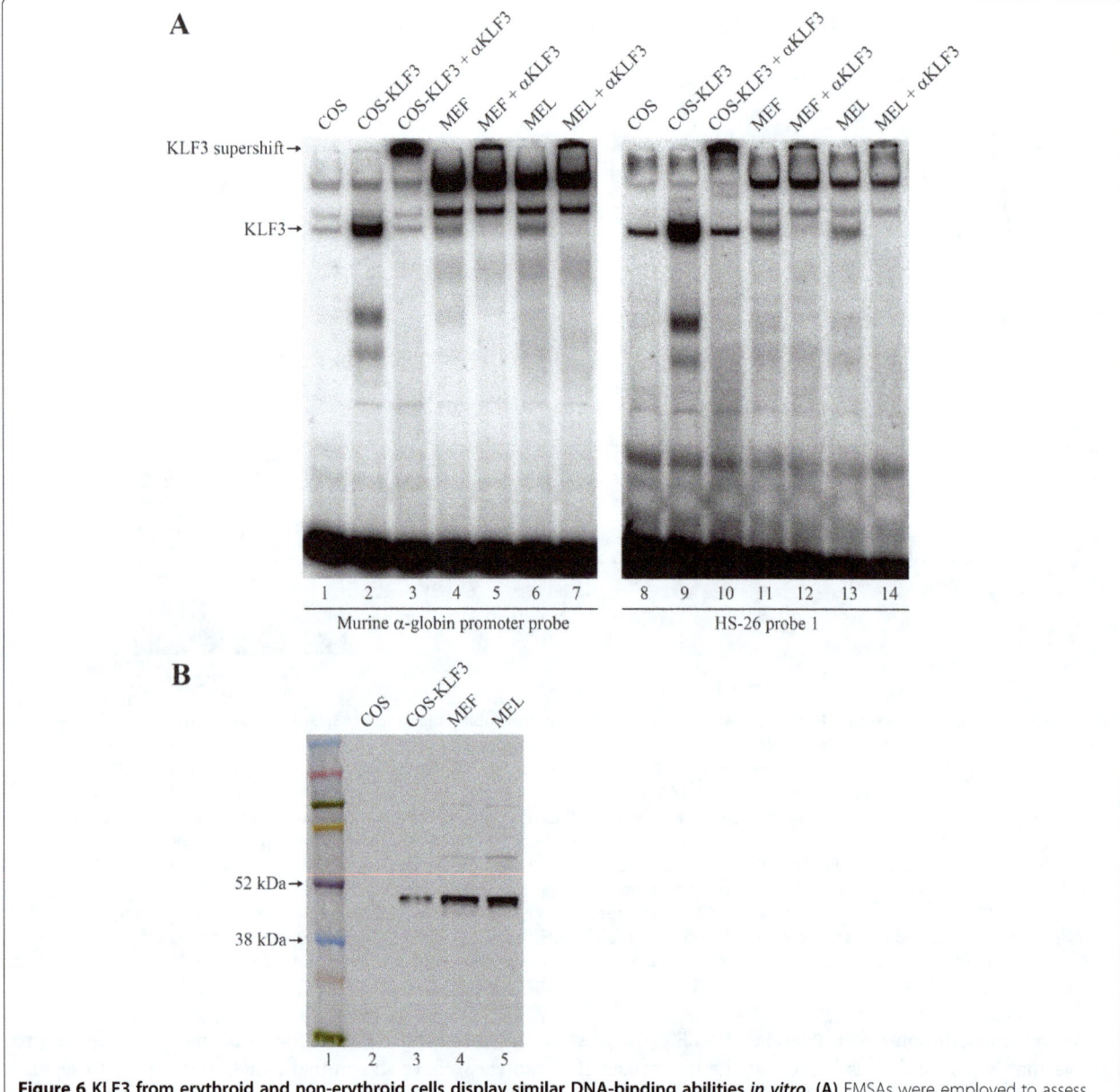

Figure 6 KLF3 from erythroid and non-erythroid cells display similar DNA-binding abilities *in vitro*. (A) EMSAs were employed to assess the binding of KLF3 to the murine *α-globin* promoter (lanes 1–7) and a site in the *HS-26* element from Figure 5 (lanes 8–14). Nuclear extracts were harvested from non-erythroid MEF (lanes 4, 5, 11 and 12) or erythroid MEL cells (lanes 6, 7, 13 and 14). Nuclear extracts from mock transfected COS-7 cells (lanes 1 and 8) or cells expressing KLF3 (lanes 2, 3, 9 and 10) were included as negative and positive controls respectively. The identity of KLF3 was confirmed by specific antibody supershifts (lanes 3, 5, 7, 10, 12 and 14). **(B)** Western blot demonstrating the relative amounts of KLF3 in MEF (lane 4) and MEL (lane 5) nuclear extracts used in the EMSAs in (A). As negative and positive controls, COS and COS-KLF3 nuclear extracts have been included (lanes 2 and 3) at 20-fold lower relative amounts than in (A) to facilitate visualization. A size ladder is shown in lane 1.

may play a role in *α-globin* regulation in human erythroid cells. Indeed, the related factor KLF4 has been shown to positively regulate the human *α-globin* promoter in reporter assays and to drive the endogenous *HBA* gene in K562 cells [36].

The up-regulation of GFP expression in Line3::*Klf3*$^{-/-}$ mice shows that KLF3 can functionally repress the transgenic *α-globin* regulatory sequences *in vivo*, and may function as an epigenetic modifier of transgene expression.

KLF3 mediates repression of its target genes by binding the co-repressor CtBP [4], which in turn recruits a repressive complex that includes several epigenetic modifiers, such as LSD1, G9A, EUHMT, PC2, HDAC1, and HDAC2 [37,38]. These factors facilitate histone methylation, demethylation and deacetylation, and are responsible for the addition of repressive epigenetic marks and gene silencing. It is possible that the absence of KLF3 in Line3::*Klf3*$^{-/-}$ erythrocytes prevents CtBP from being recruited to the transgene, and it is this that allows the rewriting of epigenetic marks permissive for transcription, resulting in the up-regulation of GFP expression. Indeed, the Line3 mice have frequently been used in ENU mutagenesis screens for modulators of variegated expression, and these screens have predominantly culminated in the identification of epigenetic modifiers, including HDAC1 [39-43].

Another possible explanation for the lack of de-repression of the endogenous *a-globin* gene in red blood cells is that the locus is already fully open and maximally expressed, so significant further de-repression cannot occur. In contrast, the transgene contains only a limited subset of regulatory sequences, and may therefore be expressed at lower levels allowing its up-regulation in the absence of KLF3. To circumvent this, we examined regulation in murine embryonic fibroblasts, as *a-globin* mRNA expression is limited to low but detectable levels in this cell type. In these non-erythroid cells, we identified a modest but significant increase in *a-globin* gene expression in the absence of KLF3. Moreover, in support of a role for CtBP in the regulation of the *a-globin* locus, we note that another group have observed a similar de-repression (4-fold) of *a-globin* gene expression from microarray analysis of $Ctbp^{-/-}$ murine embryonic fibroblasts [44].

Both the human and mouse *a-globin* loci lie in an open chromosomal region, surrounded by a number of actively expressed genes and in non-erythroid cells these loci retain the hallmarks of constitutively accessible chromatin [45]. This contrasts significantly with the more isolated *β-globin* gene cluster, where in non-erythroid cells a silent heterochromatic state is established and maintained. It therefore appears that the *a-globin* locus employs different silencing mechanisms to prevent expression in non-red blood cells. In the case of the human locus, this is achieved by targeted recruitment of the repressive polycomb complex, PRC2, to CpG islands in the promoter regions [45]. However, these CpG islands have been significantly eroded in the murine *a-globin* locus (Figure 2) and recruitment of PRC2 has not been detected, most likely due to loss of polycomb recruitment sites [46]. The mechanism of *a-globin* gene silencing in non-erythroid tissue in the mouse therefore remains unclear. Here we suggest that KLF3 participates in this silencing and may do so not through direct interaction with the *a-globin* proximal promoter but via distal regulatory regions such as *HS-26*. In erythroid cells, *HS-26* is an enhancer element that loops to the *a-globin* promoter and is required for appropriate regulation of expression [17]. In non-erythroid cells, such looping is disrupted and occurs at a much lesser frequency [47]. Whilst these observations allude to the functional importance of the *HS-26* element, it should be noted that loss of *HS-26* only modestly deregulates *a-globin* expression in erythroid cells and has not been reported to perturb non-erythroid silencing [48,49].

Thus it is likely that correct tissue-specific control of the locus is achieved by a complex interplay between multiple *cis*-acting regulatory regions and positively- and negatively-acting *trans* factors such as KLF3 and KLF1.

Conclusions

Excessive α-globin expression can be detrimental to cells and thus it is important that mechanisms exist to limit its expression. Collectively, the findings presented here suggest that the broadly expressed transcriptional repressor KLF3 may have a role in silencing the *a-globin* locus in some but not all contexts, and in particular in non-erythroid tissues. These results complement the previous observation that the KLF3 co-repressor CtBP is also required for the appropriate control of *a-globin* expression in non-erythroid cells [44].

Competing interests

The authors declare that they have no competing interests.

Authors' contributions

RCMP, APWF and MC designed the study and wrote the manuscript. DV, KGRQ, DRH and EW coordinated and oversaw experiments, and assisted in manuscript preparation. RCMP performed FACS analysis. APWF, KSM and GEM conducted EMSA experiments. DV, WFL, BW, JB and KSM performed ChIP studies. CMA generated MEF cell lines. APWF and WFL conducted qRT-PCR. All authors have read, contributed to, and approved the final manuscript.

Acknowledgements

This work is supported by funding from the Australian National Health and Medical Research Council and the Australian Research Council.

Author details

[1]School of Biotechnology and Biomolecular Sciences, University of New South Wales, Sydney, NSW 2052, Australia. [2]The Roslin Institute, University of Edinburgh, Easter Bush Campus, Midlothian EH25 9RG, UK. [3]MRC Molecular Haematology Unit, Weatherall Institute of Molecular Medicine, University of Oxford, John Radcliffe Hospital, Headington, Oxford OX3 9DS, UK. [4]La Trobe Institute for Molecular Science, La Trobe University, Melbourne, Victoria 3086, Australia.

References

1. McConnell BB, Yang VW: **Mammalian Kruppel-like factors in health and diseases.** *Physiol Rev* 2010, **90**(4):1337–1381.
2. Pearson RC, Funnell AP, Crossley M: **The mammalian zinc finger transcription factor Kruppel-like factor 3 (KLF3/BKLF).** *IUBMB life* 2011, **63**(2):86–93.
3. Pearson R, Fleetwood J, Eaton S, Crossley M, Bao S: **Kruppel-like transcription factors: A functional family.** *Int J Biochem Cell Biol* 2008, **40**(10):1996–2001.
4. Turner J, Crossley M: **Cloning and characterization of mCtBP2, a co-repressor that associates with basic Kruppel-like factor and other mammalian transcriptional regulators.** *Embo J* 1998, **17**(17):5129–5140.
5. Funnell AP, Norton LJ, Mak KS, Burdach J, Artuz CM, Twine NA, Wilkins MR, Power CA, Hung TT, Perdomo J, Koh P, Bell-Anderson KS, Orkin SH, Fraser ST, Perkins AC, Pearson RC, Crossley M: **The CACCC-binding protein KLF3/BKLF represses a subset of KLF1/EKLF target genes and is required for proper erythroid maturation in vivo.** *Mol Cell Biol* 2012, **32**(16):3281–3292.
6. Funnell AP, Mak KS, Twine NA, Pelka GJ, Norton LJ, Radziewic T, Power M, Wilkins MR, Bell-Anderson KS, Fraser ST, Perkins AC, Tam PP, Pearson RC, Crossley M: **Generation of Mice Deficient in both KLF3/BKLF and KLF8**

Reveals a Genetic Interaction and a Role for These Factors in Embryonic Globin Gene Silencing. *Mol Cell Biol* 2013, 33(15):2976–2987.

7. Bell-Anderson KS, Funnell AP, Williams H, Mat Jusoh H, Scully T, Lim WF, Burdach JG, Mak KS, Knights AJ, Hoy AJ, Nicholas HR, Sainsbury A, Turner N, Pearson RC, Crossley M: Loss of Kruppel-like factor 3 (KLF3/BKLF) leads to upregulation of the insulin-sensitizing factor adipolin (FAM132A/CTRP12/C1qdc2). *Diabetes* 2013, 62(8):2728–2737.

8. Sue N, Jack BH, Eaton SA, Pearson RC, Funnell AP, Turner J, Czolij R, Denyer G, Bao S, Molero-Navajas JC, Perkins A, Fujiwara Y, Orkin SH, Bell-Anderson K, Crossley M: Targeted disruption of the basic Kruppel-like factor gene (Klf3) reveals a role in adipogenesis. *Mol Cell Biol* 2008, 28(12):3967–3978.

9. Himeda CL, Ranish JA, Pearson RC, Crossley M, Hauschka SD: KLF3 regulates muscle-specific gene expression and synergizes with serum response factor on KLF binding sites. *Mol Cell Biol* 2010, 30(14):3430–3443.

10. Vu TT, Gatto D, Turner V, Funnell AP, Mak KS, Norton LJ, Kaplan W, Cowley MJ, Agenès F, Kirberg J, Brink R, Pearson RC, Crossley M: Impaired B cell development in the absence of Kruppel-like factor 3. *J Immunol* 2011, 187(10):5032–5042.

11. Turchinovich G, Vu TT, Frommer F, Kranich J, Schmid S, Alles M, Loubert JB, Goulet JP, Zimber-Strobl U, Schneider P, Bachl J, Pearson R, Crossley M, Agenès F, Kirberg J: Programming of marginal zone B-cell fate by basic Kruppel-like factor (BKLF/KLF3). *Blood* 2011, 117(14):3780–3792.

12. Funnell AP, Maloney CA, Thompson LJ, Keys J, Tallack M, Perkins AC, Crossley M: Erythroid Kruppel-like factor directly activates the basic Kruppel-like factor gene in erythroid cells. *Mol Cell Biol* 2007, 27(7):2777–2790.

13. Siatecka M, Bieker JJ: The multifunctional role of EKLF/KLF1 during erythropoiesis. *Blood* 2011, 118(8):2044–2054.

14. Tallack MR, Perkins AC: KLF1 directly coordinates almost all aspects of terminal erythroid differentiation. *IUBMB life* 2010, 62(12):886–890.

15. Nuez B, Michalovich D, Bygrave A, Ploemacher R, Grosveld F: Defective haematopoiesis in fetal liver resulting from inactivation of the EKLF gene. *Nature* 1995, 375(6529):316–318.

16. Perkins AC, Sharpe AH, Orkin SH: Lethal beta-thalassaemia in mice lacking the erythroid CACCC-transcription factor EKLF. *Nature* 1995, 375(6529):318–322.

17. Vernimmen D, De Gobbi M, Sloane-Stanley JA, Wood WG, Higgs DR: Long-range chromosomal interactions regulate the timing of the transition between poised and active gene expression. *Embo J* 2007, 26(8):2041–2051.

18. Shyu YC, Wen SC, Lee TL, Chen X, Hsu CT, Chen H, Chen RL, Hwang JL, Shen CK: Chromatin-binding in vivo of the erythroid kruppel-like factor, EKLF, in the murine globin loci. *Cell Res* 2006, 16(4):347–355.

19. Tallack MR, Whitington T, Yuen WS, Wainwright EN, Keys JR, Gardiner BB, Nourbakhsh E, Cloonan N, Grimmond SM, Bailey TL, Perkins AC: A global role for KLF1 in erythropoiesis revealed by ChIP-seq in primary erythroid cells. *Genome Res* 2010, 20(8):1052–1063.

20. Vernimmen D, Marques-Kranc F, Sharpe JA, Sloane-Stanley JA, Wood WG, Wallace HA, Smith AJ, Higgs DR: Chromosome looping at the human alpha-globin locus is mediated via the major upstream regulatory element (HS −40). *Blood* 2009, 114(19):4253–4260.

21. Drissen R, Palstra RJ, Gillemans N, Splinter E, Grosveld F, Philipsen S, de Laat W: The active spatial organization of the beta-globin locus requires the transcription factor EKLF. *Genes Dev* 2004, 18(20):2485–2490.

22. Preis JI, Downes M, Oates NA, Rasko JE, Whitelaw E: Sensitive flow cytometric analysis reveals a novel type of parent-of-origin effect in the mouse genome. *Curr Biol* 2003, 13(11):955–959.

23. Reddel RR, De Silva R, Duncan EL, Rogan EM, Whitaker NJ, Zahra DG, Ke Y, McMenamin MG, Gerwin BI, Harris CC: SV40-induced immortalization and ras-transformation of human bronchial epithelial cells. *Int J Cancer* 1995, 61(2):199–205.

24. Burdach J, Funnell AP, Mak KS, Artuz CM, Wienert B, Lim WF, Tan LY, Pearson RC, Crossley M: Regions outside the DNA-binding domain are critical for proper in vivo specificity of an archetypal zinc finger transcription factor. *Nucleic Acids Res* 2014, 42(1):276–289.

25. Eaton SA, Funnell AP, Sue N, Nicholas H, Pearson RC, Crossley M: A Network of Kruppel-like Factors (Klfs): Klf8 is repressed by Klf3 and activated by Klf1 in vivo. *J Biol Chem* 2008, 283(40):26937–26947.

26. Hancock D, Funnell A, Jack B, Johnston J: Introducing undergraduate students to real-time PCR. *Biochem Mol Biol Educ Bimonthly Publication Int Union Biochem Mol Biol* 2010, 38(5):309–316.

27. Crossley M, Whitelaw E, Perkins A, Williams G, Fujiwara Y, Orkin SH: Isolation and characterization of the cDNA encoding BKLF/TEF-2, a major CACCC-box-binding protein in erythroid cells and selected other cells. *Mol Cell Biol* 1996, 16(4):1695–1705.

28. Perdomo J, Verger A, Turner J, Crossley M: Role for SUMO modification in facilitating transcriptional repression by BKLF. *Mol Cell Biol* 2005, 25(4):1549–1559.

29. Anguita E, Hughes J, Heyworth C, Blobel GA, Wood WG, Higgs DR: Globin gene activation during haemopoiesis is driven by protein complexes nucleated by GATA-1 and GATA-2. *Embo J* 2004, 23(14):2841–2852.

30. Robinson JT, Thorvaldsdottir H, Winckler W, Guttman M, Lander ES, Getz G, Mesirov JP: Integrative genomics viewer. *Nat Biotechnol* 2011, 29(1):24–26.

31. Mak KS, Burdach J, Norton LJ, Pearson RCM, Crossley M, Funnell APW: Repression of chimeric transcripts emanating from endogenous retrotransposons by a sequence-specific transcription factor. *Genome Biol* 2014, 15:4.

32. Miller IJ, Bieker JJ: A novel, erythroid cell-specific murine transcription factor that binds to the CACCC element and is related to the Kruppel family of nuclear proteins. *Mol Cell Biol* 1993, 13(5):2776–2786.

33. Deisseroth A, Hendrick D: Human alpha-globin gene expression following chromosomal dependent gene transfer into mouse erythroleukemia cells. *Cell* 1978, 15(1):55–63.

34. Biggin MD: Animal transcription networks as highly connected, quantitative continua. *Dev Cell* 2011, 21(4):611–626.

35. Li XY, MacArthur S, Bourgon R, Nix D, Pollard DA, Iyer VN, Hechmer A, Simirenko L, Stapleton M, Luengo Hendriks CL, Chu HC, Ogawa N, Inwood W, Sementchenko V, Beaton A, Weiszmann R, Celniker SE, Knowles DW, Gingeras T, Speed TP, Eisen MB, Biggin MD: Transcription factors bind thousands of active and inactive regions in the Drosophila blastoderm. *PLoS Biol* 2008, 6(2):e27.

36. Marini MG, Porcu L, Asunis I, Loi MG, Ristaldi MS, Porcu S, Ikuta T, Cao A, Moi P: Regulation of the human HBA genes by KLF4 in erythroid cell lines. *Br J Haematol* 2010, 149(5):748–758.

37. Shi Y, Sawada J, Sui G, Affar EB, Whetstine JR, Lan F, Ogawa H, Luke MP, Nakatani Y, Shi Y: Coordinated histone modifications mediated by a CtBP co-repressor complex. *Nature* 2003, 422(6933):735–738.

38. Kagey MH, Melhuish TA, Wotton D: The polycomb protein Pc2 is a SUMO E3. *Cell* 2003, 113(1):127–137.

39. Daxinger L, Harten SK, Oey H, Epp T, Isbel L, Huang E, Whitelaw N, Apedaile A, Sorolla A, Yong J, Bharti V, Sutton J, Ashe A, Pang Z, Wallace N, Gerhardt DJ, Blewitt ME, Jeddeloh JA, Whitelaw E: An ENU mutagenesis screen identifies novel and known genes involved in epigenetic processes in the mouse. *Genome Biol* 2013, 14(9):R96.

40. Blewitt ME, Vickaryous NK, Hemley SJ, Ashe A, Bruxner TJ, Preis JI, Arkell R, Whitelaw E: An N-ethyl-N-nitrosourea screen for genes involved in variegation in the mouse. *Proc Natl Acad Sci U S A* 2005, 102(21):7629–7634.

41. Chong S, Vickaryous N, Ashe A, Zamudio N, Youngson N, Hemley S, Stopka T, Skoultchi A, Matthews J, Scott HS, de Kretser D, O'Bryan M, Blewitt M, Whitelaw E: Modifiers of epigenetic reprogramming show paternal effects in the mouse. *Nat Genet* 2007, 39(5):614–622.

42. Ashe A, Morgan DK, Whitelaw NC, Bruxner TJ, Vickaryous NK, Cox LL, Butterfield NC, Wicking C, Blewitt ME, Wilkins SJ, Anderson GJ, Cox TC, Whitelaw E: A genome-wide screen for modifiers of transgene variegation identifies genes with critical roles in development. *Genome Biol* 2008, 9(12):R182.

43. Whitelaw NC, Chong S, Morgan DK, Nestor C, Bruxner TJ, Ashe A, Lambley E, Meehan R, Whitelaw E: Reduced levels of two modifiers of epigenetic gene silencing, Dnmt3a and Trim28, cause increased phenotypic noise. *Genome Biol* 2010, 11(11):R111.

44. Grooteclaes M, Deveraux Q, Hildebrand J, Zhang Q, Goodman RH, Frisch SM: C-terminal-binding protein corepresses epithelial and proapoptotic gene expression programs. *Proc Natl Acad Sci U S A* 2003, 100(8):4568–4573.

45. Garrick D, De Gobbi M, Samara V, Rugless M, Holland M, Ayyub H, Lower K, Sloane-Stanley J, Gray N, Koch C, Dunham I, Higgs DR: The role of the polycomb complex in silencing alpha-globin gene expression in nonerythroid cells. *Blood* 2008, 112(9):3889–3899.

46. Lynch MD, Smith AJ, De Gobbi M, Flenley M, Hughes JR, Vernimmen D, Ayyub H, Sharpe JA, Sloane-Stanley JA, Sutherland L, Meek S, Burdon T, Gibbons RJ, Garrick D, Higgs DR: An interspecies analysis reveals a key role for unmethylated CpG dinucleotides in vertebrate Polycomb complex recruitment. *Embo J* 2012, 31(2):317–329.

47. Zhou GL, Xin L, Song W, Di LJ, Liu G, Wu XS, Liu DP, Liang CC: Active chromatin hub of the mouse alpha-globin locus forms in a transcription factory of clustered housekeeping genes. *Mol Cell Biol* 2006, 26(13):5096–5105.

48. Bouhassira EE, Kielman MF, Gilman J, Fabry MF, Suzuka S, Leone O, Gikas E, Bernini LF, Nagel RL: **Properties of the mouse alpha-globin HS-26: relationship to HS-40, the major enhancer of human alpha-globin gene expression.** *Am J Hematol* 1997, **54**(1):30–39.

49. Anguita E, Sharpe JA, Sloane-Stanley JA, Tufarelli C, Higgs DR, Wood WG: **Deletion of the mouse alpha-globin regulatory element (HS −26) has an unexpectedly mild phenotype.** *Blood* 2002, **100**(10):3450–3456.

Screening and analysis of PoAkirin1 and two related genes in response to immunological stimulants in the Japanese flounder (*Paralichthys olivaceus*)

Chang-Geng Yang[1,2], Xian-Li Wang[3], Bo Zhang[4], Bing Sun[1], Shan-Shan Liu[1] and Song-Lin Chen[1*]

Abstract

A member of the NF-κB signaling pathway, PoAkirin1, was cloned from a full-length cDNA library of Japanese flounder (*Paralichthys olivaceus*). The full-length cDNA comprises a 5′UTR of 202 bp, an open reading frame of 564 bp encoding a 187-amino-acid polypeptide and a 521-bp 3′UTR with a poly (A) tail. The putative protein has a predicted molecular mass of 21 kDa and an isoelectric point (pI) of 9.22. Amino acid sequence alignments showed that PoAkirin1 was 99% identical to the *Scophthalmus maximus* Akirin protein (ADK27484). Yeast two-hybrid assays identified two proteins that interact with PoAkirin1: PoHEPN and PoC1q. The cDNA sequences of *PoHEPN* and *PoC1q* are 672 bp and 528 bp, respectively. Real-time quantitative reverse-transcriptase polymerase chain reaction analysis showed that bacteria could induce the expressions of PoAkirin1, PoHEPN and PoC1q. However, the responses of PoHEPN and PoC1q to the bacterial challenge were slower than that of PoAkirin1. To further study the function of PoAkirin1, recombinant PoAkirin1 and PoHEPN were expressed in *Escherichia coli* and would be used to verify the PoAkirin1-PoHEPN binding activity. These results identified two proteins that potentially interact with PoAkirin1 and that bacteria could induce their expression.

Keywords: Akirin, Japanese flounder, NF-κB, Yeast two-hybrid assay, Immunity, HEPN, C1q

Background

Biological processes are primarily performed and controlled by proteins. Therefore, clarifying the biological functions of proteins and their biological response mechanisms at the cellular level has become the main objective of proteomics research. Protein-protein interactions play a crucial role in various biological functions, including the formation of polymer structure, cell signal transduction, gene regulation, and metabolic pathways. In the post-genome sequencing era, protein interaction bridges the gap between prediction of the relationship between the proteins and the annotation of important genes. Thus, comprehensive analysis of protein-protein interactions is crucial for the full understanding of proteomics [1]. Studies of protein-protein interactions not only can reveal the protein function on the molecular level but also are critical for understanding growth, development, differentiation and apoptosis and other crucial life activities, such studies also provide an important theoretical basis for disease mechanisms, disease treatment, disease prevention and drug development. The yeast two-hybrid system is a simple, but powerful, tool for detecting interactions between proteins and has been widely applied in many research areas. Recently, the yeast two-hybrid system has been used to study the large-scale interaction group in viruses, bacteria, Drosophila, and *Caenorhabditis elegans* [2-7].

Nuclear factor κB (NF-κB) is a nuclear transcription factor that plays a key role in the regulation of apoptosis, viral replication, cancer, inflammation and the regulation of the expression of other related genes. In particular, NF-κB can be activated by a variety of stimulatory factors, including cytokines,

* Correspondence: chensl@ysfri.ac.cn
[1]Yellow Sea Fisheries Research Institute, Chinese Academy of Fisheries Sciences, Qingdao 266071, China
Full list of author information is available at the end of the article

lymphokines, UV, pharmaceutical preparations, and growth and stress factors. Such activation of NF-κB is part of the stress response. Although many members in the NF-κB signaling pathway have been identified in the past 20 years, a highly conserved protein Akirin, a member in the NF-κB signaling pathway, was recently identified in a study of the immune defense system at 2008 [8]. This 20–25-kDa protein participates in the regulation of gene expression in many physiological processes, including the insect and mammalian innate immune response [8], cancer, insect reproduction and arthropods growth [9,10]. Knockout of the Akirin gene led to embryonic lethality of mice and caused death or reduced growth in Drosophila, ticks, and nematodes. Consequently, Akirin is considered important in animal development [11]. Akirin cannot directly combine with DNA, but interacts with the promoter or assisting factors that inhibit the transcription of genes encoding such proteins as the 14-3-3 protein and the helix-loop-helix transcription factor, Twist [12]. Research on fish Akirin has been limited to the analysis of gene structure and function in several species [13-15]. Furthermore, Akirin's interaction mechanism requires further investigation.

The Japanese flounder (*Paralichthys olivaceus*) is economically important and is widely cultured in Europe and China. However, flounder diseases have a serious impact on the aquaculture industry. Recently, to explore the molecular mechanisms of disease resistance and host-pathogen interactions in this species, Nam's team [16,17] constructed a cDNA library from an immune stimulated Japanese flounder. In addition, many immune-related genes including those encoding STAT, Nramp (natural resistance associated macrophage protein), MHCIIA and IIB (major histocompatibility complex class II) have also been investigated [18-20].

As an important protein required for NF-κB-dependent gene regulation in the immune response, little is known about Akirin's function, interacting proteins, and regulation mechanism. In this study, we

Table 1 Oligonucleotide primers used in this study

Name	Sequence	Purpose used
AKI-F-S1	ATGGCCTGCGGAGCGACGTT	ORF amplification
AKI-F-A1	TCAGGAGACATAACTAGCAGGCCG	ORF amplification
β-actin-s1	GCTGTGCTGTCCCTGTA	RT-PCR
β-actin-a1	GAGTAGCCACGCTCTGTC	RT-PCR
GAP-S1	CAACGGCGACACTCACTCCTC	RT-PCR
GAP-A1	TCGCAGACACGGTTGCTGTAG	RT-PCR
TUBA-S1	TGACATCACAAACGCCTGCTTC	RT-PCR
TUBA-A1	GCACCACATCTCCACGGTACAG	RT-PCR
AKI-R-S1	AGGACCAGCCCTCGTTCACACT	RT-PCR
AKI-R-A1	TCCGTATCTTCGCATGATCTGGT	RT-PCR
HEPN-R-S1	TACAAGGACAATGGTGGGGG	RT-PCR
HEPN-R-A1	GGCAAGGGCTGAGATGGAG	RT-PCR
C1q-R-S1	CTCCAGAAAACGAAGCAGGC	RT-PCR
C1q-R-A1	TGTCGCACATCATCAAGTGAAC	RT-PCR
AKI-F-A7	GCGTCGACGGAGACATAACTAGCAG	Plasmid construction
AKI-T-A1	GCGGTACCGGAGACATAACTAGCAG	Plasmid construction
AD-AKI-S1	GTGAATTCATGGCCTGCGGAGCGACGTT	Plasmid construction
AD-AKI-A1	CAGCTCGAGTCAGGAGACATAACTAGCAG	Plasmid construction
C1q-S1	GCCGAATTCAAGGGGGCACCAGGTCTTAA	Plasmid construction
C1q-A1	CAGGTCGACTCAGGCCGTGGGGAAGACGA	Plasmid construction
HEPN-S3	GTGAATTCGGCACGAGGCTCAGGTGGCA	Plasmid construction
HEPN-A2	CAGCTCGAGAGCCTCCTCTTTGTTTGGCC	Plasmid construction
HEPN-A3	CAGCTCGAGGGAAGGTTGACCGTGCCTTT	Plasmid construction
HEPN-S5	AGTGAATTCTCTGGCAGCTCCATCTCAGC	Plasmid construction
HEPN-A4	CAGCTCGAGCTATTTCACATATGCCTCAAC	Plasmid construction
HEPN-ET-S1	CCGAATTCAGGCTCAGGTGGCATCCT	Plasmid construction
HEPN-ET-A1	GTGCTCGAGCTATTTCACATATGCCTCA	Plasmid construction

Table 2 Primer names and restriction sites used to construct the expression vectors

Name	Restriction site	Vector
AKI-F-A7	*Sal*I	pGBKT7-PoAkirin1
AKI-T-A1	*Kpn*I	
AD-AKI-S1	*Eco*RI	pGADT7-PoAkirin1
AD-AKI-A1	*Xho*I	
C1q-S1	*Eco*RI	pGBKT7- PoC1q
C1q-A1	*Sal*I	
HEPN-S3	*Eco*RI	pGADT7- PoHEPN-M1
HEPN-A2	*Xho*I	
HEPN-S3	*Eco*RI	pGADT7- PoHEPN-M2
HEPN-A3	*Xho*I	
HEPN-S5	*Eco*RI	pGADT7- PoHEPN-M3
HEPN-A4	*Xho*I	
HEPN-ET-S1	*Eco*RI	pET30a-PoHEPN
HEPN-ET-A1	*Xho*I	

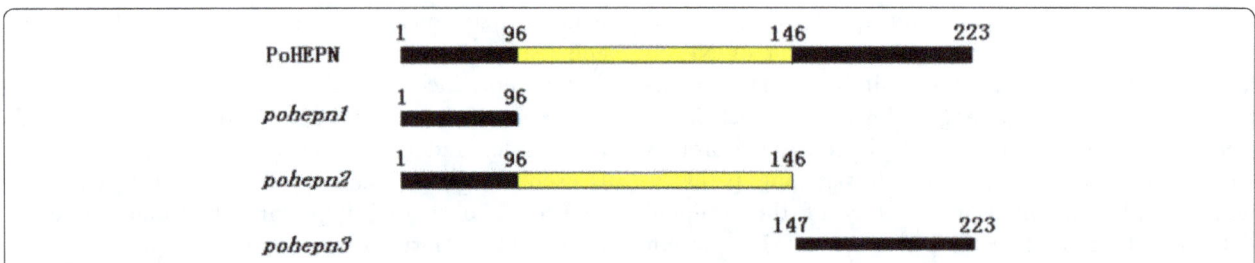

Figure 1 Sketch map of the *PoHEPN* deleted segment. PoHEPN: normal *PoHEPN*; *pohepn1*: deletion of 96-223aa; *pohepn2*: deletion of 146-223aa; *pohepn3*: deleted 1-146aa.

screened for Akirin interacting proteins and analyzed the interaction mechanism using the yeast two-hybrid system and a cDNA library of flounder immune tissue. Two possible interacting proteins were identified: PoHEPN (a higher eukaryotes and prokaryotes nucleotide-binding domain (HEPN) protein) and PoC1q (complement component C1, q subunit). The expression profiles of PoAkirin1, PoHEPN and PoC1q were also analyzed by a bacterial challenge test. This study increases our understanding of the Akirin family, and provides a theory of flounder immunity and disease resistance mechanisms.

Methods

Ethics statement

The Yellow Sea Fisheries Research Institute's animal care and use committee waived the need for ethical approval, as this is not required in China.

Experimental animals

Japanese flounders with an average weight of 200 g were obtained from Haiyang Fisheries Company in Yantai and raised in a breeding tank with seawater (20°C). For cloning and tissue expression analysis, RNA was extracted from 13 tissues (brain, gill, skin, muscle, fin, heart, liver,

```
   1    GGAGTACCCATACGACGTACCAGATTACGCTCATCTGGCCGTGGAGGCCAGTGAATTCGG
  61    CACGAGGGTTGTGAAGAGGAAGAAAGACTCAACTCTTCCGTGGGACGCGACACCGACCAA
 121    CCGCCGCCGTTACACCCCGGTTCCTGCGGGTCTTTTTTCCCCCTCCAGCTCTGTTTGTTC
 181    AGCCCGTCTCGTCTCAGCGATCATGGCCTGCGGAGCGACGCTAAAGCGGTCGATGGAGTT
   1                                M   A   C   G   A   T   L   K   R   S   M   E   F
 241    TGAGGCCCTCCTCAGTCCCCAGTCTCCCAAGCGGAGAAGGTGCAATCCACTACCGGGGAC
  14     E   A   L   L   S  P   Q   S   P   K   R   R   R   C   N  P   L   P   G   T
 301    TCCTGGAACTCCGTCCCCGCAAAGATGCAACCTCCGTCCGCCGGTGGACAGCCCCACGCA
  34     P   G   T   P   S   P   Q   R   C   N   L   R   P   P   V   D   S   P   T   H
 361    CTCGATGTCCCCCCCGGCCATAGGAGGCGAGCACCGGCTCACCCCAGTGCAGATCTTCCA
  54     S   M   S   P   P   A   I   G   G   E   H   R   L   T   P   V   Q   I   F   Q
 421    GAACCTCCGACAGGAGTACAGTCGGATCCAGAGGCGGCGACAGCTGGAGGGGGCTTTCAA
  74     N   L   R   Q   E   Y   S   R   I   Q   R   R   R   Q   L   E   G   A   F   N
 481    CCAGACTGAGGCCTGTAGCTCCAGTGACGCCCCCAGCCCCAGCTCATCCATCAATGCTCC
  94     Q   T   E   A   C   S   S   S   D   A   P   S   P   S   S   I   N   A   P
 541    CAGCTCCCCACCAGGTGCCTCAAGGAAGGACCAGCCCTCGTTCACACTGAAGCAGGTGAG
 114     S   S   P   P   G   A   S   R   K   D   Q   P   S   F   T   L   K   Q   V   S
 601    CTACCTGTGCGAGCGCCTGCTCAAAGACCATGAGGAGAAGATACGGGAGGAGTACGAACA
 134     Y   L   C   E   R   L   L   K   D   H   E   E   K   I   R   E   E   Y   E   Q
 661    GATCCTTAACACAAAACTTGCAGAACAATATGAATCTTTTGTGAAATTCACACAAGACCA
 154     I   L   N   T   K   L   A   E   Q   Y   E   S   F   V   K   F   T   Q   D   Q
 721    GATCATGCGAAGATACGGAGCCCGGCCTGCTAGTTATGTCTCCTGAACTCACGTGGATTT
 174     I   M   R   R   Y   G   A   R   P   A   S   Y   V   S   *
 781    AGAACCCAGCTTCCTCTCACAAGACGGCTGCTCTTCCACTGCCCCCTCCCTCCACTCGAT
 841    TTTCATTTTATTCCCCTATTTTTTTCCCCTCCTCCATCTCAAACTGTTCGGGTGCCATGG
 901    TTTATCCAATTTTTTAATCCAAAAGTTGATTTCTTAACTTTAAAACTATTGCTGCTGGCC
 961    AAAAGTTAGAATCTGGGAGATGAACTGTGTATGAAGCCCGTCCCTCTTTTTTTGCCCTTT
1021    TTCTTGTTTATATGCTCTTTTTCTTGAAATCCTAATGAAAGCAGATCTGTGTCAAATCTC
1081    TTGCCAGCGCTGTAGAAAAAAAGCTTATTGGCTCATATGATCATTCTGTATAACCTTCCA
1141    GCAGCTGTATTTAATGTCTCCAGCTTCCCACGTCGTCCTACATTTCTCACCTCCCCCTCC
1201    TTCCCTCCTCACTCTTTACTCGAATGTGGATTTTTCTCCCTTTTGAAAAAAATAAAAAGT
1261    TCTGTGATTAAAAAAAAAAAAAAAAAAA
```

Figure 2 Nucleotide sequence (above) and deduced amino acid sequence (below) of *PoAkirin1*. Nucleotides are numbered from the first base at the 5'end. Amino acids, shown as one letter abbreviations, are numbered from the initiating methionine. The predict NLS site is boxed, the AATAAA box is underlined, and the poly(A) region is double-underlined. The stop codon is marked by an asterisk.

spleen, eye, pituitary, kidney, head kidney, and intestine) from three individuals. For the bacterial challenge experiment, RNA was extracted from three tissues (liver, spleen, and kidney) from three individuals.

Bacterial challenge

The bacterium *Vibrio anguillarum*, which is pathogenic in Japanese flounders, was cultured at 28°C to mid-logarithmic growth on 2216E medium, centrifuged to

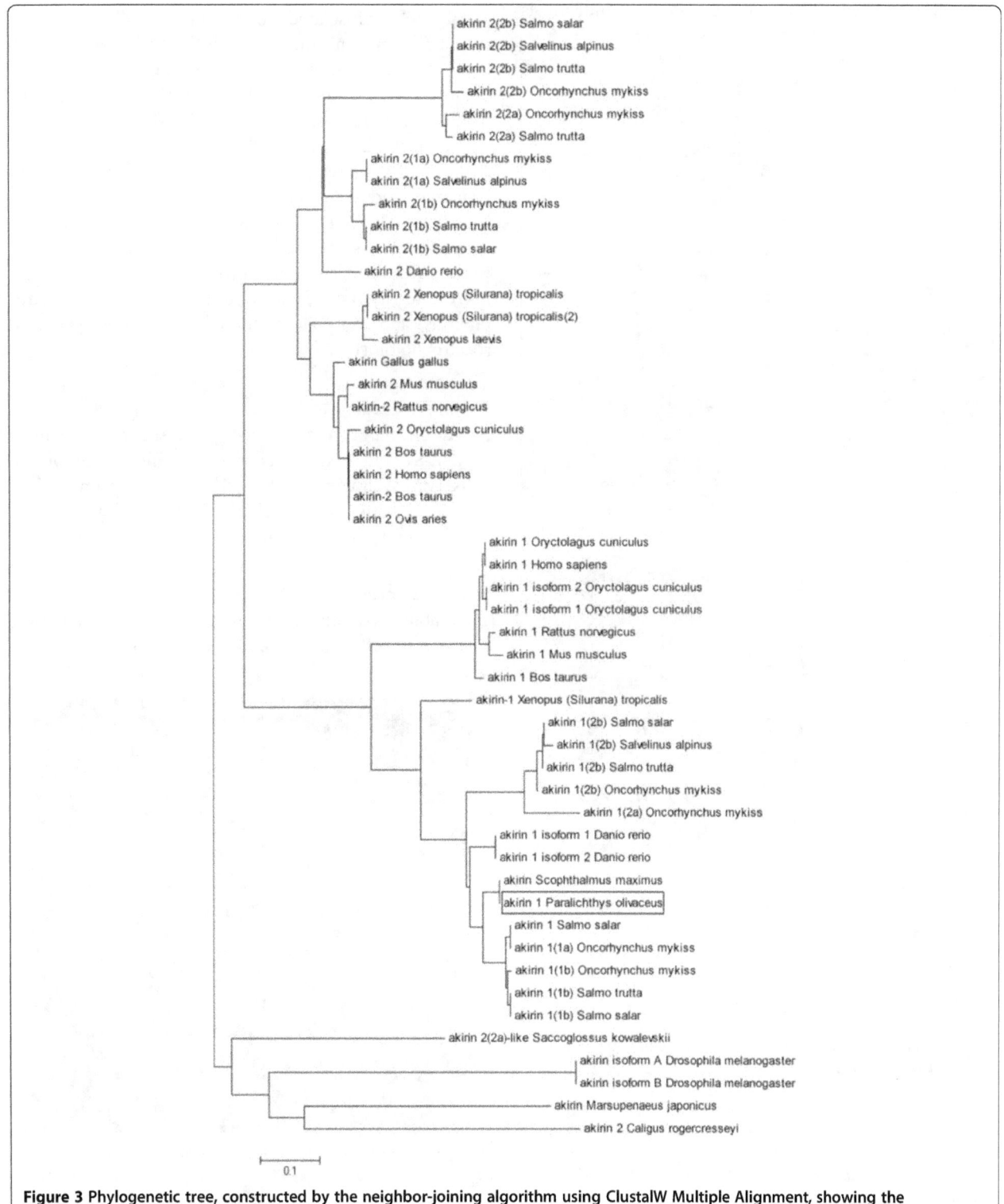

Figure 3 Phylogenetic tree, constructed by the neighbor-joining algorithm using ClustalW Multiple Alignment, showing the relationship between Akirin proteins. The PoAkirin1 protein is boxed.

collect the bacteria and suspended in 0.9% saline [21,22]. A cell counter was used to measure the number of bacteria in the suspension. A final concentration of 7×10^6 cfu of *V. anguillarum* was used for each injection, and 0.9% saline was used as the negative control. At 6 h, 12 h, 24 h, 48 h, 72 h and 96 h post-injection, three individuals

from each time point were sacrificed and tissues were used for RNA extraction. For the negative control, tissues were taken 12 h after the saline injection.

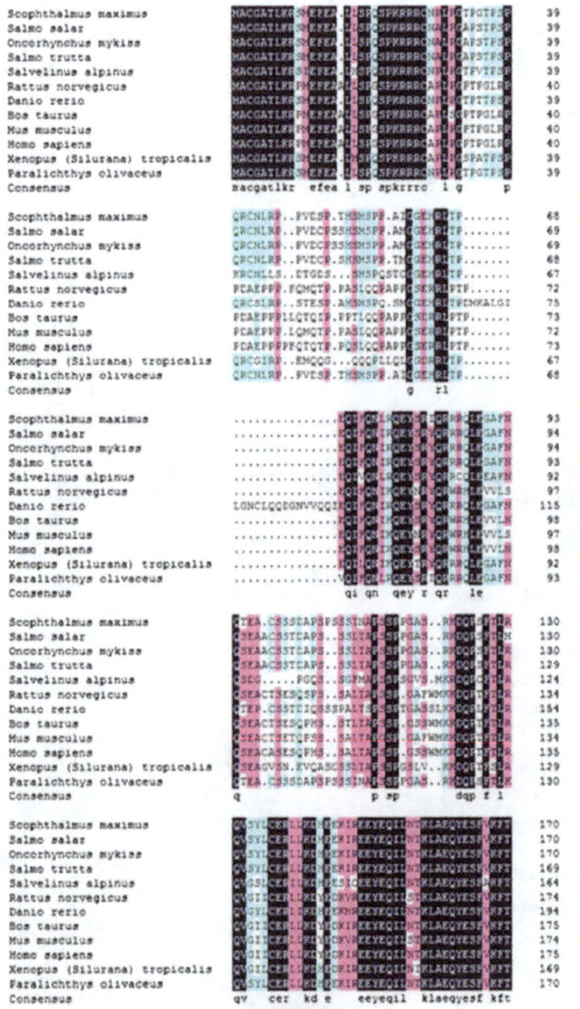

RNA extraction

Total RNA was isolated from 500 mg powdered fish tissues by homogenization in 5 mL TRIzol (Invitrogen), held at room temperature for 5 min. An aliquot of chloroform (1 mL) was added to each extract, and the resulting mixture was centrifuged (10 min, 13,000 g). The aqueous layer was transferred to a clean tube, and the RNA firstly precipitated by the addition of 3 mL isopropanol, and then pelleted by centrifugation (15 min, 13,000 g). The RNA pellet was washed twice with 75% ethanol and re-suspended in diethylpyrocarbonate-treated water. After DNA removal (Turbo DNA-free kit, Ambion), RNA integrity was detected using agarose gel electrophoresis, and the concentration of RNA was quantified spectrophotometrically.

Cloning of PoAkirin1

Based on the turbot Akirin1 gene sequence, primers (AKI-R-S1 and AKI-R-A1) were designed to amplify a conservative fragment. The 5′ and 3′ fragments of PoAkirin1 gene were amplified from the flounder full-length cDNA Library (a mix of liver, spleen and kidney).

Sequence analysis

Translation was performed using DNASTAR software. The conserved domain analysis and BLAST analysis

Figure 4 Alignment of amino acid sequences of PoAkirin1 and other Akirin1 proteins. Identical residues are shown in white letters with a black background. Sequence accession nos.: *Scophthalmus maximus*, ADK27484.1; *Salmo salar*, NP_001161992; *Oncorhynchus mykiss*, ACV49717; *Salmo trutta*, ACV49710; *Salvelinus alpines*, ACV49696; *Danio rerio*, NP_001025225; *Rattus norvegicus*, NP_001107272; *Bos taurus*, DAA31047; *Mus musculus*, AAH03291; *Homo sapiens*, AAI19746; *Xenopus (Silurana) tropicalis*, NP_001016080; *Oryctolagus cuniculus*, XP_002708554.

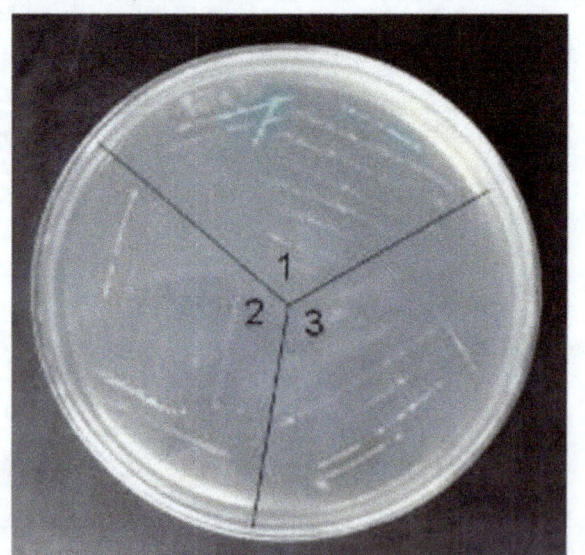

Figure 5 Transcription activation of PoAkirin1. The *PoAkirin1* gene was fused to the GAL4 DNA binding domain (GAL4-DB) in the vector pGBKT7. The positive control plasmid was pGBKT7-53. The negative control plasmid was pGBKT7-Lam.

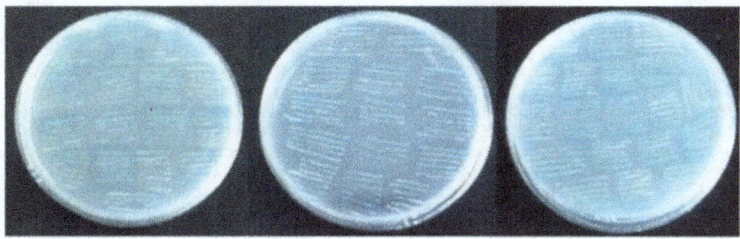

Figure 6 Forty-nine positive clones that may interact with the PoAkirin1 from the screening library.

was performed at http://blast.ncbi.nlm.nih.gov/Blast.cgi, containing blastn, blastp and tblastp. The PSORT II server (http://psort.ims.u-tokyo.ac.jp) was used to predict the putative nuclear localization signal (NLS). The alignment of Akirins from different species was performed using the ClustalW alignment program, and the phylogenetic tree was constructed on the basis of the proportion of the amino acid differences (p-distances) determined by the neighbor-joining method [23] using MEGA 3 software [24]. The following proteins were used in the alignment: AAF50569 [*Drosophila melanogaster*], AAN12062 [*D. melanogaster*], ADK27484 [*Scophthalmus maximus*], BAI49701 [*Marsupenaeus japonicus*], ADK26453 [*Gallus gallus*], NP_001161992 [*Salmo salar*], ACV49724 [*Oncorhynchus mykiss*], ACV49723 [*O. mykiss*], ACV49722 [*O. mykiss*], ACV49721 [*O. mykiss*], ACV49720 [*O. mykiss*], ACV49719 [*O. mykiss*], ACV49718 [*O. mykiss*], ACV49717 [*O. mykiss*], ACV49716 [*Salmo trutta*], ACV49715 [*S. trutta*], ACV49714 [*S. trutta*], ACV49712 [*S. trutta*], ACV49710 [*S. trutta*], ACV49708 [*S. salar*], ACV49706 [*S. salar*], ACV49704 [*S. salar*], ACV49702 [*S. salar*], ACV49700 [*Salvelinus alpinus*], ACV49697 [*S. alpinus*], ACV49696 [*S. alpinus*], ADK39312 [*Caligus rogercresseyi*], NP_001107272 [*Danio rerio*], NP_001007187 [*D. rerio*], NP_988914 [*Xenopus (Silurana) tropicalis*], NP_001085484 [*Xenopus laevis*], NP_001025225 [*Rattus norvegicus*], DAA31047 [*Bos taurus*], DAA26175 [*B. taurus*], XP_002715780 [*Oryctolagus cuniculus*], XP_002714617 [*O. cuniculus*], XP_002708555 [*O. cuniculus*], XP_002708554 [*O. cuniculus*], XP_002736520 [*Saccoglossus kowalevskii*], AAH97074 [*D. rerio*], AAH03291 [*Mus musculus*], AAH61612 [*X. tropicalis*], CAM16479 [*M. musculus*],

AAI19746 [*Homo sapiens*], AAH05051 [*H. sapiens*], NP_001103557 [*B. taurus*], NP_001016080 [*X. tropicalis*], NP_001035003 [*R. norvegicus*], and AEO17042 [*Ovis aries*].

Quantitative real-time RT-PCR (RT-qPCR)

The RT-qPCR protocol adhered to the 'Minimum Information for Publication of Quantitative Real-time PCR experiment guidelines [25]. The relative mRNA steady-state level was measured by RT-qPCR. The total RNA from different tissues was prepared using the TRIzol reagent. The cDNA was synthesized from each RNA sample (500 ng) using a PrimeScript® RT reagent kit (Takara, China), following manufacturer's protocol. RT-qPCR was conducted on an Applied Biosystems 7500 Real-Time PCR System with SYBR® Premix Ex Taq™ (Takara, China). Genes encoding β-actin (for normal tissue types and infected kidney), a-tubulin (for infected spleen) and glyceraldehyde-3-phosphate dehydrogenase (for infected liver) were selected for normalization, as their expressions have been reported to be stable [26-28]. The RT-qPCR primer pairs were designed to generate a product size of 150–250 bp and a Tm of 60 ±1°C. cDNA from 13 normal and infected tissues was chosen for the detection of PoAkirin1, PoHEPN and PoC1q expression, using the gene-specific primers AKI-R-S1, AKI-R-A1 (for PoAkirin1); HEPN-R-S1, HEPN-R-A1 (for PoHEPN); and C1q-R-S1, C1q-R-A1 (for PoC1q) (Table 1). The primers β-actin-s1, β-actin-a1, GAP-S1, GAP-A1, and TUBA-S1 and TUBA-A1 were used to amplify the β-actin, a-tubulin, and glyceraldehyde-3-phosphate dehydrogenase fragments, respectively (Table 1). PCR efficiency (E) of

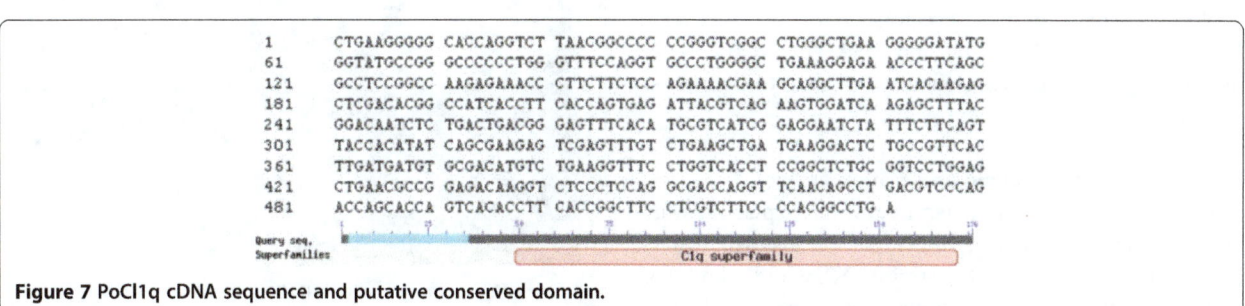

Figure 7 PoCl1q cDNA sequence and putative conserved domain.

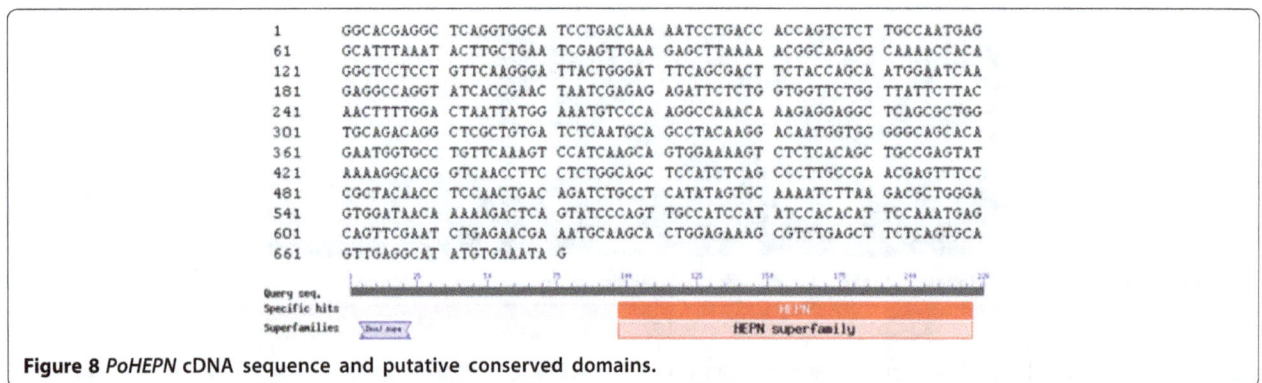

```
  1    GGCACGAGGC TCAGGTGGCA TCCTGACAAA AATCCTGACC ACCAGTCTCT TGCCAATGAG
 61    GCATTTAAAT ACTTGCTGAA TCGAGTTGAA GAGCTTAAAA ACGGCAGAGG CAAAACCACA
121    GGCTCCTCCT GTTCAAGGGA TTACTGGGAT TTCAGCGACT TCTACCAGCA ATGGAATCAA
181    GAGGCCAGGT ATCACCGAAC TAATCGAGAG AGATTCTCTG GTGGTTCTGG TTATTCTTAC
241    AACTTTTGGA CTAATTATGG AAATGTCCCA AGGCCAAACA AAGAGGAGGC TCAGCGCTGG
301    TGCAGACAGG CTCGCTGTGA TCTCAATGCA GCCTACAAGG ACAATGGTGG GGGCAGCACA
361    GAATGGTGCC TGTTCAAAGT CCATCAAGCA GTGGAAAAGT CTCTCACAGC TGCCGAGTAT
421    AAAAGGCACG GTCAACCTTC CTCTGGCAGC TCCATCTCAG CCCTTGCCGA ACGAGTTTCC
481    CGCTACAACC TCCAACTGAC AGATCTGCCT CATATAGTGC AAAATCTTAA GACGCTGGGA
541    GTGGATAACA AAAAGACTCA GTATCCCAGT TGCCATCCAT ATCCACACAT TCCAAATGAG
601    CAGTTCGAAT CTGAGAACGA AATGCAAGCA CTGGAGAAAG CGTCTGAGCT TCTCAGTGCA
661    GTTGAGGCAT ATGTGAAATA G
```

Figure 8 *PoHEPN* cDNA sequence and putative conserved domains.

these primers were between 92 and 110% and correlation coefficient (R2) ranged from 0.991 to 0.998.

RT-PCR was carried out with a 1 μl cDNA sample, 10 μl SYBR® Premix Ex Taq™, 0.4 μl ROX Reference Dye II, 0.4 μl PCR forward/reverse primers (10 μM) and 7.8 μl nuclease-free water. The thermo-cycling conditions for the reaction were as follows: 95°C for 30 s, followed by 40 cycles of 95°C for 15 s and 61°C for 34 s. The reaction was carried out with triplicate with duplicates of each sample. Data (normalized Ct values) from the treated and control tissues templates were compared, and the 2-ΔΔCT method was selected for relative quantification. All data were expressed as the mean ± S.D. and analyzed by one-way analysis of variance to determine significant differences between samples, using SPSS 16.0. Values were considered statistically significant when $P < 0.05$ or $P < 0.01$.

Plasmid construction

Using the sequences of the open reading frames (ORFs) of the PoAkirin1, PoHEPN and PoC1q genes, primers were designed to construct pGEX4T-1-PoAkirin1 (for expression in *E. coli*), pET30a-PoHEPN (for expression in *E. coli*), pGBKT7-PoAkirin1 (for Yeast two-hybrid assay), pGADT7-PoAkirin1 (for Yeast two-hybrid assay), pGBKT7-PoC1q (for Yeast two-hybrid assay), pGADT7-PoHEPN-M1 (for Yeast two-hybrid assay), pGADT7-PoHEPN-M2 (for Yeast two-hybrid assay), and pGADT7-PoHEPN-M3 (for Yeast two-hybrid assay) vectors. The primer sequences and restriction sites corresponding to the vectors are shown in Table 1 and Table 2. The mutation vectors pGADT7-PoHEPN-M1, pGADT7-PoHEPN-M2, and pGADT7-PoHEPN-M3 mutation vectors were constructed according to the model in Figure 1.

Yeast two-hybrid screening and mating assays

The pGBKT7-PoAkirin1 plasmid and DNA plasmids for cDNA library clones were individually transformed into *Saccharomyces cerevisiae* strain Y2HGold and *S. cerevisiae* strain Y187, following the manufacturer's

instructions in the Matchmaker™ Gold Yeast Two-Hybrid System User Manual. Approximately 1×10^7 library clones were screened by yeast mating with selection by growth for   3–5 days at 30°C on agar media lacking Leu and Trp, but with X-α-galactosidase and Aureobasidin A.

Recombinant protein expression in *Escherichia coli*

The recombinant plasmids pGEX4T-1-PoAkirin1, pET30a-PoHEPN, pGEX4T-1, and pET30a were transformed into *E. coli* BL21(DE3) competent cells. After growth at 37°C, 1.0 mM IPTG was added to the transformed cells and incubation continued for 4 h. Sodium dodecyl sulfate polyacrylamide gel electrophoresis (SDS-PAGE) analysis of total proteins was used to detect recombinant protein expression. No IPTG and cells transformed with pGEX-4 T-1 or pET30a and induced with IPTG were used as negative controls.

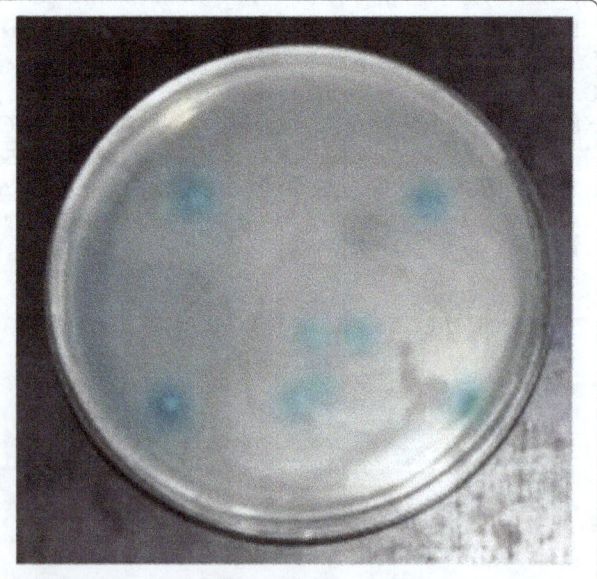

Figure 9 Yeast rotary verify to PoC1q and PoAkirin1.

Screening and analysis of PoAkirin1 and two related genes in response to immunological stimulants...

159

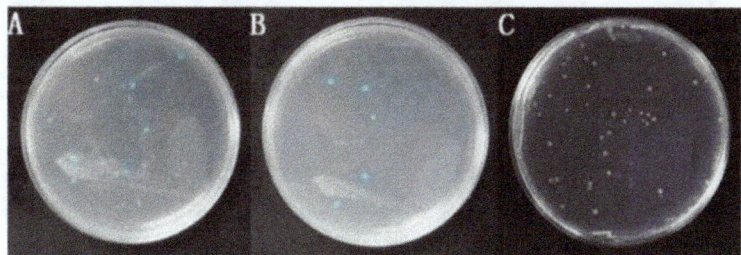

Figure 10 Growth yeast transformed with the *PoHEPN* deletion segment and PoAkirin1 in SD/–Leu/–Trp/x-α-gal. **A**: *pohepn1*; **B**: *pohepn2*; **C**: *pohepn3*.

Western blotting analysis

The protein concentrations of the samples were determined using an Enhanced BCA Protein Assay Kit (Beyotime Institute of Biotechnology, China). Recombinant PoHEPN and PoAkirin1 were serially diluted (to 5 ng) and separated by 12% SDS-PAGE and transferred to a BioTrace NT Nitrocellulose Transfer Membrane (PALL, USA). The membranes were blocked and then incubated with a 1 μg/mL dilution of the primary antibody (monoclonal antibodies against His or GST (glutathione-S-transferase; Uscn Life Science Inc., Wuhan, China)). After washing, the membranes were incubated with the secondary antibody, horseradish peroxidase-conjugated goat anti-mouse IgG (1:1,000 dilutions) (Beyotime Institute of Biotechnology, China). Reactive proteins were detected using chemiluminescence (ECL Western Blotting Analysis System, Thermo Fisher Scientific Inc., IL, USA).

Results

Cloning and sequence analysis of PoAkirin1

The PoAkirin1 cDNA acquired from the full-length cDNA Library of Japanese flounder comprised a 5′UTR of 202 bp, an ORF of 564 bp encoding a polypeptide of amino acids and a 521-bp 3′UTR with a poly (A) tail (Figure 2). Phylogenetic analysis showed that the PoAkirin1 gene clustered together with fish Akirin1 genes (Figure 3, 4).

The PoAkirin1 cDNA encodes a putative protein of 187 amino acids with a predicted molecular mass of 21 kDa and an isoelectric point (pI) of 9.22. A putative NLS was predicted in the N-terminus of the protein (Figure 2). The sequence of PoAkirin1 has been submitted to GenBank with the accession number KC190111.

Cloning and testing the bait construct for autoactivation

Bait plasmid (100 ng; pGBKT7-PoAkirin1) was transformed into yeast strainY2H Gold and selected on medium SD/Trp (lacking Trp). No autoactivation of fish Akirin1 expression was observed (Figure 5).

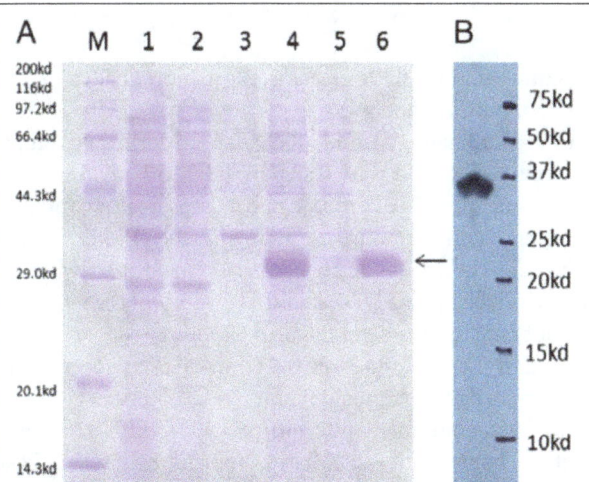

Figure 11 SDS-PAGE and Western blot analysis of PoHEPN protein. A: SDS-PAGE analysis of PoHEPN protein expression. M: Protein MW marker (Broad); 1: BL21(DE3)T1R induced whole cell; 2: BL21(DE3)T1R induced Supernatant; 3: BL21(DE3)T1R induced precipitate; 4: pET30-HEPN induced whole cell; 5: pET30-HEPN induced whole cell; 6: pET30-HEPN induced whole cell. **B**: Western blot analysis of PoHEPN protein.

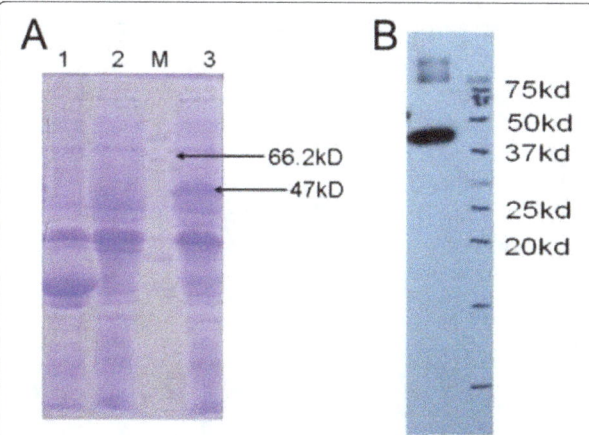

Figure 12 SDS-PAGE and Western blot analysis of PoAkirin1 protein. A: SDS-PAGE analysis of PoAkirin1 protein expression. 1: pGEX4T-1 induced; 2: pGEX4T-1- PoAkirin1 non-induced; 3: pGEX4T-1- PoAkirin1 induced; M: Protein MW marker. **B**: Western blot analysis of PoAkirin1.

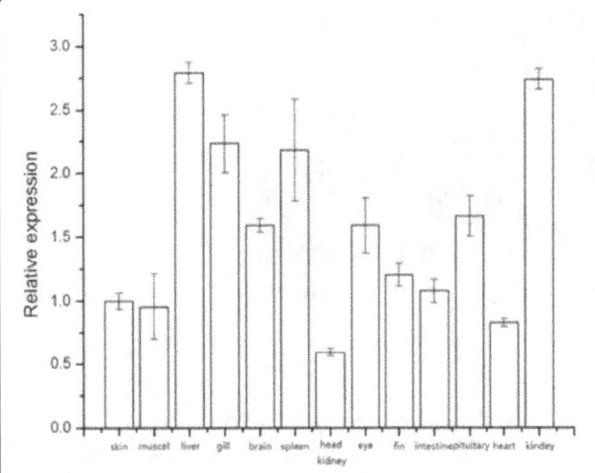

Figure 13 Expression pattern of *PoAkirin1* in different tissues of the Japanese flounder detected by quantitative RT-PCR. *PoAkirin1* mRNA was expressed in all tissues detected. The β-actin gene was used as an internal control to calibrate the cDNA template for all the samples. All data are expressed as the mean ± S.D. (n = 3).

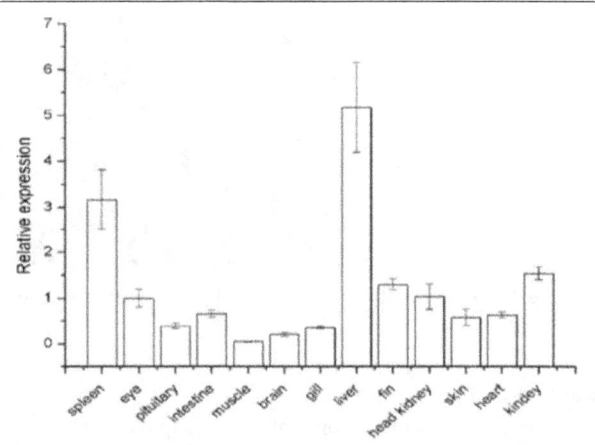

Figure 15 Expression pattern of *PoC1q* in different tissues of Japanese flounder detected by quantitative RT-PCR. *PoC1q* was expressed in all tissues, with higher expression levels in the liver. The β-actin gene was used as an internal control to calibrate the cDNA templates for all the samples. Data are expressed as the mean ± S.D. (n = 3).

Yeast two-hybrid screening of a Japanese flounder library using bait pGBKT7-PoAkirin1

Using PoAkirin1 as the bait to screen interactions in the library, 49 positive clones were selected. Yeast plasmids extracted from positive clones were transformed into *E. coli* TOP10 and sequenced. DNAMAN was used to analyze the sequence information and BLASTP at the National Center for Biotechnology Information (http://ncbi.nlm.nih.gov/blast) was used to perform the alignments. The results showed that there were seven possible interacting proteins (CD63, HEPN, C1q, T-complex protein 1, voltage-dependent anion channel, asparaginyl endopeptidase and Chaperonin) among the 49 positive clones (Figure 6). To verify the interaction of the

proteins, we chose the possible immune-related protein (C1q) (Figure 7) and the protein predicted in the nucleus (HEPN) (Figure 8) to perform yeast rotary and domain verification tests. The results showed that the yeast grew with a blue color in the yeast rotary verification test using PoC1q and PoAkirin1, and the yeast expressing PoHEPN with a 96–223 aa segment deleted and PoAkirin with a 146–223 aa segment deleted were blue when grown on SD/−Leu/−Trp/x-α-gal. (Figures 1, 9, and 10).

PoAkirin1 and PoHEPN proteins expression in *E. coli* and western blotting analysis

The recombinant plasmid pGEX-4T-1-Akirin and empty plasmid pGEX-4T-1 were transformed into *E. coli*. After induction by 1 mM IPTG for 4 h, bacterial lysates contained a protein band with a Mr of approximately 47 kDa, as anticipated. The lysates from bacteria transformed with the empty vector pGEX-4 T-1 and that had no IPTG induction did not contain this band (Figures 12A).

Similarly, after induction by 1 mM IPTG(Isopropyl β-D-1-Thiogalactopyranoside) for 4 h, bacterial lysates with the recombinant plasmid pET30a-HEPN showed a protein band with an Mr of approximately 31 kDa, which corresponded to the predicted size. This band was not present in the negative control bacterial lysates (Figures 11A).

The recombinant PoHEPN and PoAkirin1 contain His and GST tags; therefore, western blotting was used to detect the expressed proteins using His and GST monoclonal antibodies (Figures 11B, 12B).

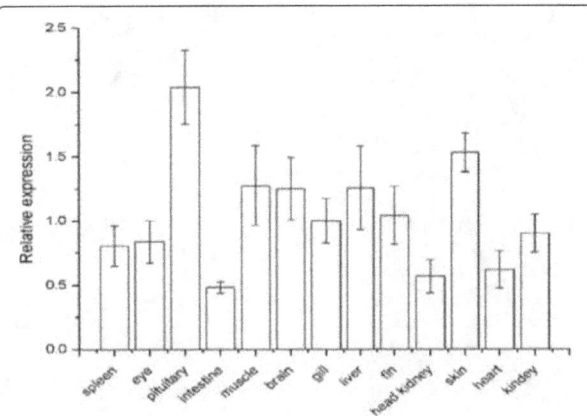

Figure 14 Expression pattern of *PoHEPN* in different tissues of the Japanese flounder detected by quantitative RT-PCR. *PoHEPN* mRNA was expressed in all tissues. The β-actin gene was used as an internal control to calibrate the cDNA templates for all samples. Data are expressed as the mean ± S.D. (n = 3).

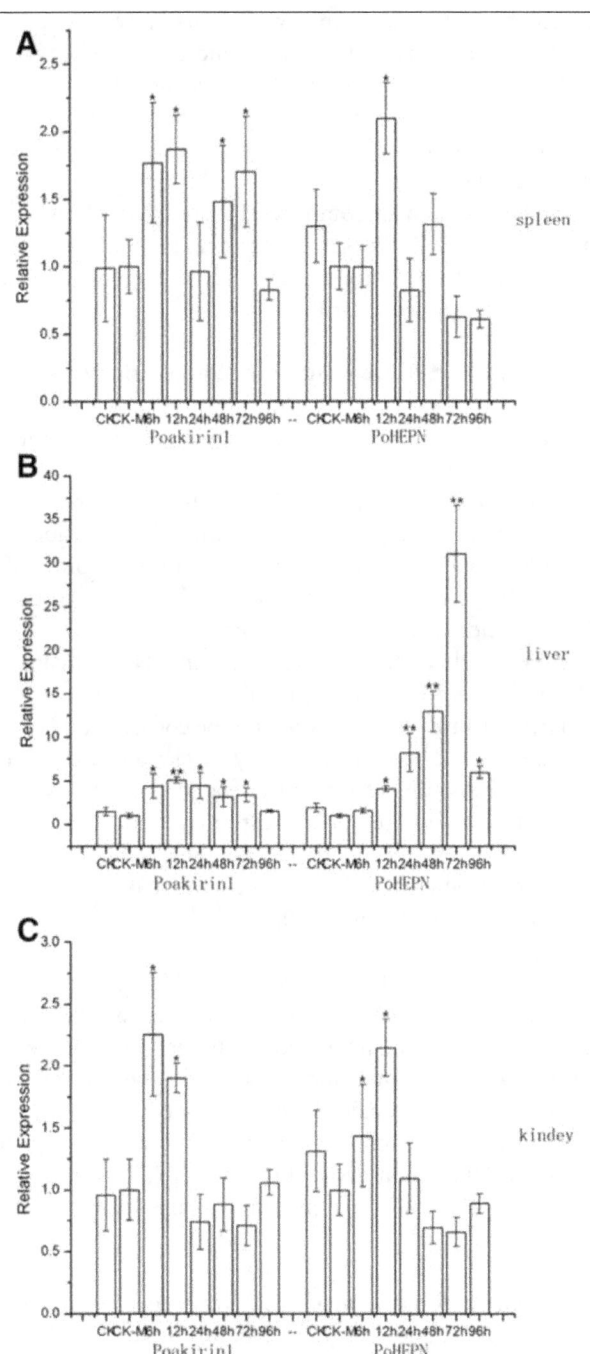

Figure 16 Quantitative RT-PCR analyses of the expression level of *PoAkirin1* and *PoHEPN* mRNA in the spleen (A), liver (B), and kidney (C) after injection with *Vibrio anguillarum*. Figures show the relative expression levels of *PoAkirin1* and *PoHEPN* at 6 h, 12 h, 24 h, 48 h, 72 h and 96 h post-injection. CK-M represents the tissues at 12 h after saline solution injection, and CK means live fish under normal conditions. Data are expressed as the mean ± S.D. (n = 3). Significant differences are indicated with an asterisk at P < 0.05 and two asterisks at P < 0.01.

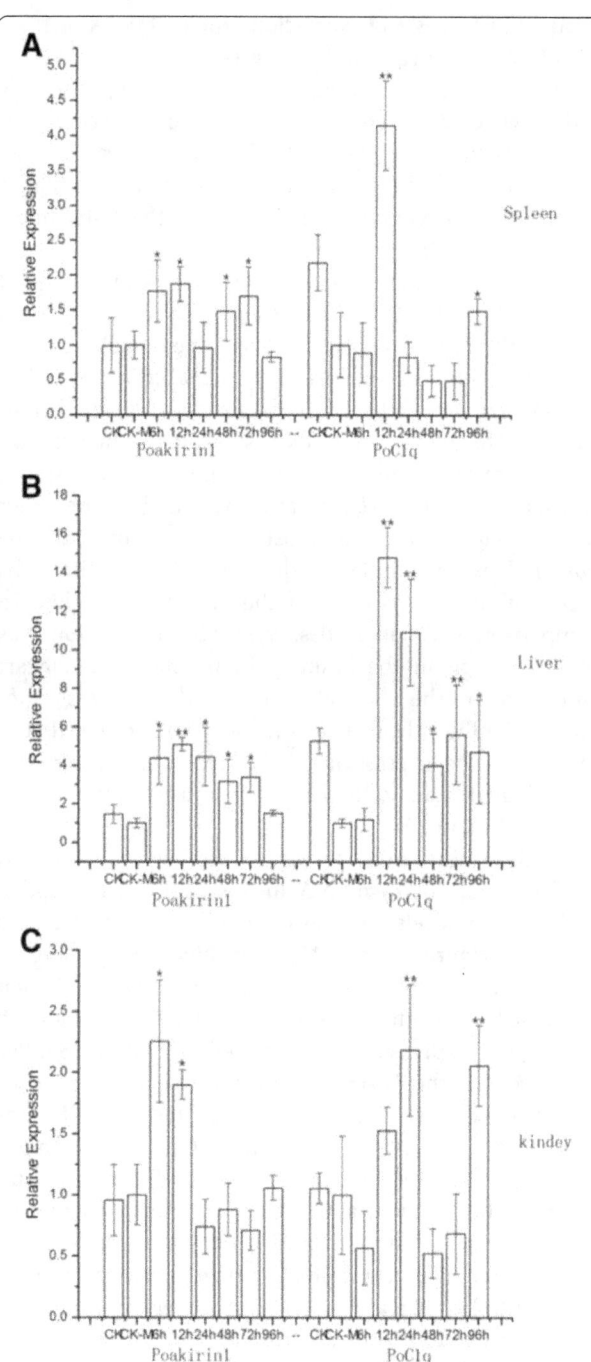

Figure 17 Quantitative RT-PCR analysis of the expression levels of *PoAkirin1* and *PoC1q* mRNA, relative to β-actin mRNA, in the spleen (A), liver (B), and kidney (C) after injection with *Vibrio anguillarum*. Graphs show the relative expression levels of *PoAkirin1* and *PoC1q* at 6 h, 12 h, 24 h, 48 h, 72 h and 96 h post-injection. CK-M represents the tissues at 12 h after saline solution injection, and CK represents live fish under normal conditions. Data are expressed as the mean ± S.D. (n = 3). Significant differences are indicated with an asterisk at P < 0.05 and two asterisks at P < 0.01.

Sequence analysis and expression profile of PoAkirin1, PoHEPN, and Poc1q in various tissues

Yeast two-hybrid screening identified a 672-bp *PoHEPN* cDNA, which encoded a protein with a conserved HEPN domain at the C-terminus. The *PoC1q* gene was 528 bp, and analysis of the predicted protein indicated that it contained a conserved C1q domain at the C-terminus (Figures 7 and 8).

RT-qPCR was used to quantify the expressions of *PoAkirin1*, *PoHEPN* and *PoC1q* mRNA in different flounder tissues in normal and pathogen-challenged individuals. The mRNA transcripts of *PoAkirin1*, *PoHEPN* and *PoC1q* were detected in all tissues of normal individuals. The *PoAkirin1* mRNA showed the highest expression in liver, and was highly expressed in the spleen and kidney. *PoAkirin1* was also expressed in the other tissues (Figure 13). In most tissues examined, the *PoHEPN* expression levels did not differ significantly, with slightly higher levels in the pituitary and skin. In comparison with other tissues, *PoC1q* expression was relatively high in the brain, gills, pituitary, skin, heart and eyes. In the liver and spleen, the expression of *PoC1q* significantly higher compared with other tissues (p < 0.01). The expression in the liver was 25 and 100 times higher than in the brain and muscle, respectively. (Figures 14 and 15).

To determine the expression profile of *PoAkirin1*, *PoHEPN* and *PoC1q* mRNA in the challenged flounder, their mRNA levels were examined in three tissues. As shown in Figures 16 and 17, after infection by the bacteria, *PoAkirin1* expression initially increased and then decreased in the three tissues (liver, spleen and kidney); the highest expression was detected at 6–24 h. Similar to *PoAkirin1*, the *PoHEPN* expression could also be induced by bacteria, but the response time was 6 h later than that of *PoAkirin1* and the significant increment started at 12 h. *PoC1q* expression was induced at 12 and 24 h by the bacteria, and showed a relatively large increase. In the kidney and spleen, the expression levels of PoC1q were significantly higher than those of the normal control at 24 and 12 h, respectively. Significantly higher levels of *PoC1q* were also detected at 12 and 24 h in the liver (Figures 16 and 17).

Discussion

The HEPN domain and the sacsin protein

Spastic ataxia of Charlevoix-Saguenay (SACS) originated from the Lac-Saint-Jean region, Quebec, in Canada. The sacsin gene, the pathogenic gene of SACS, consists of nine exons, including a gigantic exon of over 12.8 kbp, which is the biggest exon in vertebrates. The gene encodes the sacsin protein, 65% certain to be located in the in the nucleus. By SMART [29] analysis, this protein has

a recognizable DnaJ conserved domain, which is involved in the interaction of sacsin and Hsp70. The sacsin protein has the other three domains, including a UBQ region at the N-terminus, a HEPN domain at the C-terminus and another DnaJ region upstream of the HEPN domain.

The HEPN domain comprises 110 amino acid residues at C-terminus of the sacsin protein. In some invertebrates, bacteria and archaea, sacsin homologous proteins have this conserved domain.

Proteins with HEPN Domains are of three main types:

a) A single HEPN domain. In many types of bacterium, the protein is usually followed by nucleotidyltransferase (NT), which is often in a different reading frame, with both ORFs overlapping. Only two exceptions were found: in the genomes of *Pyrococcus furiosus* and *P. abyssi*, the HEPN protein does not exist after the NT protein.

b) Two HEPN domains, the N-terminal NT domain and C-terminal HEPN domain.

c) HEPN proteins with a variety of conserved domains. These occur mainly in fish and mammals (monkeys, rats, mice and humans). The HEPN domain is located at the C terminus of the protein, adjacent to the DnaJ domain. However, conserved HEPN domains in lower eukaryotes have not been found [30].

The function of the HEPN domain in sacsin is not clear and is difficult to predict. HEPN has an important role in stabilizing nucleotide binding in complexes formed with the DnaJ domain and may be involved in determining the specificity. In diseases caused by sacsin mutations, sacsin mutant proteins have an incomplete DnaJ and HEPN conserved domain [31]. Thus, the functions of HEPN and sacsin proteins require further study.

C1q

Complement component C1 is composed of three subunits: C1q, C1r and C1s [32]. C1q, in C1, combined with immune complexes (IC) or other non-immune complexes, activates the classical pathway of complement activation. The C1q complement system, activated via the classic, bypass and mannose-binding lectin activation pathways, is composed of 30 types of protein. C1q has two functional units: a collagen-like region (CLR) near the N-terminus and a globular region (GR) at C-terminus. The CLR of C1q, with C1r and C1s, forms the C1 macromolecules (C1qC1r2C1s2). After the recognizing and binding to the GR through the IC, C1r and C1s are activated, thereby initiating the classical pathway. When complement is activated, a number of complement proteins can be cleaved into a variety of small cell surface fragments that are recognized by complement receptors. The primary function of the cell surface complement receptors is to promote the natural immune system to remove the foreign proteins, cell debris and microorganisms from the circulatory system [33,34].

PoAkirin1 interacts with the N-terminus of PoHEPN

The HEPN conserved domain is predicted to locate in the 97–223 amino acids region of PoHEPN. Therefore, in the mutation experiment of PoHEPN segment deletion, we constructed three vectors comprising 1-96aa (pohepn1), 1-146aa (pohepn2), and 147-224aa (pohepn3), to verify whether PoHEPN can interact with PoAkirin1, where the binding site of PoHEPN with PoAkirin1 is, whether the interaction is associated with the conserved sequences, and whether the incomplete sequence of PoHEPN retains its binding activity. The result showed that the region of PoHEPN that binds with PoAkirin1 is located in the 1-96aa region, before the HEPN domain, and is unrelated with the HEPN domain. This suggests that the nucleoprotein PoAkirin1 has no transcriptional activation activity and must be combined with other proteins (transcription factors or other DNA-binding proteins), such as PoHEPN, which can bind DNA and initiate expression of immune-related genes as a complex. In this process, PoAkirin1 is likely to play a specific-enhancer role; however, this hypothesis should be tested by further experimentation.

PoAkirin1 and PoHEPN expression in *E. coli* and western blotting analysis

Prokaryotic fusion protein expression vectors for PoAkirin1 and PoHEPN were constructed and the recombinant PoHEPN and PoAkirin1 were successfully expressed in *E. coli*. The recombinant proteins were verified by western blotting. The recombinant proteins will be separated and purified. These recombinant proteins will be useful for further investigation of the function of PoAkirin1, such as to verify the binding activity of PoAkirin1 with PoHEPN, changes in protein expression levels, in situ hybridization, crystal structure, and protein activity.

Expression profiles of PoAkirin1, PoHEPN, PoC1q

To investigate the expression profile of PoAkirin1, the mRNA expressions of PoHEPN and PoC1q in different tissues from normal and *V. anguillarum* challenged fish were examined at different time points in three important immune organs: the kidney, spleen and liver. The results showed that PoAkirin1 was highly expressed in all the three tissues. The expression profile suggested that PoAkirin1 might be involved in growth, development and, especially, immunity. In most normal tissues, the

expression of PoHEPN did not differ significantly. It was only slightly higher in the pituitary and skin than in other tissues. PoC1q was hardly expressed in muscle, slightly expressed in brain, gills, pituitary, skin, heart and eyes, but highly expressed in the liver, spleen, kidney and head kidney. In the liver and spleen, the expression level is significantly different compared with other tissues ($p < 0.01$). The expression in liver is 25 times higher than in the brain and 100 times higher than in the muscle. The reason for this tissue-specific expression may relate to their different functions. The sacsin protein has mainly been associated with nerve function [31,35]. Therefore, although it is expressed in various tissues and organs, its expression is higher in the pituitary and the skin. However, for the C1q protein, as a subunit of complement C1, its distribution is mainly in the immune tissues, where it is involved in a series of signal transductions.

The expressions of the PoAkirin1, PoHEPN, and PoC1q gene were all induced by bacterial challenge, where the expression of PoAkirin1 reached its highest point reached after 12–24 h. Interestingly, however, the responses to the bacterial challenge of PoHEPN and PoC1q were slower than that of PoAkirin1, which may indicate that PoHEPN and PoC1q are located downstream in the response to bacterial challenge, and that the change in expression of PoAkirin1 leads directly to changes in the expressions of PoHEPN and PoC1q. However, the specific relationship between the upstream and downstream proteins in this signaling pathway requires further study.

Conclusions

In this paper, we cloned the Akirin1 homologous gene, PoAkirin1, from Japanese flounder and identified two proteins that potentially interact with PoAkirin1. The expression patterns of PoAkirin1 and the genes encoding the interacting proteins are closely associated with immunity. Bacteria can induce the expression of these genes; therefore, the function of this protein merits further investigation, particularly in the context of protecting fish populations.

Competing interests
The authors declare that they have no competing interests.

Authors' contributions
S-LC obtained the project. S-LC and C-GY designed the study. C-GY, X-LW, BZ, BS, and S-SL carried out the molecular genetic studies, participated in the sequence alignment and drafted the manuscript. C-GY, X-LW, and BZ carried out the immunoassays. BS and S-SL participated in the statistical analysis. S-LC revised the manuscript. All authors read and approved the final manuscript.

Acknowledgments
This work was supported by grants from the State 863 High-Technology R&D Project of China (2012AA092203, 2012AA10A408), the 973 National Basic Research Program of China (2010CB126303), and the Taishan Scholar Project Fund of Shandong, China.

Author details

[1]Yellow Sea Fisheries Research Institute, Chinese Academy of Fisheries Sciences, Qingdao 266071, China. [2]Yangtze River Fisheries Research Institute, Chinese Academy of Fishery Sciences, Wuhan 430223, China. [3]Translational Center for Stem Cell Research, Tongji Hospital, Stem Cell Research Center, Tongji University School of Medicine, Shanghai 200065, China. [4]Bohai Sea Fisheries Research Institute of Tianjin, Tianjin, China.

References

1. Auerbach D, Thaminy S, Hottiger MO, Stagljar I: The post-genomic era of interactive proteomics: facts and perspectives. *Proteomics* 2002, **2**(6):611–623.
2. Ito T, Tashiro K, Muta S, Ozawa R, Chiba T, Nishizawa M, Yamamoto K, Kuhara S, Sakaki Y: Toward a protein-protein interaction map of the budding yeast: A comprehensive system to examine two-hybrid interactions in all possible combinations between the yeast proteins. *Proc Natl Acad Sci USA* 2000, **97**(3):1143–1147.
3. Rain JC, Selig L, De Reuse H, Battaglia V, Reverdy C, Simon S, Lenzen G, Petel F, Wojcik J, Schachter V, *et al*: The protein-protein interaction map of Helicobacter pylori. *Nature* 2001, **409**(6817):211–215.
4. Terradot L, Durnell N, Li M, Ory J, Labigne A, Legrain P, Colland F, Waksman G: Biochemical characterization of protein complexes from the Helicobacter pylori protein interaction map: strategies for complex formation and evidence for novel interactions within type IV secretion systems. *Mol Cell Proteomics* 2004, **3**(8):809–819.
5. Giot L, Bader JS, Brouwer C, Chaudhuri A, Kuang B, Li Y, Hao YL, Ooi CE, Godwin B, Vitols E, *et al*: A protein interaction map of Drosophila melanogaster. *Science* 2003, **302**(5651):1727–1736.
6. Ito T, Chiba T, Ozawa R, Yoshida M, Hattori M, Sakaki Y: A comprehensive two-hybrid analysis to explore the yeast protein interactome. *Proc Natl Acad Sci USA* 2001, **98**(8):4569–4574.
7. Li S, Armstrong CM, Bertin N, Ge H, Milstein S, Boxem M, Vidalain PO, Han JD, Chesneau A, Hao T, *et al*: A map of the interactome network of the metazoan C. elegans. *Science* 2004, **303**(5657):540–543.
8. Goto A, Matsushita K, Gesellchen V, El Chamy L, Kuttenkeuler D, Takeuchi O, Hoffmann JA, Akira S, Boutros M, Reichhart JM: Akirins are highly conserved nuclear proteins required for NF-kappaB-dependent gene expression in drosophila and mice. *Nat Immunol* 2008, **9**(1):97–104.
9. de la Fuente J, Almazan C, Blas-Machado U, Naranjo V, Mangold AJ, Blouin EF, Gortazar C, Kocan KM: The tick protective antigen, 4D8, is a conserved protein involved in modulation of tick blood ingestion and reproduction. *Vaccine* 2006, **24**(19):4082–4095.
10. Almazan C, Blas-Machado U, Kocan KM, Yoshioka JH, Blouin EF, Mangold AJ, de la Fuente J: Characterization of three Ixodes scapularis cDNAs protective against tick infestations. *Vaccine* 2005, **23**(35):4403–4416.
11. Marshall A, Salerno MS, Thomas M, Davies T, Berry C, Dyer K, Bracegirdle J, Watson T, Dziadek M, Kambadur R, *et al*: Mighty is a novel promyogenic factor in skeletal myogenesis. *Exp Cell Res* 2008, **314**(5):1013–1029.
12. Komiya Y, Kurabe N, Katagiri K, Ogawa M, Sugiyama A, Kawasaki Y, Tashiro F: A novel binding factor of 14-3-3beta functions as a transcriptional repressor and promotes anchorage-independent growth, tumorigenicity, and metastasis. *J Biol Chem* 2008, **283**(27):18753–18764.
13. Macqueen DJ, Kristjansson BK, Johnston IA: Salmonid genomes have a remarkably expanded akirin family, coexpressed with genes from conserved pathways governing skeletal muscle growth and catabolism. *Physiol Genomics* 2010, **42**(1):134–148.
14. Macqueen DJ, Johnston IA: Evolution of the multifaceted eukaryotic akirin gene family. *BMC Evol Biol* 2009, **9**:34.
15. Yang CG, Wang XL, Wang L, Zhang B, Chen SL: A new Akirin1 gene in turbot (Scophthalmus maximus): molecular cloning, characterization and expression analysis in response to bacterial and viral immunological challenge. *Fish Shellfish Immunol* 2011, **30**(4–5):1031–1041.
16. Nam BH, Hirono I, Aoki T: Bulk isolation of immune response-related genes by expressed sequenced tags of Japanese flounder Paralichthys olivaceus leucocytes stimulated with Con A/PMA. *Fish Shellfish Immunol* 2003, **14**(5):467–476.
17. Nam BH, Yamamoto E, Hirono I, Aoki T: A survey of expressed genes in the leukocytes of Japanese flounder, Paralichthys olivaceus, infected with Hirame rhabdovirus. *Dev Comp Immunol* 2000, **24**(1):13–24.
18. Park EM, Kang JH, Seo JS, Kim G, Chung J, Choi TJ: Molecular cloning and expression analysis of the STAT1 gene from olive flounder. Paralichthys olivaceus. *BMC Immunol* 2008, **9**:31.
19. Chen SL, Wang ZJ, Xu MY, Gui JF: Molecular identification and expression analysis of natural resistance associated macrophage protein (Nramp) cDNA from Japanese flounder (Paralichthys olivaceus). *Fish Shellfish Immunol* 2006, **20**(3):365–373.
20. Xu TJ, Chen SL, Ji XS, Tian YS: MHC polymorphism and disease resistance to Vibrio anguillarum in 12 selective Japanese flounder (Paralichthys olivaceus) families. *Fish Shellfish Immunol* 2008, **25**(3):213–221.
21. Zhang YX, Chen SL: Molecular identification, polymorphism, and expression analysis of major histocompatibility complex class IIA and B genes of turbot (Scophthalmus maximus). *Mar Biotechnol (NY)* 2006, **8**(6):611–623.
22. Chen S, Xu M, Ji X, Yu G: Cloning and characterization of natural resistance associated macrophage protein (Nramp) cDNA from red sea bream (Chrysophrys major). *Fish Shellfish Immunol* 2004, **17**:305–313.
23. Saitou N, Nei M: The neighbor-joining method: a new method for reconstructing phylogenetic trees. *Mol Biol Evol* 1987, **4**(4):406–425.
24. Tamura K, Dudley J, Nei M, Kumar S: MEGA4: Molecular Evolutionary Genetics Analysis (MEGA) software version 4.0. *Mol Biol Evol* 2007, **24**(8): 1596–1599.
25. Bustin SA, Benes V, Garson JA, Hellemans J, Huggett J, Kubista M, Mueller R, Nolan T, Pfaffl MW, Shipley GL, *et al*: The MIQE guidelines: minimum information for publication of quantitative real-time PCR experiments. *Clin Chem* 2009, **55**(4):611–622.
26. Zheng WJ, Sun L: Evaluation of housekeeping genes as references for quantitative real time RT-PCR analysis of gene expression in Japanese flounder (Paralichthys olivaceus). *Fish Shellfish Immunol* 2011, **30**(2):638–645.
27. Martin-Antonio B, Jimenez-Cantizano RM, Salas-Leiton E, Infante C, Manchado M: Genomic characterization and gene expression analysis of four hepcidin genes in the redbanded seabream (Pagrus auriga). *Fish Shellfish Immunol* 2009, **26**(3):483–491.
28. Kurobe T, Yasuike M, Kimura T, Hirono I, Aoki T: Expression profiling of immune-related genes from Japanese flounder Paralichthys olivaceus kidney cells using cDNA microarrays. *Dev Comp Immunol* 2005, **29**(6):515–523.
29. Letunic I, Goodstadt L, Dickens NJ, Doerks T, Schultz J, Mott R, Ciccarelli F, Copley RR, Ponting CP, Bork P: Recent improvements to the SMART domain-based sequence annotation resource. *Nucleic Acids Res* 2002, **30**(1):242–244.
30. Grynberg M, Erlandsen H, Godzik A: HEPN: a common domain in bacterial drug resistance and human neurodegenerative proteins. *Trends Biochem Sci* 2003, **28**(5):224–226.
31. Engert JC, Berube P, Mercier J, Dore C, Lepage P, Ge B, Bouchard JP, Mathieu J, Melancon SB, Schalling M, *et al*: ARSACS, a spastic ataxia common in northeastern Quebec, is caused by mutations in a new gene encoding an 11.5-kb ORF. *Nat Genet* 2000, **24**(2):120–125.
32. Lepow IH, Naff GB, Todd EW, Pensky J, Hinz CF: Chromatographic resolution of the first component of human complement into three activities. *J Exp Med* 1963, **117**:983–1008.
33. Kishore U, Leigh LE, Eggleton P, Strong P, Perdikoulis MV, Willis AC, Reid KB: Functional characterization of a recombinant form of the C-terminal, globular head region of the B-chain of human serum complement protein, C1q. *Biochem J* 1998, **333**(Pt 1):27–32.
34. Lu J, Wu X, Teh BK: The regulatory roles of C1q. *Immunobiology* 2007, **212**(4–5):245–252.
35. Bouchard JP, Barbeau A, Bouchard R, Bouchard RW: Autosomal recessive spastic ataxia of Charlevoix-Saguenay. *Can J Neurol Sci* 1978, **5**(1):61–69.

Néstor-Guillermo Progeria Syndrome: a biochemical insight into Barrier-to-Autointegration Factor 1, alanine 12 threonine mutation

Nicolas Paquet[1†], Joseph K Box[1†], Nicholas W Ashton[1], Amila Suraweera[1], Laura V Croft[1], Aaron J Urquhart[1], Emma Bolderson[1], Shu-Dong Zhang[2], Kenneth J O'Byrne[1] and Derek J Richard[1*]

Abstract

Background: Premature aging syndromes recapitulate many aspects of natural aging and provide an insight into this phenomenon at a molecular and cellular level. The progeria syndromes appear to cause rapid aging through disruption of normal nuclear structure. Recently, a coding mutation (*c.34G > A* [p.A12T]) in the Barrier to Autointegration Factor 1 (*BANF1*) gene was identified as the genetic basis of Néstor-Guillermo Progeria syndrome (NGPS). This mutation was described to cause instability in the BANF1 protein, causing a disruption of the nuclear envelope structure.

Results: Here we demonstrate that the BANF1 A12T protein is indeed correctly folded, stable and that the observed phenotype, is likely due to the disruption of the DNA binding surface of the A12T mutant. We demonstrate, using biochemical assays, that the BANF1 A12T protein is impaired in its ability to bind DNA while its interaction with nuclear envelope proteins is unperturbed. Consistent with this, we demonstrate that ectopic expression of the mutant protein induces the NGPS cellular phenotype, while the protein localizes normally to the nuclear envelope.

Conclusions: Our study clarifies the role of the A12T mutation in NGPS patients, which will be of importance for understanding the development of the disease.

Keywords: Progeria, Nuclear envelope, Aging

Background

Aging is a natural process that affects all organisms, although the precise mechanisms of its progression remain poorly understood. As such, human premature aging syndromes, which recapitulate many aspects of natural aging, may allow us to further investigate this phenomenon at the molecular and cellular level. These syndromes largely result from heritable genetic alterations that mainly affect DNA repair proteins, or proteins associated with the nuclear periphery [1,2]. For instance, one category of premature aging syndromes, known as Human Progeroid syndromes (or laminopathies), are caused by mutations in

nuclear lamins or other proteins of the nuclear envelope. As a result, cells from these patients are characterized by nuclear envelope dysfunction, altered nuclear activity, impaired structural dynamics and aberrant cell signaling [3]. These irregularities may manifest in premature aging, as well as conditions such as neuropathy.

Recently, two unrelated patients who exhibited several Hutchinson-Gilford Progeria syndrome-like phenotypes were described [4,5]. These patients, however, did not present with many of the symptoms common to those with known human Progeroid syndromes. For instance, neither patient showed signs of ischemia or atherosclerosis, both fundamental phenotypes of Hutchinson-Gilford Progeria syndrome. Cognitive function was also identified as normal in both patients. Moreover, the age of the patients studied was not consistent with the current understanding of known human Progeroid syndromes, with

* Correspondence: derek.richard@qut.edu.au

†Equal contributors

[1]School of Biomedical Science, Institute of Health and Biomedical Innovation at the Translational Research Institute, Queensland University of Technology, Brisbane, QLD, Australia

Full list of author information is available at the end of the article

these patients being much older than the average life span of progeroid patients. This condition was named Néstor–Guillermo Progeria Syndrome (NGPS). Whole-genome and exome sequencing of both affected patients identified them as homozygous for a mutation in the Barrier-to-Autointegration Factor 1 (*BAF1* or *BANF1*) gene [5]. This mutation (*c.34G > A* [p.Ala12Thr]) results in the expression of a BANF1 protein where alanine 12 is mutated to a threonine residue.

BANF1 encodes a protein consisting of 89 amino acid residues with a molecular weight of approximately 10 kDa [6]. Nuclear magnetic resonance and crystallographic studies have determined that BANF1 may form a homodimer, which is the active state required to bind chromatin [7,8]. During G1, S and G2 phases of the cell cycle, BANF1 is known to predominantly associate with the nuclear envelope, where it interacts with the Lamina associated polypeptides Emerin-MAN1 (LEM) domain of the nuclear scaffold proteins MAN1 [9], Emerin [10] and LAP2 [11,12]. Here, BANF1 regulates organization of the chromatin structure at the nuclear envelope by condensing DNA via a looping mechanism [13], as well as by binding histone H3 and histone linker H1.1 [14]. The interaction between BANF1, chromatin and protein from the lamina is tightly regulated, allowing proper nuclear assembly and chromatin organization during cell cycle progression [15,16]. In addition to these roles, BANF1 has been proposed to regulate the transcription of specific genes [17], to suppress the integration of retroviruses within the genome [6,18,19], and to regulate specific developmental signals [15,20].

Despite the available data on BANF1 biology, the contribution of the A12T mutation to the development of Nestor-Guillermo Progeria Syndrome is poorly understood. In their original study of the disease, Puente *et al.* observed reduced levels of mutant BANF1 in patient fibroblasts, and thus proposed that the mutation may affect protein stability [5]. It was therefore assumed that the observed phenotype of these patients was due to physiologically low levels of BANF1 [5]. In the present study, we aimed to further decipher the role of the BANF1 A12T mutation in the molecular processes leading to the development of the disease. To do so, we used a series of biochemical and molecular tools to understand the defect resulting from this genetic mutation.

Results

In order to characterize the effect of the A12T mutation on the BANF1 protein and NGPS phenotype, we initially purified recombinant His-tagged BANF1 wild type (WT) and A12T proteins from *E. coli* using a protocol adapted from Harris *et al.* [19], although unlike Harris *et al*, we retained the N-terminal His-tag. As previously described, recombinant WT BANF1 was found in inclusion bodies

(as was the A12T mutant), indicating aggregation of the protein in a higher-ordered complex, and thus necessitating denaturation and subsequent refolding of the protein during purification. Using this method, both wild type and A12T BANF1 displayed similar purification characteristics (Figure 1A). BANF1 forms a stable dimer in solution with a dimerization interface formed by helix α3, the C-terminus of α5 and part of the loop linking α1 to the helical turn [7]. Although alanine 12 is not located at the dimerization interface of BANF1, we experimentally tested whether A12T mutation may influence this interface indirectly. To do this, we detected both mutant and wild type protein as they eluted from a size exclusion column (Figure 1B). Here, both wild type and A12T recombinant BANF1 eluted with a profile consistent with a mixture of monomeric and dimeric BANF1, supporting that this mutation does not disrupt the dimerization of the protein *in vitro*.

As alanine 12 of BANF1 is positioned in a loop immediately following helix α1, we hypothesized that mutation of this residue to a bulky β-branched threonine could influence the structure of the BANF1 protein. To assess this, we analyzed recombinant WT and A12T BANF1 by circular dichroism (CD), a standard biophysical technique used to study the secondary structure content of proteins in solution. CD spectra of both wild type and A12T BANF1 showed typical α-helical profiles, with minima in the near far- UV at a wavelength of 220 nm (Figure 1C). By this method we detected a spectrum for wild type BANF1 that was consistent with published data [19]. Furthermore, overlay of the A12T BANF1 spectrum indicated the secondary structure of the protein was not affected as a result of mutation.

To explore whether A12T mutation may alter the BANF1 structure on a smaller scale, we obtained the crystal structure of wild type BANF1 bound to DNA from the Protein Data Bank (PDB ID 2BZF), and used Phyre2 [21] (http://www.sbg.bio.ic.ac.uk/~phyre2) to predict the three-dimensional structure of BANF1 following A12T mutation. Modeling was subsequently confirmed using I-TASSER [22] (http://zhanglab.ccmb.med.umich.edu/I-TASSER/). Consistent with the CD spectra data, Phyre2 model prediction of the A12T mutant did not suggest major modifications to the structure of BANF1, although interestingly did indicate a potential alteration in the position of amino acids essential for DNA binding (Figure 2A). Based on the crystal structure, the N terminus of BANF1 helix α1 is important for contacting DNA and establishing hydrogen bonds between Gln5, Lys6 and the nucleotide phosphates [7]. In our model, we therefore predict that the BANF1 A12T mutation may displace the side chain of Lys6 from its original position (Figure 2B), preventing the formation of hydrogen bonds between this residue and the phosphodiester backbone

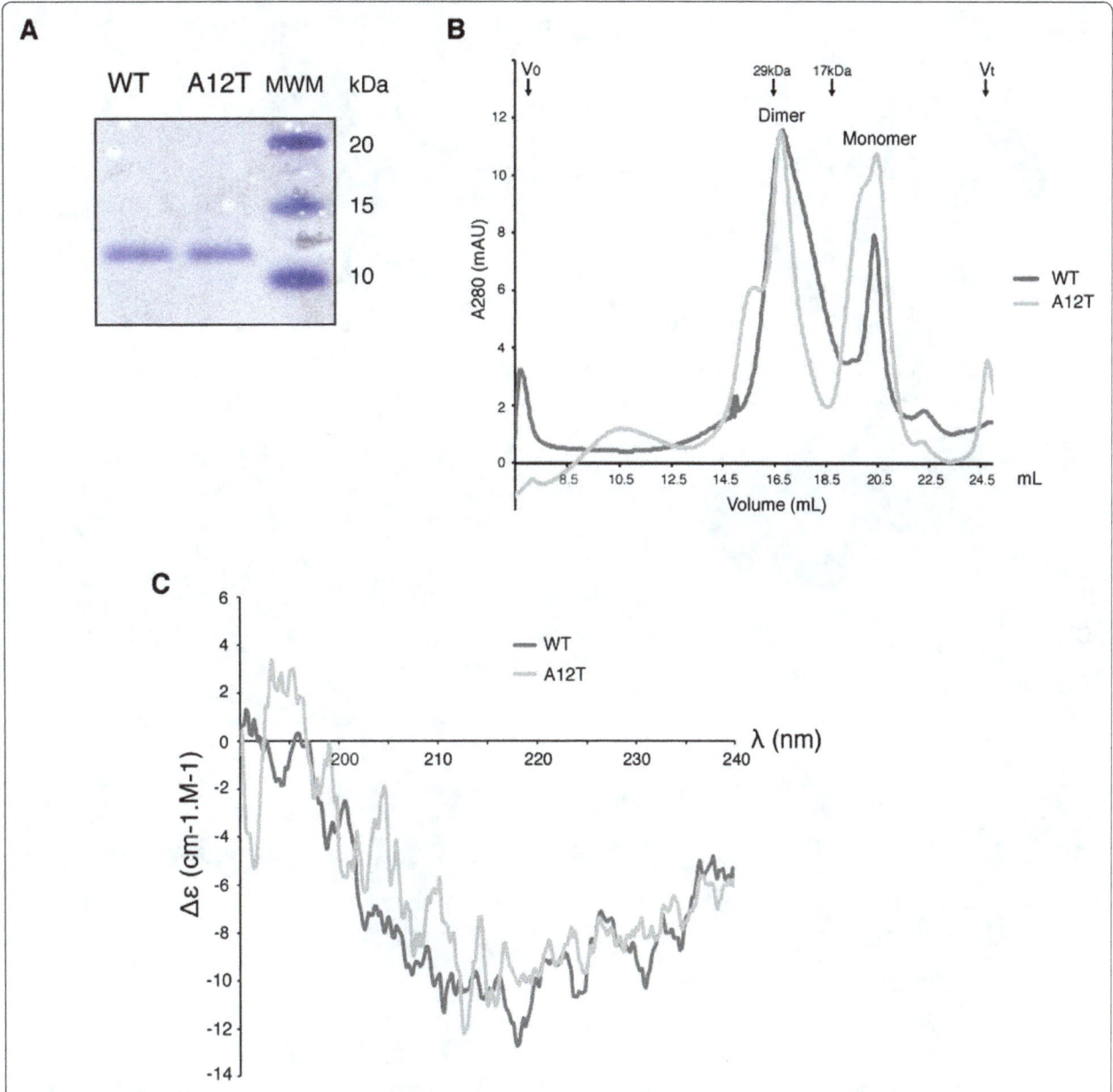

Figure 1 Characterization of WT and A12T Barrier to Autointegration 1 proteins. (A): Coomassie blue staining of 500 ng of recombinant wild type and A12T HexaHis-tagged BANF1 run on a Nu-Page gel. **(B)**: Gel filtration chromatography of wild type and A12T mutant BANF1 proteins. Absorbance at 280 nm was plotted against the elution volume. Both proteins eluted in two predominant peaks, which are consistent with dimeric and monomeric BANF1. Vo indicates the void volume of the column and Vt indicates the termination volume. Arrows indicate the elution volume of the protein standards carbonic anhydrase (29 kDa) and Myoglobin (17 kDa), which were used to calibrate the column prior to BANF1 filtration. **(C)**: A12T BANF1 secondary structure is not modified when compared to the WT. CD spectroscopy of wild-type and A12T BANF1 protein. The mean residue ellipticity Δε of the indicated protein is plotted against the wavelength, and is the results of 3 independent measurements. Δε is in cm^{-1}.M^{-1}.

and thus indicating a possible DNA binding defect. Interestingly, Lys6 is also known to be buried in a pocket formed by the carbonyl groups of Gly21, Ile26 and Leu23 [7]. In our prediction, the Lys6 ε-amino group sits outside this pocket, which is likely to have structural and functional consequences (Figure 2C). In addition, our model also predicts the side chains of Glu13 are displaced, potentially affecting the formation of the salt bridge between this residue and Lys18 [23] (Figure 2D). Both Lys6 and Glu13 are important for the stabilization of helix α1 and the loop connecting α1 and α3, which brings positively charged Lys6 and Arg8 to the DNA binding site.

To understand further the instability previously described for A12T BANF1, we expressed 3x FLAG-tagged WT or A12T BANF1 in HeLa cells from a CMV promoter. Consistent with the observations of Puente *et al.*

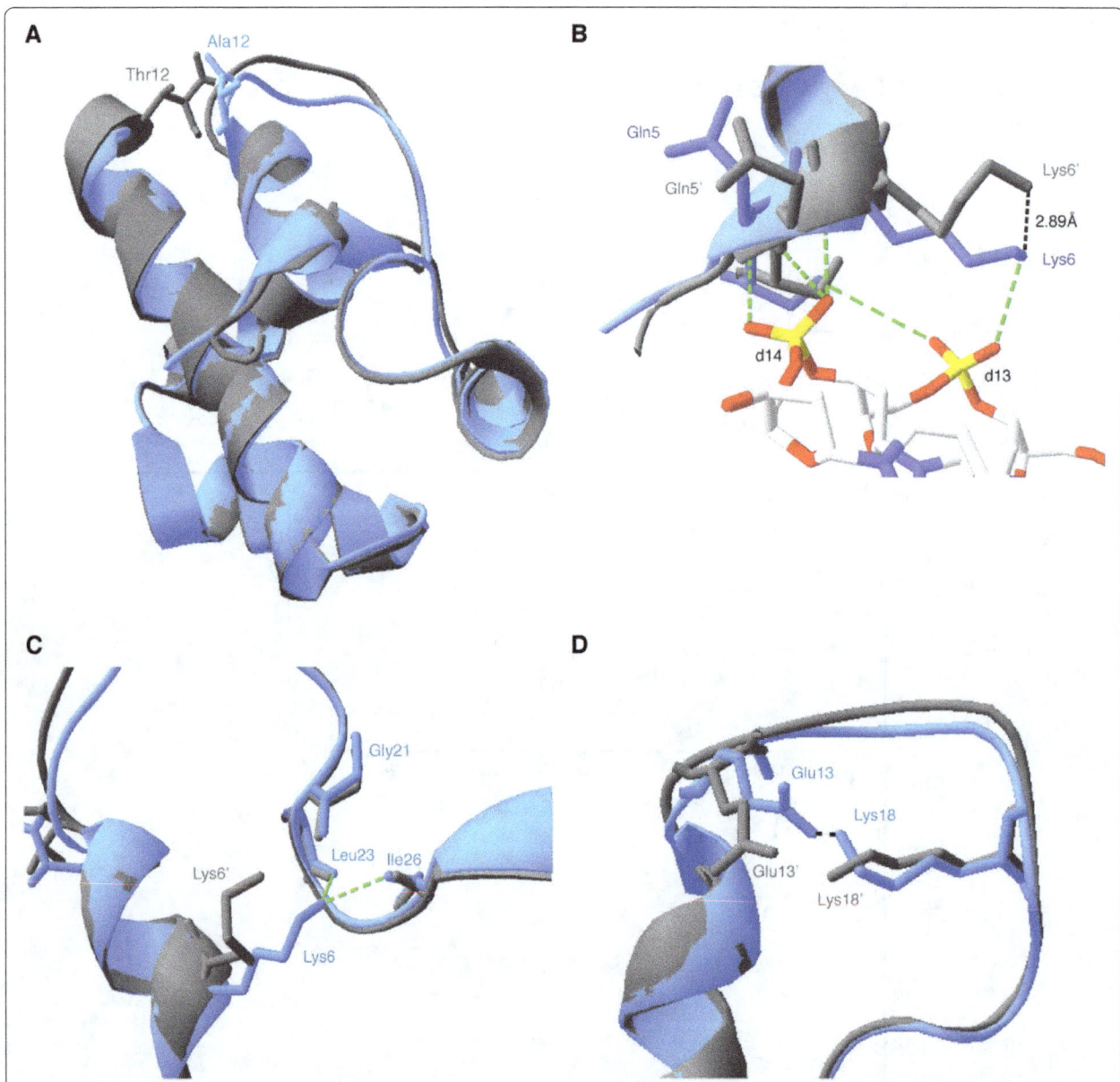

Figure 2 Structural model of the BANF1 A12T protein. (A-D): BANF A12T modeling predicts fine changes in the position of key amino acids. WT and A12T sequences were used to generate a model using Phyre2 software. The resulting models were fitted on the existing crystal structure of BANF1 bound to DNA (PDB: 2BZF). **(A)**: Superposition of ribbon diagrams of monomeric WT (Blue) and A12T (gray) BANF1, indicating modification of the loop connecting helix α1 and α3. **(B)**: Ribbons diagrams of WT (Blue) and A12T (Gray) BANF1, indicating the hydrogen bonds between Lys6 and Gln5 with DNA nucleotides d13 and d14 DNA (as depicted in PDB: 2BZF). Lys6' and Gln5' from BANF1 A12T are seen displaced from their WT counterpart. **(C)**: Ribbon diagrams of WT (Blue) and A12T (Gray) BANF1. Schematics indicate that in the WT, Lys6 protrudes into the pocket formed by Gly21, Ile26 and Leu23. Lys6' sits outside this pocket in the A12T mutant. **(D)**: Ribbon diagrams of WT (Blue) and A12T (Gray) BANF1, showing the salt bridge between Glu13 and Lys18 in the WT. Glu13' and Lys18' are displaced in the BANF1 A12T prediction.

[5], we observed minimal levels of the A12T mutant protein when compared to the wild type protein, as determined by immunoblotting using the same BANF1 antibody as in their study (Figure 3A). To confirm that we had indeed detected ectopically expressed FLAG tagged BANF1, we stripped the nitrocellulose membrane of the BANF1 antibody and re-probed with an antibody against the FLAG tag (Sigma, F1804 SL11063). Unexpectedly, this indicated equal expression of both the wild type and mutant BANF1.

These data raise the possibility that alanine 12 is required for antibody recognition of BANF1, such that reduced antigenicity may result from A12T mutation. To test this further, 500 ng of recombinant BANF1 WT and A12T was run on a SDS page gel and transferred to a nitrocellulose membrane for immunoblotting (Figure 3B).

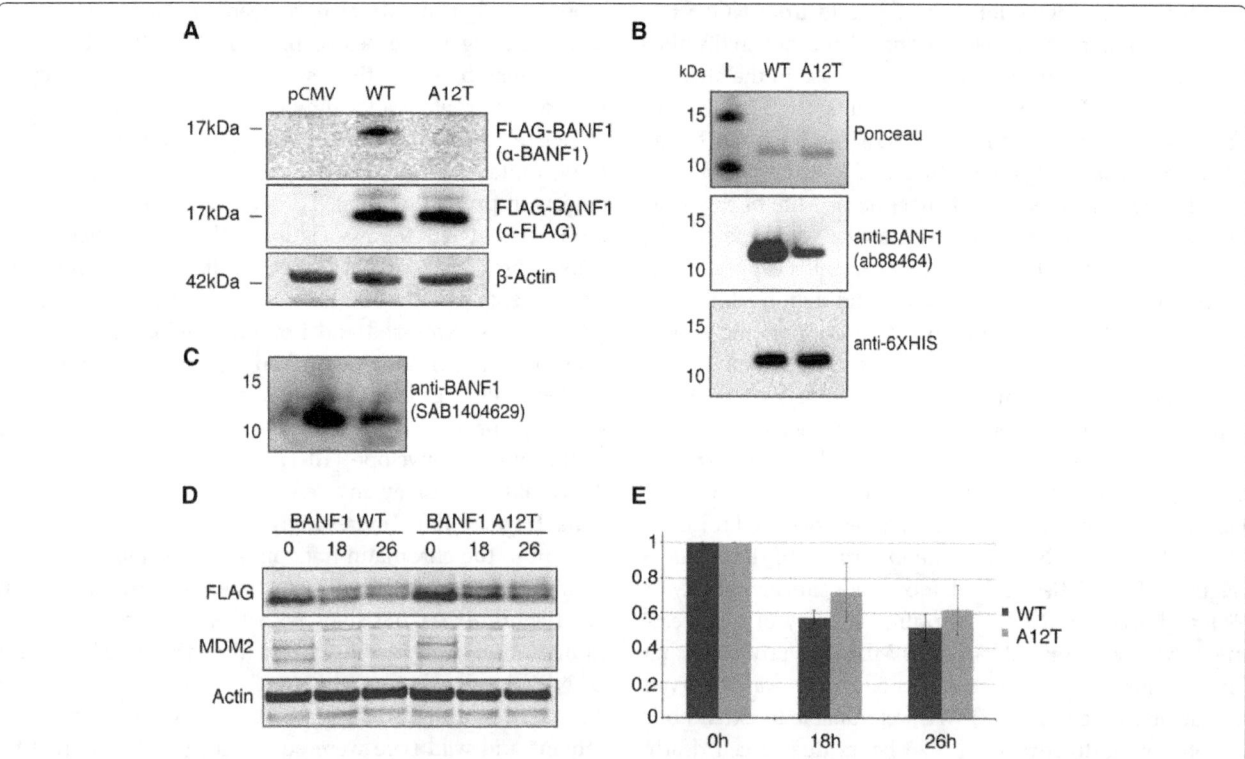

Figure 3 Alteration of BANF1 A12T antigenicity and stability. (A): The anti-BANF1 antibody (Abcam: ab88464) does not recognize exogenous A12T BANF1 in cell lysates. Cell extracts of U2OS cells overexpressing WT or A12T 3x FLAG-tagged BANF1 or an empty vector (pCMV-AN-3DDK) were resolved on a Nu-page gel and immunoblotted using an anti-BANF1 antibody. Membranes were stripped and immunoblotted using an anti-FLAG antibody. A β-actin antibody was used as an internal protein loading control. **(B)**: Recombinant A12T BANF1 has a lower antigenicity than WT BANF1. 500 ng of recombinant HexaHis tagged WT and A12T BANF1 were run on a Nu-page gel and subsequently transferred to a nitrocellulose membrane. The membrane was stained using Ponceau red as a loading control, then blotted with an anti-BANF1 antibody (Abcam ab88464). After visualization, the membrane was stripped and immunoblotted using an anti HexaHis antibody. **(C)**: Recombinant A12T BANF1 has a lower antigenicity than WT BANF1. BANF1 was visualized by blotting with an anti-BANF1 antibody (SAB1404629). **(D)**: Wild type and A12T mutant 3x FLAG BANF1 protein stability. Cells were incubated with cycloheximide (50 µg/ml) for the indicated time periods and cell lysates harvested for western blot analysis. Protein levels were assessed using anti-FLAG antibodies to detect FLAG- tagged BANF1 and anti-actin antibodies for protein loading. MDM2 degradation is shown as a positive control for the cycloheximide treatment. **(E)**: Quantification of **(D)**. Band signal intensity was quantified using ImageJ and standardized against the protein level a t = 0. Error bars represent the standard deviation (SD) from at least three independent experiments.

The loading of equal quantities of proteins was verified by staining of the membrane using Ponceau red. The immunoblot was then probed using an anti-BANF1 antibody (Abcam ab88464). Similar to our findings using overexpressed WT and A12T BANF1 in cells, the BANF1 antibody failed to recognize the A12T recombinant protein to a level comparable to that observed for the wild type protein. To confirm this result, the membrane was stripped of antibody and then re-probed using an anti-HexaHis antibody (Abcam ab1187) to detect the N-terminal His tag of the recombinant BANF1 proteins. Consistent with the Ponceau staining, the anti-HexaHis antibody demonstrated that both proteins were present in similar amounts (Figure 3B). In an attempt to confirm these findings, further immunoblots using a BANF1 antibody supplied by Sigma (SAB1404629) were performed, and again the BANF1 antibody demonstrated a lack of recognition of the mutant protein (Figure 3C). This experiment

was repeated several times with proteins from different purification batches and consistently suggested a reduced antigenicity of the mutant protein when anti-BANF1 antibody was used. Interestingly, a similar alteration of antigenicity of BANF1 as a result of epitope mutation has previously been reported [18]. In this study by Lin and Engelman, an anti-BANF1 antibody raised against a peptide encompassing amino acids 4 to 20 did not recognize K6A and K18A BANF1 mutant proteins. These findings may be explained by the consideration that BANF1 is a small and structured protein and that the antigenic regions recognized by the antibody are likely to be composed of only a few amino acids. The mutation of one of these residues, or a residue impacting the structure of this epitope, may therefore interfere with antibody binding.

The observation that A12T BANF1 may be of reduced antigenicity compared to the wild type protein is of importance, as using the same BANF1 antibody, Puente *et al.*

detected a lower signal for BANF1 in cells from NGPS patients [5]. The authors then discussed the possibility that the pathology observed is due to a decrease in the stability of BANF1 A12T. In light of their results however, the hypothesis of a reduced half-life for A12T BANF1 may be inadequate. To investigate whether A12T BANF1 is indeed unstable, we investigated whether the half-life of this protein is reduced in cells compared to the wild type. To do this, we transfected HeLa cells with plasmids expressing 3x FLAG-tagged WT or A12T BANF1 and then treated these cells with cycloheximide to block transcription and subsequent protein synthesis. Cells were then harvested 18 and 26 hours post treatment, lysates were separated by SDS-PAGE and proteins transferred to nitrocellulose. Ectopically expressed BANF1 was then detected by Western blot analysis using the anti-FLAG antibody. As shown in Figure 3D-E, similar decay patterns were detected for both WT and A12T BANF1 over the experimental time course (Figures 3D-E). These data demonstrate similar stability of WT and A12T BANF1. To confirm efficacy of our treatment, we also immunoblotted for MDM2, a protein with a known short-life; as expected, loss of protein was observed by our first time point [24]. We were unable to extend our cycloheximide treatment past 26 hours due to cell death. Taken together, our results provide strong evidence that the BANF1 A12T mutant is stable and that the phenotype seen in the NGPS patients is likely to be due to an altered function of BANF1.

As mentioned previously, our structural modeling indicated the presence of a bulky threonine residue, in place of alanine residue 12, might disrupt the DNA binding pocket of BANF1. To investigate this further, we performed DNA mobility shift assays. Escalating concentrations of recombinant purified WT and A12T BANF1 were incubated with a 21 nucleotide double-stranded DNA oligonucleotide. The DNA probe was labeled with a 5' FAM and the interaction between BANF1 and the DNA observed as retardation in the migration of the complex through a polyacrylamide gel. As previously described, BANF1 WT has a high affinity for double-stranded DNA and bundles DNA in a highly ordered nucleoprotein complex [25]. The kinetics of binding were first studied after incubation for 5 minutes at 37°C, however due to the rate of binding, we were unable to observe intermediate complexes. The reaction was then performed at 4°C for 30 minutes and this allowed us to observe the lower order complexes. Interestingly, the BANF1 A12T mutant exhibited a marked defect in DNA binding compared to the wild type protein (Figure 4A,B). To confirm our observation we conducted DNA mobility shift assay in the same conditions using a 4.5 kb double stranded DNA plasmid. Consistent with our previous observations, A12T BANF1 exhibited a decreased affinity for longer DNA substrates (Figure 4C).

Several other mutations have been reported to affect the DNA binding properties of BANF1 [19,23,26]. They can be classified based on the severity of the binding defect, as well as their ability to dimerize and to bind Lamina-associated proteins [20,26]. However, to date, none of these mutants have solely impacted DNA binding versus defects in protein binding. To determine if the A12T mutation also affects the association with known nuclear envelope proteins, we performed co-immunoprecipitation experiments. For this, 3x FLAG tagged wild type or A12T BANF1 was expressed and immunoprecipitated in HeLa cells. Immunoblotting was then performed using antibodies against Emerin, Lamin and Histone H3, proteins that have previously been shown to interact with BANF1 at the nuclear envelope [10-12,14]. Interestingly, BANF1 A12T did not display any defects in binding to these proteins (Figure 4D). Interestingly, BANF1 A12T was still present in the chromatin fraction of a subcellular fractionation, consistent with its interaction with histone protein H3 (Figure 4E). Consistent with these observations, immunofluorescence demonstrated that the A12T mutant BANF1 localized normally to the nuclear envelope in U2OS cells. Interestingly, although expression levels of mutant and wild type were equivalent (as determined by immunoblotting with the FLAG antibody), the majority of cells expressing the A12T mutant demonstrated nuclear envelope aberrations consistent with that observed in NGPS patients (Figure 5A,B,C). Together, our data indicates that the A12T mutation of BANF1, found in Nestor-Guillermo Progeria syndrome, causes a disruption of the DNA binding surface, inhibiting its normal interaction with double stranded DNA. The BANF1 A12T mutant however localizes normally to the nuclear envelope, where it interacts with nuclear envelope proteins and chromatin.

Discussion

The identification of two Progeria individuals with a single point mutation in BANF1 is important for our understanding of these syndromes. Interestingly, Puente et al. reported that although mRNA levels of A12T mutant BANF1 were found at similar levels to wild type patients, BANF1 A12T protein was detected at a much lower level. This was originally interpreted as a result of protein instability. In our study, we sought to understand the mechanism through which this may occur. Interestingly, our study suggested that the A12T mutant was not unstable, and that the lower levels of the protein observed were merely an artifact of antigenicity alterations towards the BANF1 antibody, as a result of the A12T mutation. We therefore reasoned that this mutation might affect protein function in other ways that could explain the NGPS phenotype. Our structural modeling of BANF1, predicted that the mutation of alanine 12 to a bulky threonine, could disrupt the BANF1 DNA-binding pocket and thus disrupt

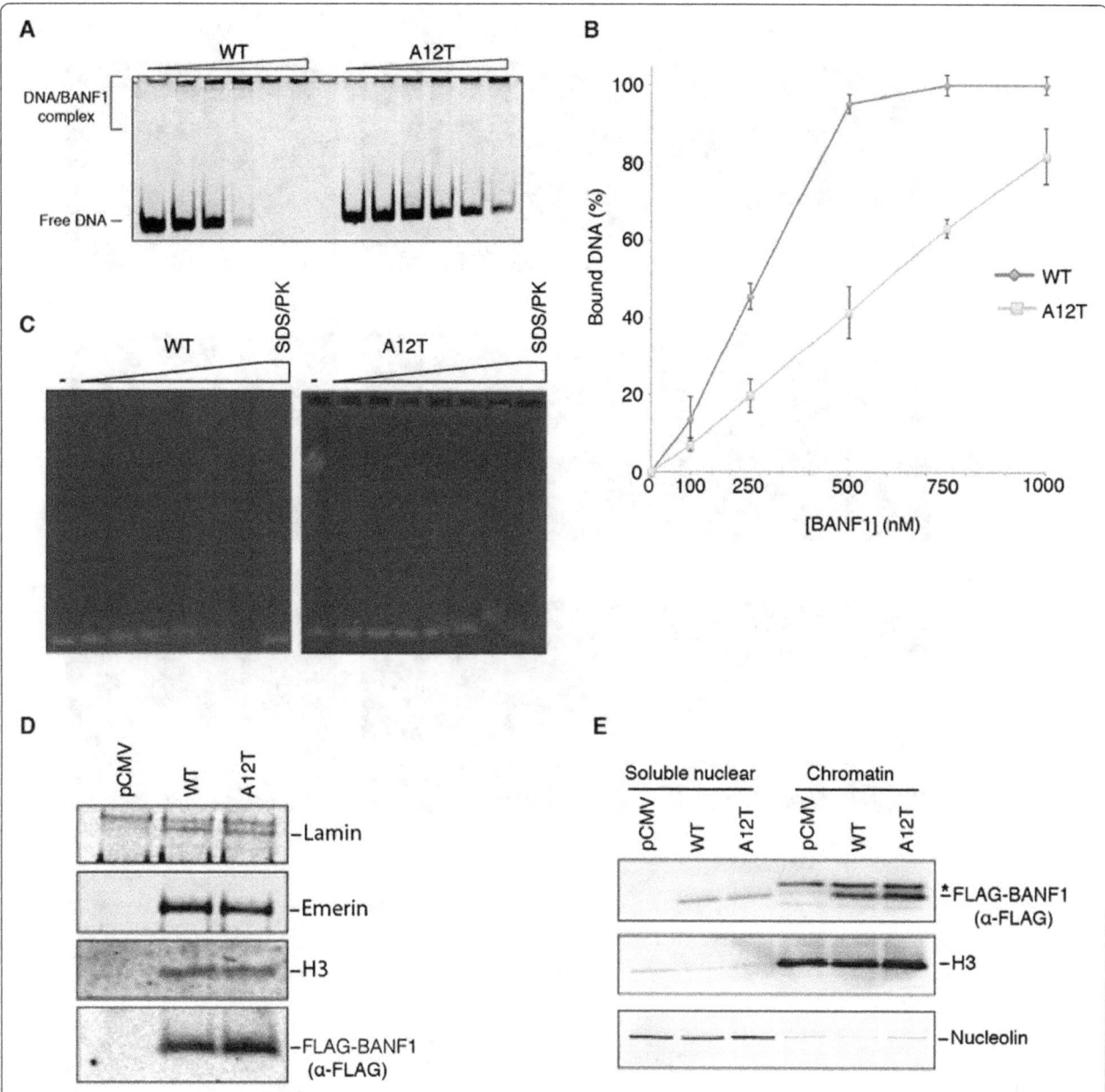

Figure 4 Alteration of BANF1 A12T DNA binding. (A): A12T BANF1 has a reduced affinity for short double-stranded DNA. WT and A12T BANF1 (0, 0.1, 0.25, 0.5, 0.75, 1, μM) was incubated for 30 min at 4°C with 10 nM of dsDNA that was labeled with a 5' FAM label. The FAM label was visualized using a Starion FLA-9000 image scanner. **(B)**: Quantification of **(A)**. Intensity of the signal was quantified using MultiGauge software (Fujifilm). Error bars represent the standard deviation (SD) from at least three independent experiments. **(C)**: A12T BANF1 has a reduced affinity for long double-stranded DNA. WT and A12T BANF1 (0, 0.1, 0.25, 0.5, 1, 2 and 4 μM) was incubated for 30 min at 4°C with 150 ng of double stranded DNA plasmid. The binding was visualized on an agarose gel following staining with Ethidium bromide. **(D)**: The interaction between WT or A12T BANF1 and known partners was tested by co-immunoprecipitation. HeLa cells were transfected with the indicated vectors and exogenous protein expressed for 24 hours Total protein was then extracted and treated with Benzonase to degrade genomic DNA. M2 magnetic FLAG beads were used to immunoprecipitate 3x FLAG BANF1 and eluent probed using specific antibodies against Lamin, Emerin and Histone H3. **(E)**: A12T BANF1 nuclear distribution is similar to that of WT BANF1. HeLa cells were transiently transfected with the indicated vectors prior to sub-cellular fractionation. Western blotting of fractions was performed using an anti-FLAG antibody (to detect exogenous BANF1), anti-H3 (chromatin fraction loading control) and anti-nucleolin (soluble nuclear loading control). *Bleed-through from antiH3 channel.

the interaction of BANF1 with DNA. We confirmed by EMSA that the A12T BANF1 was indeed perturbed in its ability to bind to DNA, suggesting that the modeling was correct. Further, our modeling and CD spectra analysis

suggested that this was the only disruption to the BANF1 structure. Moreover, we found that the A12T mutant, like WT BANF1, localized to the nuclear envelope and interacted with lamin and histone H3. Further studies using

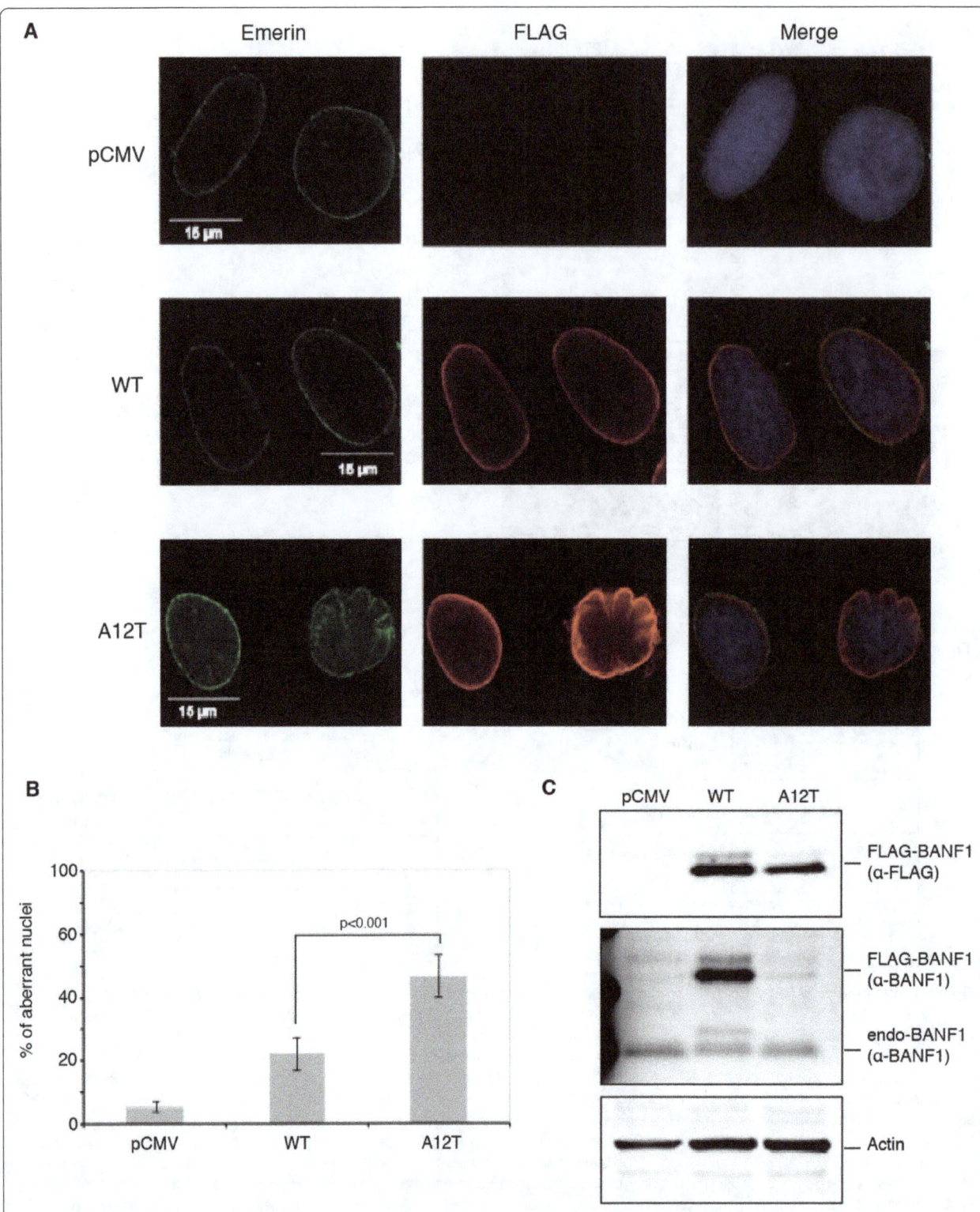

Figure 5 Nuclear envelope localization of the BANF1 A12T protein. (A): A12T BANF1 localization to the nuclear envelope is unaltered from the WT protein. U2OS cells were transfected with constructs expressing WT or A12T 3x FLAG BANF1. Soluble proteins were extracted with detergent and cells permeabilised. BANF1 distribution was analyzed by immunofluorescence using an anti-FLAG antibody and a Deltavision PDV microscope. Aberrant nuclear envelope conformation can be seen in cells expressing A12T BANF1. Emerin visualization is representative of the nuclear membrane. **(B)**: Quantification of the aberrant nuclei seen in the BANF1 A12T overexpression. **(C)**: Immunoblot showing the expression levels of FLAG-tagged BANF1 WT and A12T in comparison to BANF1 endogenous levels.

nuclear magnetic resonance (NMR) or crystallographic methods would be needed to gain a more detailed visualization of A12T structural modulations.

Although we cannot exclude that the NGPS phenotypes result from an undiscovered role of BANF1, we suggest that the DNA binding deficiency observed in the BANF1 A12T mutant contributes to the cellular phenotypes observed in NGPS. BANF1 has many proposed roles within the cell that requires it to bind to DNA. BANF1 has been shown to simultaneously bind to the nuclear membrane, LEM domain containing protein, LAP2 and DNA *in vitro*, implicating BANF1 as having a crucial role in tethering the chromatin to the nuclear envelope (12). Supporting this, both LAP2 and another LEM-domain, BANF1-interacting protein, Emerin, interact with other major structural components of the nuclear envelope called Lamins. Lamins play a key role in nuclear structure and assembly and mutations in Lamin genes also lead to a human progeria syndrome, Hutchinson–Gilford Progeria Syndrome (HGPS)(2). In HGPS the disruption to the lamina organization, induced by the mutant Lamin proteins leads to areas of weakness around the nuclear envelope, this can cause the chromatin to herniate, pushing out the destabilized membrane. Interestingly, while we were unable to assess NGPS patient cells endogenously expressing mutant BANF1, ectopic expression of the A12T BANF1 did result in herniation of the nuclear envelope, consistent with what has been reported for the patients ((5) Figure 5). It has been proposed that BANF1, through direct binding to LEM proteins and indirect binding to Lamins, may link chromatin to the inner nuclear envelope and this appears to be the case in *C. elegans* (15). In support of this, herein we have shown that mutation of A12T in BANF1 disrupts the DNA binding of BANF1, leading to the disruption of the nuclear envelope.

Conclusions

Our study now clarifies the role of the BANF1 A12T mutation in NGPS, providing insight into the disease process. Our study has important implications for the treatment of NGPS patients and has provided new mechanistic insights into the function of BANF1 and the nuclear envelope in aging.

Methods

Ethics approval

All experimental procedures are approved by the Institutional Biosafety Sub-Committee of the University of Queensland, Brisbane, Australia.

Plasmids

A pCMV6-AN-3DDK vector containing the BANF1 CDS cloned into the AsiSI and MluI restriction sites was purchased from Origene. The BANF1 CDS was further sub-cloned into pEX-N-His (Origene) using the AsiSI and MluI restriction sites. Enzymes were purchased from New England Biolabs.

A12T mutations were introduced into both BANF1 vectors by site-directed mutagenesis, using the primers A12T F: CCGAGACTTCGTGA**C**AGAGCCCA, and A12T R: CCATGGGCTCTGTCACGAAGTCT. PCR was conducted as per: AccuPrime Pfx polymerase (0.02 U.μl^{-1}; Life Technologies), 1x AccuPrime Pfx polymerase reaction mix (Life Technologies), primers (0.3 μM) and template (0.8 ng.μl^{-1}), then cycling 19x at: 94°C (20 s), 57°C (30 s), 68°C (6 min 30 s). This was followed by Dpn1 (New England Biolabs) digestion (0.8 U.μl^{-1}, 2 h, 37°C) and transformation by heat-shock into chemically competent α-select *E. coli* (Bioline). Successful mutagenesis was confirmed by DNA sequencing (Australian Genome Research Facility) using the primer VP1.5: GGACTTTCCAA AATGTCG. Primers were purchased from Sigma-Aldrich.

BANF1 purification

Plasmids expressing HexaHis-tagged WT or A12T BANF1 were transformed into BL21 (DE3) pLys *E. coli*. Cells were grown at 37°C and protein expression induced with 1 mM IPTG. *E. coli* were harvested 3 h after induction by centrifugation and stored overnight at -80°C. Cell pellet was resuspended in 8 mL of lysis buffer (25 mM HEPES pH 7.5, 150 mM NaCl) per g of cells, and sonicated. Cell lysates were centrifuged for 30 min at 17,00 rpm and the supernatant discarded. The pellet fraction containing HexaHis BANF1 was solubilized in buffer (25 mM HEPES pH 7.5, 150 mM NaCl, 25 mM imidazole) containing 6 M guanidinium chloride, and kept under agitation for 1 h at 4°C. The lysate was then further centrifuged and the clarified supernatant incubated with HIS-Select® *Nickel Affinity* Gel for 2 h at 4°C, under agitation. The affinity gel was extensively washed with the solubilization buffer and the protein was eluted from the beads in buffer K (20 mM KH$_2$PO$_4$, pH 7.4, 0.5 mM EDTA, 10% glycerol, 0.01% IGEPAL) complemented with 300 mM KCl and 250 mM Imidazole. Eluents were supplemented with 100 mM DTT and incubated for 2 h at 40°C to reduce any remaining disulfide bonds.

Protein was then concentrated on a 10 kDa cut-off *Microsep*™ centrifugal device (Pall corporation) to a volume of 250 μL and loaded on a *Superose 6 10/300* GL size exclusion chromatography column (GE healthcare) run with K buffer containing 300 mM KCl. High molecular weight fractions containing BANF1 were discarded and fractions containing monomeric BANF1 at near homogeneity pooled, concentrated and stored at -80°C.

Protein model

Amino acids sequences from BANF1 WT and A12T were used to generate three-dimensional model with Phyre2 (http://www.sbg.bio.ic.ac.uk/~phyre2). Modeling was subsequently confirmed using I-TASSER (http://zhanglab.ccmb.med.umich.edu/I-TASSER/). Models generated were visualized and analyzed using Swiss PDB viewer [27].

Cell lines

U2OS and HeLa cells were cultured in RPMI 1640 medium (Sigma-Aldrich) containing 10% FCS and maintained in a humidified incubator at 37°C/5% CO_2.

Transfection of FLAG-tagged Banf1 constructs

U2OS and HeLa cells were transfected with pCMV6-AN-3DDK, WT or A12T 3x FLAG-tagged BANF1 constructs using Lipofectamine™ 2000 (Invitrogen) as described by the manufacturer. Expression of the 3x FLAG-tagged BANF1 constructs was determined 24 h post-transfection by immunoblotting with an anti-FLAG antibody. Cellular fractionation and immunofluorescence was also carried out 24 hours post transfection.

Antibodies

Primary antibodies used are as follows: ant-BANF1 (Abcam: ab88464, monoclonal, Sigma: SAB1404629, monoclonal), anti-HexaHis (Abcam ab1187), anti-FLAG M2 (Sigma, F3165), anti-Histone H3 (Cell Signaling, 9715), anti-nucleolin (Cell Signaling, 12247S) and anti-Emerin (Cell Signaling, 5430S). Fluorescent secondary antibodies used are: Donkey anti-Mouse 800 nm (LiCor; IRDye 800CW 926-32212), Donkey anti-Rabbit (LiCor; IRDye 680LT 926-28023) and Alexa Fluor 488 and 594 (Molecular Probes).

Cycloheximide block

HeLa cells were seeded in 6 cm dishes (300, 000 cells per dish) and the following day transfections were carried out using Lipofectamine 2000 as per the manufacturers instructions using 2 μg plasmid DNA per dish. Cycloheximide (Sigma-Aldrich) was added to the dishes 24 h post transfection at a final concentration of 50 μg/ml and cells were incubated for the indicated amount of time. For the t = 0 h time point, cells were harvested immediately after addition of cycloheximide.

Cellular fractionation

Cells were separated into cytoplasmic, membrane bound, soluble nuclear, chromatin and cytoskeletal fractions using the Subcellular Protein Fractionation Kit for cultured cells (Thermo Scientific), according to the manufacturer's instructions. Protein concentrations were estimated using a Bicinchoninic acid assay (Sigma) and subsequently 10 μg of the soluble nuclear and chromatin fractions were separated on a 4-12% SDS-PAGE gel (Invitrogen) and immunoblotted with the indicated antibodies.

Western blot

Proteins were resolved on 4-12% gradient Nu-PAGE gels (Life Technologies) and transferred to nitrocellulose membrane. Membranes were blocked in 2% v/v fish gelatin (Sigma) in PBS-T for 30 min and incubated with the anti BANF1 antibody diluted in 1% fish gelatin in PBS-T overnight at 4°C. Membranes were washed in PBS-T, incubated with secondary antibodies (LiCor) and scanned on an Odyssey infrared imaging system (LiCor). Where necessary, membranes were stripped using a mild stripping buffer (15 g L^{-1} glycine pH 2.2, 1 g L^{-1} SDS, 1% Tween20) and reprobed with the appropriate antibodies.

Synthetic DNA substrates

All oligonucleotides were purchased from Integrated DNA Technology (IDT), forward: 5'FAM (carboxyfluorescein) - CTCTCCCTTCGCTCCTTTCCTCT, reverse: AGAGGAAAGGAGCGAAGGGAGAG.

All nucleotides were purified on 12% polyacrylamide, 7 M urea gels prior to further use. For dsDNA annealing, equimolar amounts of corresponding oligonucleotides were mixed in annealing buffer (50 mM Tris pH 7.5, 100 mM NaCl, 10 mM $MgCl_2$), heated at 95°C for 10 minutes and slowly cooled to room temperature. The substrates were then purified on native 10% polyacrylamide gels.

Concentrations were determined using the OD_{260} and the molar extinction coefficient of the oligonucleotides (with $\varepsilon_{fluoresceine}$ = 13,700 L/mol.cm).

Electrophoretic mobility shift assay

Reaction were carried out in 10 μL of buffer (10 mM Tris HCl pH 7.0, 20 mM NaCl, 100 ng/mL BSA, 5 mM DTT) with 10 nM of FAM labeled DNA duplex with various concentration of WT or A12T HexaHis BANF1. Proteins and DNA were incubated for 30 min at 4°C. Reactions were resolved on 7% polyacrylamide gels in 0.5x TBE buffer run at 4°C for 90 min at 90 V.

Gels were scanned using a Starion FLA-9000 image scanner (Fujifilm) and quantified using MultiGauge software (Fujifilm).

Long substrate assays were performed in 10 μL of buffer with 150 ng of empty pEX-N-His (Origene) with various concentration of WT or A12T HexaHis BANF1. Proteins and DNA were incubated for 30 min at 4°C. Reactions were resolved on 0.6% agarose gels in 1x TBE running buffer and post stained using Ethidium bromide. Images were taken using Bio-Rad's *Gel Doc* system.

Immunofluorescence

U2OS cells were seeded the day prior to transfection with the 3x FLAG-tagged BANF1 constructs. Following transfection, the cells were grown for 24 h and were subsequently treated for 5 min on ice with extraction buffer

(20 mM Hepes, 20 mM NaCl, 5 mM MgCl$_2$, 1 mM ATP, 0.1 mM sodium orthovanadate, 1 mM sodium fluoride, protease inhibitor cocktail (Roche), 0.5% IGEPAL (same chemically as obsolete Nonidet P-40), pH 7.5), to remove the soluble proteins. The cells were then fixed with 4% paraformaldehyde and subsequently permeabilized with 0.2% Triton-X for 5 min and then blocked with 3% BSA for 1 h. Cells were then incubated with the indicated primary antibodies for 1 h hour at RT, washed and counterstained with the corresponding Alexa Fluor conjugated secondary antibodies for 1 h at RT. DNA was counterstained with DAP1 (Sigma, D9564). Images were captured using a DeltaVision deconvolution microscope and the figures were assembled using Adobe Photoshop CS6.

Competing interests
The authors declare that they have no competing interests.

Authors' contributions
NP was involved in the design, coordination and implementation of the study, drafted the manuscript and carried out the biochemical studies. JKB also carried out the biochemical studies. NWA performed the site-directed mutagenesis and assessed 3x FLAG A12T BANF1 antigenicity and in-cell protein associations. AS carried out the immunofluorescence studies. LVC carried out the cycloheximide studies. AJU carried out the cloning and western blot analysis of the Banf1 expression constructs. SZ was involved in identifying the Banf1 protein for use in this study. DJR, KO and EB conceived the study, participated in its design and coordination and helped to draft the manuscript. All authors read and approved the final manuscript.

Acknowledgments
This work was supported by an ARC project grant (DP 120103099) and by a Queensland Health Senior Clinical Research Fellowship. D.J.R was funded by an ARC future fellowship. N.W.A was supported by the Marylyn Mayo scholarship awarded by Cancer Council Queensland.

Author details
[1]School of Biomedical Science, Institute of Health and Biomedical Innovation at the Translational Research Institute, Queensland University of Technology, Brisbane, QLD, Australia. [2]Center for Cancer Research and Cell Biology, School of Medicine, Dentistry and Biomedical Sciences, Queen's University Belfast, Lisburn Road 97, Belfast, UK.

References
1. Ramirez CL, Cadinanos J, Varela I, Freije JM, Lopez-Otin C: **Human progeroid syndromes, aging and cancer: new genetic and epigenetic insights into old questions.** *Cell Mol Life Sci* 2007, **64**:155–170.
2. Kudlow BA, Kennedy BK, Monnat RJ Jr: **Werner and Hutchinson-Gilford progeria syndromes: mechanistic basis of human progeroid diseases.** *Nat Rev Mol Cell Biol* 2007, **8**:394–404.
3. Hutchison CJ: **The role of DNA damage in laminopathy progeroid syndromes.** *Biochem Soc Trans* 2011, **39**:1715–1718.
4. Cabanillas R, Cadinanos J, Villameytide JA, Perez M, Longo J, Richard JM, Alvarez R, Duran NS, Illan R, Gonzalez DJ, Lopez-Otin C: **Nestor-Guillermo progeria syndrome: a novel premature aging condition with early onset and chronic development caused by BANF1 mutations.** *Am J Med Genet A* 2011, **155A**:2617–2625.
5. Puente XS, Quesada V, Osorio FG, Cabanillas R, Cadinanos J, Fraile JM, Ordonez GR, Puente DA, Gutierrez-Fernandez A, Fanjul-Fernandez M, Levy N, Freije JM, Lopez-Otin C: **Exome sequencing and functional analysis identifies BANF1 mutation as the cause of a hereditary progeroid syndrome.** *Am J Hum Genet* 2011, **88**:650–656.
6. Lee MS, Craigie R: **A previously unidentified host protein protects retroviral DNA from autointegration.** *Proc Natl Acad Sci U S A* 1998, **95**:1528–1533.
7. Bradley CM, Ronning DR, Ghirlando R, Craigie R, Dyda F: **Structural basis for DNA bridging by barrier-to-autointegration factor.** *Nat Struct Mol Biol* 2005, **12**:935–936.
8. Cai M, Huang Y, Suh JY, Louis JM, Ghirlando R, Craigie R, Clore GM: **Solution NMR structure of the barrier-to-autointegration factor-Emerin complex.** *J Biol Chem* 2007, **282**:14525–14535.
9. Mansharamani M, Wilson KL: **Direct binding of nuclear membrane protein MAN1 to emerin in vitro and two modes of binding to barrier-to-autointegration factor.** *J Biol Chem* 2005, **280**:13863–13870.
10. Lee KK, Haraguchi T, Lee RS, Koujin T, Hiraoka Y, Wilson KL: **Distinct functionnal domains in emerin bind lamin A and DNA-bridging protine BAF.** *J Cell Sci* 2001, **114**:4567–4573.
11. Cai M, Huang Y, Ghirlando R, Wilson KL, Craigie R, Clore GM: **Solution structure of the constant region of nuclear envelope protein LAP2 reveals two LEM-domain structures: one binds BAF and the other binds DNA.** *EMBO J* 2001, **20**:4399–4407.
12. Shumaker DK, Lee KK, Tanhehco YC, Craigie R, Wilson KL: **LAP2 binds to BAF-DNA complexes: requirement for the LEM domain and modulation by variable regions.** *EMBO J* 2001, **20**:1754–1764.
13. Skoko D, Li M, Huang Y, Mizuuchi M, Cai M, Bradley CM, Pease PJ, Xiao B, Marko JF, Craigie R, Mizuuchi K: **Barrier-to-autointegration factor (BAF) condenses DNA by looping.** *Proc Natl Acad Sci U S A* 2009, **106**:16610–16615.
14. Montes De Oca R, Lee KK, Wilson KL: **Binding of barrier to autointegration factor (BAF) to histone H3 and selected linker histones including H1.1.** *J Biol Chem* 2005, **280**:42252–42262.
15. Margalit A, Segura-Totten M, Gruenbaum Y, Wilson KL: **Barrier-to-autointegration factor is required to segregate and enclose chromosomes within the nuclear envelope and assemble the nuclear lamina.** *Proc Natl Acad Sci U S A* 2005, **102**:3290–3295.
16. Asencio C, Davidson IF, Santarella-Mellwig R, Ly-Hartig TB, Mall M, Wallenfang MR, Mattaj IW, Gorjanacz M: **Coordination of kinase and phosphatase activities by Lem4 enables nuclear envelope reassembly during mitosis.** *Cell* 2012, **150**:122–135.
17. Wang X, Xu S, Rivolta C, Li LY, Peng GH, Swain PK, Sung CH, Swaroop A, Berson EL, Dryja TP, Chen S: **Barrier to autointegration factor interacts with the cone-rod homeobox and represses its transactivation function.** *J Biol Chem* 2002, **277**:43288–43300.
18. Lin CW, Engelman A: **The Barrier-to-Autointegration Factor Is a Component of Functional Human Immunodeficiency Virus Type 1 Preintegration Complexes.** *J Virol* 2003, **77**:5030–5036.
19. Harris D, Engelman A: **Both the structure and DNA binding function of the barrier-to-autointegration factor contribute to reconstitution of HIV type 1 integration in vitro.** *J Biol Chem* 2000, **275**:39671–39677.
20. Furukawa K, Sugiyama S, Osouda S, Goto H, Inagaki M, Horigome T, Omata S, McConnell M, Fisher PA, Nishida Y: **Barrier-to-autointegration factor plays crucial roles in cell cycle progression and nuclear organization in Drosophila.** *J Cell Sci* 2003, **116**:3811–3823.
21. Kelley LA, Sternberg MJ: **Protein structure prediction on the Web: a case study using the Phyre server.** *Nat Protoc* 2009, **4**:363–371.
22. Zhang Y: **I-TASSER server for protein 3D structure prediction.** *BMC Bioinformatics* 2008, **9**:40.
23. Umland TC, Wei SQ, Craigie R, Davies DR: **Structural basis of DNA bridging by Barrier-to-Autointegration Factor.** *Biochemistry* 2000, **39**:9130–9138.
24. Pan Y, Haines DS: **The pathway regulating MDM2 protein degradation can be altered in human leukemic cells.** *Cancer Res* 1999, **59**:2064–2067.
25. Zheng R, Ghirlando R, Lee MS, Mizuuchi K, Krause M, Craigie R: **Barrier-to-autointegration factor (BAF) bridges DNA in a discrete, higher-order nucleoprotein complex.** *Proc Natl Acad Sci U S A* 2000, **97**:8997–9002.
26. Segura-Totten M, Kowalski AK, Craigie R, Wilson KL: **Barrier-to-autointegration factor: major roles in chromatin decondensation and nuclear assembly.** *J Cell Biol* 2002, **158**:475–485.
27. Guex N, Peitsch MC: **SWISS-MODEL and the Swiss-PdbViewer: an environment for comparative protein modeling.** *Electrophoresis* 1997, **18**:2714–2723.

Identification of nucleotides and amino acids that mediate the interaction between ribosomal protein L30 and the SECIS element

Abby L Bifano[1], Tarik Atassi[1], Tracey Ferrara[1,2] and Donna M Driscoll[1,3*]

Abstract

Background: Ribosomal protein L30 belongs to the L7Ae family of RNA-binding proteins, which recognize diverse targets. L30 binds to kink-turn motifs in the 28S ribosomal RNA, L30 pre-mRNA, and mature L30 mRNA. L30 has a noncanonical function as a component of the UGA recoding machinery that incorporates selenocysteine (Sec) into selenoproteins during translation. L30 binds to a putative kink-turn motif in the Sec Insertion Sequence (SECIS) element in the 3' UTR of mammalian selenoprotein mRNAs. The SECIS also interacts with SECIS-binding protein 2 (SBP2), an essential factor for Sec incorporation. Previous studies showed that L30 and SBP2 compete for binding to the SECIS in vitro. The SBP2:SECIS interaction has been characterized but much less is known about how L30 recognizes the SECIS.

Results: Here we use enzymatic RNA footprinting to define the L30 binding site on the SECIS. Like SBP2, L30 protects nucleotides in the 5' side of the internal loop, the 5' side of the lower helix, and the SECIS core, including the GA tandem base pairs that are predicted to form a kink-turn. However, L30 has additional determinants for binding as it also protects nucleotides in the 3' side of the internal loop, which are not protected by SBP2. In support of the competitive binding model, we found that purified L30 repressed UGA recoding in an in vitro translation system, and that this inhibition was rescued by SBP2. To define the amino acid requirements for SECIS-binding, site-specific mutations in L30 were generated based on published structural studies of this protein in a complex with its canonical target, the L30 pre-mRNA. We identified point mutations that selectively inhibited binding of L30 to the SECIS, to the L30 pre-mRNA, or both RNAs, suggesting that there are subtle differences in how L30 interacts with the two targets.

Conclusions: This study establishes that L30 and SBP2 bind to overlapping but non-identical sites on the SECIS. The amino acid requirements for the interaction of L30 with the SECIS differ from those that mediate binding to the L30 pre-mRNA. Our results provide insight into how L7Ae family members recognize their cognate RNAs.

Keywords: RNA-binding protein, L30, SECIS-binding protein 2, Selenocysteine, Selenoprotein

Background

Eukaryotic ribosomal protein L30 is a component of the large ribosomal subunit. L30 has no prokaryotic ortholog but the gene is essential in yeast [1]. Cryo-electron microscopy studies of the wheat germ and canine 80S ribosomes revealed that L30 is located in a eukaryotic-specific bridge between the large and small subunits [2,3]. The interaction of L30 with the 60S ribosome is mediated primarily through binding of the protein to a kink-turn motif in helix 58 of the large rRNA [2,4]. L30 also binds to a kink-turn in the 5' untranslated region (UTR) of its cognate pre-mRNA and the mature spliced mRNA to auto-regulate its own expression at the level of pre-mRNA splicing or mRNA translation, respectively [5-8].

The repertoire of L30 functions was expanded by the discovery that the protein is involved in the mechanism that recodes the UGA stop codon as selenocysteine (Sec) during selenoprotein synthesis [9]. In humans,

* Correspondence: driscod@ccf.org
[1]Department of Cellular & Molecular Medicine, Lerner Research Institute, Cleveland Clinic, 9500 Euclid Avenue NC10, Cleveland, OH 44195, USA
[3]Department of Molecular Medicine, Cleveland Clinic Lerner College of Medicine of Case Western Reserve University, Cleveland, OH, USA
Full list of author information is available at the end of the article

there are 25 known selenoprotein genes, whose products play critical roles in anti-oxidant defense, thyroid hormone metabolism, immunity, and development [10,11]. Sec incorporation at the UGA/Sec codon is dependent on a stem-loop structure, the Selenocysteine Insertion Sequence (SECIS) element, which is found in the 3' UTR of eukaryotic selenoprotein mRNAs. This structure consists of two helices separated by an internal loop with an apical loop or bulge at the top (see Figure 1A). The core of the SECIS contains a quartet of non-Watson-Crick base pairs, including two sheared G•A tandem base pairs which are characteristic of kink-turn motifs. Based on structure probing and computer modeling, Walczak et al. proposed a three-dimensional structure of the SECIS in which the RNA is kinked at the internal loop, exposing the sheared G•A tandem base pairs in the SECIS core to the solvent [12]. The SECIS core, which is essential for Sec incorporation, is required for binding of two proteins, L30 and SECIS-binding protein 2 (SBP2) [9,13-16]. *In vitro* studies support a model in which the two proteins bind to the SECIS element independently, and likely sequentially [9]. SBP2 has been shown to recruit the Sec-tRNA[Sec]:EFSec complex, bind to the ribosome, and induce a conformational change in the A site [17-19]. However, the exact functions of SBP2 and L30 in UGA recoding have not been fully defined, and alternative models have been proposed [9,17,18].

A prerequisite for elucidating the mechanism of Sec incorporation is a detailed understanding of the molecular basis for protein:SECIS interactions. L30 and SBP2 are both members of the L7Ae family of RNA-binding proteins. In addition to the founding member archaeal ribosomal protein L7Ae, this family includes other eukaryotic ribosomal proteins as well as proteins involved in RNA processing, ribonucleoprotein assembly, and termination of protein synthesis [20,21]. The L7Ae family members share a similar RNA-binding domain and characteristically bind to kink-turn motifs in their cognate RNA. However, the kink-turn does not represent a single structural motif [22]. Each protein in the L7Ae family has a unique RNA-binding specificity, which allows it to distinguish its cognate RNA from other kink-turn containing transcripts in the cell. Indirect evidence from our lab suggests that the SECIS core is part of a noncanonical kink-turn, which may explain why the SECIS is bound by L30 and SBP2, but not by other proteins in the L7Ae family [9].

We, and others have characterized the SBP2:SECIS interaction at the molecular level, although no structural studies on the complex have been performed to date. Based on RNA footprinting experiments, SBP2 binds to both sides of the SECIS core, as well as to the 5' strand of the internal loop and lower helix [14]. The upper helix and a large internal loop in the SECIS may also be

important for recognition by SBP2 [23]. Mutational analysis of SBP2 revealed that the L7Ae motif is necessary but not sufficient to mediate SECIS-binding and that additional amino acids in a K-rich region N-terminal to this motif are required [24,25]. The SBP2:SECIS interaction is critical for human health as mutations in either the SBP2 binding site or in the SBP2 RNA-binding domain result in a reduction in selenoprotein synthesis and a variety of phenotypes [26-28].

In contrast to SBP2, much less is known about how L30 recognizes the SECIS. We previously showed that L30 binds to SECIS elements from multiple selenoprotein mRNAs [9], but the actual L30 binding site has not been defined. Compared to SBP2, L30 is a relatively small protein (11 kDa) and it lacks a K-rich region. Structural studies of L30 in a complex with its cognate target, the stem-loop from the L30 pre-mRNA, have defined the RNA-binding interface of the protein [29,30]. However, it is not known whether the same amino acids mediate binding to the SECIS. In this study, we used RNA footprinting and site-directed mutagenesis to identify nucleotides and amino acids that are important for the L30:SECIS interaction.

Results

Defining the L30 binding site by RNA footprinting

Eukaryotic SECIS elements are stem-loop structures containing two highly conserved motifs that are essential for Sec incorporation, namely the SECIS core and the AAR motif in an apical bulge or loop. The SECIS element from the rat Phospholipid Hydroperoxide Glutathione Peroxidase (PHGPx) mRNA is shown in Figure 1A. We previously showed that binding of L30 to the PHGPx SECIS was abrogated by mutations in the sheared G•A tandem base pairs in the SECIS core, but not by deletion of the AAR motif [9]. In order to define the L30 binding site on the SECIS, we used enzymatic RNA footprinting. The ^{32}P-labeled PHGPx SECIS RNA was incubated in the presence or absence of purified rat L30. The native RNA and RNA:protein complexes were then partially digested with different ribonucleases and analyzed by electrophoresis. Regions of cleavage and protection were identified by comparing samples with RNA sequencing reactions (G and C + U) and alkali ladders. A schematic illustrating the results is shown in Figure 1A and a gel representative of 3 independent experiments is shown in Figure 1B.

The cleavage results with the native RNA are consistent with the published structure of the PHGPx SECIS, which was determined by enzymatic and chemical probing [31]. RNase T1, which cleaves after single-stranded G bases, cleaved the native PHGPx SECIS RNA in the apical region, both strands of the SECIS core, and the 3' side of the internal loop (Figure 1B). Similar results were

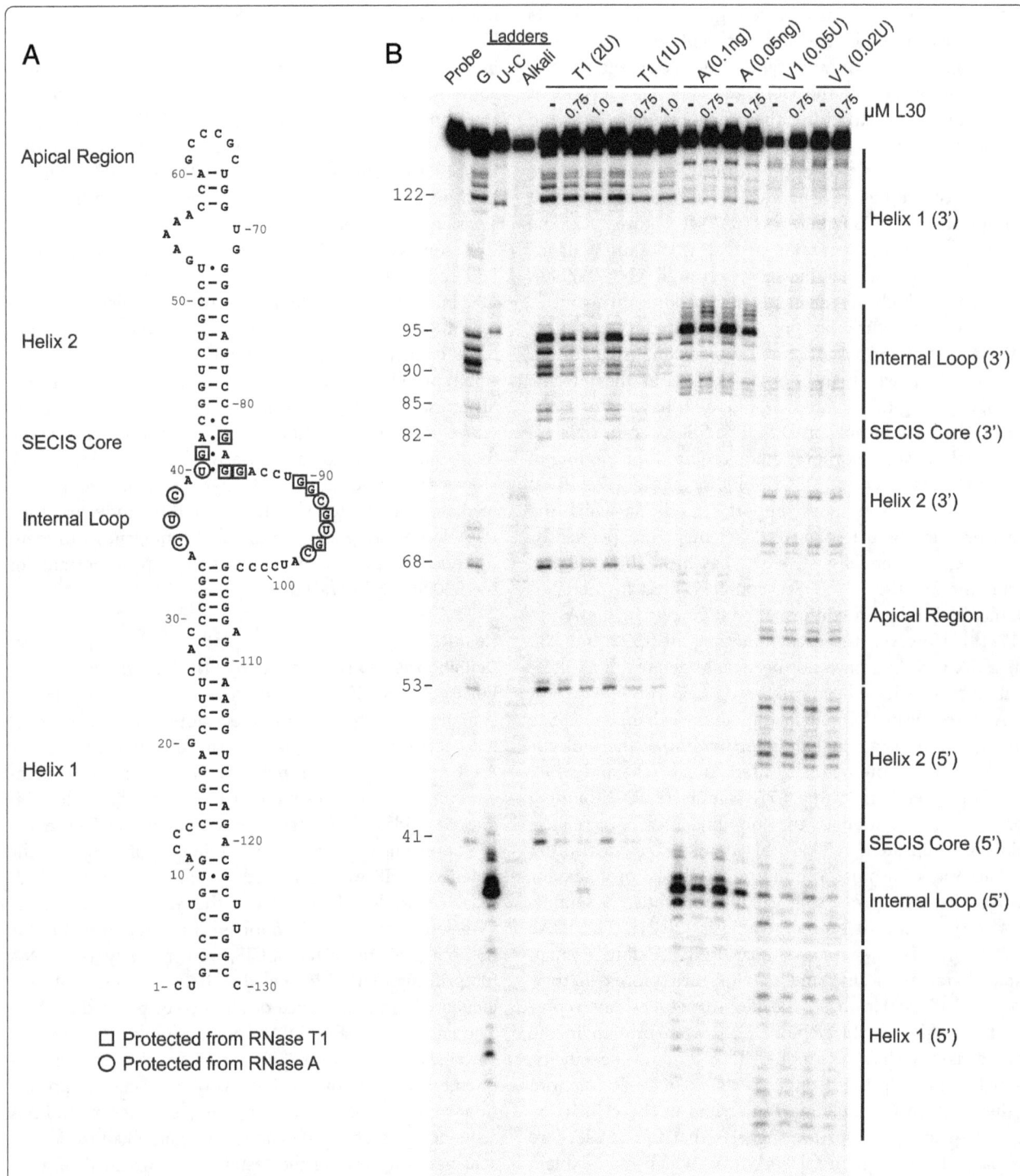

Figure 1 RNA footprinting of the L30:SECIS complex. A. The structure of the SECIS element from the rat PHGPx mRNA is shown. Boxes and circles indicate nucleotides that are protected from cleavage by RNase T1 and RNase A, respectively. **B**. The 5' end-labeled PHGPx SECIS was incubated in the absence or presence of L30 (0.75 or 1.5 µM). The reactions were then partially digested with RNase T1, A, or V1 as indicated. The products were analyzed by denaturing gel electrophoresis. The sequencing (G and C + U) and alkali ladders are shown in the left lanes. The numbers to the left of the gel indicate the positions of G nucleotides using the numbering in (A). The bars on the right indicate the different regions of the SECIS element. The gel is a representative example from 3 independent experiments.

obtained when the SECIS was partially digested with RNase A, which cleaves at single-stranded C and U bases. Cleavage by RNase A was detected in the 5′ side of the SECIS core, the apical loop, as well as both sides of the internal loop. Both RNase A and T1 cleaved at the base of helix 1 (nucleotides 121-128) suggesting that this region may breathe due to imperfect base pairing. RNase V1, which cleaves in double-stranded regions, cleaved at multiple positions in helix 1 and helix 2. We also observed faint V1 digestion in the large internal loop, which suggests that this region may occasionally form an alternate structure.

When the SECIS was incubated with 0.75 or 1.5 μM L30, there was a marked reduction in cleavage by RNase T1 on both sides of the SECIS core (G41, G82, G84) and along the 3′ side of the internal loop 2 (G85, G90, G91, G93, G95). We were unable to achieve full protection by increasing the amount of L30 in the binding reaction, most likely due to the high on/off rate of the L30:SECIS complex as determined by Surface Plasmon Resonance (data not shown). G53 and G68 in the apical region and G122, G125, G127, and G128 at the base of helix 1 were not reproducibly protected by L30 from RNase T1 cleavage.

When the L30:SECIS complexes were partially digested with RNase A, there was a reduction in cleavage in the 5′ side of the internal loop and SECIS core (bases 36-38 and 40) and the 3′ side of the internal loop (bases 92, 94, 96). Other bases in the internal loop (C87, C88, U89) and the base of helix 1 were still cleaved by RNase A in the presence of L30. Similarly, the binding of L30 did not protect nucleotides in helix 1 or helix 2 from RNase V1 cleavage. Taken together, our results show that binding of L30 protects nucleotides in the SECIS core and in the 5′ and 3′ sides of the internal loop. We

previously showed that SBP2 interacts with both sides of the SECIS core, the 5′ strand of the internal loop, and the 5′ strand of helix 1, but not with the 3′ side of the internal loop [14]. Thus the two proteins bind to similar but not identical regions of the SECIS, as illustrated in Figure 2A.

L30 represses UGA recoding in vitro and this inhibition is rescued by SBP2

The fact that the SBP2 and L30 binding sites overlap is consistent with our earlier finding that purified recombinant L30 and SBP2 compete for binding in vitro to an isolated SECIS in the absence of other factors [9]. To test whether this competition could occur during translation, we used a luciferase reporter construct and a modified rabbit reticulocyte lysate (RRL) system. This assay has been previously validated to be specific for UGA recoding [32]. The Luc/UGA/PHGPx reporter RNA contains a UGA/Sec codon at position 258 in the open reading frame and the PHGPx SECIS in the 3′ UTR. Since RRL contains very little SBP2, we added recombinant SBP2-CT, which represents the C-terminal half of the protein and encodes all known functions of SBP2. Translation assays were supplemented with purified 70 nM SBP2-CT to be within the linear range of the assay. Reactions were performed in the absence or presence of increasing amounts of L30, and the products were analyzed for luciferase activity.

As shown in Figure 3A (top panel), the addition of exogenous L30 inhibited recoding of Luc/UGA/PHGPx in a dose-dependent manner. This effect is codon-specific and SECIS-dependent as the inhibition was not observed when the reactions were primed with a Luc/UGU/PHGPx reporter construct, which contains a UGU/Cys

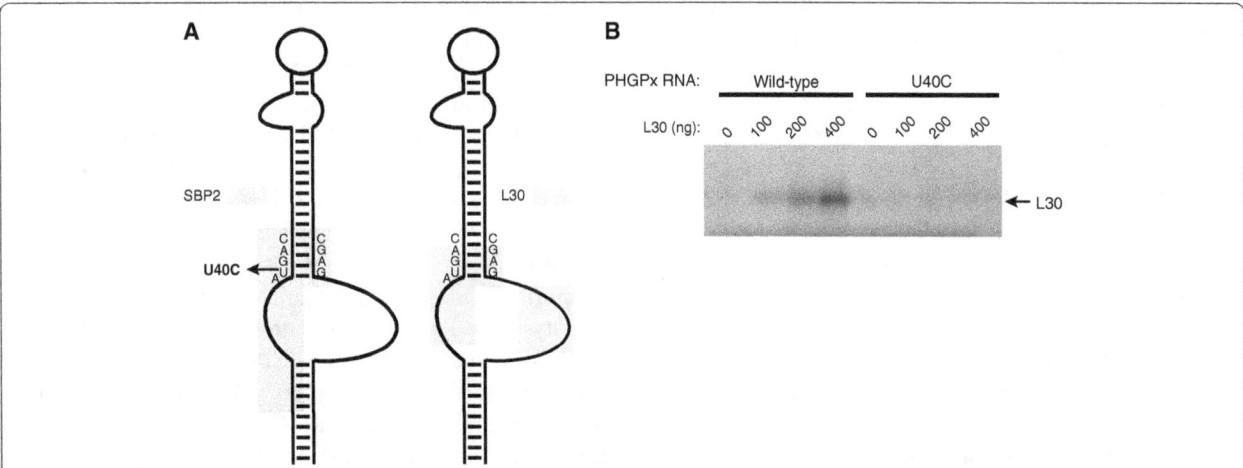

Figure 2 The U40C mutation abrogates L30 binding. A. Schematic illustrating the binding sites of SBP2 [14] and L30 (this work) on the SECIS, as determined by RNA footprinting. The position of the U40C point mutation is indicated. **B.** UV cross-linking experiments were performed using the ^{32}P-labeled wild-type PHGPx SECIS or the U40C mutant RNA, which were incubated with increasing amounts of purified L30 as indicated. After RNase digestion, the products were analyzed by SDS-PAGE and autoradiography.

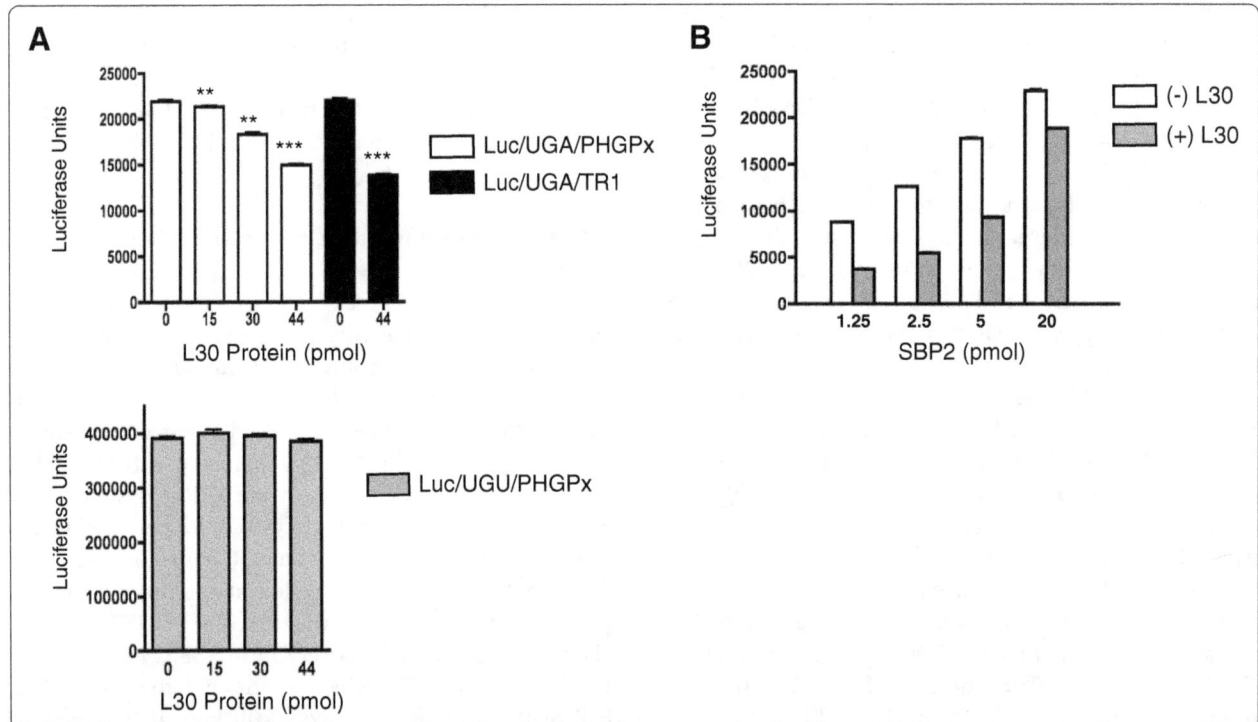

Figure 3 Repression of UGA recoding by L30 is rescued by SBP2. A. In vitro UGA recoding assays were performed using a luciferase reporter RNA that contains UGA (top panel) or UGU (bottom panel) at position 258 of the coding region, fused to either the PHGPx or TR1 SECIS element as indicated. Translation assays were performed in the presence of increasing amounts of L30 as indicated. The products were analyzed for luciferase activity using a luminometer, and the results are expressed as means ± SEM. Statistical significance is indicated by ** ($p < 0.01$) and *** ($p < 0.001$). **B**. UGA recoding assays with the luc/UGA/TR1 reporter construct were performed with a constant amount of L30 (44 pmol/reaction) and increasing amounts of SBP2 as indicated. The products were analyzed as described in (A).

codon (Figure 3A, bottom panel) or a normal luciferase RNA that lacks a SECIS element (data not shown). The addition of L30 also repressed UGA recoding directed by the SECIS element from Thioredoxin Reductase 1 (TR1) (Figure 3A, top panel). We hypothesized that the exogenous L30 protein interacted with the SECIS and prevented binding of SBP2, which is limiting in our translation system. To test this competitive model, we added increasing amounts of SBP2 to the translation reaction while keeping the amount of L30 constant. As shown in Figure 3B, the inhibitory effect of L30 on recoding from the TR1 SECIS was rescued by the addition of SBP2. Similar results were obtained using the Luc/UGA/PHGPx construct (data not shown). Taken together, these results demonstrate that SBP2 and L30 can functionally compete in an in vitro translation system.

Mutations in L30 inhibit SECIS-binding

The L7Ae family members are functionally diverse and recognize a variety of targets. A number of these RNA: protein complexes have been analyzed at the structural level [29,30,33-36]. An emerging theme from these studies is that the L7Ae family members bind to their different cognate RNAs in a similar manner. Therefore, we designed site-directed mutations based on the solution

structures of yeast L30 in complex with the stem-loop from the L30 pre-mRNA target, which were solved by NMR spectroscopy and x-ray crystallography [29,30]. An induced fit model was proposed in which L30 folds into an α/β/α sandwich, with the three loops at the end of the sandwich in direct contact with the RNA [29,30]. We generated alanine point mutations in these functionally important regions, including L29 in the first loop of the α/β/α sandwich, L35 and K36 in the α2 region, and K87, Y89, and V91 in the α4–β4 loop (Figure 4A). We also mutated M108 and E110 in the C-terminal region, as these residues are present in L30 sequences from higher eukaryotes and SBP2, but not in S12, an L7Ae family member that does not bind the SECIS [9].

The wild-type and mutant L30 proteins were analyzed for purity by SDS-PAGE (Figure 4B) and for RNA-binding activity using RNA Electrophoretic Mobility Shift Assays (REMSA). The rat PHGPx SECIS element and the stem-loop structure from the yeast L30 pre-mRNA (subsequently referred to as L30 RNA) were used as ^{32}P-labeled probes. As shown in Figure 5A, the wild-type protein bound to the SECIS in a dose-dependent manner with an apparent K_D of ~ 0.49 µM, which is comparable to what was previously reported [9]. The affinity of L30 for the L30 RNA was several-fold higher with a K_D of 0.17 µM

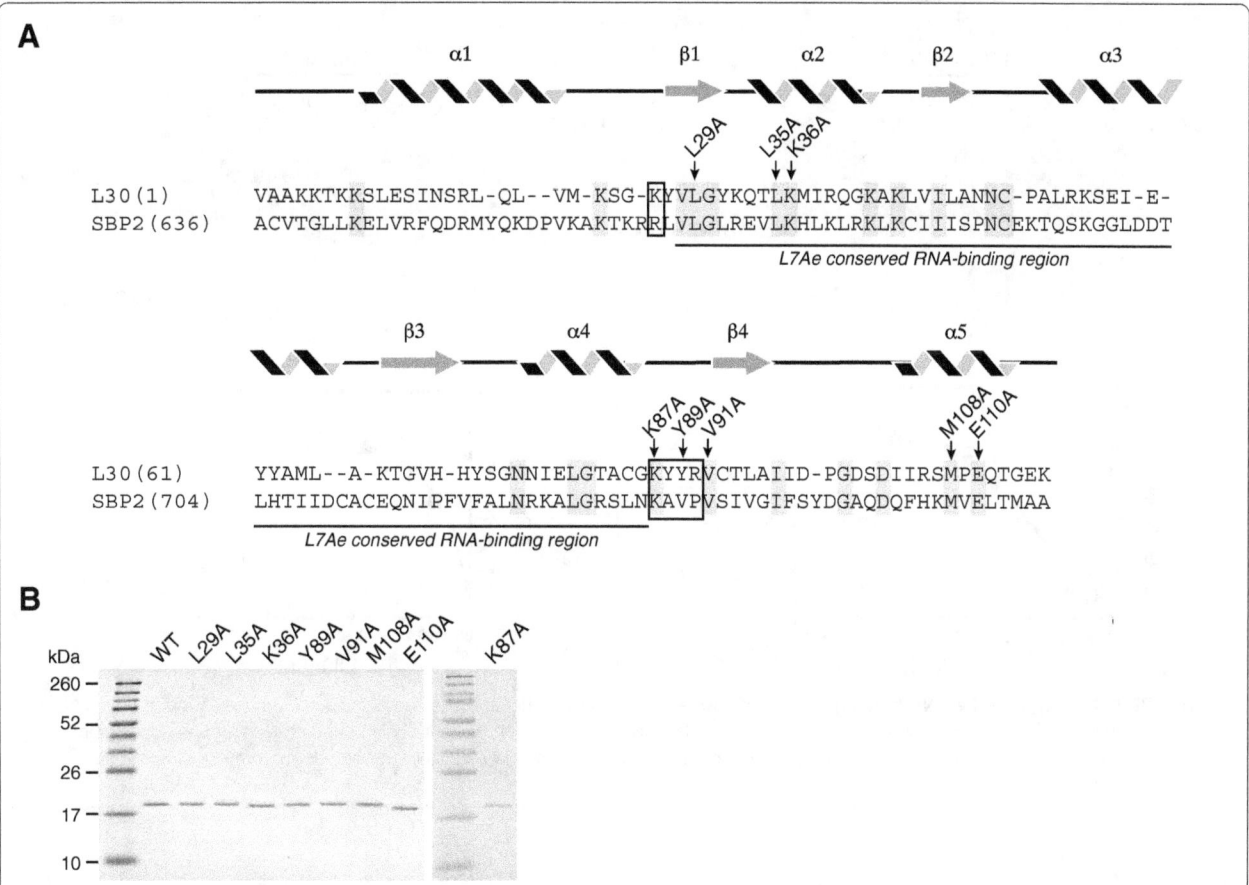

Figure 4 Mutational analysis of L30. A. A schematic illustrating the primary sequences and predicted secondary structures of L30 and SBP2. The position numbers refer to the rat protein sequences. The L7Ae conserved RNA-binding domain is underlined and the conserved signature amino acid motifs for L30 and SBP2 are boxed as described in [37]. Arrows indicate the amino acids in L30 that were mutated to alanine. **B**. The wild-type and mutant L30 protein were expressed in bacteria, purified and analyzed by SDS-PAGE and Coomassie Blue staining. The molecular weight markers are shown in the left lane of each gel.

(Figure 5B). We performed a survey of the mutant proteins, using a protein concentration at which 50% of the probe was bound by the wild-type L30. Representative REMSAs are shown in Figure 6A, and the graphs of the results from 5 and 3 independent experiments, respectively, are presented in Figure 6B.

We found that the amino acid requirements for L30 binding to the SECIS and the L30 RNA are similar but not identical. The Y89A mutant protein was the most defective, with 32% and 37% binding to the SECIS and L30 RNA, respectively, compared to the wild-type protein. We also identified two mutations that selectively inhibited binding of L30 to one target RNA but not the other. As shown in Figure 6B, the K87 mutation in rat L30 reduced binding to the SECIS by 40% but had little effect on the L30:L30 RNA interaction. In contrast the K36A mutant protein was more impaired in its ability to interact with the L30 RNA (reduced by 60%) than the SECIS (reduced by 22%). There was a slight (14%) reduction in binding of the L35A mutant to both the SECIS and L30 RNA, but

only the latter result was statistically significant (Figure 6B). The other mutants, L29A, V91A, M108A, and E110A, were comparable to the wild-type protein with respect to their ability to bind to both targets.

We also analyzed several mutant L30 proteins for their ability to inhibit UGA recoding in our in vitro translation system. As shown in Figure 7, the L29A and M108A mutant proteins, which have wild-type levels of SECIS-binding activity, reduced UGA recoding by ~50%, similar to wild-type L30. However, the K36A and Y89A mutants, that are defective in SECIS-binding, only reduced UGA recoding by 9% and 4% respectively Thus, there was a correlation between the ability of L30 to bind to the SECIS and repress recoding in vitro.

A naturally occurring SECIS mutation inhibits L30 binding
One interesting finding from the footprinting experiments was that L30 protected U40 in the SECIS core. A previous study identified a naturally occurring U to C point mutation at this position in the Selenoprotein N

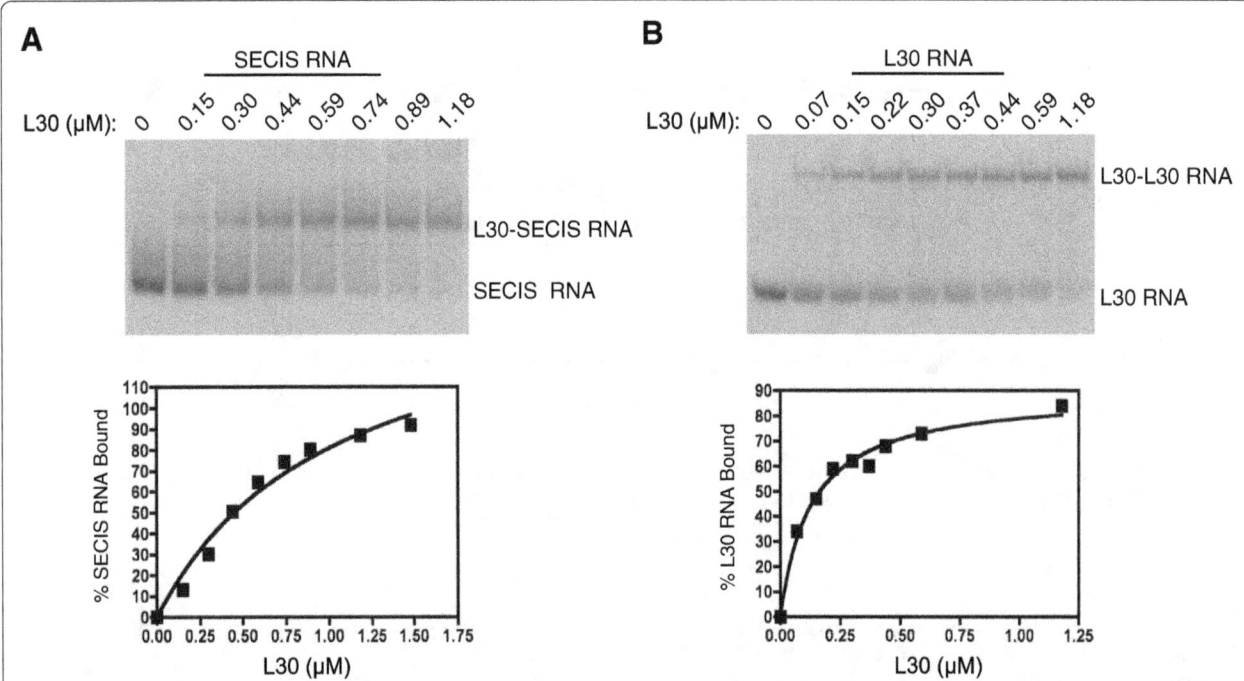

Figure 5 REMSA analysis of the RNA-binding activity of L30. A. The ^{32}P-labeled PHGPx SECIS was incubated with increasing amounts of L30 as indicated. The complexes were analyzed by native gel electrophoresis (top panel). The percent SECIS bound at each protein concentration is shown graphically (bottom panel). **B**. REMSA analysis was performed as described in (A) except that the stem-loop from the L30 pre-mRNA (L30 RNA) was used as the probe.

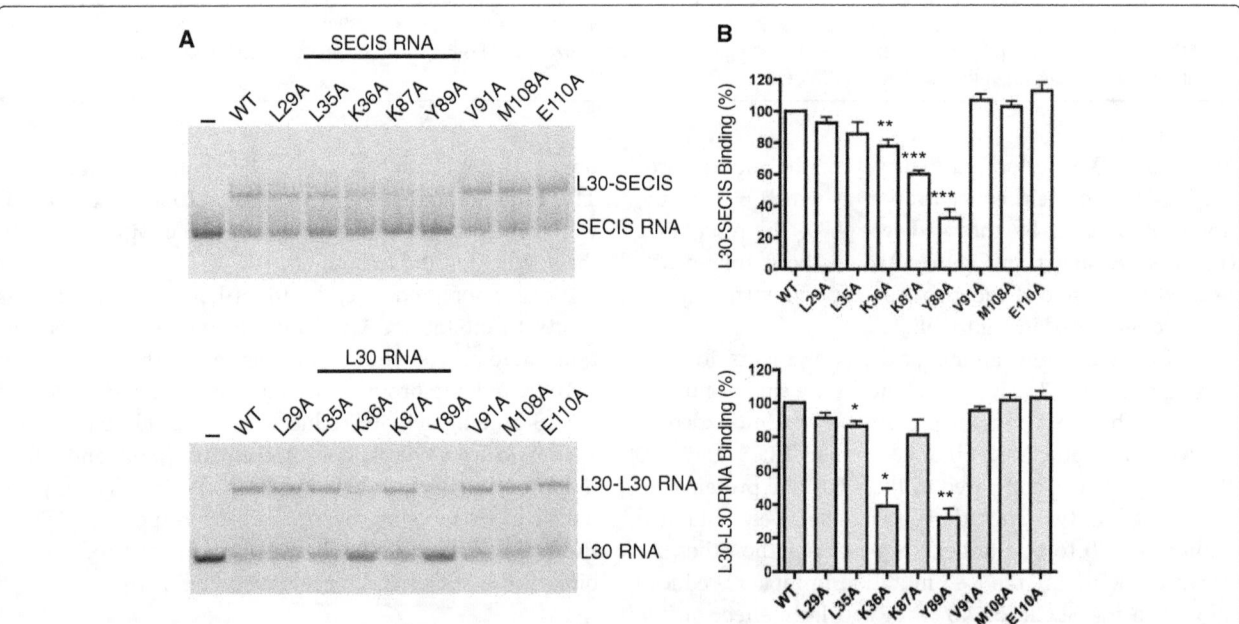

Figure 6 Point mutations in L30 affect RNA-binding activity. A. Representative REMSA analysis of the ^{32}P-labeled PHGPx SECIS (top panel) or L30 RNA (bottom panel). The RNA probes were incubated in the absence or presence of wild-type and mutant L30 proteins as indicated. The positions of the free probes and protein:RNA complexes are indicated. **B**. Graphical representation of REMSA results for L30 binding to the SECIS (top panel) or L30 RNA (bottom panel) from 5 or 3 independent experiments, respectively. The results are expressed relative to the activity of the wild-type protein, which is expressed as 100%. Statistical significance is shown by asterisks, with * ($p < 0.05$), ** ($p < 0.01$) and *** ($p < 0.001$).

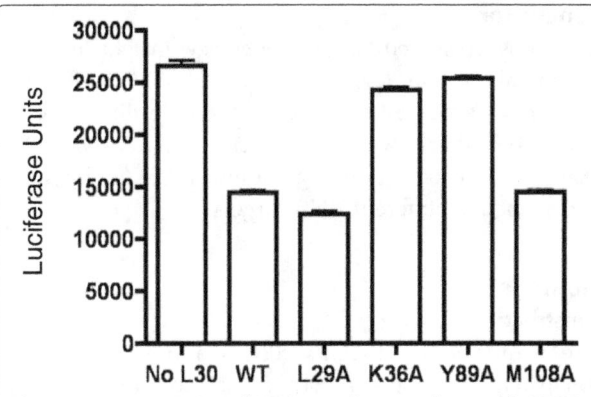

Figure 7 L30 repression of UGA recoding correlates with SECIS-binding. UGA recoding assays were performed using the luc/UGA/TR1 reporter construct as described in the legend to Figure 4A. Reactions contained either no L30 or 44 pmol of wild-type or mutant L30 proteins. Results are expressed as means ± SEM.

(SelN) SECIS element from a patient with a mild form of rigid spine muscular dystrophy [16]. SelN expression was impaired in this patient, and this mutation was shown to abrogate binding of SBP2 to the SelN and PHGPx SECIS elements [13,16]. We hypothesized that the mutant RNA may also be defective in the L30:SECIS interaction, as L30 and SBP2 have overlapping binding sites (Figure 2A). The ^{32}P-labeled wild-type and U40C mutant SECIS RNAs were incubated with increasing amounts of purified L30 protein. The RNA:protein complexes were UV cross-linked, digested with RNase A, and analyzed by SDS-PAGE electrophoresis and autoradiography. As shown in Figure 2B, L30 cross-linked to the wild-type SECIS but not to the U40C mutant RNA. Thus, defects in selenoprotein synthesis in patients with mutant SECIS elements may not be solely due to an impaired SBP2:SECIS interaction.

Discussion

As members of the L7Ae family, L30 and SBP2 share a similar RNA-binding domain. However, there are important differences in how the two proteins interact with the SECIS. Our RNA footprinting experiments revealed that the binding site of L30 centers on the SECIS core and internal loop. L30 protects nucleotides in the SECIS core, including the Gs in the two, sheared G•A tandem base pairs that form the putative kink-turn motif. The protein also protects many of the nucleotides in the internal loop. Thus, the L30 binding site described here overlaps with the regions that have been shown to be protected by SBP2 binding, including the SECIS core and 5' side of the internal loop [14]. In addition to sharing these common nucleotide requirements, L30 and SBP2 each have unique determinants for binding to the SECIS. We show here that L30 protects the 3' side of the internal loop, which is distinct from the known SBP2

binding site [14]. Likewise, SBP2 protects nucleotides in the upper part of helix 1, which are not part of the L30 binding site.

The fact that the binding sites of L30 and SBP2 overlap provides a molecular explanation for our earlier finding that the two proteins cannot interact simultaneously with the SECIS [9]. Our results also have implications for interpreting defects in SBP2:SECIS interactions that are associated with human disease. A naturally occurring point mutation in a highly conserved nucleotide in the SelN SECIS that was previously shown to disrupt SBP2 binding and selenoprotein synthesis also inhibited binding of L30. Thus, it is important to keep in mind that mutant SECIS elements may be defective in more than one function.

To date, the SECIS element is the only known cognate RNA for SBP2. In contrast, L30 binds to several other kink-turn containing targets, including the L30 pre-mRNA, L30 mature mRNA, and 28S ribosomal RNA. As the structure of the SBP2:SECIS complex has not been solved, we turned to the NMR and co-crystal structures of the yeast L30 protein in a complex with the L30 pre-mRNA to guide our mutational analysis. We chose to design site-specific mutations in regions defined as the RNA-binding interface for the L30:L30 pre-RNA interaction. Interestingly, the crystal structure of a 60S ribosomal subunit from the fresh water ciliate *Tetrahymena thermophila* was recently published, and it is the first such structure from an organism that synthesizes selenoproteins [4]. Using the published coordinates, we utilized several molecular modeling programs to analyze the structure of L30 on the ribosome where it is bound to helix 58 of the 26S rRNA (unpublished observations). *T. Thermophila* L30 appears to use a similar RNA-binding interface to bind to a kink-turn in helix 58, as the yeast protein uses to bind to the L30 pre-mRNA. Thus, we expected that some of the same amino acids might also be important for the L30:SECIS interaction. Indeed, the Y89A mutant protein was defective in binding to both the SECIS and the L30 RNA, its canonical target. This residue is equivalent to F85 in the yeast protein, where it is the most prominent amino acid contact in the L30:L30 pre-mRNA complex [29,30]. Based on mutagenesis studies, an aromatic group is required at this position for binding of the yeast protein to the L30 RNA [29], and a similar requirement may be true for the L30:SECIS interaction.

Unlike Y89, mutagenesis of K36 and K87 had selective effects on the interaction of L30 with the two target RNAs. The K36A mutation had a greater inhibitory effect on binding of the protein to the L30 RNA than to the SECIS. The equivalent amino acid in the yeast protein, K32, did not directly contact the RNA in the NMR and crystal structures of the L30:L30 RNA complex [30].

However, it was proposed that this highly conserved amino acid could play an important role by neutralizing the negatively charged phosphate backbone [29,30]. We also found that mutagenesis of L35, which is also in the α2 helical region, modestly reduced binding inhibition to both targets. This result was only statistically significant for the L30:L30 RNA interaction, however.

In contrast, we found that K87 is important for the L30:SECIS interaction, but is dispensable for binding to the L30 RNA. This was a particularly interesting result as K87 is part of a signature motif that has been identified for L30 [37]. It is well established that each L7Ae family member has a unique RNA-binding specificity. However the basis for this selectivity has not been well understood as the proteins share a relatively conserved RNA-binding domain. An elegant study by Gagnon et al. recently identified five signature amino acids flanking this region that are unique to each family member [37]. One amino acid is N-terminal to the RNA-binding domain whereas the other four residues comprise a motif that is C-terminal to the domain. The C-terminal motif in particular is quite different between family members, with respect to the structure and chemical properties of the conserved amino acids. Functional evidence was provided for the L7Ae and 15.5 kDa proteins, showing that the unique conserved amino acids are necessary although not sufficient to mediate specific binding to the appropriate cognate RNA. The consensus N-terminal amino acid for L30 and SBP2 is a basic residue, K26 in rat L30 and R665 in rat SBP2 (Figure 5A). K87, which is important for the L30:SECIS interaction, is the first amino acid in KYYR, the C-terminal signature motif for L30 [37]. This residue is also highly conserved in SBP2 (K733 in rat SBP2, see Figure 5), where it is part of this protein's signature motif KAVP [37]. Interestingly, K is not present in the first position of the signature motifs for other eukaryotic L7Ae family members, including L7Ae, 15.5 kDa protein, rpL7a, Nhu2p, and Rpp38p [37]. The unique signature motifs that have evolved in L30 and SBP2 may explain the ability of these proteins to bind to the SECIS, which contains a large internal loop and a non-canonical kink-turn motif. As discussed above, the binding of L30 to the SECIS and the L30 RNA depends on Y89, which is the third amino acid in the KYYR motif. SBP2 has a V at this position, and mutational analysis of human SBP2 showed that this amino acid is important for the SBP2:SECIS interaction [38]. The fact that the signature motifs for L30 and SBP2 are not identical is consistent with the unique RNA-binding specificities of the two proteins, as L30 has been shown to bind to multiple cognate RNAs whereas the only known target for SBP2 is the SECIS element. An important future direction will be to solve the structures of the L30:SECIS and SBP2:SECIS complexes.

Conclusions

The study presented here provides new insight into how ribosomal protein L30 recognizes the SECIS element. Our findings suggest that there are subtle differences in how L30 interacts with its different cognate RNAs. The results expand our knowledge of how L7Ae family members recognize different RNA targets.

Methods

Constructs

The wild-type rat PHGPx SECIS and U40C mutant constructs were previously described [13]. The rat PHGPx SECIS elements of 129 nucleotides and 102 nucleotides, which were used for RNA footprinting and REMSA analysis respectively, are described in [39]. The stem-loop L30 pre-mRNA was generated by annealing primers (Additional file 1: Table S1) in 10 mM NaCl and slow cooling the sample from 95°C to room temperature. The luc/UGA^{258}/PHGPx and luc/UGA^{258}/TR1 constructs were described in [40]. The L30 cDNA was amplified from rat liver cDNA (BD Bioscience) by polymerase chain reaction (PCR) and the products cloned into the Champion™ pET200 Directional TOPO° vector (Life Technologies) downstream of an N-terminal His$_6$ tag. Mutants of L30 were constructed using the QuikChange Site-directed mutagenesis method (Stratagene), using the appropriate mutagenic primers (Additional file 1: Table S1). All constructs were verified by Sanger sequencing.

RNA synthesis

Plasmid DNAs were linearized and used as templates for in vitro transcription. For UV cross-linking and REMSA analyses, RNAs were synthesized in the presence of [α^{32}P] UTP (800 Ci/mmole; Perkin Elmer Easy Tides) using the RiboMAX™ Large Scale RNA Production Systems (T7) (Promega). Transcripts were purified using organic extraction followed by gel filtration using Micro Bio-Spin 30 Columns (BioRad). L30 RNA was gel purified and resuspended in 10 mM Tris HCl, pH 7.5. Unlabeled RNAs were prepared using the AmpliScribe T7-Flash Transcription kit (Epicentre) and purified as described above. For RNase footprinting experiments, cold synthetic transcripts were dephosphorylated with SuperSAP (Affymetrix), purified, and resuspended in nuclease-free water. Dephosphorylated transcripts were end-labeled in the presence of [γ^{32}P] ATP (3000 Ci/mmole; Perkin Elmer Easy Tides) and T4 PNK (NEB) at 20 units/pmole RNA. The transcripts were gel purified on 8% acrylamide (19:1)/7M urea gels and eluted in 10 mM Tris HCl, pH 7.5, 1 mM EDTA, pH 8, 300 mM NaAc, pH 5.5 at 4°C overnight. Purified RNA was stored in 10 mM Tris HCl, pH 7.5 at -20°C.

Protein purification

Recombinant rat L30 was expressed in BL21 Star (DE3) cells (Invitrogen). IPTG was added to 1 mM and the culture was grown at 37°C for 3 hr. The cells were harvested at 4°C, and the pellets washed with 1X PBS and frozen on dry ice. Purification buffers (PB) contained 20 mM HEPES-HCl, pH 8.0 and the indicated amounts of NaCl and imidazole. The frozen cell pellet was re-suspended in 20 ml of PB/0.2 M NaCl/20 mM imidazole, to which 60 kU rlysozyme (Novagen), 500 U of Benzonase Nuclease (Novagen) and two Complete, Mini, EDTA-free Protease Inhibitor Cocktail tablets (Roche) were added. The cells were further lysed by sonication on ice. The insoluble material was removed by two centrifugations of the lysate at ~16,000 × g for 15 min at 4°C. The cell lysate was filtered through a 0.45 μM filter and passed across a HisTrap TM FF crude column (GE Healthcare) using an AKTA Purifier (UPC-900) (GE Healthcare). The resin was sequentially washed with PB/0.2 M NaCl/20 mM imidazole, PB/1 M NaCl/20 mM imidizole, and PB/0.2M NaCl/40 mM imidazole. The protein was eluted with PB/0.2 M NaCl/250 mM imidazole and collected in 1 or 2 ml fractions. Fractions containing the pure protein were combined and dialyzed against 1 L of PB/0.2 M NaCl buffer for 2 hr at 4°C, which was then repeated with fresh buffer for an additional 2 hr at 4°C. The protein was dialyzed against PB/0.2 M NaCl containing 50% glycerol for 18 hr at 4°C. The dialyzed protein was aliquoted and stored at –70°C. Recombinant SBP2-CT was purified as previously described [40].

RNase footprinting

End-labeled ^{32}P-labeled PHGPx SECIS RNA was heated to 95°C and slow cooled to room temperature. The RNA (2.5 nM) was incubated in L30 binding buffer (30 mM Tris HCl, pH 8.0, 75 mM KCl, 5 mM MgCl$_2$, 1 mM DTT, 0.04 μg/μL BSA (NEB), 10% glycerol, and 50 ng/μL yeast tRNA) with or without rat L30 protein (0.75 or 1.5 μM as indicated) at 30°C for 15 min. Reactions were cooled to room temperature over a 2 min period and then placed at 22°C for 2-5 min. The indicated amounts of RNase T1, A, or V1 (Ambion) were added to the appropriate samples and incubated at 22°C for 5 min. Enzymatic reactions were quenched with 20 μL Inactivation/Precipitation buffer (Ambion) and purified according to manufacturer's directions. Samples were resuspended in 10 μL of dye-less loading buffer (95% formamide, 18 mM EDTA, and 0.025% SDS), heat-denatured at 95°C for 5 min, and separated in a denaturing 8% (19:1) polyacrylamide/7 M urea gel. The dried gels were visualized with a phosphorimager or on film.

Sequencing ladders were prepared by incubating end-labeled ^{32}P-labeled PHGPx SECIS RNA (2.5 nM) in 1X Sequencing Buffer (Ambion) supplemented with 50 ng/μL yeast tRNA. The RNAs were incubated at 50°C for 5

min, cooled to 22°C and the indicated amounts of RNase T1, A, or V1 added. The samples were incubated, quenched, and purified as described above. Alkali ladders were prepared by incubating end-labeled ^{32}P-labeled PHGPx SECIS RNA (2.5 nM) in 100 mM NaOH, 2 mM EDTA, pH 8.0, and 2 μg/μL yeast tRNA at 37°C for 3 min, to which 0.2 M Tris HCl, pH 8.0 (final) was added. The samples were frozen on dry ice and combined with an equal volume of denaturing loading buffer (10 M urea, 1X TBE).

UV cross-linking

The ^{32}P-labeled PHGPx SECIS or U40C mutant RNAs (10 fmol) were incubated in buffer containing 0.7X PBS, 11 mM DTT, 250 ng/μL yeast tRNA and RNAguard (Amersham). The indicated amounts of L30 were added last and the reactions incubated at 37°C for 30 min. The samples were irradiated on ice at 254 nm for 10 min in Costar 96-well polystyrene plates (Corning, Inc) using a Bio-Rad GS Genelinker. The RNA was digested with 20 U of RNase A (Fermentas) at 37°C for 1 hr. The samples were separated on a 15% SDS–PAGE (37.5:1) gel. The gels were dried and visualized using a phosphorimager.

UGA recoding assays

In vitro transcribed RNAs (100 ng) were added to an *in vitro* translation reaction (25 μL) containing rabbit reticulocyte lysate (Promega), complete amino acid mix (Promega), Protector RNase Inhibitor (Roche), and 70 nM purified recombinant SBP2-CT protein. Purified recombinant L30 protein was added as indicated. Reactions were incubated at 30°C for 30 min and placed on ice for 15 min. Aliquots of the translation products (2.5 μL) were then added to the luciferase substrate (50 μL) in a well of a Microlite 1 microtiter plate (Thermoscientific). Luminescence was measured in 10 sec intervals using a Victor[3] Multilabel Counter (Perkin Elmer).

RNA Electrophoretic Mobility Shift Assays (REMSA)

The ^{32}P-labeled SECIS RNA (10 fmol) was incubated in L30 binding buffer (30 mM Tris HCl, pH 8.0, 75 mM KCl, 5 mM MgCl$_2$, 1 mM DTT, 0.04 μg/μL BSA (NEB), 10% glycerol, 50 ng/μL yeast tRNA, and 0.2 U/μL Protector RNase Inhibitor (Roche)) with the indicated final concentration of L30 protein at 30°C for 15 min. Samples were transferred to ice and then separated in non-denaturing polyacrylamide gels, either 6% (19:1) or 8% (29:1), in 0.5X TBE gel at 4°C. The dried gels were visualized with a Phosphorimager or on film. The ^{32}P-labeled yeast L30 RNA was incubated in 350 mM KCl, 30 mM Tris HCl, 8.0, and 10 mM DTT at 60°C and allowed to slow cool to room temperature as previously described [41]. The binding reactions were then set-up and processed as described above. The reactions were separated in a non-denaturing 8% (29:1) polyacrylamide/ 0.5X TBE gel at 4°C.

Statistical analysis

Data were analyzed using GraphPad Prism software. The results are expressed as means +/- standard error of the mean (S.E.M). Statistical significance is indicated as * ($P < 0.05$), ** ($P < 0.01$), or *** ($P < 0.001$).

Abbreviations

PHGPx: Phospholipid Hydroperoxide Glutathione Peroxidase; REMSA: RNA Electrophoretic Mobility Shift Assay; RRL: Rabbit reticulocyte lysate; SBP2: SECIS-binding protein 2; Sec: Selenocysteine; SECIS: Selenocysteine insertion sequence; SelN: Selenoprotein N; TR1: Thioredoxin Reductase 1.

Competing interests

The authors declare they have no competing interests.

Authors' contributions

AB, TF, and DD conceived and designed the experiments, and analyzed data. AB, TF, and TA performed experiments. AB and DD wrote the manuscript. All authors read and approved the final manuscript.

Acknowledgements

This work was supported by National Institutes of Health grants DK085391 and HL29582 (to D.M.D.). We would like to thank Jodi Bubenik and Angela Miniard for reading the manuscript.

Author details

[1]Department of Cellular & Molecular Medicine, Lerner Research Institute, Cleveland Clinic, 9500 Euclid Avenue NC10, Cleveland, OH 44195, USA. [2]Present address: Department of Human Medical Genetics, University of Colorado, Aurora, CO, USA. [3]Department of Molecular Medicine, Cleveland Clinic Lerner College of Medicine of Case Western Reserve University, Cleveland, OH, USA.

References

1. Dabeva MD, Warner JR: The yeast ribosomal protein L32 and its gene. *J Biol Chem* 1987, **262**(33):16055–16059.
2. Halic M, Becker T, Frank J, Spahn CM, Beckmann R: Localization and dynamic behavior of ribosomal protein L30e. *Nat Struct Mol Biol* 2005, **12**(5):467–468.
3. Chandramouli P, Topf M, Menetret JF, Eswar N, Cannone JJ, Gutell RR, Sali A, Akey CW: Structure of the mammalian 80S ribosome at 8.7 A resolution. *Structure* 2008, **16**(4):535–548.
4. Klinge S, Voigts-Hoffmann F, Leibundgut M, Arpagaus S, Ban N: Crystal structure of the eukaryotic 60S ribosomal subunit in complex with initiation factor 6. *Science* 2011, **334**(6058):941–948.
5. Vilardell J, Warner JR: Ribosomal protein L32 of Saccharomyces cerevisiae influences both the splicing of its own transcript and the processing of rRNA. *Mol Cell Biol* 1997, **17**(4):1959–1965.
6. Vilardell J, Chartrand P, Singer RH, Warner JR: The odyssey of a regulated transcript. *RNA* 2000, **6**(12):1773–1780.
7. Dabeva MD, Warner JR: Ribosomal protein L32 of Saccharomyces cerevisiae regulates both splicing and translation of its own transcript. *J Biol Chem* 1993, **268**(26):19669–19674.
8. Macias S, Bragulat M, Tardiff DF, Vilardell J: L30 binds the nascent RPL30 transcript to repress U2 snRNP recruitment. *Mol Cell* 2008, **30**(6):732–742.
9. Chavatte L, Brown BA, Driscoll DM: Ribosomal protein L30 is a component of the UGA-selenocysteine recoding machinery in eukaryotes. *Nat Struct Mol Biol* 2005, **12**(5):408–416.
10. Bellinger FP, Raman AV, Reeves MA, Berry MJ: Regulation and function of selenoproteins in human disease. *Biochem J* 2009, **422**(1):11–22.
11. Papp LV, Lu J, Holmgren A, Khanna KK: From selenium to selenoproteins: synthesis, identity, and their role in human health. *Antioxid Redox Signal* 2007, **9**(7):775–806.
12. Walczak R, Carbon P, Krol A: An essential non-Watson-Crick base pair motif in 3'UTR to mediate selenoprotein translation. *RNA* 1998, **4**(1):74–84.
13. Lesoon A, Mehta A, Singh R, Chisolm GM, Driscoll DM: An RNA-binding protein recognizes a mammalian selenocysteine insertion sequence element required for cotranslational incorporation of selenocysteine. *Mol Cell Biol* 1997, **17**(4):1977–1985.
14. Fletcher JE, Copeland PR, Driscoll DM, Krol A: The selenocysteine incorporation machinery: interactions between the SECIS RNA and the SECIS-binding protein SBP2. *RNA* 2001, **7**(10):1442–1453.
15. Copeland PR, Fletcher JE, Carlson BA, Hatfield DL, Driscoll DM: A novel RNA binding protein, SBP2, is required for the translation of mammalian selenoprotein mRNAs. *EMBO J* 2000, **19**(2):306–314.
16. Allamand V, Richard P, Lescure A, Ledeuil C, Desjardin D, Petit N, Gartioux C, Ferreiro A, Krol A, Pellegrini N, et al: A single homozygous point mutation in a 3'untranslated region motif of selenoprotein N mRNA causes SEPN1-related myopathy. *EMBO Rep* 2006, **7**(4):450–454.
17. Caban K, Copeland PR: Selenocysteine insertion sequence (SECIS)-binding protein 2 alters conformational dynamics of residues involved in tRNA accommodation in 80 S ribosomes. *J Biol Chem* 2012, **287**(13):10664–10673.
18. Caban K, Kinzy SA, Copeland PR: The L7Ae RNA binding motif is a multifunctional domain required for the ribosome-dependent Sec incorporation activity of Sec insertion sequence binding protein 2. *Mol Cell Biol* 2007, **27**(18):6350–6360.
19. Zavacki AM, Mansell JB, Chung M, Klimovitsky B, Harney JW, Berry MJ: Coupled tRNA(Sec)-dependent assembly of the selenocysteine decoding apparatus. *Mol Cell* 2003, **11**(3):773–781.
20. Koonin EV, Bork P, Sander C: A novel RNA-binding motif in omnipotent suppressors of translation termination, ribosomal proteins and a ribosome modification enzyme? *Nucleic Acids Res* 1994, **22**(11):2166–2167.
21. Klein DJ, Schmeing TM, Moore PB, Steitz TA: The kink-turn: a new RNA secondary structure motif. *EMBO J* 2001, **20**(15):4214–4221.
22. Goody TA, Melcher SE, Norman DG, Lilley DM: The kink-turn motif in RNA is dimorphic, and metal ion-dependent. *RNA* 2004, **10**(2):254–264.
23. Clery A, Bourguignon-Igel V, Allmang C, Krol A, Branlant C: An improved definition of the RNA-binding specificity of SECIS-binding protein 2, an essential component of the selenocysteine incorporation machinery. *Nucleic Acids Res* 2007, **35**(6):1868–1884.
24. Takeuchi A, Schmitt D, Chapple C, Babaylova E, Karpova G, Guigo R, Krol A, Allmang C: A short motif in Drosophila SECIS Binding Protein 2 provides differential binding affinity to SECIS RNA hairpins. *Nucleic Acids Res* 2009, **37**(7):2126–2141.
25. Bubenik JL, Driscoll DM: Altered RNA binding activity underlies abnormal thyroid hormone metabolism linked to a mutation in selenocysteine insertion sequence-binding protein 2. *J Biol Chem* 2007, **282**(48):34653–34662.
26. Azevedo MF, Barra GB, Naves LA, Ribeiro Velasco LF, Godoy Garcia Castro P, de Castro LC, Amato AA, Miniard A, Driscoll D, Schomburg L, et al: Selenoprotein-related disease in a young girl caused by nonsense mutations in the SBP2 gene. *J Clin Endocrinol Metab* 2010, **95**(8):4066–4071.
27. Schoenmakers E, Agostini M, Mitchell C, Schoenmakers N, Papp L, Rajanayagam O, Padidela R, Ceron-Gutierrez L, Doffinger R, Prevosto C, et al: Mutations in the selenocysteine insertion sequence-binding protein 2 gene lead to a multisystem selenoprotein deficiency disorder in humans. *J Clin Invest* 2010, **120**(12):4220–4235.
28. Di Cosmo C, McLellan N, Liao XH, Khanna KK, Weiss RE, Papp L, Refetoff S: Clinical and molecular characterization of a novel selenocysteine insertion sequence-binding protein 2 (SBP2) gene mutation (R128X). *J Clin Endocrinol Metab* 2009, **94**:4003–4009.
29. Mao H, White SA, Williamson JR: A novel loop-loop recognition motif in the yeast ribosomal protein L30 autoregulatory RNA complex. *Nat Struct Biol* 1999, **6**(12):1139–1147.

30. Chao JA, Williamson JR: **Joint X-ray and NMR refinement of the yeast L30e-mRNA complex.** *Structure* 2004, **12**(7):1165–1176.

31. Fagegaltier D, Lescure A, Walczak R, Carbon P, Krol A: **Structural analysis of new local features in SECIS RNA hairpins.** *Nucleic Acids Res* 2000, **28**(14):2679–2689.

32. Mehta A, Rebsch CM, Kinzy SA, Fletcher JE, Copeland PR: **Efficiency of mammalian selenocysteine incorporation.** *J Biol Chem* 2004, **279**(36):37852–37859.

33. Vidovic I, Nottrott S, Hartmuth K, Luhrmann R, Ficner R: **Crystal structure of the spliceosomal 15.5kD protein bound to a U4 snRNA fragment.** *Mol Cell* 2000, **6**(6):1331–1342.

34. Moore T, Zhang Y, Fenley MO, Li H: **Molecular basis of box C/D RNA-protein interactions; cocrystal structure of archaeal L7Ae and a box C/D RNA.** *Structure* 2004, **12**(5):807–818.

35. Hamma T, Ferre-D'Amare AR: **Structure of protein L7Ae bound to a K-turn derived from an archaeal box H/ACA sRNA at 1.8 A resolution.** *Structure* 2004, **12**(5):893–903.

36. Li L, Ye K: **Crystal structure of an H/ACA box ribonucleoprotein particle.** *Nature* 2006, **443**(7109):302–307.

37. Gagnon KT, Zhang X, Qu G, Biswas S, Suryadi J, Brown BA 2nd, Maxwell ES: **Signature amino acids enable the archaeal L7Ae box C/D RNP core protein to recognize and bind the K-loop RNA motif.** *RNA* 2010, **16**(1):79–90.

38. Allmang C, Carbon P, Krol A: **The SBP2 and 15.5 kD/Snu13p proteins share the same RNA binding domain: identification of SBP2 amino acids important to SECIS RNA binding.** *RNA* 2002, **8**(10):1308–1318.

39. Copeland PR, Driscoll DM: **Purification, redox sensitivity, and RNA binding properties of SECIS-binding protein 2, a protein involved in selenoprotein biosynthesis.** *J Biol Chem* 1999, **274**(36):25447–25454.

40. Budiman ME, Bubenik JL, Miniard AC, Middleton LM, Gerber CA, Cash A, Driscoll DM: **Eukaryotic initiation factor 4a3 is a selenium-regulated RNA-binding protein that selectively inhibits selenocysteine incorporation.** *Mol Cell* 2009, **35**(4):479–489.

41. White SA, Hoeger M, Schweppe JJ, Shillingford A, Shipilov V, Zarutskie J: **Internal loop mutations in the ribosomal protein L30 binding site of the yeast L30 RNA transcript.** *RNA* 2004, **10**(3):369–377.

Mechanical stimulation of human tendon stem/progenitor cells results in upregulation of matrix proteins, integrins and MMPs, and activation of p38 and ERK1/2 kinases

Cvetan Popov[1], Martina Burggraf[1], Ludwika Kreja[2], Anita Ignatius[2], Matthias Schieker[1] and Denitsa Docheva[1*]

Abstract

Background: Tendons are dense connective tissues subjected periodically to mechanical stress upon which complex responsive mechanisms are activated. These mechanisms affect not only the development of these tissues but also their healing. Despite of the acknowledged importance of the mechanical stress for tendon function and repair, the mechanotransduction mechanisms in tendon cells are still unclear and the elucidation of these mechanisms is a key goal in tendon research. Tendon stem/progenitor cells (TSPC) possess common adult stem cell characteristics, and are suggested to actively participate in tendon development, tissue homeostasis as well as repair. This makes them an important cell population for tendon repair, and also an interesting research target for various open questions in tendon cell biology. Therefore, in our study we focused on TSPC, subjected them to five different mechanical protocols, and investigated the gene expression changes by using semi-quantitative, quantitative PCR and western blotting technologies.

Results: Among the 25 different genes analyzed, we can convincingly report that the tendon-related genes - fibromodulin, lumican and versican, the collagen I-binding integrins - α1, α2 and α11, the matrix metalloproteinases - MMP9, 13 and 14 were strongly upregulated in TSPC after 3 days of mechanical stimulation with 8% amplitude. Molecular signaling analyses of five key integrin downstream kinases suggested that mechanical stimuli are mediated through ERK1/2 and p38, which were significantly activated in 8% biaxial-loaded TSPC.

Conclusions: Our results demonstrate the positive effect of 8% mechanical loading on the gene expression of matrix proteins, integrins and matrix metalloproteinases, and activation of integrin downstream kinases p38 and ERK1/2 in TSPC. Taken together, our study contributes to better understanding of mechanotransduction mechanisms in TPSC, which in long term, after further translational research between tendon cell biology and orthopedics, can be beneficial to the management of tendon repair.

Keywords: Tendon stem/progenitor cells, Mechanical stimulation, Tendon-related genes, Collagen-binding integrins, Matrix metalloproteinases

* Correspondence: denitsa.docheva@med.uni-muenchen.de
[1]Department of Surgery, Experimental Surgery and Regenerative Medicine, Ludwig-Maximilians-University (LMU), Nussbaumstr. 20, D-80336 Munich, Germany
Full list of author information is available at the end of the article

Background

Tendons are able to transmit forces with minimal deformation or energy loss due to their unique hierarchically organized structure. The tendon extracellular matrix (ECM) is mainly composed of collagens (type 1, 3-6) and various proteoglycans, whilst the tendon cellular content is dominated by tenocytes. Within the tendon cell niche, Bi et al., [1] have reported the existence of a novel cell population possessing classical stem cell features such as self-renewing and multipotentiality. These cells were named tendon stem/progenitor cells (TSPC). TSPC are very closely related to the better known mesenchymal stem cells isolated from bone marrow, however they convey features distinguishable from other stem cells, namely the expression of tendon-related genes and the ability to form tendon-like tissue in vivo. Furthermore, it is suggested that TSPC are essential during tendon development and repair, and if their functions are disturbed, they can contribute to the progression of tendon pathologies [1]. Thus, TSPC represent a very important cell type for in-depth investigation of tendon cell behaviour, and their easy isolation and cultivation in vitro makes them useful and powerful tools for tendon researchers.

Within the tendon tissue, tendon cells interact with each other and with the proteins from the ECM [2]. These interactions are essential for the cells to sense and respond to mechanical loading, which in turn influences tendon metabolism [3]. The cells react to mechanical stimuli through complex mechanotransduction processes that can regulate the anabolic (ECM synthesis) and catabolic (matrix metalloproteinases expression and ECM degradation) pathways. In normal conditions, these processes are balanced and resulting in the maintenance of tendon homeostasis. However, changes in the equilibrium may lead to tendon pathology due to tissue degradation because of augmented ECM remodelling [4-6].

A major factor of the mechanotransduction process is the mechanical deformation of the ECM, which can affect cell actin cytoskeleton and thereby alter cell shape, motility and function. Mechanical forces can be transmitted by focal adhesion sites and cell-cell junctions [4,6]. The core components of focal adhesions are the integrin receptors; transmembrane heterodimers that can be activated by changes in the ECM or actin cytoskeleton and are mediating "outside-in" and "inside-out" signalling between the cell and the ECM [7]. Integrin signaling is initiated at the focal adhesion sites, which are membrane-associated platforms consisting of clustered, ECM-bound integrins as well as various enzymes, kinases, cytoskeletal and adaptor proteins (e.g. focal adhesion kinase, FAK, paxillin, p130cas) in the cytoplasm. Integrin adhesion triggers "outside-in" signaling which frequently synergizes with growth factor-dependent cascades and activates downstream proteins such as extracellular signal-regulated protein kinases 1 and 2 (ERK), p38 mitogen-activated protein kinases and c-Jun N-terminal kinases (Jnk) [2]. Furthermore, integrin-mediated anchorage and signaling can also regulate cell survival processes through the activation of protein kinase B (Akt) survival pathway [2,8].

So far, there are only few studies reporting on the phenotypic responses of tendon-derived cells to mechanical stress [5,9-12]. In particular, Fong et al., [10] and Mackley et al., [9] applied a microarray technology and studied the effects of mechanical load on the transcriptome of rat palmar flexor tendon cells and mouse embryonic tendon fibroblasts, respectively. These studies have suggested some candidate genes, such as transforming growth factors and cytoskeletal adaptor proteins, to be involved in the tendon cell mechanotransduction. However, the tendon mechanobiology is still not fully understood and it needs further scientific exploration. Here, we focussed on human TSPC from Achilles tendon and stimulated them with three different mechanical magnitudes for one or three days. As an experimental approach, we applied low cost analysis, by semi-quantitative and quantitative PCR, of 25 selected genes. To our knowledge, we can present for the first time novel data on the effect of mechanical stimulation on the expression of genes that are: 1) essential matrix components of the TSPC niche; 2) integrin receptors, which establish the necessary cell-matrix interactions and translate mechanical stimuli in signalling cascades; 3) matrix metalloproteinases, which are downstream targets of the integrin signalling, and in turn can remodel the TSPC niche; and 4) five key kinases from the integrin-mediated signalling pathways.

Results

Stem cell characteristics and tenogenic profile of TSPC

TSPC were successfully isolated from human Achilles tendon biopsies [13] and their phenotype was re-validated in vitro based on the expression of stem cell surface markers (Figure 1A), tendon-related genes (Figures 1B and 2B) and the ability to undergo two-lineage differentiation (Figure 1C).

First, immunofluorescent staining for CD146, Nestin and STRO-1 stem cell markers demonstrated their ubiquitous expression in TSPC (Figure 1A). Next, quantitative PCR results confirmed the expression of the transcription factor Scleraxis and the tendon-specific gene tenomodulin in TSPC (Figure 1B). Additionally, the cells expressed collagen type 1 and 3, and several proteoglycans like cartilage oligomeric matrix protein (COMP), decorin, tenascin C, biglycan, fibromodulin, lumican and versican (Figure 2B), which are known to be highly expressed in tendons. Finally, we subjected TSPC to two different differentiation protocols, namely adipogenic and osteogenic stimulation (Figure 1C). Our results after 21 days of stimulation, demonstrated that TSPC could successfully commit towards

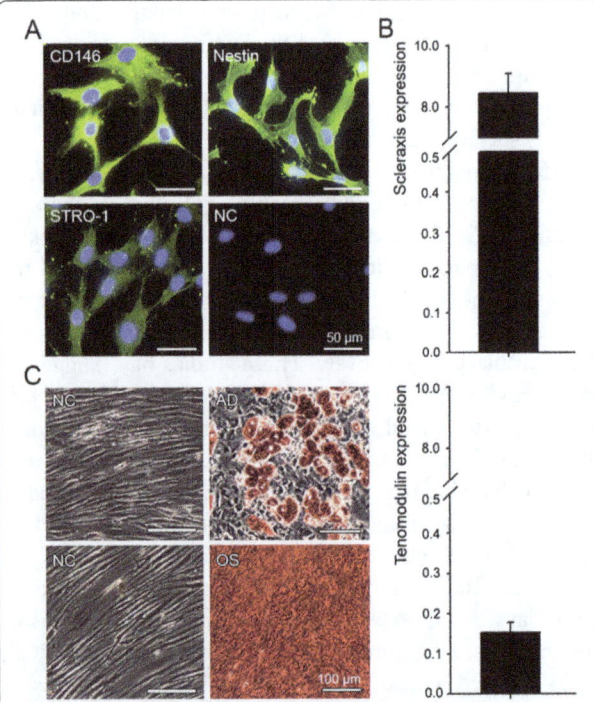

Figure 1 Characterization of the human TSPC. (A) Expression of CD146, Nesitn and STRO-1 stem cell markers demonstrated by immunocytochemistry; NC – negative control, cells incubated only with secondary antibody. Bar 50 μm. **(B)** Quantitative PCR analyses for Scleraxis and tenomodulin gene expression in TSPC, demonstrated as a ratio to HPRT housekeeping gene. **(C)** Adipogenic and osteogenic stimulation of TSPC for 21 days. Adipogenic (AD) and osteogenic (OS) differentiation visualized by Oil Red-O and Alizarin red staining, correspondingly; NC – negative control, unstimulated cells. Bar 100 μm. Data is representative of 3 donors, each used in 3 independent experiments.

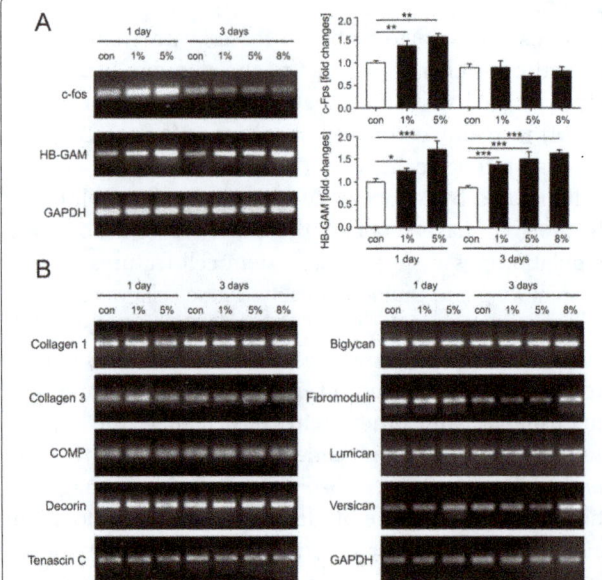

Figure 2 Expression of mechanoresponsive and extracellular matrix genes upon mechanical stimulation of TSPC. (A) RT-PCR for c-fos and HB-GAM expression changes after 1 and 3 days stimulation with 1, 5 and 8% mechanical stress. PCR densitometric quantification is shown as fold changes to the unstimulated controls; *p < 0.1, **p < 0.05, ***p < 0.01. **(B)** RT-PCR analysis for collagen 1 and 3, and tendon-related proteoglycans (COMP, decorin, tenascin C, biglycan, fibromodulin, lumican and versican). Data is representative of 3 donors, each used in 3 independent experiments.

adipocyte and osteoblast lineages visualized by Oil Red O (for lipid droplets) and Alizarin Red staining (for calcified matrix). Taken together, the above data reconfirmed the classical TSPC features.

Mechanical loading of TSPC

In order to validate that TSCP have been subjected to mechanical stimulation, we analyzed the expression levels of well-known mechano-responsive genes such as c-fos and heparin-binding growth-associated molecule HB-GAM (Figure 2A). PCR analysis of c-fos demonstrated increased gene expression in TSPC stimulated mechanically for 1 day. Densitometric evaluation of the PCR bands showed that c-fos was significantly upregulated with 1.4 and 1.6 folds when stimulated with 1% or 5% loading strain, respectfully. At day 3, no effect of the stretching on c-fos expression was observed, as the gene expression levels remained similar to the non-stimulated cells. HB-GAM expression was clearly changed in response to the applied mechanical loading at day 1 and 3. The densitometric evaluation of the PCR bands

demonstrated that at day 1, the HB-GAM expression was increased with 1.2 and 1.7 folds when stimulated with 1 and 5% strain, correspondingly. At day 3, a dose-dependent trend was observed and upon stimulation with 1, 5 and 8% strain, HB-GAM expression changed significantly with 1.4, 1.5 and 1.6 folds in comparison to the non-stimulated cells. Due to the observed expression changes in the tested mechano-regulated gene, we validated that TSCP were successfully stimulated with the applied axial stretching.

Effect of mechanical loading on the expression of extracellular matrix genes

Next, we analyzed the effect of loading on the gene expression of ECM proteins that are highly expressed in the tendon tissue (Figure 2B). In particular, we investigated the expression of nine different genes - collagen type 1 and 3, COMP, tenascin C, decorin, biglycan, fibromodulin, lumican and versican. Our results showed that mRNA expression levels of the ECM genes have not been apparently altered by the mechanical stimulation and it mostly remained similar to that of non-stimulated controls at day 1 and day 3. Only in the case of fibromodulin, lumican and versican, we observed an increased expression at day 3 when TSPC were stimulated with 8% strain. These results suggested that 8% mechanical loading over a longer period have a positive effect on certain ECM genes expressed by the tendon cells. Therefore, we propose that in order to achieve a successful mechanical stimulation in TSPC, higher mechanical strain applied for extended time period is required.

Analysis of integrin expression in response to mechanical stimulation

In order to sense and translate the applied external mechanical signals, cells implicate mechanoreceptors on their surface, such as the integrins. An individual integrin receptor consists of two non-covalently bound subunits – α and β [7]. In our study, we analyzed the expression changes of eight alpha (α1-6, α11 and αV) and two beta (β1 and β3) integrin subunits (Figure 3). RT-PCR analysis demonstrated a slight increase of integrin α3, α4, α5, α6 and αV subunit expression at day 3 in comparison to day 1 (Figure 3A). However, this difference was independent from the applied mechanical stress since it was also observed in the non-stimulated TSPC. Next, we analyzed the expression changes in the collagen-binding integrins α1, α2 and α11 by quantitative PCR (Figure 3B). We found that when stimulated for 3 days, TSPC upregulated integrin α1 with 1.3 and 1.5 folds upon 5 and 8% stress, correspondingly. The expression of integrin α2 was upregulated with 1.2 and 1.3 folds when stimulated with 1% strain at

day 1 or day 3, respectively. The highest increase of integrin α2 (1.9 folds) was observed when TSPC were stimulated with 8% strain for 3 days. Mechanical loading of TSPC resulted in 1.3 folds increase in integrin α11 expression at day 1 when the cells were stimulated with 1% strain and at day 3, when stimulated with 8% strain. Regarding, integrin β-subunits, we did no observe any pronounced changes in between the control and mechanically stimulated TSPC. The expression levels of integrin β1 and β3 was similar also between the different days of stimulation. Taken together, our results suggest that 8% biaxial mechanical loading applied for 3 days modulated the expression levels of the three collagen I-binding alpha-subunits.

Expression of matrix metalloproteinases upon mechanical stimulation

Next, we examined the effect of mechanical loading on the gene expression changes of matrix metalloproteinases responsible for collagen degradation (MMP1, 2, 3, 9, 13 and 14). The expression of MMP1, 2 and 3 in TSPC did not respond to the mechanical stimulation as their gene levels remained similar to the non-stimulated cells at day 1 and 3 (Figure 4A and B). In contrast, the expression of MMP9, 13 and 14 increased when cells were stretched for 3 days (Figure 4B). The MMP9 expression was affected significantly by the 5% mechanical loading for 3 days as the gene levels increased with approximately 100 folds. When stimulated with 1, 5 or 8% loading strain, TSPC clearly upregulated MMP13 and 14, but only at day 3. In particular, MMP13 was increased with 2 folds at any of the applied loading strain. Similarly, MMP14 expression was elevated with at least 1.5 folds at day 3 in comparison to the non-stimulated TSPC. In conclusion, our data suggested that biaxial mechanical stimulation of TSPC for 3 days period of time upregulates the MMP9, 13 and 14 and is independently of the applied strain magnitude.

Validation of gene expression changes on protein level upon 8% mechanical loading of TSPC

Based on the obtained mRNA data, we performed western blotting analyses for the major candidate genes as we investigated: 1) the collagen binding integrins α1, α2 and α11; 2) the activation of five key integrin downstream kinases and 3) the expression of MMP9, 13 and 14 responsible for the collagen degradation. For this analysis, TSPC were stimulated with 8% biaxial loading for 3 days, since this was the best condition according to our mRNA screening. Similar to the mRNA results (Figure 3B), stimulation with 8% mechanical strain resulted in significantly higher protein production for all three integrins. However, the increase in the protein expression was more pronounced for integrin α2 (1.5 folds) and α11 (1.9 folds) (Figure 4A). Next, we analyzed the

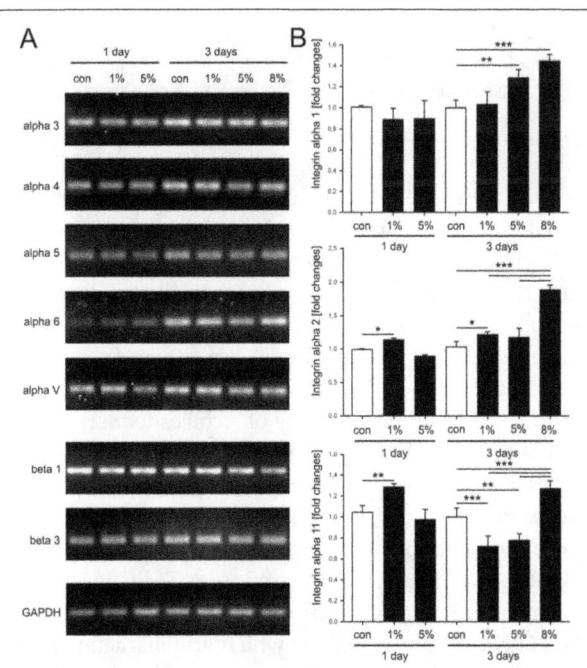

Figure 3 Integrin expression changes after mechanical stimulation of TSPC. (A) RT-PCR analysis of integrin alpha 3, 4, 5, 6 and V, and beta 1 and 3 in TSPC stimulated with 1, 5 and 8% strain for 1 and 3 days. **(B)** Quantitative PCR analysis for integrin alpha 1, 2 and 11 (fold change to the non-stimulated TSPC at day 1). Data is representative of 3 donors each used in 3 independent experiments; *p < 0.1, **p < 0.05, ***p < 0.01.

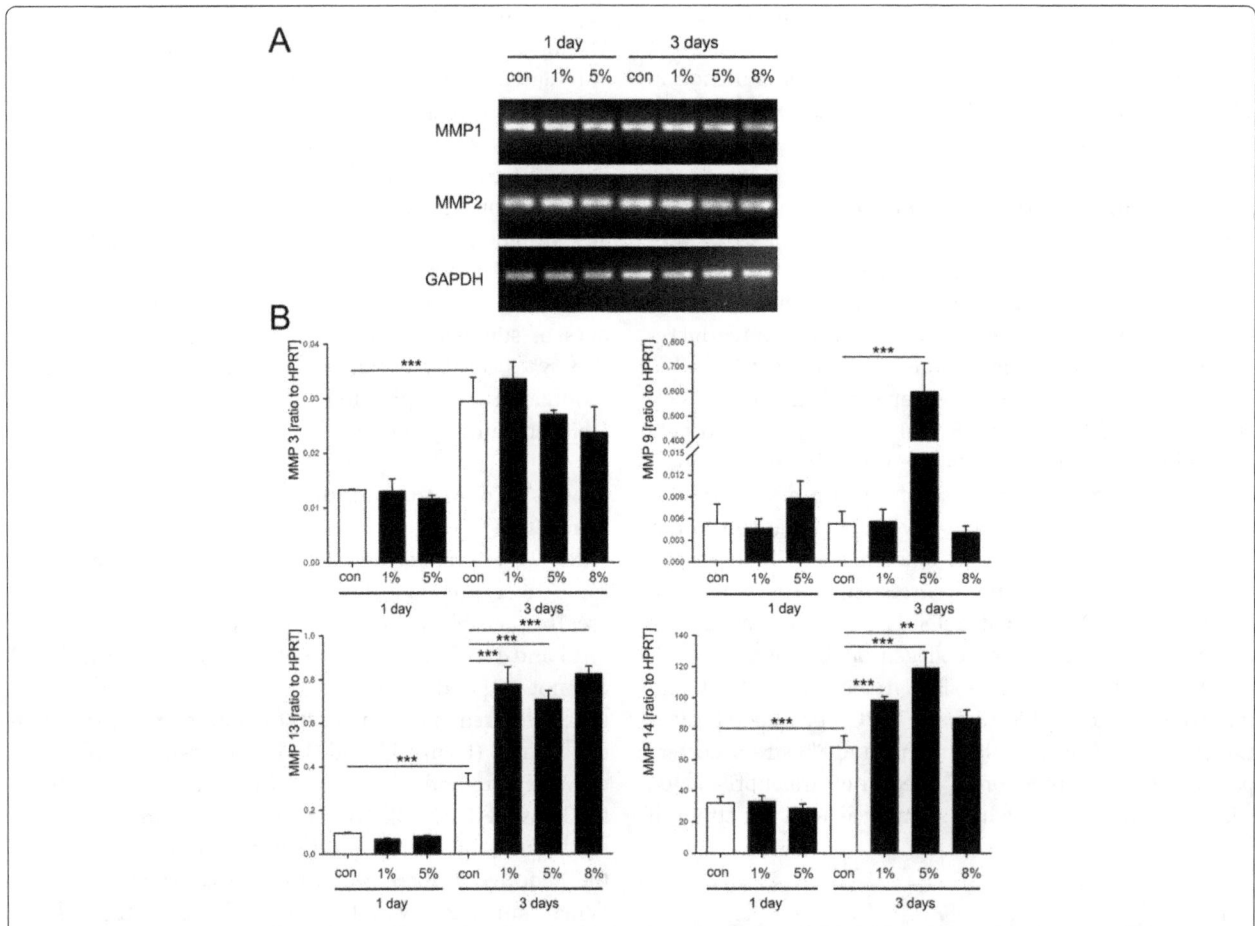

Figure 4 Gene expression of matrix metalloproteinases upon mechanical stimulation of TSPC. (A) RT-PCR analysis for MMP1 and MMP2. **(B)** Quantitative PCR analysis for MMP3, MMP9, MMP13 and MMP14 (ratio to HPRT housekeeping gene). Data is representative of 3 donors, each used in 3 independent experiments; *p < 0.1, **p < 0.05, ***p < 0.01.

changes in the activity of five integrin downstream kinases FAK, ERK, Akt, p38 and Jnk (Figure 5B). We found that upon mechanical loading, the activity of ERK (1.5 folds) and p38 (1.4 folds) kinases significantly increased, whereas the activity of Jnk (1.6 folds) was significantly reduced in the mechanically loaded TSPC. The other two kinases – FAK and Akt were not influenced by the mechanical loading. The analysis of MMP protein levels (Figure 5C) demonstrated that upon mechanical stimulation the expression of MMP9, 13 and 14 were significantly increased as the most pronounced changes were observed in the expression of MMP13 (1.4 folds) and 14 (1.8 folds) in comparison to the non-stimulated TSPC.

Discussion

Despite of the acknowledged importance of mechanical loading for tendon development and healing, the effect of different loading magnitudes on tendon cells has not been studied in details. Thus, with our study we shade a light on the changes in the gene expression of tendon-related matrix proteins, integrins and MMPs occurring

after different mechanical loadings – 1, 5 and 8%, applied for one or three days. As a cell source, we used TSPC isolated from human Achilles tendon [13], because they are not only an important cell type in tendon development and repair, but also a potential clinical cell source for tendon regeneration. A limitation of the present study was the use of only three TSPC biological replicates due to the difficult obtainability of Achilles tendon biopsies from young and healthy patient donors. Thus, in future studies an increased cohort of TSPC donors has to be investigated.

In previous publication, we have already reported the isolation and initial characterization of TSPC [13]. These cells express common stem cell surface markers such as CD146, Nestin and STRO-1, which are characteristic for stem cells from different tissue sources [14,15]. For example, CD146 was shown as marker of stem cells localized in the vascular wall [16], Nestin was found predominantly in multipotent cells from the central nervous system [17] and STRO-1 was expressed on mesenchymal stem cells in bone marrow [18]. Bi et al., [1] and Tempfer et al., [19]

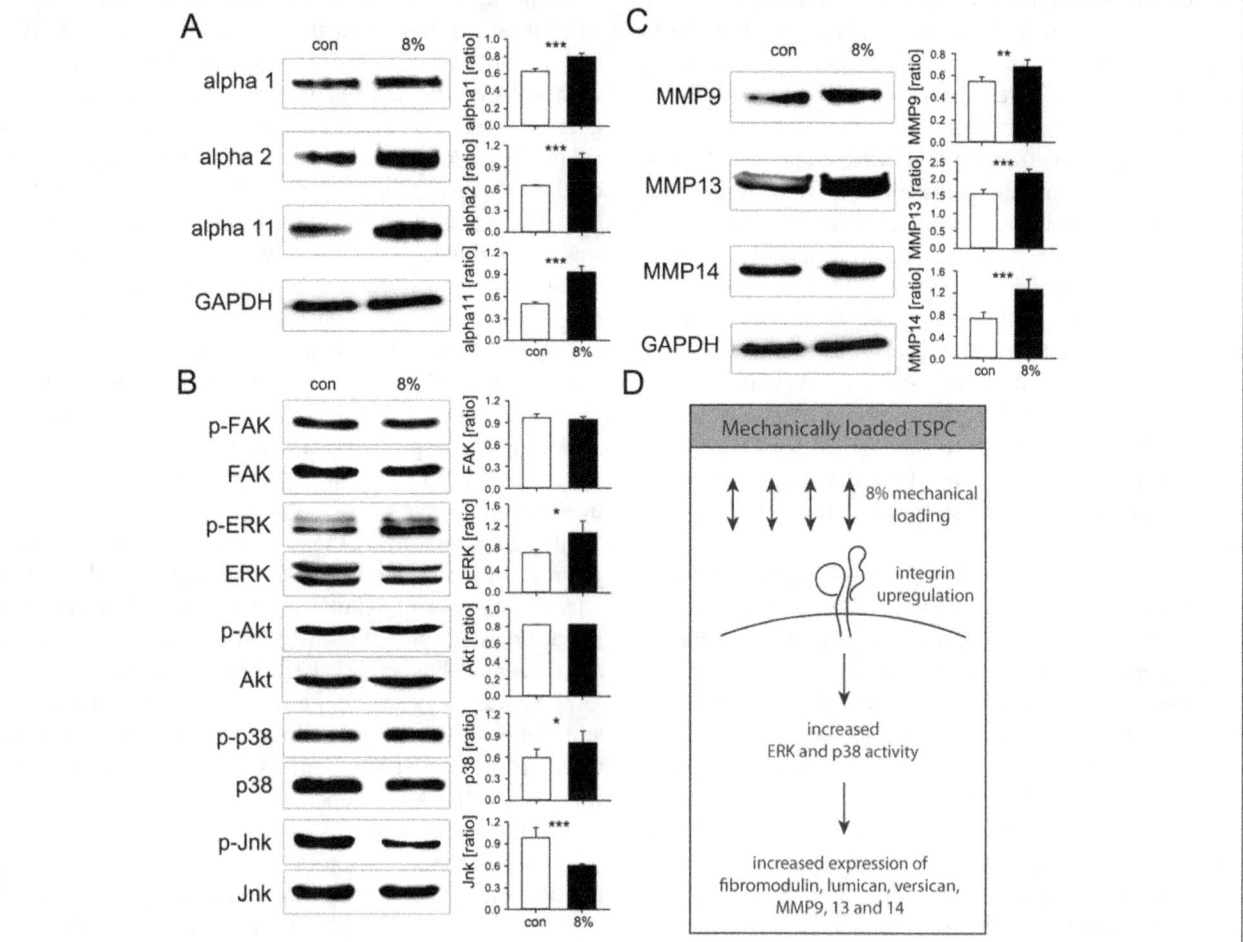

Figure 5 Western blotting analysis for collagen-binding integrins, integrin-downstream kinases and matrix metalloproteinases upon 8% mechanical stimulation of TSPC. (A) Collagen I-binding integrins α1, α2 and α11 (ratio to GAPDH protein expression); **(B)** Phosphorylated and total levels of FAK, ERK, Akt, p38 and Jnk (ratio phospho-/total protein); **(C)** MMP9, 13 and 14 (ratio to GAPDH). **(D)** Schematic summary of the changes occurring in TSPC upon 8% biaxial mechanical loading. Data is representative of 3 donors, each used in 3 independent experiments; *p < 0.1, **p < 0.05, ***p < 0.01.

have also determined the expression of these three markers in TSPC. Additionally to the expression of stem cell-related markers, TSPC preserve the expression of the typical tendon lineage genes namely Scleraxis [20] and tenomodulin [21]. Here, we have reconfirmed the TSPC characteristics by: 1) the expression of CD146, Nestin and STRO-1; 2) the expression of Scleraxis and tenomodulin; and 3) the ability of these cells to commit to adipogenic and osteogenic lineages. Hence, we concluded that TSPC represent a good model system to study the behavior of the tendon-derived cells under mechanical stress in vitro.

In vivo, the whole tendon unit can be subjected to tissue stretching that normally does not exert 4%. This stretching is considered as a maximum of the physiological range. In some rare occasions, the applied forces can surpass that range. Then, this can result in formation of microscopic tears in the collagen fibers. Extreme cases, when the generated forces exerted 8-10% tissue stretching

or prolonged sub-physiological loadings can cause tissue rupture and thereon tendon failure [6]. However, studies have shown that avian flexor and rabbit Achilles tendons can withstand stretching up to 14% and 16%, respectively [6,22] suggesting that the tendon strain might be underestimated. With regards to the cells within the tendon tissue, it has been proposed that up to 10% mechanical stretching of tendon fibroblast is within their physiological range [22]. The cell response to mechanical stimuli depends mainly on the type of loading (static or cyclic, uniaxial or biaxial), stretching magnitude, frequency and duration [22]. Several studies, performed with tendon fibroblasts obtained from various species have used mechanical stimulation (cyclic uniaxial or biaxial) in the range of 4 to 10% stretching [22]. Therefore, we stimulated TPSC from three different biological replicates with three magnitudes located in the low (1%), middle (5%) and high (8%) region of the physiological range. We applied the mechanical stress for two different time periods, one and three days, in order to study short and long-term effects. To confirm the successful mechanical loading of TSPC, we first investigated the expression of short-term and long-term mechanically stimulated genes c-fos and HB-GAM. The transcription factor c-fos is established short-term mechanically-regulated gene, which peaks after 30 min upon mechanical stress [23]. HB-GAM gene was upregulated in hMSC upon cyclic mechanical stimulation as the gene expression was detected after longer periods of mechanical loading (48 hours) [24]. Our results clearly demonstrated that upon mechanical stimulation the expression levels of both mechano-regulated gene markers c-fos and HB-GAM were increased, as in the case of HB-GAM the increased gene expression was detected in each of the two different time points and it was in correlation to the strength of the biaxial mechanical load.

Then, we analyzed the changes in the expression levels of collagen 1 and 3 genes, which are key matrix proteins in the TSPC niche. Few articles have suggested that mechanical stimulation has a positive effect on the collagen 1 expression. For example, in human tendon fibroblasts from anterior cruciate ligament, 4% and 8-10% cyclic uniaxial stimulation resulted in the induction of collagen 1 gene expression [25]. Howard et al., [26] reported an increase of collagen 1 and fibronectin expression in human ligament fibroblasts when stimulated with 5% cyclic biaxial stretch, but observed no effect on collagen 1 expression by 10% stretching [22]. Interestingly, Hsieh et al., [27] suggested that different tendon types can be influenced differently by the mechanical loading. In human anterior cruciate ligaments, 7.5% biaxial stretch increased collagen 1 and decreased collagen 3 expression. In contrast, in medial collateral ligaments the same mechanical loading led only to elevation of the collagen 3 expression, while collagen 1 expression remain unchanged [22]. In TSPC, our PCR results demonstrated no changes in the expression levels of collagen 1 and 3 at the three different magnitudes and at the two different time points. Hence, we can conclude that TSPC isolated from human Achilles tendon do not elevate the mRNA levels of collagen 1 and 3 when mechanically stimulated. Taking into account the complex post-translational modification of collagens, it remains to be further clarified if mechanical stress might influence the protein levels of collagen 1 and 3 in TSPC.

Next, we studied the expression levels of proteoglycans that are characteristic for the tendon tissue. We detected an increased expression of fibromodulin, lumican and versican when TSPC were stimulated with 8% axial loading, while the levels of COMP, decorin, tenascin C and biglycan remained unchanged. The observed increase of fibromodulin, lumican and versican RNA levels upon mechanical loading can be explained by the function of these proteoglycans in the tendon tissue. Fibromodulin and lumican were found to be important in the fibrilogenesis and maturation of the collagen fibers [28], while versican is involved in cell adhesion, proliferation and migration as well as in ECM assembly [29]. Little is known about the alterations in proteoglycan expression levels in tendons, triggered by mechanical loading. Only few studies have reported that decorin and biglycan expression was not influenced by mechanical stimulation [5,30,31], while versican was significantly upregulated [31]. Taken together, our results are in line with the above literature and demonstrated that high levels of mechanical stress can upregulate the gene expression of fibromodulin, lumican and versican. It will be of great interest to further explore how these proteoglycans affect the TSPC functions upon mechanical stress, which can be the focus of follow up studies.

Integrins can convert the mechanical signals into cytoplasmic signals via outside-in signaling cascade, which affects a number of cellular processes including cell proliferation, migration and differentiation, gene expression, matrix remodeling, cytoskeletal dynamics and cell survival [3,7]. Subramony et al., [32] have reported that mechanical stimulation of hMSC can trigger the upregulation of integrin $\alpha2$, $\alpha5$ and $\beta1$ after 7 or 14 days of 1% axial stimulation. The authors concluded that the increase in integrin expression is important to ensure the sufficient numbers of surface receptors that are necessary to sense and translate the mechanical stimuli in signaling cascade in MSC [32]. With regards to TSPC, to our knowledge, this study delivers the first data on the effect of mechanical stimulation on the RNA expression of different integrins responsible for collagen ($\alpha1$, $\alpha2$ and $\alpha11$), fibronectin ($\alpha3$, $\alpha4$, $\alpha5$ and αV) and laminin ($\alpha6$) binding. The RT-PCR analysis performed for integrin $\alpha3$, $\alpha4$, $\alpha5$, $\alpha6$ and αV demonstrated increased integrin expression at day 3. However, that expression did not dependent on the magnitude of the mechanical stimulation. In contrast, the expression of collagen-binding integrins were strongly upregulated at day 3 on mRNA and protein levels when TSPC were stimulated with 8% mechanical stress, suggesting that this mechanical loading was optimal to trigger molecular response in TSPC. Taken together, we are the first to report integrin expression changes occurring in TSPC upon mechanical stimulation.

Based on these initial findings, we investigated further the integrin-orchestrated signaling cascade in TPSC triggered by the mechanical stimulation. Involvement of several different kinase pathways from the signaling cascade of integrin-transmitted mechanical stimuli was already demonstrated in cell, delivered from muscles [33,34], cartilage [35-37] and bone [38,39] tissues. A common tendency between all musculoskeletal tissues was the activation of ERK signaling pathway upon mechanical loading. With regards to other kinases, the published data is contradictive. For example, Zhou et al., [35] and De Croos et al., [36] found that mechanical stress in human and bovine chondrocytes results in increased Jnk phosphorylation, whereas Xu et al., [37] did not confirmed this observation in rat chondrocytes, in which no changes in the phospho-Jnk levels were detected. Others have demonstrated that integrin-dependent transmission of the mechanical signaling in the osteoblasts and chondrocytes was mediated through increased FAK [38], p38, Jnk and Akt activity [40-42]. Matsui et al.,[41] suggested that the activation of the different kinases might correlate to the magnitude of the mechanical stress: when MC3T3 cells were treated with lower cyclic mechanical loading ERK phosphorylation was elevated, whereas when MC3T3 were subjected to higher mechanical stretch Jnk and p38 active forms were also increased. Our protein results from TSPC treated with 8% mechanical loading clearly demonstrated a significant increase in ERK and p38 phosphorylation. This result suggests that 8% biaxial mechanical loading is an optimal magnitude of mechanical loading for TSPC. In contrast to MC3T3, TSPC stimulated with 8% mechanical loading had decreased Jnk activity, a finding that needs to be further investigated.

Finally, we analyzed the expression of the matrix metalloproteinases. It is known that binding of the integrins to their ECM ligands results in increased expression of various MMPs, which participate in the important cell processes such as cell migration as well as matrix remodeling [43]. Increased MMP expression upon mechanical loading without or with presence of other factors has been previously demonstrated in tenocytes [44,45], chondrocytes [46] and osteoblasts [47]. This effect, however, depends strongly on the cell type, the duration and the magnitude of the applied stress. For example, Archambault J et al. [48] found an increased MMP-1 and MMP-3 expression in rabbit Achilles tendon cells when stimulated with IL-1β and 5% cyclic biaxial stretching. Oppositely, Sun et al. [49] reported that in human synovial cells that were stimulated with 2% cyclic stretch, the expression levels of MMP-1 and MMP-13 decreased. Our analysis clearly demonstrated that 8% biaxial mechanical stress of TSCP for 3 days significantly increased MMP9, 13 and 14 expressions in comparison to the non- and to 1 day stimulated cells.

Conclusions

To our knowledge, our study is the first one to address the effect of different mechanical loadings on the expression of selected genes in human TSPC isolated from Achilles tendon. Here, we showed that biaxial mechanical stress induces the expression of the proteoglycans fibromodulin, lumican and versican; collagen-binding integrin receptors α1, α2 and α11; and MMP9, 13 and 14

via ERK and p38 kinase activation (Figure 5D). These proteins play an important role for TSPC's niche composition, cell survival, mechanosignaling and matrix remodeling. Furthermore, we established an efficient experimental protocol for the mechanical stimulation of human TSPC and we can propose 8% mechanical strain for 3 days as an optimal setup that promotes mechanoresponse and gene expression changes in these cells. Taken together, we believe that our study contributes to a better understanding of the TSPC and their response to mechanical stimuli, and it can serve as an experimental model for further in-depth analysis of the mechanotransduction mechanisms in tendon cells.

Methods

Cell isolation and cultivation

Achilles tendon biopsies derived from three young and healthy human patients (male, age 28 ± 5 years), who had undergone surgical operations due to lower extremity accidents in the Surgical Clinic of Ludwig-Maximilians-University in Munich (LMU). TSPC were isolated and initially characterized by Kohler et al., [13]. The procedure was approved by the Ethical Commission of the LMU Medical Faculty (grant No. 166-08) and written informed consent was obtained from all donors. Briefly, TSPC were isolated as tendon tissue (without the paratenon) was minced into small pieces and enzymatically digested overnight with 0.15% collagenase II (Worthington, USA). Then the digested tissue was filtered (100 μm nylon mesh) and centrifuged for 10 min. The cell pellet was suspended in DMEM/Ham's F-12 supplemented with stabile glutamine, 1% MEM-Amino-acids (Biochrom, Germany), 10% FBS and 1% L-ascorbit-acid-2-phosphate (Sigma-Aldrich, Germany) [13]. The cell suspension was plated on polystyren dishes and the obtained TSPC were expanded in a humidified incubator at constant 37°C and 5% CO2 and then used in different experiments at passages 4-6 or stored in liquid nitrogen tank.

Immunocytochemistry

TSPC at passage 6 were plated and cultured on 20 μg/ml collagen 1-coated glass slides (BD Bioscience, USA) for 48 h. Then, the cells were fixed with 4% paraformaldehyde (Merck, Germany), permeabilized with Triton X100 (Sigma-Aldrich) and blocked with 3% BSA (PAA, USA). Primary antibodies against CD146 (Millipore, USA), Nestin (Proteintech, USA) and STRO-1 (R&D Systems, USA) were applied overnight at 4°C. Next, secondary Alexa Flour 488-conjugated antibodies and DAPI were used (all Life technologies, USA). As negative control were used cell-seeded slides which were incubated only with secondary Alexa Flour 488-conjugated antibodies and DAPI. Photomicrographs were taken with Axiocam MRm camera on Axioskope 2 microscope (Carl Zeiss, Germany). Staining procedures were reproduced at least twice.

Cell differentiation

TSPC were stimulated at passage 5 towards adipogenic and osteogenic lineages. For adipogenic differentiation, TSPC were plated in 6-well dishes (1×10^5 cells/cm^2). Cells were stimulated for 21 days using the induction medium composed of DMEM-high glucose medium, 10% FBS, 1 μM dexamethasone, 0.2 mM indomethacin, 0.1 mg/ml insulin and 1 mM IBMX (all Sigma-Aldrich). The extent of adipogenic differentiation was evaluated by standard Oil Red-O staining.

For osteogenic differentiation, 3.5×10^4 cells/cm^2 were seeded in 6-well dishes and cultivated in osteogenic medium composed of DMEM high glucose, 10% FBS, 10 mM β-glycerophosphate, 50 μM L-ascorbic-acid-2-phosphate and 100 nM dexamethasone (all Sigma). After 21 days, Alizarin Red staining was performed using the Osteogenic Quantification kit (Millipore). After each differentiation experiment, photomicrographs were taken with AxiocamICc3 camera mounted on AxiovertS100 microscope (Carl Zeiss).

Mechanical stimulation

Biaxial mechanical stimulation of the TSPC from three different donors at passage 4-6 was performed in a six-station stimulation apparatus driven by eccentric motor [50]. For this, 1×10^5 cells were plated and cultured for 4 days on FBS-coated flexible silicone dishes (60 mm × 30 mm). Then, triplicate dishes were stretched cyclically in the long axis at a frequency of 1Hz and a magnitude of 1%, 5% or 8% (corresponding to 1×10^4, 5×10^4 or 8×10^4 μstrain) continuous for 60 min (1 day) or intermittent on 3 consecutive days for 60 min/day (3 days). In parallel, non-stimulated cells, plated on flexible silicone dishes, were used as controls on day 1 and day 3, respectively. Directly after stimulation, cells were lysed and subjected to mRNA isolation.

Reverse transcriptase and quantitative PCR analysis

Total RNA was extracted directly after stimulation with RNeasy Mini Kit (Qiagen, Germany). Then, total RNA concentration was determined by Nanodrop (Thermo scientific, USA) and its integrity was verified by 28S/18S rRNA ratio on agarose gel electrophoresis (Peqlab, Germany). For cDNA synthesis, 1 μg total RNA and AMV First-Strand cDNA Synthesis Kit (Life technologies) were used. RT-PCR was performed with Taq DNA Polymerase (Life technologies) in MGResearch instrument (BioRad, Germany). Primer pairs and PCR conditions are listed in Additional file 1: Table S1. For c-Fos and HB-GAM, PCR bands were quantified densitometrically using the BioCapt software (Vilber Lourmat, Germany). Values were normalized to GAPDH and results reported as fold change to the none-stimulated cells.

For quantitative PCR, LightCycler Fast Start DNA Master SYBR Green kit (Roche, Germany) and target-specific, company designed and validated primer kits for Scleraxis, tenomodulin, α1, α2, α11, MMP3, MMP9, MMP13, MMP14, HPRT and GAPDH (Search-LC, Germany) were used. The quantitative PCR was performed in LightCycler1.5 instrument (Roche) equipped with LightCycler 3.5.3 software. Crossing points for each sample and inter-run calibrator were determined by the second derivative maximum method and relative quantification was performed using the comparative ΔΔCt method with efficiency correction according to the manufacturer's protocol. The relative gene expression was calculated as a ratio to GAPDH or HPRT, depending on the expression levels (high or low) of the targeted genes. GAPDH and HPRT were selected for normalization due to reported stability upon mechanical loading [51,52]. Detailed information about qPCR template and conditions can be found in Additional file 2: Table S2. All PCR results have been reproduced three independent times.

Western blot analysis

Total protein from TSPC (none-stimulated or with 8% mechanical loading) at passage 6 was isolated according to Alberton et al., [53]. In brief, adherent cells were lysed in 1x Cell Culture Lysis Reagent (Promega, 25 mM Tris-phosphate pH 7.8, 2, mM DTT, 2 mM 1,2-diaminocyclo-hexane-N,N,N',N'-tetraacetic acid, 10% glycerol, 1% Triton X-100). Total protein was quantified with Micro BCA protein assay kit (Pierce, USA). Aliquots of 20 μg were denatured at 99°C for 5 min and loaded on SDS-PAGE gels. Then, proteins were transferred onto PVDF membrane, blocked with 5% skim milk (Merck) and incubated in primary anti-human antibodies: integrins α2 (BD Bioscience) and α11 (R&D Systems), phospho-FAK (Life technologies), total FAK, total and phospho-ERK1/2; p38, Jnk and MMP9 (all Cell Signaling, USA), MMP13 and MMP14 (Thermo scientific, USA) and GAPDH loading control (Merck) overnight at 4°C. Secondary HRP-conjugated antibodies (Rockland, USA or Cell Signaling) were applied for 1 h at room temperature. Western blots were visualized with ECL solution (GE Healthcare, USA) as photomicrographs were taken on ImageQuant LAS 4000 mini (GE healthcare, USA) as bent size was quantify by using the machine software.

Statistics

Statistical evaluation was performed using the GraphPad Prism 5 software (GraphPad Software, USA). N = 3 means that the results of three different TSPC donors were pull together as for each donor we used a mean value of three independent experiments (N = 3). Graphs and bar charts show mean values and standard deviation. Unpaired *t*-test was used for each condition versus the control (non-stimulated TSPC). A p-value <0.05 was considered statistically significant (*p < 0.05; **p < 0.01, ***p < 0.001).

Abbreviations

TSPC: Tendon stem/progenitor cells; MMP: Matrix metalloproteinases; ERK1/2: Extracellular-signal-regulated kinases 1 and 2; p38: p38 Mitogen-activated protein kinases; ECM: Extracellular matrix; JNK: c-Jun N-terminal kinases; Akt: protein kinase B; COMP: Cartilage oligomeric matrix protein; FAK: Focal adhesion kinase; hMSC: Human mesenchymal stem cells; MSC: Mesenchymal stem cells; IBMX: 3-Isobutyl-1-methylxanthine and FBS, fetal bovine serum.

Competing interest

The authors declare that they have no competing interests.

Authors' contributions

CP: collection and/or assembly of data, data analysis and interpretation, manuscript writing; MB and LK: collection and/or assembly of data; AI and MS: final approval of manuscript; DD: conception and design, data analysis and interpretation. All authors read and approved the final manuscript.

Acknowledgements

We acknowledge the support of the German Research Foundation (Grants PO1718/1-1 and DO1414/1-1) and Medical Faculty of the Ludwig-Maximilians-University (FöFoLe Grant Reg. Nr-668). Prof. Matthias Schieker present address is at Amgen GmbH, Munich, Germany. We thank Manuela Mißbach for technical assistance.

Author details

[1]Department of Surgery, Experimental Surgery and Regenerative Medicine, Ludwig-Maximilians-University (LMU), Nussbaumstr. 20, D-80336 Munich, Germany. [2]Institute of Orthopaedic Research and Biomechanics, University of Ulm, Helmholtzstr. 14, D-89081 Ulm, Germany.

References

1. Bi Y, Ehirchiou D, Kilts TM, Inkson CA, Embree MC, Sonoyama W, et al. Identification of tendon stem/progenitor cells and the role of the extracellular matrix in their niche. Nat Med. 2007;13(10):1219–27.
2. Docheva D, Popov C, Alberton P, Aszodi A. Integrin signaling in skeletal development and function. Birth defects research Part C, Embryo today: reviews. 2014;102(1):13–36.
3. Mammoto A, Mammoto T, Ingber DE. Mechanosensitive mechanisms in transcriptional regulation. J Cell Sci. 2012;125(Pt 13):3061–73.
4. Kjaer M. Role of extracellular matrix in adaptation of tendon and skeletal muscle to mechanical loading. Physiol Rev. 2004;84(2):649–98.
5. Maeda E, Shelton JC, Bader DL, Lee DA. Differential regulation of gene expression in isolated tendon fascicles exposed to cyclic tensile strain in vitro. J Appl Physiol. 2009;106(2):506–12.
6. Wang JH. Mechanobiology of tendon. J Biomech. 2006;39(9):1563–82.
7. Docheva D, Popov C, Mutschler W, Schieker M. Human mesenchymal stem cells in contact with their environment: surface characteristics and the integrin system. J Cell Mol Med. 2007;11(1):21–38.
8. Popov C, Radic T, Haasters F, Prall WC, Aszodi A, Gullberg D, et al. Integrins alpha2beta1 and alpha11beta1 regulate the survival of mesenchymal stem cells on collagen I. Cell death & disease. 2011;2:e186.
9. Mackley JR, Ando J, Herzyk P, Winder SJ. Phenotypic responses to mechanical stress in fibroblasts from tendon, cornea and skin. Biochemical j. 2006;396(2):307–16.
10. Fong KD, Trindade MC, Wang Z, Nacamuli RP, Pham H, Fang TD, et al. Microarray analysis of mechanical shear effects on flexor tendon cells. Plast Reconstr Surg. 2005;116(5):1393–404. discussion 1405-1396.
11. Scott A, Cook JL, Hart DA, Walker DC, Duronio V, Khan KM. Tenocyte responses to mechanical loading in vivo: a role for local insulin-like growth factor 1 signaling in early tendinosis in rats. Arthritis Rheum. 2007;56(3):871–81.
12. Androjna C, Spragg RK, Derwin KA. Mechanical conditioning of cell-seeded small intestine submucosa: a potential tissue-engineering strategy for tendon repair. Tissue Eng. 2007;13(2):233–43.
13. Kohler J, Popov C, Klotz B, Alberton P, Prall WC, Haasters F, et al. Uncovering the cellular and molecular changes in tendon stem/progenitor cells attributed to tendon aging and degeneration. Aging cell. 2013;12(6):988–99.
14. Russell KC, Phinney DG, Lacey MR, Barrilleaux BL, Meyertholen KE, O'Connor KC. In vitro high-capacity assay to quantify the clonal heterogeneity in trilineage potential of mesenchymal stem cells reveals a complex hierarchy of lineage commitment. Stem Cells. 2010;28(4):788–98.
15. Kolf CM, Cho E, Tuan RS. Mesenchymal stromal cells. Biology of adult mesenchymal stem cells: regulation of niche, self-renewal and differentiation. Arthritis res ther. 2007;9(1):204.
16. Covas DT, Panepucci RA, Fontes AM, Silva Jr WA, Orellana MD, Freitas MC, et al. Multipotent mesenchymal stromal cells obtained from diverse human tissues share functional properties and gene-expression profile with CD146+ perivascular cells and fibroblasts. Exp Hematol. 2008;36(5):642–54.
17. Toma JG, Akhavan M, Fernandes KJ, Barnabe-Heider F, Sadikot A, Kaplan DR, et al. Isolation of multipotent adult stem cells from the dermis of mammalian skin. Nat Cell Biol. 2001;3(9):778–84.
18. Lin G, Liu G, Banie L, Wang G, Ning H, Lue TF, et al. Tissue distribution of mesenchymal stem cell marker Stro-1. Stem Cells Dev. 2011;20(10):1747–52.
19. Tempfer H, Wagner A, Gehwolf R, Lehner C, Tauber M, Resch H, et al. Perivascular cells of the supraspinatus tendon express both tendon- and stem cell-related markers. Histochem Cell Biol. 2009;131(6):733–41.
20. Brent AE, Schweitzer R, Tabin CJ. A somitic compartment of tendon progenitors. Cell. 2003;113(2):235–48.
21. Docheva D, Hunziker EB, Fassler R, Brandau O. Tenomodulin is necessary for tenocyte proliferation and tendon maturation. Mol Cell Biol. 2005;25(2):699–705.
22. Wang JH, Thampatty BP, Lin JS, Im HJ. Mechanoregulation of gene expression in fibroblasts. Gene. 2007;391(1–2):1–15.
23. Ying B, Fan H, Wen F, Xu D, Liu D, Yang D, et al. Mechanical strain-induced c-fos expression in pulmonary epithelial cell line A549. Biochem Biophys Res Commun. 2006;347(1):369–72.
24. Liedert A, Kassem M, Claes L, Ignatius A. Mechanosensitive promoter region in the human HB-GAM gene. Biochem Biophys Res Commun. 2009;387(2):289–93.
25. Yang G, Crawford RC, Wang JH. Proliferation and collagen production of human patellar tendon fibroblasts in response to cyclic uniaxial stretching in serum-free conditions. J Biomech. 2004;37(10):1543–50.
26. Howard PS, Kucich U, Taliwal R, Korostoff JM. Mechanical forces alter extracellular matrix synthesis by human periodontal ligament fibroblasts. J Periodontal Res. 1998;33(8):500–8.
27. Hsieh AH, Tsai CM, Ma QJ, Lin T, Banes AJ, Villarreal FJ, et al. Time-dependent increases in type-III collagen gene expression in medical collateral ligament fibroblasts under cyclic strains. Journal of orthopaedic research: official publication of the Orthopaedic Research Society. 2000;18(2):220–7.
28. Yoon JH, Halper J. Tendon proteoglycans: biochemistry and function. Journal of musculoskeletal & neuronal interactions. 2005;5(1):22–34.
29. Wight TN. Versican: a versatile extracellular matrix proteoglycan in cell biology. Curr Opin Cell Biol. 2002;14(5):617–23.
30. Juncosa-Melvin N, Matlin KS, Holdcraft RW, Nirmalanandhan VS, Butler DL. Mechanical stimulation increases collagen type I and collagen type III gene expression of stem cell-collagen sponge constructs for patellar tendon repair. Tissue Eng. 2007;13(6):1219–26.
31. Desmoulin GT, Hewitt CR, Hunter CJ. Disc strain and resulting positive mRNA expression from application of a noninvasive treatment. Spine. 2011;36(14):E921–8.
32. Subramony SD, Dargis BR, Castillo M, Azeloglu EU, Tracey MS, Su A, et al. The guidance of stem cell differentiation by substrate alignment and mechanical stimulation. Biomaterials. 2013;34(8):1942–53.
33. Burkholder TJ. Mechanotransduction in skeletal muscle. Frontiers in bioscience : a journal and virtual library. 2007;12:174–91.
34. Sakamoto K, Goodyear LJ. Invited review: intracellular signaling in contracting skeletal muscle. J Appl Physiol (1985). 2002;93(1):369–83.

35. Zhou Y, Millward-Sadler SJ, Lin H, Robinson H, Goldring M, Salter DM, et al. Evidence for JNK-dependent up-regulation of proteoglycan synthesis and for activation of JNK1 following cyclical mechanical stimulation in a human chondrocyte culture model. Osteoarthritis and cartilage / OARS, Osteoarthritis Research Society. 2007;15(8):884–93.

36. De Croos JN, Dhaliwal SS, Grynpas MD, Pilliar RM, Kandel RA. Cyclic compressive mechanical stimulation induces sequential catabolic and anabolic gene changes in chondrocytes resulting in increased extracellular matrix accumulation. Matrix biology : journal of the International Society for Matrix Biology. 2006;25(6):323–31.

37. Xu HG, Li ZR, Wang H, Liu P, Xiang SN, Wang CD, et al. Expression of ectonucleotide pyrophosphatase-1 in end-plate chondrocytes with transforming growth factor beta 1 siRNA interference by cyclic mechanical tension. Chin Med J. 2013;126(20):3886–90.

38. Toma CD, Ashkar S, Gray ML, Schaffer JL, Gerstenfeld LC. Signal transduction of mechanical stimuli is dependent on microfilament integrity: identification of osteopontin as a mechanically induced gene in osteoblasts. Journal of bone and mineral research : the official journal of the American Society for Bone and Mineral Research. 1997;12(10):1626–36.

39. Plotkin LI, Mathov I, Aguirre JI, Parfitt AM, Manolagas SC, Bellido T. Mechanical stimulation prevents osteocyte apoptosis: requirement of integrins, Src kinases, and ERKs. American journal of physiology Cell physiology. 2005;289(3):C633–43.

40. Kong D, Zheng T, Zhang M, Wang D, Du S, Li X, et al. Static mechanical stress induces apoptosis in rat endplate chondrocytes through MAPK and mitochondria-dependent caspase activation signaling pathways. PLoS One. 2013;8(7):e69403.

41. Matsui H, Fukuno N, Kanda Y, Kantoh Y, Chida T, Nagaura Y, et al. The expression of Fn14 via mechanical stress-activated JNK contributes to apoptosis induction in osteoblasts. The Journal of biological chemistry. 2014;289(10):6438–50.

42. Danciu TE, Adam RM, Naruse K, Freeman MR, Hauschka PV. Calcium regulates the PI3K-Akt pathway in stretched osteoblasts. FEBS Lett. 2003;536(1–3):193–7.

43. Munshi HG, Stack MS. Reciprocal interactions between adhesion receptor signaling and MMP regulation. Cancer Metastasis Rev. 2006;25(1):45–56.

44. Yang G, Im HJ, Wang JH. Repetitive mechanical stretching modulates IL-1beta induced COX-2, MMP-1 expression, and PGE2 production in human patellar tendon fibroblasts. Gene. 2005;363:166–72.

45. Wang T, Lin Z, Day RE, Gardiner B, Landao-Bassonga E, Rubenson J, et al. Programmable mechanical stimulation influences tendon homeostasis in a bioreactor system. Biotechnol Bioeng. 2013;110(5):1495–507.

46. Fujisawa T, Hattori T, Takahashi K, Kuboki T, Yamashita A, Takigawa M. Cyclic mechanical stress induces extracellular matrix degradation in cultured chondrocytes via gene expression of matrix metalloproteinases and interleukin-1. J Biochem. 1999;125(5):966–75.

47. Sasaki K, Takagi M, Konttinen YT, Sasaki A, Tamaki Y, Ogino T, et al. Upregulation of matrix metalloproteinase (MMP)-1 and its activator MMP-3 of human osteoblast by uniaxial cyclic stimulation. J Biomed Mater Res B Appl Biomater. 2007;80(2):491–8.

48. Archambault J, Tsuzaki M, Herzog W, Banes AJ. Stretch and interleukin-1beta induce matrix metalloproteinases in rabbit tendon cells in vitro. Journal of orthopaedic research : official publication of the Orthopaedic Research Society. 2002;20(1):36–9.

49. Sun HB, Yokota H. Reduction of cytokine-induced expression and activity of MMP-1 and MMP-13 by mechanical strain in MH7A rheumatoid synovial cells. Matrix biology : journal of the International Society for Matrix Biology. 2002;21(3):263–70.

50. Neidlinger-Wilke C, Wilke HJ, Claes L. Cyclic stretching of human osteoblasts affects proliferation and metabolism: a new experimental method and its application. Journal of orthopaedic research : official publication of the Orthopaedic Research Society. 1994;12(1):70–8.

51. Liu J, Zou L, Wang J, Zhao Z. Validation of beta-actin used as endogenous control for gene expression analysis in mechanobiology studies. Stem Cells. 2009;27(9):2371–2.

52. Kopf J, Petersen A, Duda GN, Knaus P. BMP2 and mechanical loading cooperatively regulate immediate early signalling events in the BMP pathway. BMC Biol. 2012;10:37.

53. Alberton P, Popov C, Pragert M, Kohler J, Shukunami C, Schieker M, et al. Conversion of human bone marrow-derived mesenchymal stem cells into tendon progenitor cells by ectopic expression of scleraxis. Stem Cells Dev. 2012;21(6):846–58.

Nop17 is a key R2TP factor for the assembly and maturation of box C/D snoRNP complex

Marcela B Prieto[1], Raphaela C Georg[1,2], Fernando A Gonzales-Zubiate[1], Juliana S Luz[1,3] and Carla C Oliveira[1*]

Abstract

Background: Box C/D snoRNPs are responsible for rRNA methylation and processing, and are formed by snoRNAs and four conserved proteins, Nop1, Nop56, Nop58 and Snu13. The snoRNP assembly is a stepwise process, involving other protein complexes, among which the R_2TP and Hsp90 chaperone. Nop17, also known as Pih1, has been shown to be a constituent of the R_2TP (Rvb1, Rvb2, Tah1, Pih1) and to participate in box C/D snoRNP assembly by its interaction with Nop58. The molecular function of Nop17, however, has not yet been described.

Results: To shed light on the role played by Nop17 in the maturation of snoRNP, here we analyzed the interactions domains of Nop58 – Nop17 – Tah1 and the importance of ATP to the interaction between Nop17 and the ATPase Rvb1/2.

Conclusions: Based on the results shown here, we propose a model for the assembly of box C/D snoRNP, according to which R_2TP complex is important for reducing the affinity of Nop58 for snoRNA, and for the binding of the other snoRNP subunits.

Keywords: box C/D snoRNP assembly, R_2TP complex, Nop17

Background

Box C/D snoRNP complexes are involved in pre-rRNA cleavage and in 2′-O-methylation of nucleotides at specific positions in rRNAs, snRNAs, and other RNAs during maturation [1]. In yeast, these complexes are formed by snoRNAs that contain conserved sequences (boxes C, D, C′ and D′), and four core proteins, Nop1, Nop56, Nop58 and Snu13. In addition to these proteins, some other factors may associate with specific snoRNP, such as the U3 snoRNP [2]. snoRNP complexes are conserved from archaea to eukaryotes, although in the latter they are more complex [3].

The assembly of snoRNP is initiated in the nucleoplasm and completed in the cajal bodies in mammalian cells, whereas in yeast, the final steps of assembly and maturation of snoRNPs are considered to occur in the nucleolus, a compartment where the snoRNPs also catalyze the rRNA modifications [4,5].

During snoRNP assembly, Snu13 binds RNA by recognizing a conserved RNA secondary structure that is present in box C/D snoRNAs, as well as in the U4 snRNA [6]. Due to its affinity to RNA, the human orthologue of Snu13, 15.5kD, has been shown to be the first core snoRNP subunit to bind box C/D snoRNAs [7]. Although the later steps of assembly are less well defined, it has been shown that Nop1 and Nop58 bind the snoRNAs independently, whereas Nop56 depends on Nop1 for binding the complexes [3]. Due to the many structural rearrangements that occur during assembly, chaperones may be important for the maturation of snoRNPs.

Nop17, also known as Pih1, has been shown to strongly interact with Nop58 [8] and with the chaperone Hsp90 [9]. Through its interaction with Hsp90, Nop17 was identified as part of a complex named R_2TP (Rvb1, Rvb2, Tah1, Pih1) [9,10]. Rvb1 and Rvb2 are ATP dependent helicases, belonging to the class of AAA^+ ATPases [11], that form heterohexamers *in vitro* and participate in processes ranging from DNA repair, transcription, chromatin remodeling, ribosomal RNA processing, to small nucleolar RNP formation [12]. The human orthologues of Rvb1/Rvb2, TIP48/TIP49, have also been shown to be involved in box C/D snoRNP assembly [7].

Tah1, a TPR (tetratricopeptide repeat)-containing protein associated with heat-shock protein Hsp90 has been

---

* Correspondence: ccoliv@iq.usp.br
[1]Department of Biochemistry, Institute of Chemistry, University of São Paulo, Av. Prof. Lineu Prestes 748, 05508-000 São Paulo, SP, Brazil
Full list of author information is available at the end of the article

shown to bind directly to Hsp90 [12,13]. Nop17 binds to the Hsp90-Tah1 complex and has been proposed to control Hsp90 ATPase activity [13]. The structures of the interaction regions of Tah1-Hsp90 and Tah1-Nop17 have been determined, and a model was proposed, according to which the Tah1 TPR domain adopts a highly folded structure, whereas the C-terminal region of Tah1 only folds upon its interaction with Nop17 [14].

Nop17 interacts with proteins involved in various cellular processes [8,15,16], probably helping the assembly of different complexes. The role played by Nop17 in snoRNP assembly depends on its interaction with Hsp90 as part of the R_2TP complex. Despite the studies on the interactions between the R_2TP complex and Hsp90, and the determination of the structure of the complex, the molecular function of Nop17 remains elusive.

Rsa1, although not a subunit of the R_2TP complex, has also been shown to be involved in box C/D snoRNP formation through its interaction with Snu13 and with Nop17 [17]. It has been proposed that Rsa1 binds immature snoRNP particles and is released upon assembly of the mature protein subunits of the complexes for their active conformation [18].

R_2TP also interacts with the prefoldin complex, which participates in protein folding, degradation and rearrangements [19], broadening the range of protein interactions of the R_2TP complex, and therefore, of Nop17. In this work, we describe further studies on the interactions between Nop17 and the R_2TP complex, and between Nop17 and the box C/D snoRNP core subunits. Through the analysis of the interaction of a Nop17 point mutant with Tah1, we were able to narrow down the interface regions of these proteins, and also analyzed the effect of the presence of ATP on the interaction of the Rvb1/2 ATPase with Nop17. In addition, we mapped the region of Nop58 involved in the interaction with Nop17. Based on the data presented here, we propose a model for the role of R_2TP in snoRNP assembly.

Results

Interactions of Nop17 within R_2TP complex

Nop17/Pih1 was identified as part of the R_2TP complex, together with Rvb1, Rvb2, and Tah1 [20]. In this complex, Nop17 has been shown to interact directly with Tah1 in pull-down assays [9], and with Rvb1 and Rvb2 in the two-hybrid system [17]. In order to analyze in more detail Nop17 interactions with R_2TP subunits, two-hybrid and pull-down assays were performed. Nop17 showed interaction with Rvb1 and Rvb2 in the two-hybrid system when fused to both domains: the lexA DNA binding domain, and the Gal4 transcription activation domain (Figure 1A). Interaction between Rvb1 and Rvb2 was also positive in both fusions. As expected, Rvb1 and Rvb2 show higher affinity for each other in the two-hybrid assay than for

Nop17 (Figure 1A). Since Rvb1 and Rvb2 are ATPases [12], to determine whether ATP binding or hydrolysis may affect the interaction between these proteins and Nop17, pull-down assays were performed with recombinant proteins in the absence or in the presence of either ATP or ADP. The results show that the interaction Nop17-Rvb2 is independent of ATP or its hydrolytic product, whereas Nop17 only interacts efficiently with Rvb1 in the absence of ATP (Figure 1B). These results are interesting because they suggest that Nop17 may interact directly with Rvb2, independently of Rvb1 or ATP. Despite the direct binding of Nop17 to Rvb1, this interaction is hindered by the presence of ATP or ADP.

Experiments with deletion mutants of Nop17 have shown that its C-terminal portion is important for the interaction with Tah1 [23]. To further analyze the region of Nop17 responsible for its interaction with Tah1 and other proteins, we performed random in vitro mutagenesis in NOP17 gene fused to lexA DNA binding domain and tested the interaction of the mutants with Gal4AD-Tah1 in the two-hybrid system. A nop17 mutant was obtained that no longer interacts with Tah1, this mutant has an asparagine to serine substitution in position 306 (N306S) (Figure 1C). Western blot results show that BD-nop17(N306S) is expressed in L40 cells, although in lower levels than BD-Nop17 (Additional file 1: Figure S1). Interestingly, a very recent report on the structure of Nop17-Tah1 interaction domains show that N306 is part of a beta sheet in Nop17 CS domain, interacting with Tah1 C-terminal region [24]. Further studies will reveal how this mutation might affect Nop17 structure in order to disrupt the interaction with Tah1. Molecular modeling analysis suggests that the asparagine to serine substitution in the position 306 might affect the intramolecular interactions in the Nop17 CS domain (data not shown).

Rsa1 and Tah1 affect Nop17 stability

Previous studies have shown that Tah1 interaction is important for Nop17 stability [20]. We therefore tested the expression levels of Nop17 in the Δtah1 strain. In addition, due to Rsa1 involvement in snoRNP formation [17,18], we also tested Nop17 in the Δrsa1 strain. Since the strains Δnop17 and Δrsa1 are temperature sensitive, we tested Nop17 levels at the permissive and restrictive growth temperatures. The results show that Nop17 levels decrease in Δrsa1 at 25°C, and are even lower at 37°C (Figure 2A). Interestingly, in Δtah1 strain, Nop17 levels are very low, regardless of the temperature of growth, suggesting that the interaction with Tah1 is more important for the stability of Nop17, and also suggesting that Nop17 may not be found in the cell in a free form, but only bound to Tah1, either in the R_2TP complex, or in a ternary complex with Hsp90. Further confirming that the

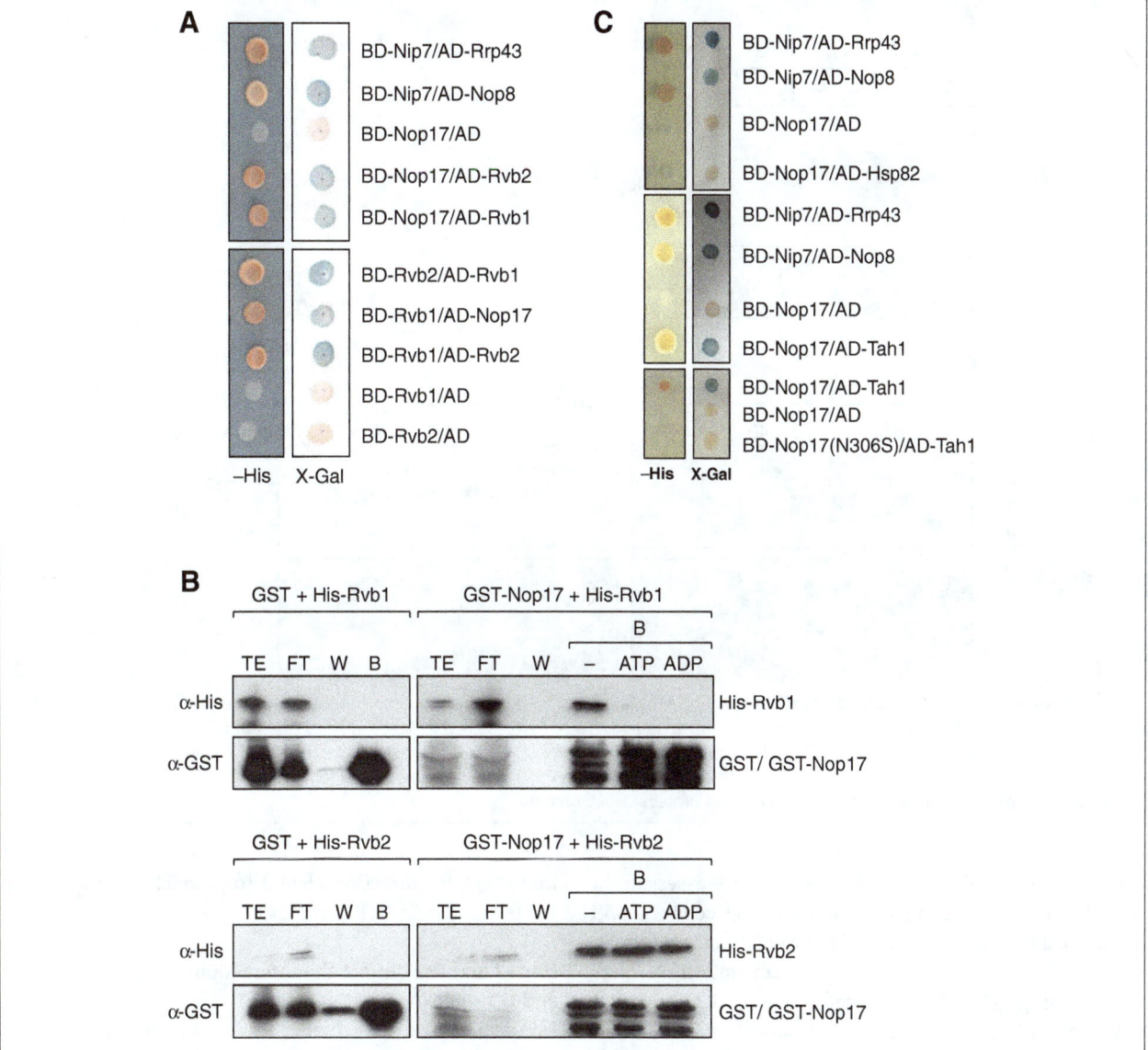

Figure 1 Interaction of Nop17 with the other subunits of the R₂TP complex. (A) Analysis of the interactions between Nop17 and Rvb1, and Rvb2 through the two-hybrid assay. BD-Nop17 interacts with both AD-Rvb1 and AD-Rvb2, as seen by the expression of the reporter genes *HIS3* and *lacZ*. BD-Rvb2 + AD-Nop17 is stronger than BD-Rvb1-AD-Nop17. BD-Nip7/AD-Rrp43 and BD-Nip7/AD-Nop8 were used as positive controls for interaction [21,22] **(B)** Pull-down assay to confirm direct interaction between Nop17 and Rvb1/2. GST or GST-Nop17 were bound to glutathione-sepharose beads, followed by the incubation with His-Rvb1, or His-Rvb2, in the absence, or presence of 1 mM ATP or ADP at 4°C for 2 hours. Fractions from total extract (TE), flow through (FT), wash (W), or bound **(B)** were separated by SDS-PAGE and subjected to western blot with anti-His or anti-GST sera. Interaction Nop17-Rvb2 is independent of ATP. **(C)** Two-hybrid assay for the analysis of Nop17-Tah1 and Nop17-Hsp90 interaction. BD-Nop17 did not interact with AD-Hsp90, whereas BD-Nop17 interacted with AD-Tah1. Mutation of Nop17 in the position 306 disrupts interaction with Tah1.

interaction with Tah1 is important for Nop17 stability, the steady-state level of the mutant Nop17(N306S) is lower than that of the wild type protein (Additional file 1: Figure S1). Interestingly, deletion of Rsa1 also leads to the destabilization of Nop17 (Figure 2B). Nop17 shows a half-life of 90 min in WT cells, but it decreases to 55 min in Δ*rsa1* strain at 37°C. These results show that Rsa1 also plays a role in Nop17 stability.

Nop17 is important for Nop58 stability

Nop17 interacts directly with Nop58 and is important for the assembly of the box C/D snoRNP particle, probably by directing Hsp90 chaperone to the particle [8,18]. Considering Δ*nop17* temperature sensitivity, and the involvement of Nop17 in targeting Hsp90-Tah1 to client proteins, those client proteins might be destabilized in the absence of Nop17 at higher temperatures. To test

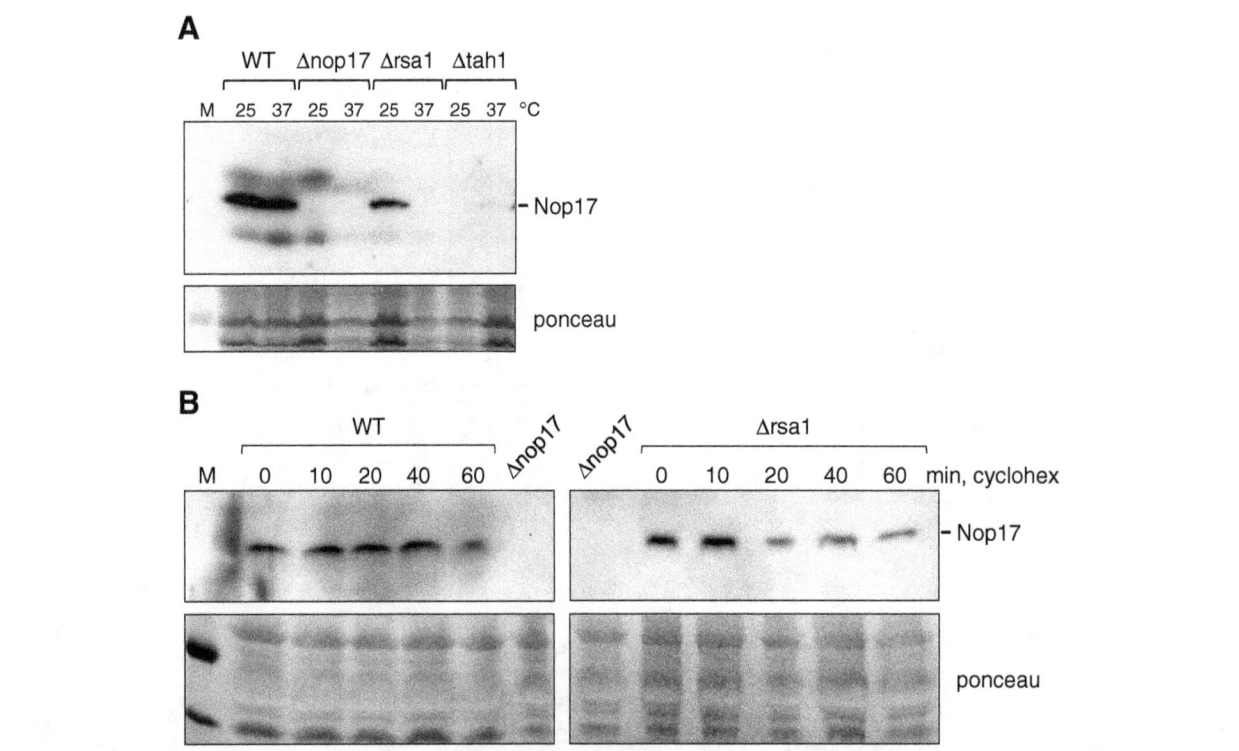

Figure 2 Absence of Rsa1 or Tah1 destabilizes Nop17. (A) Total extract from cells growing either at the permissive (25°C), or restrictive (37°C) temperature to OD_{600} 0.5 were used for western blot with serum against Nop17. Steady state levels of Nop17 do not change in wild type cells, whereas in Δ*rsa1* and Δ*tah1* strains, Nop17 levels decrease drastically. **(B)** WT and Δ*rsa1* cells were treated with cyclohexamide after incubation to OD_{600} of 0.8 at 37°C. Samples were collected at the indicated time points and subjected to western blot for the detection of Nop17. Ponceau staining of the membranes was used as control for total protein loaded on gels.

this hypothesis, ProtA-Nop58 levels were assessed in Δ*nop17* strain and compared to wild type cells. Nop58 has been shown to be unstable *in vitro* [3,25], therefore, in order to detect the protein, immunoprecipitation was performed using IgG-sepharose beads. The results show that full-length ProtA-Nop58 can be visualized in the bound fraction from wild type cells, but the protein is destabilized in Δ*nop17* strain, resulting in breakdown products that are detected in the bound fractions (Figure 3A). To test whether higher levels of the chaperone could stabilize Nop58, Hsp90 was overexpressed together with Nop58 in either wild type or Δ*nop17* strains. The results show that in the absence of Nop17, Nop58 is unstable, regardless of the overexpression of the chaperone (Figure 3B), suggesting that indeed Nop17 is required for directing Hsp90 to its client protein Nop58. These data are also in agreement with the model of Nop17 being an Hsp90 co-chaperone, responsible for inhibiting its ATPase activity, which is important for the loading of client proteins onto Hsp90 [13]. Accordingly, loss of Tah1 and Rsa1 has the same destabilizing effect on Nop58 seen in Δ*nop17* strain (Figure 3C). The effects of these latter proteins on Nop58 could be indirect, and depend upon Nop17 interaction with Nop58. These results corroborate the hypothesis of Nop17 being

important for directing Hsp90 to Nop58, and thereby, to the box C/D snoRNP complex.

Nop17 and Rsa1 affect the interaction between Nop58 and U3 snoRNA

Interestingly, despite being less stable upon depletion of the R$_2$TP complex, Nop58 shows higher affinity for box C/D snoRNA in the absence of Nop17 [8]. A similar, though weaker, effect is seen in the absence of Rsa1 (Figure 4A), suggesting that the recruitment of Hsp90 and its co-chaperones is important for the correct binding of Nop58 to the snoRNAs. A decrease in the stability of Nop58-snoRNA interaction might be required for the assembly of the other box C/D core subunits onto the complex.

To determine whether Nop58 binds snoRNA at an early or late stage of snoRNP assembly, ChIP assays were performed with different pairs of primers for the analysis of three regions of the U3 snoRNA gene. The results show that Nop58 binds U3 snoRNA co-transcriptionally in the exon 2 region, where the conserved C/D are located (Figure 4B). Corroborating the previous results, Nop58 bound snoRNA with higher affinity upon depletion of either Nop17 or Rsa1 (Figure 4B).

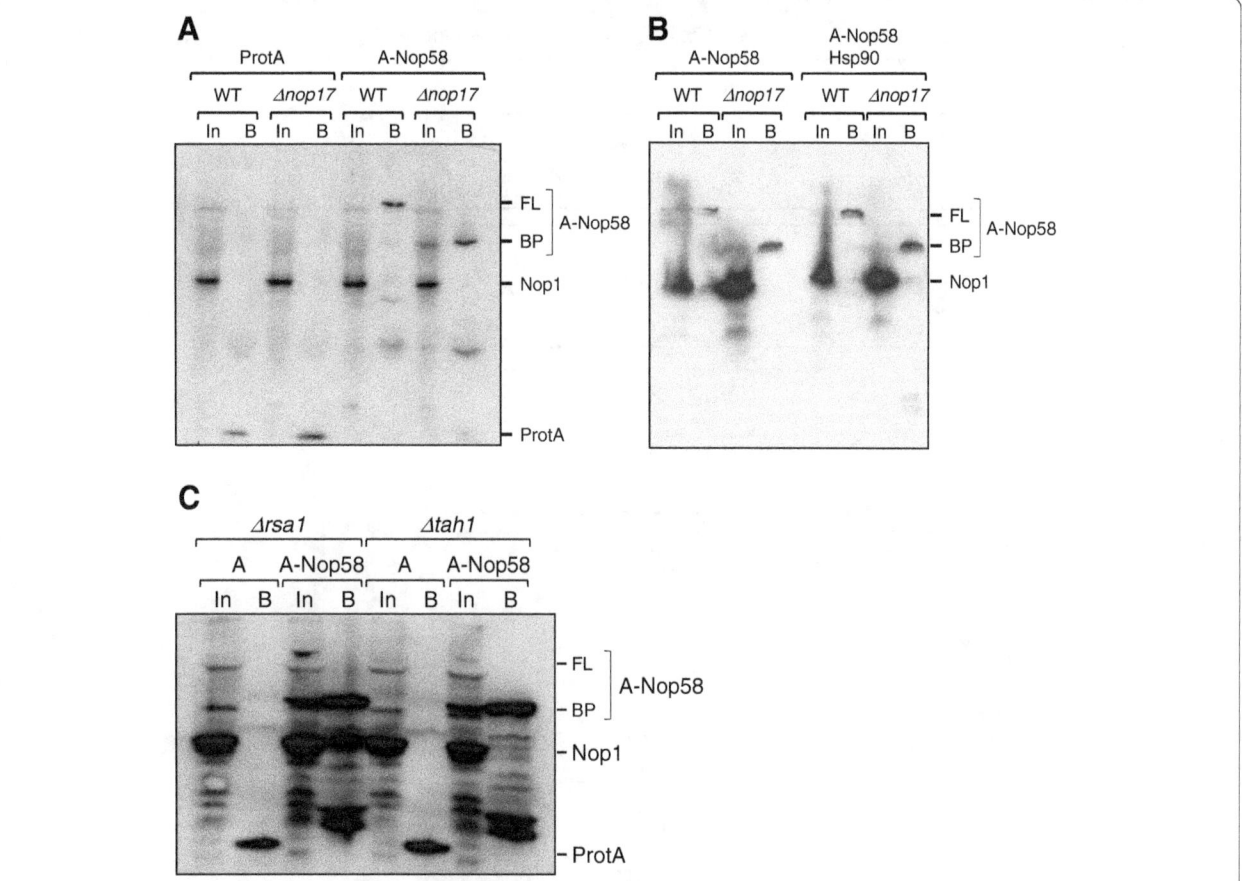

Figure 3 Nop17, Tah1 and Rsa1 affect Nop58 stability. Total extract from cells expressing either ProtA or ProtA-Nop58 were used in immunoprecipitation assays. Samples from input (In) and bound **(B)** material were analyzed by western blot for the detection of ProtA-Nop58. **(A)** Expression of ProtA or ProtA-Nop18 in strains WT and Δ*nop17*. Band corresponding to full-length (FL) ProtA-Nop58 can only be detected in samples from WT cells. In samples from Δ*nop17*, break-down products (BP) from ProtA-Nop58 are visualized. Nop1 was used as an internal control for input samples and was not co-immunoprecipitated with ProtA-Nop58 under the conditions used. **(B)** The same experiment as in **A** was performed with samples from cells overexpressing Hsp90. **(C)** Immunoprecipitation of ProtA or ProtA-Nop58 expressed in Δ*rsa1* or Δ*tah1*. Depletion of Rsa1 or Tah1 also destabilizes Nop58.

The structure of the archaeal complex Nop58/Nop1 orthologues has been determined [26]. From the structure of the archaeal complex and the analysis of protein interactions with RNA, it has been suggested that the C-terminal region of Nop5 is involved in the interaction with RNA and the protein L7Ae, the central coiled-coil region is important for the Nop5 dimerization, while the N-terminal portion interacts with Fibrillarin [27]. Taking into account the protein sequence conservation and the complexes formed in archaea and eukaryotes, the information from the archaeal complex can be used to infer that the C-terminal portion of Nop58 is responsible for its interaction with snoRNAs. As shown here and previously, Nop58 co-immunoprecipitates box C/D snoRNAs more efficiently in the absence of Nop17 or Rsa1 (Figure 4) [8], suggesting that Nop17, which interacts directly with Nop58, might compete with RNA for the interaction with the C-terminal portion of Nop58. We therefore used the two-hybrid assay to map the region of Nop58 involved in

the interaction with Nop17. The results show that Nop17 interacts with the C-terminal region of Nop58, but not with the N-terminal, or the central domains of Nop58 (Figure 5A; Additional file 2: Figure S2). Although we cannot exclude the possibility that N-terminal portions of Nop58 may not interact with Nop17 due to lower protein levels, we consider that unlikely because the region of Nop58 responsible for its instability is a highly charged KKD/E domain located at its C-terminal portion [25]. Our results therefore suggest that either Nop17 can compete with RNA for binding to Nop58, or that upon interacting with Nop17, Nop58 has its affinity for RNA decreased. To test the hypothesis of Nop17 competing with RNA for binding Nop58, we performed RNA co-immunoprecipitation assays with ProtA-Nop58 expressed in Δ*nop17* strain, and added increasing amounts of recombinant GST-Nop17 during the assay. The results show that the snoRNA U3 co-purified with ProtA-Nop58 is not released in the presence of the purified Nop17 (Figure 5B,C). These results

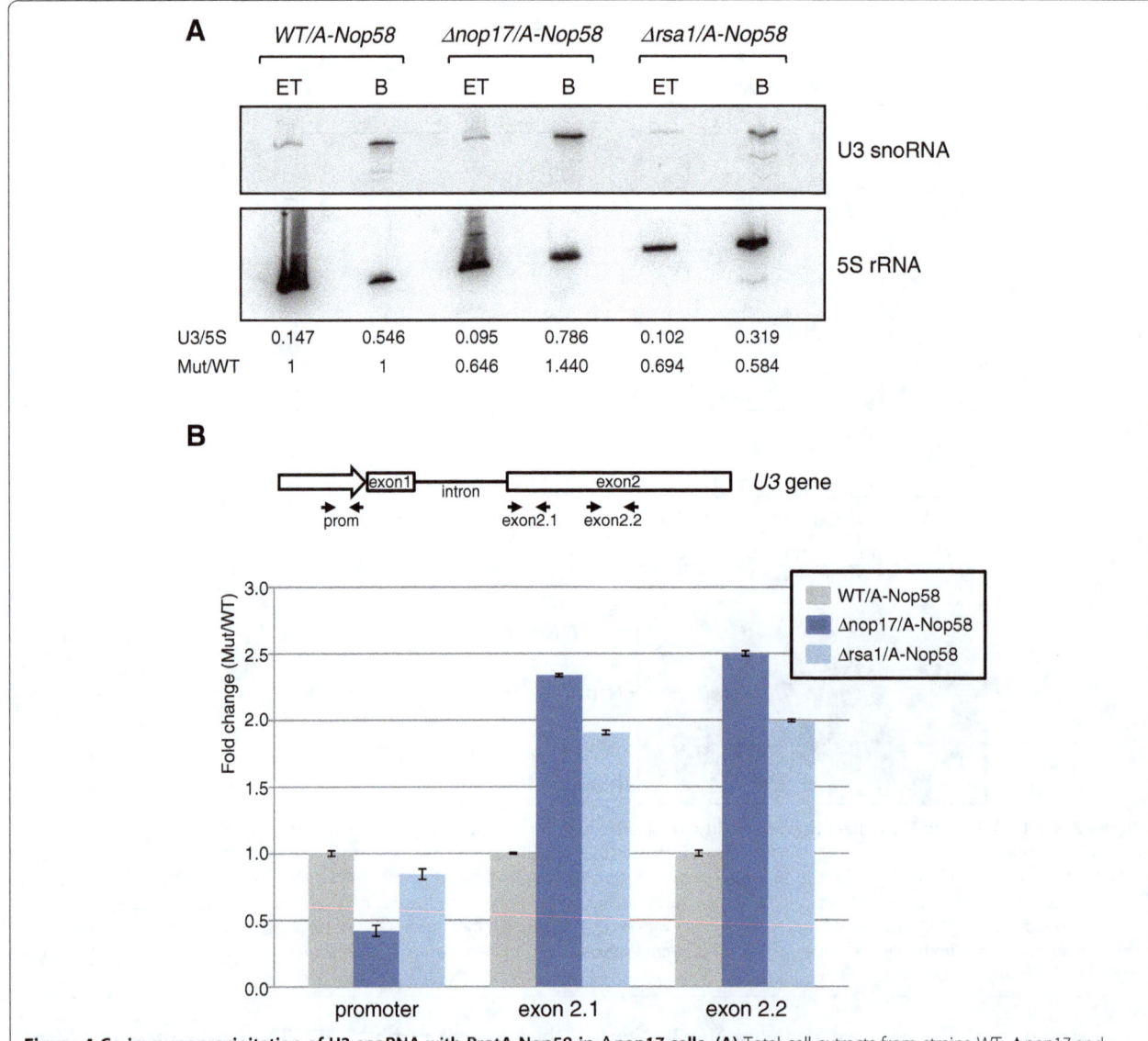

Figure 4 Co-immunoprecipitation of U3 snoRNA with ProtA-Nop58 in Δ*nop17* cells. (A) Total cell extracts from strains WT, Δ*nop17* and Δ*rsa1* were mixed with IgG-Sepharose beads for co-immunoprecipitation of snoRNAs with ProtA-Nop58. Bound RNA was detected by northern blotting using probes specific to the snoRNA U3. Membrane was washed and re-hybridized against probe specific to the 5S rRNA (internal control). TE, total extract; B, bound fraction. Bands were quantitated using Typhoon equipment and mean values of three biological replicates are shown. U3 bands were quantitated relative to 5S bands in each lane. U3/5S in mutants were then calculated relative to WT strain. **(B)** Nop58 immunoprecipitates snoRNA U3 chromatin. ChIP assay with A-Nop58 expressed in strains WT, Δ*nop17* and Δ*rsa1* was performed, followed by RT-qPCR reactions with primers for amplification of various regions of the U3 snoRNA gene. Mean values are based on three different experiments with two biological replicates.

suggest that Nop17 does not compete with RNA for binding Nop58 *in vitro*. It is also possible that only the complete R_2TP complex may affect Nop58-RNA interaction.

Nop17 interacts with other box C/D snoRNP subunits in addition to Nop58

We have previously shown that Snu13 interacts with all the other three core subunits of box C/D snoRNPs, but it does not interact with Nop17 in the two-hybrid system

[8]. To determine the snoRNP assembly step in which Nop17 is involved, we performed protein pull-down experiments with recombinant Snu13, Nop1 and Nop17. In these experiments, His-Nop1 is efficiently pulled down with GST-Snu13, but not with GST (Figure 6). Interestingly, His-Nop17 can be co-precipitated with His-Nop1 when the latter is bound to GST-Snu13 (Figure 6B). These results suggest that Nop17 can bind the heterodimer Snu13-Nop1, but not the isolated

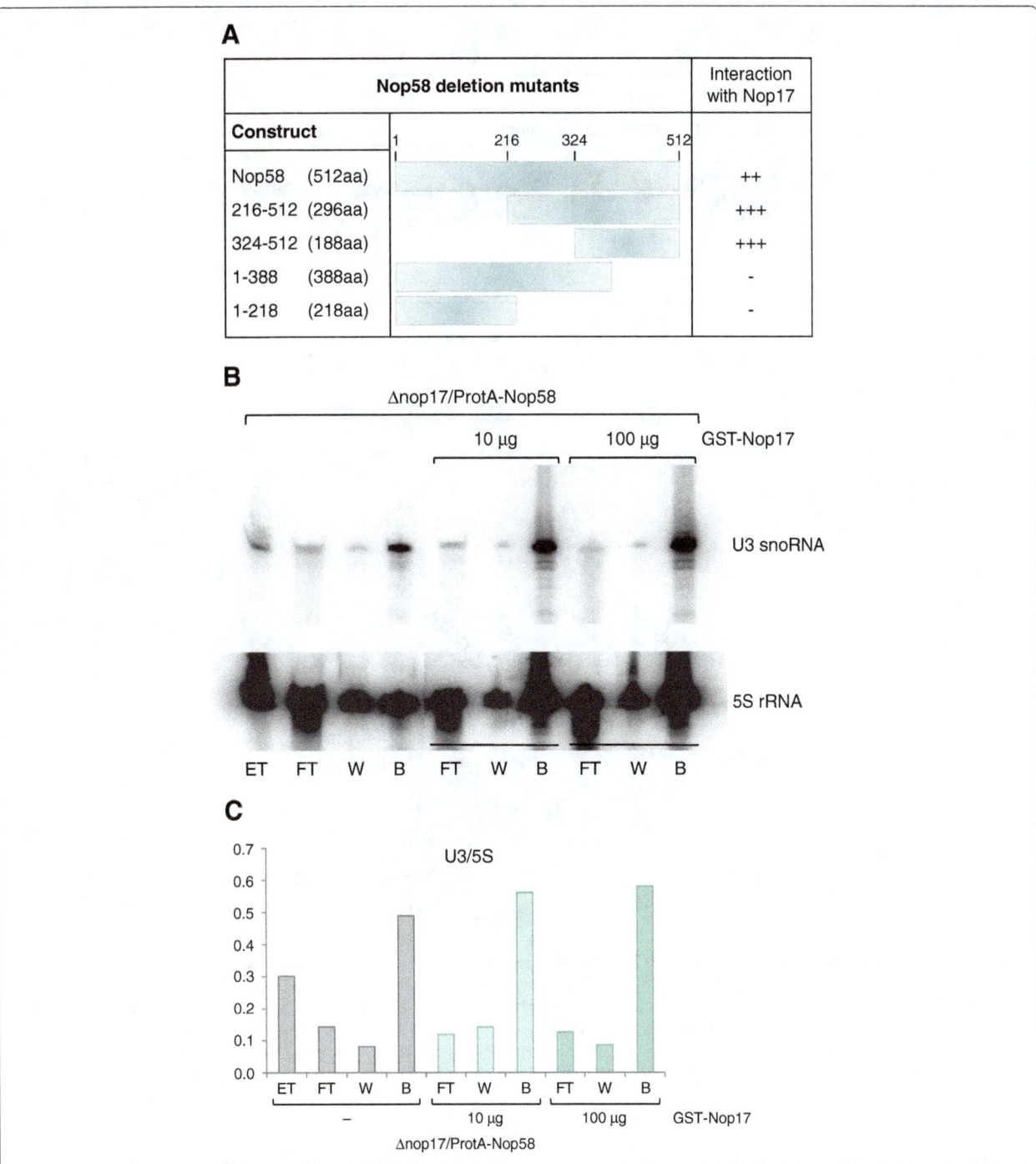

Figure 5 Nop58 interacts with Nop17 through its C-terminal portion. (A) Schematics summarizing the results of two-hybrid assay with deletion mutants of Nop58 and Nop17. Nop58(324–512) mutant interacts with Nop17. **(B)** Co-immunoprecipitation of RNA with ProtA-Nop58 in the absence or presence of different amounts of Nop17. Incubation of total extract with IgG-sepharose beads was performed for 2 h at 4°C in the presence or absence of purified GST-Nop17. Co-immunoprecipitated U3 snoRNA was detected by northern blot. 5S rRNA was used as an internal control. **(C)** Quantitation of the U3 bands corrected by 5S bands after northern hybridization.

proteins. This can be an indication that Nop17 interacts with these two proteins after they bind the snoRNA, already as part of the snoRNP assembly complex. Nop17 might be brought to the complex by its interaction with Nop58, decreasing the affinity of Nop58 for the snoRNA and allowing for the assembly of Nop1 and Snu13 onto the complex. The hydrolysis of ATP by Rvb1/2 ATPases may cause a structural rearrangement necessary for the release of the R_2TP complex and formation of the mature box C/D snoRNP.

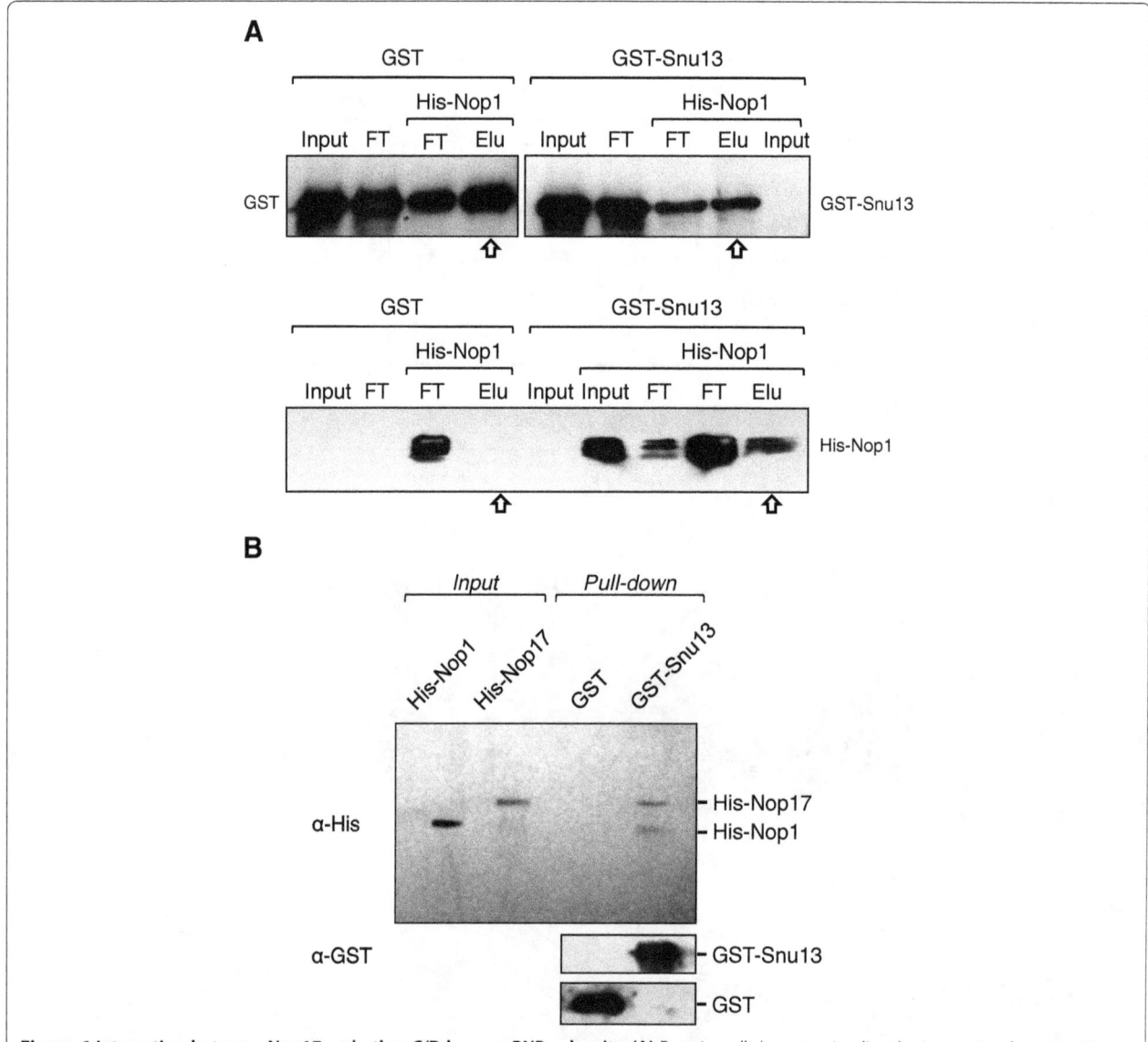

Figure 6 Interaction between Nop17 and other C/D box snoRNP subunits. (A) Protein pull-down to visualize the interaction between Nop1 and Snu13. Total extracts from *E. coli* cells expressing GST or GST-Snu13 (input) were first incubated with GST-sepharose beads. Extracts from cells expressing His-Nop1 (input) were then added to the beads. Flow through was collected (FT) after the addition of each extract, and beads were washed. Proteins were eluted with reduced glutathione (Elu). Samples from each fraction were subjected to SDS-PAGE and western blot with anti-GST and anti-His sera. Elution fractions are indicated by arrows. **(B)** Nop17 interacts with the complex Nop1/Snu13 in pull-down assays. His-Nop1 was added to either GST or GST-Snu13 immobilized in glutathione-sepharose beads. After that, His-Nop17 was added to the beads, and was pulled-down only by the GST-Snu13/His-Nop1 complex. Bands were visualized by western blot with anti-GST and anti-His sera.

Nop17 and Rsa1 affect U3 snoRNA localization

The role of R$_2$TP, and particularly of Nop17, in the assembly of box C/D snoRNP suggests that there might be an order of binding of the proteins to the snoRNA and molecular rearrangements for the formation of the mature complex. Therefore, the depletion of the R$_2$TP subunits might not only cause a mislocalization of the box C/D proteins, as has been demonstrated for Nop17 [8], but also of the snoRNAs. To test this hypothesis, we performed FISH experiments to determine the U3 snoRNA localization in the deletion strains Δ*nop17*, Δ*rsa1*, and

Δ*tah1*, compared to the wild type strain growing at the permissive or non permissive temperature. The results show that in the absence of Nop17 or Rsa1, U3 snoRNA is still concentrated in the nucleolus at 25°C, but when the cells are shifted to 37°C, the snoRNA signal becomes more disperse throughout the cells (Figure 7; Additional file 3: Figure S3). Surprisingly, however, depletion of Tah1 did not seem to affect the localization of U3 snoRNA. These results indicate that in the absence of Nop17 or Rsa1, the assembly of box C/D snoRNP is defective, mainly at the non-permissive temperature, leading to the

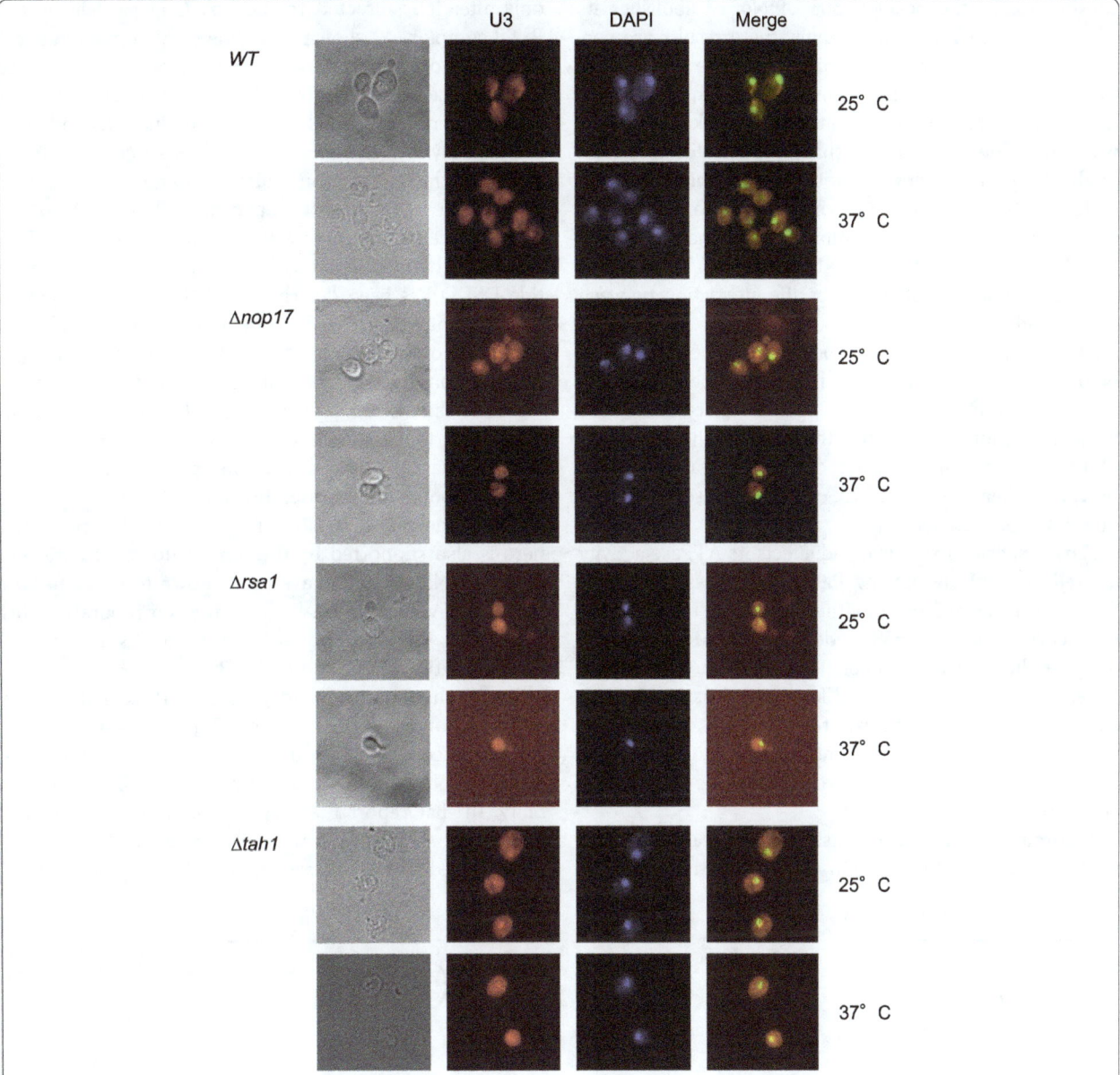

Figure 7 Depletion of R₂TP subunits affects the localization of the U3 snoRNA. FISH experiments were performed using cells that had been cultivated at 25°C or 37°C before hybridization with a fluorescent-labeled DNA oligo complementary to U3 snoRNA. DNA was labeled with DAPI.

mislocalization of the snoRNA. Interestingly, as pointed out above, absence of Nop17 has the same effect on the box C/D core proteins [8], confirming the importance of this protein for the assembly of snoRNP. Because the localization of U3 snoRNA is not strongly affected by the depletion of Tah1, these results also suggest that its function might be redundant with that of another Hsp90 co-chaperone.

Discussion

Nop17/Pih1 is a nucleolar protein [8] that has been shown to be part of the R₂TP complex, interacting with Rvb1, Rvb2, and Tah1 [9]. Rvb1 and Rvb2 are ATP dependent

helicases thought not to be present in isolated form in the cell, but instead they may form a heterohexameric complex containing three molecules of each protein [12]. This complex shows higher helicase activity *in vitro* than the isolated proteins, and undergoes nucleotide-dependent conformational changes [11,12]. Based on the results shown here, it can be hypothesized that Nop17 binds more tightly to one of the conformations of Rvb1/Rvb2 complex, thereby modulating the activity of the complex.

Nop17 has been shown to interact with Tah1 through its C-terminal region [13,14,28], it was therefore important to determine whether point mutations in the C-terminus of Nop17 affect its interaction with Tah1. As shown here, substitution in the position 306 of Nop17 abolishes its interaction with Tah1. Interestingly, these results are corroborated by the recent determination of the structure of the Nop17-Tah1 interaction domains [24]. Amino acid 306 is part of a beta sheet in the interaction pocket of Nop17 with Tah1. The amide to OH change in the N306S mutant might disrupt the interaction with the C-terminal segment of Tah1 [24]. Whereas the C-terminal region of Tah1 interacts with Nop17, its N-terminal TPR domain interacts with Hsp90 C-terminal peptide MEEVD [28]. Tah1-Hsp90 interaction might be stabilized by the interaction between Nop17 and Hsp90 [20].

As shown here, Tah1 is important for the stability of Nop17, this stabilization may be due to the interaction with the C-terminal region of Nop17. Similar results have been shown for the human orthologues of these proteins [29]. The destabilizing effect on Nop17 was not seen, however, when a Nop17-FLAG fusion was used, probably due to the protein fusion [20].

Tah1 is specific for Hsp90 and affects its ATPase activity as well as substrate binding [30]. It is interesting, therefore, that in the absence of either Nop17 or Tah1 the box C/D core subunit Nop58 is destabilized. More importantly, as shown here, the depletion of Rsa1, a protein proposed to function as a scaffold for snoRNP assembly [31], also leads to the destabilization of Nop58, confirming the importance of the R$_2$TP interaction with Hsp90 and Rsa1 for the proper assembly of the functional snoRNP complex.

According to the model of Rsa1 being a scaffold for the snoRNP assembly [31], Nop58 would bind the snoRNAs

only after it is directed to the box C/D particle by the R$_2$TP complex. As shown here, however, Nop58 binds U3 snoRNA co-transcriptionally, and binds snoRNA more stably in the absence of Nop17 [8]. We, therefore, favor a model according to which Nop58 binds snoRNA with very high affinity, but in order for the snoRNP complex to be matured, this interaction must be destabilized so that Nop58 can interact with Nop56 and allow for Nop1 to bind box D and D' (Figure 8). The structure of the PIH domain of hNop17/PIH1D1 interacting with a target peptide DSDD has been determined, and it has been shown that this interaction is dependent upon the phosphorylation state of the peptide DSDD [32]. Interestingly, Nop58 has this conserved peptide in the C-terminal portion involved in the interaction with Nop17 (positions 443–446). It remains to be determined whether the serine 444 is phosphorylated, and whether its phosphorylated state changes upon snoRNP assembly.

The model for box C/D snoRNP assembly proposed here is also supported by the observation that core box C/D snoRNP subunits have been shown to be important for snoRNA localization [33]. Further corroborating this hypothesis, the depletion of Nop17 or Rsa1 causes a mislocalization of the snoRNA U3.

During the final preparation of this article a study was published on the R$_2$TP complex [34]. That study reports the increased interaction of R$_2$TP with Nop58 in the absence of RNA, and the importance of R$_2$TP on Nop58 stability. In that report the Nop58 regions involved in the interaction with Nop17 were also mapped. In addition, they showed that the dissociation of Rvb1/2 from Nop17/

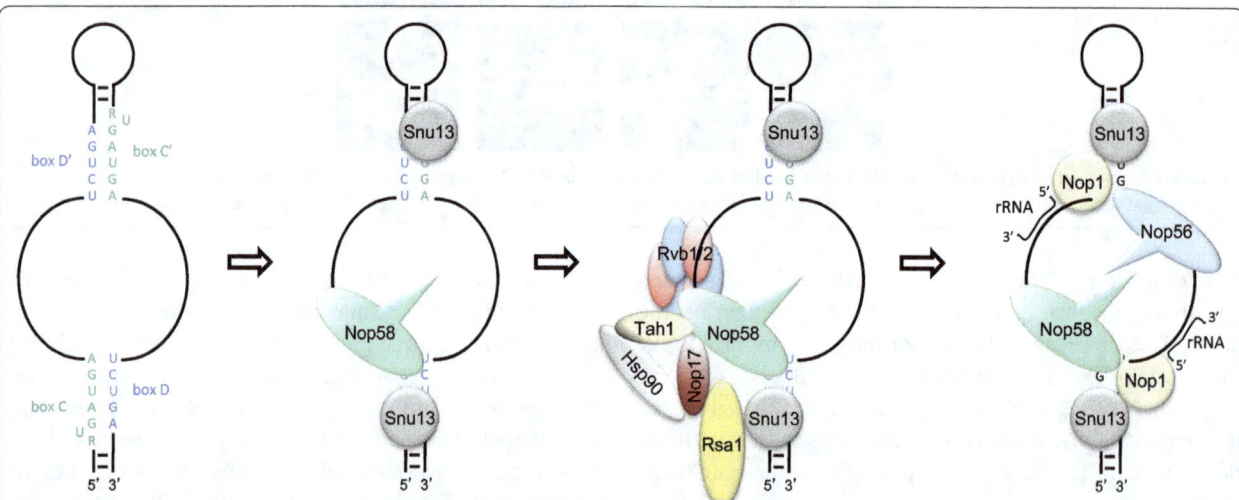

Figure 8 Model of the role of the R$_2$TP complex in the assembly of box C/D snoRNP. Snu13 binds early during the assembly, and Nop58 binds the snoRNA cotranscriptionally. The association of the R2TP complex and Hsp90 chaperone are important for the conformational change of Nop58, necessary for the binding of the Nop1 and Snu13 to form the mature snoRNP particle that will participate in rRNA maturation. In the absence of R$_2$TP subunits, Nop58 is less stable, but shows higher affinity for RNA. In addition, in the absence of R2TP, snoRNAs and core snoRNP subunits mislocalize in the cell.

Tah1 was induced by nucleotide binding rather than ATP hydrolysis. Those results are consistent with the data shown here. We show complementary data of the stronger interaction between Nop58 and RNA in the absence of R_2TP. Additionally, we show that Nop17 does not compete with RNA for binding to Nop58, confirming that RNA is not necessary for the Nop17-Nop58 interaction. The data shown here also complement those because we narrow further down the region of Nop58 responsible for the interaction with Nop17. As shown here, that region is enclosed between amino acids 389 and 512. We can therefore conclude that the 389–447 portion of Nop58 is responsible for that interaction. Interestingly, this region comprises the peptide DSDD, recently shown to interact directly with the PIH domain of Nop17 [32].

We show that the levels of full-length Nop58 decreases upon depletion not only of Nop17, but also of Rsa1 and Tah1. Interestingly, and according to the model of Nop17 directing Hsp90 to the target proteins, as shown here, the over-expression of Hsp90 in Δ*nop17* strain does not stabilize Nop58. Here we show that the interaction between Nop17 and Rvb2 is not affected by nucleotide binding or hydrolysis, contrary to the interaction with Rvb1 which is affected by the presence of ATP. Our results, therefore, corroborate and complement those recently published.

Conclusions

Nop17 interacts with the C-terminal portion of Nop58, affecting its affinity for snoRNA, Nop58 stability, and the localization of the box C/D snoRNP components, both protein and RNA moieties. Tah1 and Rsa1 affect Nop17 stability, and might therefore affect Nop58 indirectly. These results indicate a key role played by Nop17 in snoRNP assembly, and suggest a stepwise process that requires molecular rearrangements of the proteins for the binding of all subunits and formation of the mature snoRNP.

Methods
Plasmid constructions

The plasmids used in this study are listed on Table 1 and the cloning strategies described below. The genes RVB1, RVB2 and HSC82 were amplified from *S. cerevisiae* genomic DNA. For recombinants proteins with expression in bacteria (fused to glutathione-S-transferase (GST) and histidine tag (His_6)), the PCR products (RVB1 and RVB2) were first cloned into pGEM-T vector before digestion with EcoRI and SalI for cloning into the vectors pET28a and pGEX-4 T-1. The construct pGEX-NOP17 was obtained after excision of NOP17 from pET28a-NOP17 [8] with BamHI and XhoI. The genes NOP1 and SNU13 were excised from pGAD clones [8] for subcloning into the

Table 1 Plasmids used in this study

Plasmid	Characteristics	Reference
pGEX4T1	GST Tag C-terminal, Amp^R	Amersham
pGEX4T1-SNU13	GST::SNU13, Amp^R	This study
pGEX4T1-NOP17	GST::NOP17, Amp^R	This study
pET28a	His Tag N-terminal, Kan^R	Novagen
pET28a-SNU13	HIS::SNU13, Kan^R	This study
pET28a-NOP1	HIS::NOP1, Kan^R	This study
pET28a-NOP17	HIS::NOP17, Kan^R	Gonzales et al. [8]
pBTM116	lexA, DNA binding domain	Chien et al. [36]
pBTM-NOP17	lexA::NOP17, TRP1	Gonzales et al. [8]
pBTM-NOP17(N307S)	lexA::NOP17(N307S), TRP1	This study
pGAD	GAL4AD, transcription activation domain	James et al. [37]
pGAD-TAH1	GAL4AD::TAH1, LEU2	This study
pGAD-HSC82	GAL4AD::HSC82, LEU2	This study
pGAD-RVB1	GAL4AD::RVB1, LEU2	This study
pGAD-RVB2	GAL4AD::RVB2, LEU2	This study
YCP111GAL-HA-HSC82	GAL4AD::HA::Nop58, TRP1	This study
YCP33Gal-A-NOP58	GAL::ProtA-NOP58, URA3, CEN4	Gonzales et al. [8]
pBTM-RVB1	lexA::RVB1, TRP1	This study
pBTM-RVB2	lexA::RVB2, TRP1	This study
pET28a-RVB1	HIS::RVB1, Kan^R	This study
pET28a-RVB2	HIS::RVB2, Kan^R	This study

PET28a and pGEX-4 T-1 vectors. For recombinant proteins expressed in yeast (in fusion with GAL4 and LEXA), pBTM116 and pGAD-C2, RSA1, TAH1, RVB1 and RVB2 were subcloned from pGEM-T. For Hsc82 overexpression, HSC82 PCR product was digested with BamHI and XhoI and cloned into the YCP-111-GAL-HA construct [35].

Analysis of protein stability

Cells were grown at 25°C or 37°C to OD_{600} of 0.8, before addition of cyclohexamide to the final concentration of 150 µg/ml. Samples were collected at time zero, and 10, 20 40 and 60 min after addition of cyclohexamide. Total cell extract was then prepared and protein samples were separated by SDS-PAGE and subjected to western blot.

Co-immunoprecipitation of proteins and western blot analysis

Extracts from strains *WT*, *Δnop17*, *Δrsa1* and *Δtah1* expressing ProtA-Nop58 that had been grown in galactose medium were prepared in buffer (20 mM Tris-Cl pH 8.0; 0.5 mM magnesium acetate; 0.2% Triton X-100; 150 mM potassium acetate; 1 mM DTT and 1 mM PMSF). Immunoprecipitation was performed by incubating total extract with *IgG-Sepharose* (Amersham Biosciences) for 2 hours at 4°C. Fractions corresponding to total extract (TE), flow-trough (FT), wash (W) and immunoprecipitation (IP) were collected and stored at –20°C. For protein analysis by western blot, samples were separated by SDS-PAGE and transferred to PVDF membranes (GE Healthcare). For ProtA-Nop58 and HA-Hsc82 expression in *WT* and *Δnop17*, the same experiment strategy was used.

Co-immunoprecipitation of RNAs and northern blot

RNA co-immunoprecipitation with ProtA-Nop58 expressed at 37°C in strains *WT*, *Δnop17* and *Δrsa1* was performed as described previously [8]. For the Nop58 and RNA binding assay, co-immunoprecipitation was performed as described [8], and purified GST-Nop17 was added during the incubation of total extract with IgG-sepharose beads during the immunoprecipitation. Northern hybridizations were analyzed in a Typhoon equipment (GE Healthcare

Life Sciences) and bands quantitated using ImageQuant program from Typhoon.

Chromatin immunoprecipitation and qPCR analyses

WT, *Δnop17* and *Δrsa1* cells expressing ProtA-Nop58, were fixed with 1% formaldehyde and used for chromatin immunoprecipitation, as previously described [38]. For the qPCR reactions, the *Maxima SYBER Green/ROX qPCR Master Mix* (Molecular Probes Inc., Eugene) was used in the Real Time Applied Biosystem 7500 equipment. The primers used in the reactions are listed on Table 2.

snoRNA localization by FISH

The detailed protocol used for the *in situ* hybridization in strains *WT*, *Δnop17*, *Δrsa1* and *Δtah1* is described in http://www.einstein.yu.edu/labs/robert-singer/protocols/. The fluorescence tagging system used was the Cy™3ULS Labeling Kit (Amersham Biosciences) and images were obtained with a Nikon TE300 microscope coupled to a Roper CoolSnap HQ camera. Quantitation of the fluorescence signals was performed using ImageJ 1.42q.

Recombinant proteins expression in bacteria and pull-down of proteins *in vitro*

Recombinant proteins GST-Snu13, GST-Nop17, His-Nop1, His-Nop17, His-Rvb1 and His-Rvb2 were expressed in *E. coli* strain BL21. For the Snu13-Nop1-Nop17 pulldown, the total extracts containing GST or GST-Snu13 were incubated for 1 hour at 4°C with glutathione-sepharose resin (GE Healthcare) in PBS buffer. After immobilization of the first protein, the resin was washed with buffer and incubated for 2 hours with total extract expressing His-Nop1 or His-Nop17, with three washes between the extracts. After the last wash, the proteins were eluted from the resin with PBS buffer containing 10 mM reduced glutathione. Samples of all fractions were collected (TE, FT, W and Elu), and the proteins were analyzed by western blot with anti-GST, anti-His and anti-Nop17 sera. For the pull-down between Nop17 and Rvb1 and Rvb2, the same strategy was used, with the addition ATP or ADP during the second incubation step.

Table 2 Oligonucleotides used in this study

Oligonucleotide	Sequence	Reference
5S[rev]	GCGAGGCAAATCCTGAAAATTT	Granato et al. [15]
U3 prom[for]	CGAAGGCAAATCCTGAAAATTT	This study
U3 prom[rev]	TTGACAGCAGAATACAAAGCCTTT	This study
U3 exon1[for]	TCAACCATTGCAGCAGCTTT	Coltri et al. [39]
U3 exon1[rev]	TCTGCTCCGAAATGAAAACTCTAGTA	Coltri et al. [39]
U3 exon2[for]	TCTATAGGAATCGTCACTCTTTGACTC	Coltri et al. [39]
U3 exon2[rev]	GACCAAGCTAATTTAGATTCAATTTCGG	Coltri et al. [39]
U3RevFISH	ATGGGGCTCATCAACCAAGTTGG	This study

Construction of Nop17 mutant

Nop17 mutant was obtained by random *in vitro* mutagenesis [35], using pBTM-NOP17 as a template. Interaction with Tah1 was performed in the two-hybrid system using the L40/pGAD-TAH1 strain. Two-hybrid assay was performed as described previously [35].

Additional files

Additional file 1: Figure S1. Analysis of expression of mutant Nop17 (N306S) in the cells used for two-hybrid assays. Total extracts were prepared from L40 cells, either not transformed with any plasmid (−), or transformed with pBTM-NOP17, pBTM-NOP17(N306S), pGAD-TAH1, and pGAD-RVB1, as indicated, and subjected to western blot with serum against Nop17. Endogenous Nop17, BD-Nop17 and BD-Nop17 (N306S) bands are indicated on the right. Endogenous Nop17 levels, used as loading control, do not vary between the samples, but BD-Nop17(N306S) levels are much lower than those of BD-Nop17.

Additional file 2: Figure S2. Two-hybrid assay to map the Nop58 interaction domain with Nop17. *HIS3* expression is shown on the left, while *lacZ* expression is shown on the right. Upper panel, full-length Nop58 interacts with Nop17 independently of the protein fusion (DNA binding domain – BD, or transcription activation domain – AD), and interacts also with Snu13. Middle panel, Nop58(216–512) interacts only with Nop17. Lower panel, Nop58(324–512) interacts only with Nop17, but with higher affinity than the full-length protein.

Additional file 3: Figure S3. Quantitation of snoRNA U3 signal from FISH experiments shown in Figure 7. U3 signals in linear distribution throughout the cells were quantitated by using ImageJ. Position of the nucleolus in each cell is indicated. WT cells show concentration of U3 signal in the nucleolus, independently of the temperature of growth. Δ*nop17* shows concentration of U3 in the nucleolus at the permissive temperature, but not at the restrictive temperature. Δ*rsa1* and Δ*tah1* cells show very low signal of U3, but despite that, it is possible to see the mislocalization of U3 at 37°C in Δ*rsa1*. In Δ*tah1* cells, on the other hand, U3 localization does not change much at 37°C.

Abbreviations

snoRNP: Small nucleolar ribonucleoprotein; R₂TP: Rvb1, Rvb2, Tah1, Pih1; rRNA: ribosomal RNA.

Competing interests

The authors declare that they have no competing interests.

Authors' contributions

MBP, RCG and FAG carried out two-hybrid assays. MBP and RCG performed protein pull-down experiments. MBP carried out ChIP, FISH and RNA co-immunoprecipitation assays. JSL performed RNA-protein competition experiments. CCO carried out protein co-immunoprecipitation and western blot assays. All authors participated in design of the study and drafted the manuscript. CCO conceived of the study, and participated in its design and coordination. All authors read and approved the final manuscript.

Acknowledgements

We thank Frederico Gueiros Filho for the use of the microscope for FISH experiments, and Roberto Kopke Salinas for help with molecular modeling analyses. This study was supported by grants from Fundação de Amparo à Pesquisa do Estado de São Paulo (FAPESP- 10/51842-3, 12/51200-7). During this work, M.B.P. was supported by a PhD fellowship from CAPES, and R.C.G., J.S.L. and F.A.G. were supported by FAPESP postdoctoral fellowships.

Author details

[1]Department of Biochemistry, Institute of Chemistry, University of São Paulo, Av. Prof. Lineu Prestes 748, 05508-000 São Paulo, SP, Brazil. [2]Present address: Department of Biochemistry and Molecular Biology, Institute of Biological Sciences, Federal University of Goiás, Goiânia, Brazil. [3]Present address: Department of Biological Sciences, School of Pharmacy, São Paulo State University, Araraquara, Brazil.

References

1. Kiss T. Small nucleolar RNAs: an abundant group of noncoding RNAs with diverse cellular functions. Cell. 2002;109(2):145–8.
2. Granneman S, Kudla G, Petfalski E, Tollervey D. Identification of protein binding sites on U3 snoRNA and pre-rRNA by UV cross-linking and high-throughput analysis of cDNAs. Proc Natl Acad Sci U S A. 2009;106(24):9613–8.
3. Lafontaine DL, Tollervey D. Synthesis and assembly of the box C+D small nucleolar RNPs. Mol Cell Biol. 2000;20(8):2650–9.
4. Filipowicz W, Pogacic V. Biogenesis of small nucleolar ribonucleoproteins. Curr Opin Cell Biol. 2002;14(3):319–27.
5. Gerbi SA, Borovjagin AV, Lange TS. The nucleolus: a site of ribonucleoprotein maturation. Curr Opin Cell Biol. 2003;15(3):318–25.
6. Marmier-Gourrier N, Clery A, Senty-Segault V, Charpentier B, Schlotter F, Leclerc F, et al. A structural, phylogenetic, and functional study of 15.5-kD/Snu13 protein binding on U3 small nucleolar RNA. RNA. 2003;9(7):821–38.
7. Watkins NJ, Dickmanns A, Luhrmann R. Conserved stem II of the box C/D motif is essential for nucleolar localization and is required, along with the 15.5K protein, for the hierarchical assembly of the box C/D snoRNP. Mol Cell Biol. 2002;22(23):8342–52.
8. Gonzales FA, Zanchin NI, Luz JS, Oliveira CC. Characterization of Saccharomyces cerevisiae Nop17p, a novel Nop58p-interacting protein that is involved in Pre-rRNA processing. J Mol Biol. 2005;346(2):437–55.
9. Zhao R, Davey M, Hsu YC, Kaplanek P, Tong A, Parsons AB, et al. Navigating the chaperone network: an integrative map of physical and genetic interactions mediated by the hsp90 chaperone. Cell. 2005;120(5):715–27.
10. Zhao R, Houry WA. Hsp90: a chaperone for protein folding and gene regulation. Biochem Cell Biol. 2005;83(6):703–10.
11. Huen J, Kakihara Y, Ugwu F, Cheung KL, Ortega J, Houry WA. Rvb1-Rvb2: essential ATP-dependent helicases for critical complexes. Biochem Cell Biol. 2010;88(1):29–40.
12. Gribun A, Cheung KL, Huen J, Ortega J, Houry WA. Yeast Rvb1 and Rvb2 are ATP-dependent DNA helicases that form a heterohexameric complex. J Mol Biol. 2008;376(5):1320–33.
13. Eckert K, Saliou JM, Monlezun L, Vigouroux A, Atmane N, Caillat C, et al. The Pih1-Tah1 cochaperone complex inhibits Hsp90 molecular chaperone ATPase activity. J Biol Chem. 2010;285(41):31304–12.
14. Back R, Dominguez C, Rothe B, Bobo C, Beaufils C, Morera S, et al. High-resolution structural analysis shows how Tah1 tethers Hsp90 to the R2TP complex. Structure. 2013;21(10):1834–47.
15. Granato DC, Gonzales FA, Luz JS, Cassiola F, Machado-Santelli GM, Oliveira CC. Nop53p, an essential nucleolar protein that interacts with Nop17p and Nip7p, is required for pre-rRNA processing in Saccharomyces cerevisiae. FEBS J. 2005;272(17):4450–63.
16. Goldfeder MB, Oliveira CC. Cwc24p, a novel Saccharomyces cerevisiae nuclear ring finger protein, affects pre-snoRNA U3 splicing. J Biol Chem. 2008;283(5):2644–53.
17. Boulon S, Marmier-Gourrier N, Pradet-Balade B, Wurth L, Verheggen C, Jady BE, et al. The Hsp90 chaperone controls the biogenesis of L7Ae RNPs through conserved machinery. J Cell Biol. 2008;180(3):579–95.
18. Rothe B, Back R, Quinternet M, Bizarro J, Robert MC, Blaud M, et al. Characterization of the interaction between protein Snu13p/15.5K and the Rsa1p/NUFIP factor and demonstration of its functional importance for snoRNP assembly. Nucleic Acids Res. 2014;42(3):2015–36.
19. Millan-Zambrano, G, Chavez S. Nuclear functions of prefoldin. Open Biol. 2014;4(7).
20. Zhao R, Kakihara Y, Gribun A, Huen J, Yang G, Khanna M, et al. Molecular chaperone Hsp90 stabilizes Pih1/Nop17 to maintain R2TP complex activity that regulates snoRNA accumulation. J Cell Biol. 2008;180(3):563–78.
21. Zanchin NI, Goldfarb DS. Nip7p interacts with Nop8p, an essential nucleolar protein required for 60S ribosome biogenesis, and the exosome subunit Rrp43p. Mol Cell Biol. 1999;19(2):1518–25.
22. Zanchin NI, Goldfarb DS. The exosome subunit Rrp43p is required for the efficient maturation of 5.8S, 18S and 25S rRNA. Nucleic Acids Res. 1999;27(5):1283–8.

23. Paci A, Liu XH, Huang H, Lim A, Houry WA, Zhao R. The Stability of the Small Nucleolar Ribonucleoprotein (snoRNP) Assembly Protein Pih1 in Saccharomyces cerevisiae Is Modulated by Its C Terminus. J Biol Chem. 2012;287(52):43205–14.

24. Pal M, Morgan M, Phelps SE, Roe SM, Parry-Morris S, Downs JA, et al. Structural basis for phosphorylation-dependent recruitment of Tel2 to Hsp90 by Pih1. Structure. 2014;22(6):805–18.

25. Gautier T, Berges T, Tollervey D, Hurt E. Nucleolar KKE/D repeat proteins Nop56p and Nop58p interact with Nop1p and are required for ribosome biogenesis. Mol Cell Biol. 1997;17(12):7088–98.

26. Aittaleb M, Rashid R, Chen Q, Palmer JR, Daniels CJ, Li H. Structure and function of archaeal box C/D sRNP core proteins. Nat Struct Biol. 2003;10(4):256–63.

27. Ye K, Jia R, Lin J, Ju M, Peng J, Xu A, et al. Structural organization of box C/D RNA-guided RNA methyltransferase. Proc Natl Acad Sci U S A. 2009;106(33):13808–13.

28. Jimenez B, Jimenez B, Ugwu F, Zhao R, Orti L, Makhnevych T, et al. Structure of minimal tetratricopeptide repeat domain protein Tah1 reveals mechanism of its interaction with Pih1 and Hsp90. J Biol Chem. 2012;287(8):5698–709.

29. Yoshida M, Yoshida M, Saeki M, Egusa H, Irie Y, Kamano Y, et al. RPAP3 splicing variant isoform 1 interacts with PIH1D1 to compose R2TP complex for cell survival. Biochem Biophys Res Commun. 2013;430(1):320–4.

30. Millson SH, Vaughan CK, Zhai C, Ali MM, Panaretou B, Piper PW, et al. Chaperone ligand-discrimination by the TPR-domain protein Tah1. Biochem J. 2008;413(2):261–8.

31. Rothe B, Saliou JM, Quinternet M, Back R, Tiotiu D, Jacquemin C, et al. Protein Hit1, a novel box C/D snoRNP assembly factor, controls cellular concentration of the scaffolding protein Rsa1 by direct interaction. Nucleic Acids Res. 2014;42(16):10731–47.

32. Horejsi Z, Stach L, Flower TG, Joshi D, Flynn H, Skehel JM, et al. Phosphorylation-dependent PIH1D1 interactions define substrate specificity of the R2TP cochaperone complex. Cell Rep. 2014;7(1):19–26.

33. Verheggen C, Mouaikel J, Thiry M, Blanchard JM, Tollervey D, Bordonne R, et al. Box C/D small nucleolar RNA trafficking involves small nucleolar RNP proteins, nucleolar factors and a novel nuclear domain. EMBO J. 2001;20(19):5480–90.

34. Kakihara Y, Makhnevych T, Zhao L, Tang W, Houry WA. Nutritional status modulates box C/D snoRNP biogenesis by regulated subcellular relocalization of the R2TP complex. Genome Biol. 2014;15(7):404.

35. Oliveira CC, Gonzales FA, Zanchin NI. Temperature-sensitive mutants of the exosome subunit Rrp43p show a deficiency in mRNA degradation and no longer interact with the exosome. Nucleic Acids Res. 2002;30(19):4186–98.

36. Chien CT, Bartel PL, Sternglanz R, Fields S. The two-hybrid system: a method to identify and clone genes for proteins that interact with a protein of interest. Proc Natl Acad Sci U S A. 1991;88(21):9578–82.

37. James P, Halladay J, Craig EA. Genomic libraries and a host strain designed for highly efficient two-hybrid selection in yeast. Genetics. 1996;144(4):1425–36.

38. Granato DC, Machado-Santelli GM, Oliveira CC. Nop53p interacts with 5.8S rRNA co-transcriptionally, and regulates processing of pre-rRNA by the exosome. FEBS J. 2008;275(16):4164–78.

39. Coltri PP, Oliveira CC. Cwc24p is a general Saccharomyces cerevisiae splicing factor required for the stable U2 snRNP binding to primary transcripts. PLoS One. 2012;7(9):e45678.

PERMISSIONS

All chapters in this book were first published in MB, by BioMed Central; hereby published with permission under the Creative Commons Attribution License or equivalent. Every chapter published in this book has been scrutinized by our experts. Their significance has been extensively debated. The topics covered herein carry significant findings which will fuel the growth of the discipline. They may even be implemented as practical applications or may be referred to as a beginning point for another development.

The contributors of this book come from diverse backgrounds, making this book a truly international effort. This book will bring forth new frontiers with its revolutionizing research information and detailed analysis of the nascent developments around the world.

We would like to thank all the contributing authors for lending their expertise to make the book truly unique. They have played a crucial role in the development of this book. Without their invaluable contributions this book wouldn't have been possible. They have made vital efforts to compile up to date information on the varied aspects of this subject to make this book a valuable addition to the collection of many professionals and students.

This book was conceptualized with the vision of imparting up-to-date information and advanced data in this field. To ensure the same, a matchless editorial board was set up. Every individual on the board went through rigorous rounds of assessment to prove their worth. After which they invested a large part of their time researching and compiling the most relevant data for our readers.

The editorial board has been involved in producing this book since its inception. They have spent rigorous hours researching and exploring the diverse topics which have resulted in the successful publishing of this book. They have passed on their knowledge of decades through this book. To expedite this challenging task, the publisher supported the team at every step. A small team of assistant editors was also appointed to further simplify the editing procedure and attain best results for the readers.

Apart from the editorial board, the designing team has also invested a significant amount of their time in understanding the subject and creating the most relevant covers. They scrutinized every image to scout for the most suitable representation of the subject and create an appropriate cover for the book.

The publishing team has been an ardent support to the editorial, designing and production team. Their endless efforts to recruit the best for this project, has resulted in the accomplishment of this book. They are a veteran in the field of academics and their pool of knowledge is as vast as their experience in printing. Their expertise and guidance has proved useful at every step. Their uncompromising quality standards have made this book an exceptional effort. Their encouragement from time to time has been an inspiration for everyone.

The publisher and the editorial board hope that this book will prove to be a valuable piece of knowledge for researchers, students, practitioners and scholars across the globe.

LIST OF CONTRIBUTORS

Sreerangam NCVL Pushpavalli, Arpita Sarkar, M Janaki Ramaiah, Debabani Roy and Manika Pal-Bhadra
Centre for Chemical Biology, Indian Institute of Chemical Technology, Hyderabad 500607, India

Chowdhury and Utpal Bhadra
Functional Genomics and Gene Silencing Group, Centre for Cellular and Molecular Biology, Hyderabad 500007, India

Shuang Yang, Lingjia Liu and Zhuo Yang
Medical School, Tianjin Key Laboratory of Tumor Microenvironment and Neurovascular Regulation, Nankai University, 94 Weijin Road, Tianjin 300071, China

Pengjuan Xu
Tianjin University of Traditional Chinese Medicine, Tianjin 300193, China

Marlinda Hupkes, Ana M Sotoca, José M Hendriks and Everardus J van Zoelen
Department of Cell & Applied Biology, Faculty of Science, Nijmegen Centre for Molecular Life Sciences (NCMLS), Radboud University Nijmegen, Heyendaalseweg 135, 6525 AJ, Nijmegen, The Netherlands

Koen J Dechering
Department of Cell & Applied Biology, Faculty of Science, Nijmegen Centre for Molecular Life Sciences (NCMLS), Radboud University Nijmegen, Heyendaalseweg 135, 6525 AJ, Nijmegen, The Netherlands
Merck Research Laboratories, PO Box 20, 5340 BH, Oss, The Netherlands
Current affiliation: TropI Q Health Sciences, PO Box 9101, 6500 HB, Nijmegen, The Netherlands

Ju-Long Yu, Zhi-Fang An and Xiang-Dong Liu
Department of Entomology, Nanjing Agricultural University, Nanjing 210095, China

Toshitsugu Fujita and Hodaka Fujii
Chromatin Biochemistry Research Group, Combined Program on Microbiology and Immunology, Research Institute for Microbial Diseases, Osaka University, Suita, Osaka, Japan

Carmen Sánchez-Jiménez, Isabel Carrascoso and José M Izquierdo
Centro de Biología Molecular Severo Ochoa, Consejo Superior de Investigaciones Científicas, Universidad Autónoma de Madrid (CSIC/UAM), C/ Nicolás Cabrera 1, Cantoblanco, Madrid 28049, Spain

Juan Barrero
Programa de Biología de Sistemas, Centro Nacional de Biotecnología, Consejo Superior de Investigaciones Científicas, C/Darwin 3, Cantoblanco, Madrid 28049, Spain

Linlin Liu, Junfeng Liu, Jinfeng Ni, and Yulong Shen
State Key Laboratory of Microbial Technology, Shandong University, 27 Shanda Nan Rd., Jinan 250100, P. R. China

Qunxin She
Archaea Centre, Department of Biology, University of Copenhagen, Ole MaaløesVej 5, Copenhagen N DK-2200, Denmark

Qihong Huang
State Key Laboratory of Microbial Technology, Shandong University, 27 Shanda Nan Rd., Jinan 250100, P. R. China
Archaea Centre, Department of Biology, University of Copenhagen, Ole MaaløesVej 5, Copenhagen N DK-2200, Denmark

Xiaohe Lu
Department of Ophthalmology, Zhujiang Hospital, Southern Medical University, Guangzhou 510280, Guangdong Province, China

Ning Kong
Department of Ophthalmology, Zhujiang Hospital, Southern Medical University, Guangzhou 510280, Guangdong Province, China
Department of Ophthalmology, Guangzhou Panyu Central Hospital, Guangzhou 510280, Guangdong Province, China

Bin Li
Department of Ophthalmology, Guangzhou Panyu Central Hospital, Guangzhou 510280, Guangdong Province, China

Vlasta Korenková, Vendula Novosadová, Marie Jindřichová, Lucie Langerová, David Švec, and Monika Šídová
Laboratory of Gene Expression, Institute of Biotechnology, Academy of Sciences of the Czech Republic, Prague, Czech Republic

Justin Scott
QFAB Bioinformatics, University of Queensland - St Lucia QLD, Brisbane, Australia

Robert Sjöback
TATAA Biocenter, Göthenburg, Sweden

Boaz E Aronson
Division of Gastroenterology and Nutrition, Department of Medicine, Children's Hospital Boston, and Harvard Medical School, 300 Longwood Avenue, Boston, MA 02115, USA
Academic Medical Center Amsterdam, Emma Children's Hospital, Amsterdam, the Netherlands

Kelly A Stapleton and Stephen D Krasinski
Division of Gastroenterology and Nutrition, Department of Medicine, Children's Hospital Boston, and Harvard Medical School, 300 Longwood Avenue, Boston, MA 02115, USA

Eva Stokhuijzen and Hanneke Bruijnzeel
Academic Medical Center Amsterdam, Emma Children's Hospital, Amsterdam, the Netherlands

Laurens ATM Vissers
University Medical Center Groningen, Groningen, the Netherlands

Irma F Takács, Dóra Tombácz, Beáta Berta, István Prazsák, Nándor Póka and Zsolt Boldogkői
Department of Medical Biology, Faculty of Medicine, University of Szeged, Somogyi B. st. 4, Szeged H-6720, Hungary

Wei-Long Jiang, Yu-Feng Zhang and Qing-Qing Xia
Department of Respiration, Jiangyin Hospital of Traditional Chinese Medicine Affiliated to Nanjing University of Chinese Medicine, Jiangyin City, Jiangsu Province 214400, China

Jian Zhu and Tao Fan
Department of Neurology, Jiangyin Hospital of Traditional Chinese Medicine Affiliated to Nanjing University of Chinese Medicine, Jiangyin City, Jiangsu Province 214400, China

Xin Yu
Department of Internal Medicine, Jiangyin Hospital of Traditional Chinese Medicine Affiliated to Nanjing University of Chinese Medicine, Jiangyin City, Jiangsu Province 214400, China

Feng Wang
Department of Neurology, Shanghai First People's Hospital, Shanghai Jiaotong University School of Medicine, Shanghai 200080, China

Alister PW Funnell, Wooi F Lim, Ka Sin Mak, Beeke Wienert, Gabriella E Martyn, Crisbel M Artuz, Jon Burdach, Kate GR Quinlan, Richard CM Pearson and Merlin Crossley
School of Biotechnology and Biomolecular Sciences, University of New South Wales, Sydney, NSW 2052, Australia

Douglas Vernimmen
The Roslin Institute, University of Edinburgh, Easter Bush Campus, Midlothian EH25 9RG, UK

Douglas R Higgs
MRC Molecular Haematology Unit, Weatherall Institute of Molecular Medicine, University of Oxford, John Radcliffe Hospital, Headington, Oxford OX3 9DS, UK

Emma Whitelaw
La Trobe Institute for Molecular Science, La Trobe University, Melbourne, Victoria 3086, Australia

Bing Sun, Shan-Shan Liu and Song-Lin Chen
Yellow Sea Fisheries Research Institute, Chinese Academy of Fisheries Sciences, Qingdao 266071, China

Chang-Geng Yang
Yellow Sea Fisheries Research Institute, Chinese Academy of Fisheries Sciences, Qingdao 266071, China
Yangtze River Fisheries Research Institute, Chinese Academy of Fishery Sciences, Wuhan 430223, China

Xian-Li Wang
Translational Center for Stem Cell Research, Tongji Hospital, Stem Cell Research Center,
Tongji University School of Medicine, Shanghai 200065, China

Bo Zhang
Bohai Sea Fisheries Research Institute of Tianjin, Tianjin, China

Nicolas Paquet, Joseph K Box, Nicholas W Ashton, Amila Suraweera, Laura V Croft, Aaron J Urquhart, Emma Bolderson, Kenneth J O'Byrne and Derek J Richard
School of Biomedical Science, Institute of Health and Biomedical Innovation at the Translational Research Institute, Queensland University of Technology, Brisbane, QLD, Australia

Shu-Dong Zhang
Center for Cancer Research and Cell Biology, School of Medicine, Dentistry and Biomedical Sciences, Queen's University Belfast, Lisburn Road 97, Belfast, UK

Abby L Bifano and Tarik Atassi
Department of Cellular & Molecular Medicine, Lerner Research Institute, Cleveland Clinic, 9500 Euclid Avenue NC10, Cleveland, OH 44195, USA

Tracey Ferrara
Department of Cellular & Molecular Medicine, Lerner Research Institute,Cleveland Clinic, 9500 Euclid Avenue NC10, Cleveland, OH 44195, USA
Department of Human Medical Genetics, University of Colorado, Aurora, CO, USA

Donna M Driscoll
Department of Cellular & Molecular Medicine, Lerner Research Institute,Cleveland Clinic, 9500 Euclid Avenue NC10, Cleveland, OH 44195, USA
Department of Molecular Medicine, Cleveland Clinic Lerner College of Medicine of Case Western Reserve University, Cleveland, OH, USA

Cvetan Popov, Martina Burggraf, Matthias Schieker and Denitsa Docheva
Department of Surgery, Experimental Surgery and Regenerative Medicine, Ludwig-Maximilians-University (LMU), Nussbaumstr. 20, D-80336 Munich, Germany

Ludwika Kreja and Anita Ignatius
Institute of Orthopaedic Research and Biomechanics, University of Ulm, Helmholtzstr. 14, D-89081 Ulm, Germany

Marcela B Prieto and Carla C Oliveira
Department of Biochemistry, Institute of Chemistry, University of São Paulo, Av. Prof. Lineu Prestes 748, 05508-000 São Paulo, SP, Brazil

Raphaela C Georg
Department of Biochemistry, Institute of Chemistry, University of São Paulo, Av. Prof. Lineu Prestes 748, 05508-000 São Paulo, SP, Brazil
Department of Biochemistry and Molecular Biology, Institute of Biological Sciences, Federal University of Goiás, Goiânia, Brazil

Juliana S Luz
Department of Biochemistry, Institute of Chemistry, University of São Paulo, Av. Prof. Lineu Prestes 748, 05508-000 São Paulo, SP, Brazil
Department of Biological Sciences, School of Pharmacy, São Paulo State University, Araraquara, BrazilYellow Sea Fisheries Research Institute, Chinese Academy of Fisheries Sciences, Qingdao 266071, China

Index

A

Aging, 2, 165, 173, 175, 197

Akirin, 152-155, 160, 164

Alpha Globin, 140

Anaphase Bridges, 1

Apoptosis, 2, 10, 12, 15, 28, 60-61, 68-71, 74-75, 90-97, 109, 132-139, 152, 198

Archaea, 76-77, 85-89, 162, 199, 203

Ask1, 132-139

Atpase, 76-77, 79-89, 199-200, 202, 208

B

Biomark, 98-103, 105-107

Box C/d Snornp Assembly, 199, 208, 212

C

C1q, 152-153, 157, 160, 162-164

Cell Growth Arrest, 16-17, 19, 22-23

Chip, 22, 25, 28-30, 38, 52, 57-58, 99, 105, 110-112, 117-119, 140, 142, 145-147, 150, 202, 204, 211

Chk2, 1, 9-13, 15

Chromatid Bridges, 1, 3, 5, 10

Chromatin

Immunoprecipitation, 19, 52, 58-59, 110, 145, 210

Collagen-binding Integrins, 188, 191, 193

Crypt Cell Proliferation, 108-109, 111-113, 115-117, 119

Cytoskeleton, 1, 7-8, 26, 60, 65-66, 189

D

Degraded Samples, 98, 101

Dna, 1-15, 17, 20, 22, 25, 41, 44, 50, 54, 61, 67, 72-89, 92, 99, 104, 106, 109, 117, 1122, 131, 134, 140, 146-148, 150, 156, 158, 163-168, 171, 175, 196, 200, 207, 211

Dna-binding Proteins, 52-53, 57, 163

Drosophila Melanogaster, 1, 13-15, 51, 157

E

Embryogenesis, 1-3, 5, 7, 9, 11-13, 15

Endothelial Cells, 26, 72, 75, 132, 135, 137-139

Exponential Pre-amplification, 98

F

Ffpe, 98, 104, 107

Fluidigm, 98-99, 105-107

G

Ganglion Cells, 90-91, 93, 95-97

Gata6, 108-119

Gene Cloning, 43, 45, 47, 49, 51

Gene Expression, 24, 26, 28, 30-31, 34, 37, 42, 50, 62, 66, 73, 75, 80, 99, 110, 112, 121, 126, 128-132, 141, 145, 147, 149-151, 153, 164, 188, 190-192, 194, 196-198

Gene Ontology, 30, 40, 42, 60, 62, 64-65, 73, 116

Gene Regulatory Networks, 60, 70

Globin Gene Regulation, 140

H

Hela Cells, 60-71, 73, 75, 167, 170-171, 174

Helicase, 12, 67, 76-77, 80, 83, 86-89, 207

Hepn, 152-154, 157, 159-160, 162-164

Hera, 76-89

Herpesvirus, 120-121, 131

Heterozygotes, 3, 6, 8-10, 12-13

High-throughput Qpcr, 98-99, 101, 103, 105, 107

Homologous Recombination Repair, 76, 88

Homozygotes, 3, 12

I

Ichip, 52-59

Icp22, 120-121, 123-125, 127-129, 131

Igf-1, 90, 97

Immunity, 139, 152, 154, 163, 177

Immunostaining, 6, 13, 95, 113

Intestinal Differentiation, 108

J

Japanese Flounder, 152-153, 159-160, 163-164

K

Kidney Development, 16, 24

Klf1, 140-141, 147, 149-150

Klf3, 140-150

L

L30, 176-180, 182-187

Locus-specific Chip, 52

M

Matrix Metalloproteinases, 188-189, 191-193, 197-198

Mechanical Stimulation, 188-191, 193-198

Microfluidics, 98, 107

Microrna Genes, 41

Mir-19a, 132-139

Mir-100, 90-97

Mirnas, 28-30, 34, 36-38, 41, 60-73, 75, 90-91, 97, 132, 134, 137

Mitosis, 1-3, 9-10, 12, 15, 175

Mkl1, 16-27

Mof, 1-15

N

Nop17, 199-211

Nuclear Envelope, 165-166, 170-173, 175

Nuclease, 76-77, 79-80, 82-84, 87-89, 158, 185

Nuclei, 1-4, 6-10, 12-13, 15, 172

Nura, 76-89

O

Oxidative Stress, 90-92, 95, 97, 137

P

Podocyte, 16-18, 22-24, 26-27

Progeria, 165-167, 169-171, 173, 175

Pseudorabies Virus, 120-121, 123, 125, 127, 129, 131

R

R2tp Complex, 199-202, 205, 207-208, 211-212

R3xfnldd-d, 52-58

Real-time Pcr, 30, 40, 49-50, 106-107, 120, 130, 132-134, 138, 142, 150, 157, 164

Retinal Ganglion, 90-91, 93, 95-97

Rice Planthopper, 43-45, 47-50

Rna Isolation, 13, 196

Rna-binding Protein, 73-74, 176

S

Secis-binding Protein 2, 176-177, 186-187

Selenocysteine, 176-177, 186-187

Selenoprotein, 67, 176-177, 183, 186

Sensory Neurons, 90

Sogatella Furcifera, 43-47, 50-51

Spdef, 108-119

Stem Cell, 41, 118, 138, 164, 188-190, 192, 197

Syncytial Embryos, 1, 7, 9-10, 13, 15

T

Tendon Stem/progenitor Cells, 188-189, 191, 193, 195, 197

Tendon-related Genes, 188-189

Tia1, 60-63, 68, 70, 73-74

Transcription Factor, 26-27, 29, 35, 40, 68, 75, 108-109, 112, 119, 147, 150, 153, 194

U

Us1 Gene, 120-125, 129-130

W

Western Blot Analysis, 2-3, 13, 61, 63, 79, 85-86, 117, 132, 159, 169

Wing Deformation, 43

Wing Length, 44, 46, 48, 50

Wingless Gene, 43, 45, 47-49, 51

Y

Yeast Two-hybrid Assay, 152, 158

www.ingramcontent.com/pod-product-compliance
Lightning Source LLC
Chambersburg PA
CBHW080405190526
45161CB00003B/138